The Role of Artificial Intelligence in Advancing Applied Life Sciences

Tamer Emara
Damietta University, Egypt

Genghui Li
Shenzhen University, China

Esraa Hassan
Kafrelsheikh University, Egypt

Abeer Saber
Damietta University, Egypt

Thanh Trinh
Phenikaa University, Vietnam

IGI Global
Scientific Publishing
Publishing Tomorrow's Research Today

Vice President of Editorial	Melissa Wagner
Managing Editor of Acquisitions	Mikaela Felty
Managing Editor of Book Development	Jocelynn Hessler
Production Manager	Mike Brehm
Cover Design	Phillip Shickler

Published in the United States of America by
IGI Global Scientific Publishing
701 East Chocolate Avenue
Hershey, PA, 17033, USA
Tel: 717-533-8845
Fax: 717-533-7115
Website: https://www.igi-global.com E-mail: cust@igi-global.com

Library of Congress Cataloging-in-Publication Data

Names: Emara, Tamer, 1983- editor | Hassan, Esraa, 1992- editor | Trinh, Thanh, 1983- editor | Li, Genghui, 1990- editor | Saber, Abeer, 1992- editor
Title: The role of artificial intelligence in advancing applied life sciences / edited by Tamer Emara, Esraa Hassan, Thanh Trinh, Genghui Li, Abeer Saber.
Description: Hershey, PA : IGI Global Scientific Publishing, [2025] | Includes bibliographical references and index. | Summary: "This book connects the theoretical concepts of AI with practical applications in life sciences"-- Provided by publisher.
Identifiers: LCCN 2024055184 (print) | LCCN 2024055185 (ebook) | ISBN 9798369392089 hardcover | ISBN 9798369392096 paperback | ISBN 9798369392102 ebook
Subjects: LCSH: Artificial intelligence--Biological applications | Artificial intelligence--Medical applications
Classification: LCC QH324.25 .R65 2025 (print) | LCC QH324.25 (ebook) | DDC 570.285--dc23/eng/20250421
LC record available at https://lccn.loc.gov/2024055184
LC ebook record available at https://lccn.loc.gov/2024055185

British Cataloguing in Publication Data
A Cataloguing in Publication record for this book is available from the British Library.

Editorial Advisory Board

Table of Contents

Detailed Table of Contents

Artificial intelligence (AI) is rapidly transforming many fields of science and technology, including the applied life sciences. This paper examines the current and potential future impacts of AI across various domains of applied life sciences, including biotechnology, agriculture, medicine, and environmental science. We review how machine learning, natural language processing, computer vision, and other AI techniques are being leveraged to accelerate research, improve data analysis, optimize processes, and enable new capabilities. The chapter also discusses challenges and limitations of applying AI in life sciences, such as data quality issues, interpretability of models, and ethical considerations. Finally, we explore emerging trends and offer perspectives on how AI may reshape applied life sciences in the coming decades. This review aims to provide researchers, policymakers and industry professionals with a comprehensive overview of AI's transformative potential in advancing applied life science fields.

This chapter explores the transformative impact of Artificial Intelligence (AI) on applied life sciences, with a focus on environmental science. It begins by tracing the theoretical foundations of AI from early pioneers like Alan Turing and John McCarthy, and then discusses its evolution through the development of expert systems and machine learning techniques. The chapter highlights AI's significant contributions to environmental monitoring, climate change prediction, and biodiversity conservation, showcasing how AI enhances our understanding and management of complex systems. It also examines the integration of AI with satellite data and sensor networks to address environmental challenges. Concluding with future directions, the chapter addresses emerging trends and ethical considerations, emphasizing AI's role in supporting sustainable development goals. This overview provides a foundational context for understanding AI's role in life sciences and sets the stage for exploring its broader applications in subsequent chapters.

Muhammad Usman Tariq, Abu Dhabi University, UAE & University College Cork, Ireland

This chapter explores the integration of Artificial Intelligence (AI) in the life sciences, highlighting its impact on research, medical care, and understanding complex biological systems. AI's capacity to process vast datasets and identify hidden patterns is revolutionising fields such as drug discovery, genomics, clinical diagnostics, and healthcare administration. It examines the historical development of AI in life sciences, from rule-based systems like MYCIN to modern data-driven approaches enhanced by deep learning, machine learning, and natural language processing. Notable applications include improved patient outcomes, accelerated drug discovery, enhanced diagnostic accuracy, and personalised medicine. The chapter also discusses AI's transformative potential across various life science sectors, especially drug discovery, where it optimises candidate selection, reduces costs, and repurposes existing drugs. Overall, the integration of AI signifies a paradigm shift in life sciences, promising greater efficiency, innovation, and responsiveness to global health challenges.

P. Venkadesh, V.S.B. College of Engineering Technical Campus, Coimbatore, India
S. V. Divya, V.S.B. College of Engineering Technical Campus, Coimbatore, India
V. Mathumitha, Bharath Institute of Higher Education and Research, Chennai, India
S. Ninisha Nels, Noorul Islam Centre for Higher Education, Kumaracoil, India
J. Balaji, Nandha College of Technology, Erode, India
Aiswariya Ramesh, V.S.B. College of Engineering Technical Campus, Coimbatore, India

Bioinformatics, where Artificial Intelligence (AI) meets biology, as transforming our understanding of life. By combining machine learning algorithm with tools like blast and cluster Omega, scientists are deciphering DNA and predicting protein structure. AI models such as CNN's and RNN's, help analyze DNA sequence, identify genetic

This chapter provides an in-depth overview of how computational intelligence enhances the detection and diagnosis of throat-related diseases using laryngeal endoscopic imaging. It focuses on advanced methods like Convolutional Neural Networks (CNNs), Generative Adversarial Networks (GANs), and transfer learning models for identifying conditions such as laryngeal cancer, vocal cord disorders, infections, and phonetic gap disorders. Traditional machine learning techniques, including Support Vector Machines (SVMs) and Random Forests, are explored for their effectiveness in disease classification and diagnostic support. The chapter also covers essential processes like image preprocessing, segmentation, and real-time diagnostic integration with endoscopic tools. Challenges such as image variability, patient differences, and processing constraints are discussed. Future directions highlight improving diagnostic accuracy, particularly in phonetic gap identification, with the goal of enhancing clinical decision-making and predicting disease progression.

5G generation has raised significant concerns among the overall public and medical network about its potential influences on human health, especially concerning the implicit risk of cancer and neurological disorders. This debate has mainly become contentious, with a few studies claiming that 5G technology poses dangerous outcomes for human health. In contrast, others find no specific affiliation between 5G and human health. The International Agency for Research on Cancer (IARC) has classified radiofrequency electromagnetic fields, which might be a pivotal

Preface

Artificial Intelligence (AI) has emerged as a transformative force in the applied life sciences, redefining how we conduct research, manage data, and address complex challenges in healthcare, agriculture, environmental science, and beyond. The rapid advancements in AI have led to groundbreaking innovations, offering new opportunities for scientific discovery, improving operational efficiencies, and enabling personalized solutions in diverse domains. This book, *The Role of Artificial Intelligence in Advancing Applied Life Sciences*, aims to provide a comprehensive exploration of AI's profound impact on various facets of life sciences.

AI in the Modern World: Transforming Life Sciences

In today's world, AI applications extend beyond traditional computational domains, revolutionizing disciplines that are crucial for human well-being and environmental sustainability. AI-driven approaches are reshaping healthcare, making early disease detection more precise, accelerating drug discovery, and personalizing patient treatment plans. In agriculture, AI facilitates smarter farming techniques, enabling better resource management, precision irrigation, and pest control strategies. Environmental sciences benefit from AI-powered climate models, biodiversity tracking, and pollution monitoring. These developments highlight the urgency and significance of AI integration in life sciences, making this book a timely and essential resource.

Target Audience

This book is designed for a broad spectrum of readers, including researchers, academicians, industry professionals, and policymakers who seek an in-depth understanding of AI's applications in life sciences. It serves as a valuable reference for students pursuing careers in biotechnology, medicine, environmental science, and agricultural engineering. Professionals working in AI, data science, and bioinformatics will also find insights into how computational techniques are shaping real-world applications in life sciences. By bridging theoretical foundations with practical applications, this book caters to both novices and experts in the field.

Overview of Chapters

This book is divided into four major parts, each addressing critical aspects of AI in life sciences:

Part I: Foundations of AI in Life Sciences

1. **The Role of Artificial Intelligence in Advancing Applied Life Sciences: An Introduction** – Provides an overview of AI's transformative impact on life sciences, covering key domains such as medicine, biotechnology, agriculture, and environmental science. It discusses major AI techniques, challenges, and future trends.
2. **A Historical Perspective on Artificial Intelligence in Applied Life Sciences** – Traces the evolution of AI, from early theoretical foundations to its current applications in life sciences, with a special focus on environmental science and sustainability.
3. **AI-Driven Innovations: Transforming Healthcare, Agriculture, and Environmental Science** – Explores how AI technologies are revolutionizing diverse fields, including genomics, clinical diagnostics, and drug discovery.
4. **AI in Bioinformatics: Unlocking the Secrets of Life** – Discusses the role of AI in deciphering genetic codes, predicting protein structures, and accelerating biomedical research.

Part II: AI in Agriculture and Environmental Science

5. **AI-Driven Approaches in Aquaculture Disease Management: Innovations, Challenges, and Future Directions** – Examines how AI techniques enhance disease detection and prevention in aquaculture, improving sustainability and food security.
6. **Drones in the Indian Agricultural Sector - A Technological Rise** – Highlights the growing role of drones in Indian agriculture and their potential for precision farming.
7. **Machine Learning for Precision Agriculture and Crop Yield Optimization: Techniques and Applications** – Discusses ML algorithms applied to yield prediction, irrigation management, and soil health analysis.

8. **Biomanufacturing: The Role of AI in Bioprocess Optimization** – Explores AI's role in optimizing biological manufacturing processes, particularly in pharmaceuticals and biofuels.
9. **Avian Species Detection Using Spectrogram and CNN** – Showcases an AI-driven approach for species identification using deep learning techniques applied to spectrogram images.

Part III: AI in Healthcare and Medical Applications

10. **Enhanced Monkeypox Skin Lesion Classification Using Pooling-based Vision Transformer Architecture** – Presents a novel AI-based method for highly accurate monkeypox detection.
11. **Role of Computational Intelligence Methods to Identify Diseases Using Laryngeal Endoscopic Images: A Survey** – Reviews AI-driven approaches for diagnosing laryngeal disorders using endoscopic imaging.
12. **5G and Human Health: Reviewing the Risks of Cancer and Neurological Disorders** – Examines the potential health risks of 5G radiation exposure using AI-driven analytical techniques.
13. **AI-Driven Laboratory Automation: Transforming Experimental Phases of Drug Discovery** – Highlights how AI enhances efficiency in laboratory automation and accelerates pharmaceutical research.
14. **The Intersection of AI, Cloud Computing, and Healthcare: A Comprehensive Overview** – Discusses how AI and cloud computing together optimize healthcare delivery and data management.
15. **Integration of Brain-Computer Interfaces in Modern Healthcare: Enhancing Patient Care and Quality of Life** – Explores AI's role in developing brain-computer interfaces for improved patient rehabilitation and assistive technologies.

Part IV: Emerging AI Applications and Future Directions

16. **AI-Powered Smart Tourism: Reimagining Travel** – Analyzes the role of AI in the tourism industry, focusing on personalized travel experiences, sustainability, and ethical considerations.

Impact and Contribution

This book is a testament to the growing interdisciplinary nature of AI, demonstrating how computational techniques are redefining life sciences. By consolidating insights from various research domains, it provides a roadmap for future advancements, bridging gaps between AI, biotechnology, agriculture, and healthcare. The discussions presented herein not only highlight AI's current capabilities but also underscore the ethical and regulatory considerations necessary for responsible AI development.

As AI continues to evolve, its applications in life sciences will only expand, offering unprecedented opportunities for innovation. We hope this book serves as an invaluable resource for researchers, professionals, and students, fostering further exploration and collaboration in this exciting domain.

Chapter 1
The Role of Artificial Intelligence in Advancing Applied Life Sciences:
An Introduction

Ketki P. Kshirsagar
https://orcid.org/0000-0002-5851-1749
Vishwakarma Institute of Technology, Pune, India

ABSTRACT

Artificial intelligence (AI) is rapidly transforming many fields of science and technology, including the applied life sciences. This paper examines the current and potential future impacts of AI across various domains of applied life sciences, including biotechnology, agriculture, medicine, and environmental science. We review how machine learning, natural language processing, computer vision, and other AI techniques are being leveraged to accelerate research, improve data analysis, optimize processes, and enable new capabilities. The chapter also discusses challenges and limitations of applying AI in life sciences, such as data quality issues, interpretability of models, and ethical considerations. Finally, we explore emerging trends and offer perspectives on how AI may reshape applied life sciences in the coming decades. This review aims to provide researchers, policymakers and industry professionals with a comprehensive overview of AI's transformative potential in advancing applied life science fields.

DOI: 10.4018/979-8-3693-9208-9.ch001

1. INTRODUCTION

Artificial Intelligence (AI) develops systems that replicate human intelligence in tasks like decision-making, learning, and problem-solving. The complexity of AI ranges across a spectrum, from simple rule-based systems to highly sophisticated, adaptive models that can process and learn from vast data sets. AI is powered by implementation tools, programming languages, and environments designed to address this complexity. Key programming languages for AI include Python, R, and Java, each valued for their libraries, efficiency, and compatibility with AI algorithms. Python, for instance, has robust libraries like TensorFlow, Keras, and PyTorch that support machine learning and deep learning.

The implementation tools used in AI encompass a wide range of software, such as TensorFlow and PyTorch for deep learning, Scikit-learn for machine learning, and OpenCV for computer vision tasks. Natural Language Toolkit (NLTK) and spaCy facilitate natural language processing tasks, while Reinforcement Learning frameworks like OpenAI Gym support the development of adaptive, goal-oriented models.

Development environments like Jupyter Notebook, Google Colab, and Anaconda provide the ecosystem for experimenting, visualizing, and iterating AI models, allowing for streamlined development and testing. Additionally, cloud platforms like Google Cloud AI, Microsoft Azure ML, and Amazon Web Services (AWS) are integral for deploying and scaling AI solutions in real-world applications due to their computational power and flexibility.

With applications in areas like agriculture, healthcare, and finance, AI systems address complex challenges like image recognition, predictive analytics, and natural language understanding, though they often require extensive computational resources, large datasets, and specialized hardware like GPUs and TPUs to handle high-volume processing. This infrastructure supports AI's diverse use cases, making it integral to transforming industries with real-time decision-making, automation, and deep insights derived from data analysis.

AI refers to the development of computer systems capable of performing tasks that typically require human intelligence. These systems aim to mimic cognitive functions like problem-solving, learning, and decision-making (George F Ludger, 2022).

At its core, AI seeks to replicate aspects of human cognition through computational means. This involves creating algorithms and models that can process information, recognize patterns, adapt to new situations, and improve their performance (Girish Kumar jha,2007) over time.

The complexity of AI arises from the diverse range of tasks it seeks to accomplish, each requiring different levels of cognitive simulation, data processing, and adaptability. At its core, AI's complexity is influenced by the depth of understanding and decision-making required, the scale of data involved, and the sophistication of the algorithms and architectures it employs.

1. **Task Complexity**: AI applications vary widely in difficulty, from simple rule-based tasks to intricate cognitive simulations. Basic automation tasks, like rule-based sorting or basic data classification, are relatively simple to implement. However, tasks requiring language comprehension, image recognition, and multi-step decision-making add layers of complexity. These require the system not only to process information but to also adapt and learn from context, past interactions, or feedback.
2. **Data Dependency**: The complexity of AI systems is heavily dependent on the scale and quality of data they require to function. Systems like machine learning models and deep learning networks rely on vast datasets to discern patterns, make predictions, and improve over time. High-dimensional data, (George F Ludger,2022) as seen in images or natural language, adds computational demands as the model needs to process complex features. Poor quality or insufficient data adds complexity by introducing noise and reducing model accuracy, necessitating sophisticated data pre-processing and augmentation techniques.
3. **Algorithmic Sophistication**: AI's underlying algorithms also contribute to its complexity. Algorithms for supervised learning (e.g., linear regression or decision trees) are generally more straightforward, but unsupervised learning and reinforcement learning algorithms are considerably more intricate. For example, deep learning models, particularly convolutional neural networks (CNNs) for image processing and recurrent neural networks (RNNs) for sequence-based tasks, require large architectures with many layers and parameters. This introduces additional challenges in terms of optimizing performance, avoiding over fitting, and handling vanishing or exploding gradients.
4. **Adaptively and Real-Time Processing**: AI applications designed for dynamic environments, such as autonomous vehicles, robots, or financial trading systems, need real-time processing and continuous learning to respond effectively to changing conditions. The need for such adaptively demands high computational power and advanced algorithms capable of updating models on the fly, adding to the overall complexity. Several studies have focused on optimizing systems using artificial neural networks for power-saving mechanisms, which could also be extended to optimizing medical devices or life sciences equipment (Emara et al., 2014; Emara, 2016; Emara, 2019).

5. **Ethical and Interpretability Challenges**: The "black-box" nature of many AI models, especially deep learning networks, raises issues of explain ability and trustworthiness. Complex models are often challenging to interpret, which limits their transparency. This interpretability issue adds a layer of complexity, especially in applications where decisions need to be explained or validated, such as in healthcare or law.

Overall, the complexity of AI encompasses not just the technical challenges but also the ethical and operational dimensions. As AI advances, managing this complexity becomes crucial to building systems that are efficient, ethical, and adaptable across a wide range of real-world applications.

6. **Turing test:** This test was proposed by (Alan Turing, 1950) paper "Computing Machinery and Intelligence," is a method for evaluating a machine's capacity to demonstrate intelligent behaviour. This test was designed to address the fundamental question: "Can machines think?"

The procedure involves a human evaluator engaging in separate text-based conversations with both a human participant and a computer system. These interactions are conducted through a text-only interface, such as a keyboard and screen, to focus on the content of the responses rather than any potential audio rendering capabilities.

During the test, both the human participant and the computer system attempt to convince the evaluator that they are human. The key participants are kept isolated from one another. If the evaluator is unable to consistently differentiate between the human and the machine based on their responses, the computer is considered to have successfully passed the test.

Turing proposed (Alan Turing, 1950) that if a significant number of evaluators cannot reliably distinguish between the computer and the human participant, it would be reasonable to conclude that the machine exhibits thought processes comparable to human cognition(no date).

This test has become a landmark concept in the field of artificial intelligence, sparking ongoing debates about the nature of intelligence, consciousness, and the potential capabilities of machines. While passing the Turing Test doesn't necessarily prove that a machine possesses genuine understanding or consciousness, it remains an influential benchmark in assessing the advancement of conversational AI systems.

7. **Roots of Artificial Intelligence:** The field of Artificial Intelligence draws upon several established academic disciplines, including philosophy, mathematics and logic, computer science, psychology and cognitive studies, as well as biology and neuroscience. These areas of study often intersect and influence one another.

For instance, there are strong connections between philosophical inquiry and logical reasoning, as well as between mathematical theory and computational practices.

By examining each of these foundational disciplines individually, we can gain deeper insights into their contributions to AI and how they have evolved to support its development. One of the primary objectives in AI research is to create systems capable of carrying out tasks that typically demand human-level intelligence. This goal drives the integration of knowledge from various fields to push the boundaries of machine capabilities.

2. AI METHODOLOGIES:

AI methodologies encompass a range of techniques and approaches used to build intelligent systems and solve complex problems (Serey, Joel, et al., 2021). Here's an overview of some key AI methodologies:

2.1. Machine Learning (ML)

Machine Learning is a subset of AI focused on enabling systems to learn from data and improve their performance over time without being explicitly programmed. Key techniques in ML include (Serey, Joel, et al., 2021):

Supervised Learning: Models are trained on labelled data to make predictions or classifications. Examples include linear regression, decision trees, and support vector machines.

Unsupervised Learning: Models identify patterns and relationships in unlabelled data. Techniques include clustering (e.g., k-means) and dimensionality reduction (e.g., principal component analysis).

Reinforcement Learning: Agents learn to make decisions by receiving rewards or penalties based on their actions within an environment. Techniques include Q-learning and deep Q-networks.

2.2. Deep Learning

Deep Learning (Wongchai et. al., 2022) is a specialized form of machine learning that uses neural networks with many layers (deep neural networks) to model complex patterns in large datasets. (Kumar, Koushal et. al., 2012) Key components include:

Convolutional Neural Networks (CNNs): Used primarily for image and video recognition tasks.

Recurrent Neural Networks (RNNs): Suitable for sequential data tasks such as time series analysis and natural language processing.

Generative Adversarial Networks (GANs): Consist of two networks (generator and discriminator) that compete to create realistic data samples.

2.3. Natural Language Processing (NLP)

NLP focuses on enabling machines to understand, interpret, and generate human language. Techniques in NLP include:

Tokenization and Parsing: Breaking down text into manageable units and analysing grammatical structure.

NamedEntity Recognition (NER): Identifying entities such as names, dates, and locations in text.

Sentiment Analysis: Determining the sentiment or emotion conveyed in text.

Machine Translation: Translating text from one language to another.

2.4. Computer Vision

Computer Vision enables machines to interpret and understand visual information from the world. Key techniques include:

Image Classification: Identifying objects or features in an image.

Object Detection: Locating and classifying multiple objects within an image.

Image Segmentation: Partitioning an image into segments for easier analysis.

2.5. Expert Systems

Expert Systems are designed to emulate the decision-making abilities of a human expert. They use:

Rule-Based Systems: Apply predefined rules to make decisions or solve problems.

Knowledge Bases: Store expert knowledge and use inference engines to derive conclusions.

2.6. Fuzzy Logic

Fuzzy Logic deals with reasoning that is approximate rather than fixed and exact. It is used in systems where binary logic is too rigid. Techniques include:

Fuzzy Sets: Represent data with varying degrees of membership.

Fuzzy Inference Systems: Apply fuzzy logic rules to make decisions based on imprecise or uncertain information.

2.7. Robotic Process Automation (RPA)

RPA involves using software robots or "bots" to automate repetitive tasks typically performed by humans. Key aspects include:

Rule-Based Automation: Automating tasks based on predefined rules and workflows.

Process Automation: Streamlining complex processes to improve efficiency and reduce errors.

2.8. Game Theory

Game Theory applies mathematical models to understand and predict strategic interactions between rational decision-makers. It is used in:

Multi-Agent Systems: Analysing interactions between multiple autonomous agents.

Strategic Decision-Making: Developing strategies based on potential actions of other agents.

2.9. Evolutionary Algorithms

Evolutionary Algorithms are optimization techniques inspired by natural selection and biological evolution. They include:

Genetic Algorithms: Use processes such as selection, crossover, and mutation to find optimal solutions.

Evolutionary Programming: Focuses on evolving strategies to solve complex problems.

2.10. Bayesian Methods

Bayesian Methods use probability to represent and update beliefs about uncertain events. Techniques include (George F Ludger, 2022).:

Bayesian Networks: Graphical models representing probabilistic relationships between variables.

Bayesian Inference: Updating probability estimates as new data becomes available.

3. ARTIFICIAL INTELLIGENCE THROUGH MACHINE LEARNING:

Artificial Intelligence, particularly through machine learning (ML), plays a transformative role in advancing applied life sciences. Here's how ML functions in this field, the types of data it relies on, and what it achieves:

3.1. How Machine Learning Works in Life Sciences

Machine learning algorithms are designed to recognize patterns in data, learn from these patterns, and make predictions or decisions based on new information. In life sciences, this typically involves training algorithms on large datasets, such as genetic sequences, medical images, and clinical records (Serey, J, 2021).

The core steps in applying ML include:

- **Data Collection**: Gathering data relevant to the life sciences application, like patient health records, genetic information, or biological images.
- **Data Pre-processing**: Cleaning, transforming, and structuring the data for analysis (e.g., normalizing genetic sequences, standardizing image sizes).
- **Model Training**: Feeding the processed data into ML algorithms, which "learn" by adjusting their internal parameters to reduce errors in predicting outcomes.
- **Testing and Validation**: Using a portion of data not seen by the model during training to evaluate its accuracy and ensure it generalizes well to new cases.
- **Deployment**: Applying the trained model to new data for real-time predictions, such as diagnosing diseases from new patient data.

3.2. Types of Data Required in Life Sciences

Machine learning in life sciences often requires diverse data types, including (Serey, Joel, et al., 2021):

- **Genomic Data**: DNA sequences, gene expression data, and protein structures used to understand genetic influences on diseases, develop personalized medicine, and identify biomarkers.
- **Medical Imaging**: Radiological scans, such as X-rays, MRIs, and CT scans, analyzed for patterns that indicate conditions like tumors, lesions, or bone fractures.

- **Clinical and EHR Data**: Electronic health records (EHRs) capturing patient histories, symptoms, treatments, and outcomes, crucial for disease prediction and risk assessment(Mohammed Yousef Shaheen,2021).
- **Sensor and Wearable Data**: Real-time data from wearable devices monitoring heart rate, activity levels, glucose levels, and other health metrics, used in preventive healthcare.
- **Experimental and Laboratory Data**: Data from bioinformatics experiments, including drug response data, microbial growth data, and protein interactions, often used in drug discovery and development.

3.3. What Machine Learning Can Achieve in Life Sciences

Machine learning offers powerful capabilities in life sciences, achieving impactful results such as:

- **Disease Diagnosis and Prediction**: ML models trained on clinical and imaging data can predict conditions like cancer, heart disease, and genetic disorders, often with high accuracy. For example, convolutional neural networks (CNNs) can analyse radiology images to identify early-stage tumours.
- **Drug Discovery and Development**: ML algorithms expedite drug discovery by analysing the molecular properties of compounds, predicting their efficacy, and identifying potential side effects (Berg, S. Nudge., 2018). This reduces the time and cost associated with developing new medications. Wei et al. (2018) introduced a two-stage data processing algorithm, which could be useful for streamlining big data analysis in life sciences, particularly in genomics and drug discovery. Additionally, the foundational concepts of Emara et al. (2023) and Trinh et al. (2021) in data management and geographically distributed systems can be applied to the management of large-scale biological datasets, helping integrate AI systems into life sciences research efficiently.
- **Personalized Medicine**: ML helps in analysing genomic and health record data to predict individual responses to treatments, enabling customized therapeutic approaches(Al Kuwaiti A, 2023).
- **Predictive Modelling of Patient Outcomes**: Models using EHRs and clinical data can forecast patient outcomes, such as recovery rates or likelihood of disease recurrence, aiding in planning personalized care.
- **Automating Laboratory Workflows**: Robots and ML algorithms automate repetitive lab tasks, such as image analysis, cell counting, and chemical screening, improving accuracy and freeing up scientists for higher-level analysis.

4. AI'S ROLE ACROSS DIVERSE FIELDS

AI has become an indispensable tool across multiple sectors, driving innovation, enhancing efficiencies, and enabling smarter decision-making. As AI technologies evolve, their impact continues to broaden, affecting industries ranging from agriculture to gaming, healthcare to aviation. This section delves into the various fields where AI plays a significant role, demonstrating its versatility and transformative potential. From revolutionizing farming practices through precision agriculture to enhancing passenger safety in aviation, AI is reshaping industries by optimizing operations, automating complex tasks, and improving overall service delivery. By examining sectors such as agriculture, healthcare, robotics, aviation, and gaming, we can better understand how AI is not only responding to challenges but also creating new opportunities for growth and development in these diverse domains.

We will explore each field in detail, highlighting how AI technologies such as machine learning, computer vision, predictive analytics, and robotics are being integrated into real-world applications, making significant strides toward solving complex, sector-specific challenges.

4.1. AI in Agriculture

AI plays a crucial role in solving agricultural challenges, with the quality of available data significantly influencing its effectiveness (Javaid, Mohd, et al,2023). AI's potential to revolutionize agrarian practices is clear, but accessing the necessary data at the farmer level remains a major challenge. By integrating image classification techniques with remote and local sensing data, AI can transform the use of farm machinery, especially for tasks like weed control, early disease detection, and crop harvesting and grading. These technologies facilitate round-the-clock monitoring of high value crops, which is vital for effective horticultural management throughout all stages of plant growth. Figure 1 illustrate eco-friendly farming in the view of AI.

Using data from precision farming software, soil sensors, drones, or even smartphone photos, AI systems can continuously monitor soil nutrient levels, comparing them to historical data that has yielded the best results for specific crops. AI can analyse the environmental effects of different types and amounts of fertilizers to determine the optimal dose that maximizes yield while minimizing environmental harm, contributing to eco-friendly farming (Javaid, Mohd, et al,2023).

Figure 1. AI in eco-friendly farming

In recent years, (Javaid, Mohd, et al,2023) increasing pollution and erratic weather patterns have created additional challenges for farmers, particularly in determining the best time to plant crops. This is where AI proves invaluable, providing insights into how weather, sunlight, wind, and rainfall affect planting cycles. With AI's help, farmers can use weather forecasts to plan optimal planting schedules.

Advances in computer vision, mechatronics, AI, and machine learning (ML) have enabled the development of remote sensing technologies for identifying and managing plants, weeds, pests, and diseases. AI also facilitates precision seeding and fertilization. Farmers can minimize waste, improve product quality, and ensure faster market access with AI tools. Automated tractors, guided by GPS, prepare fields for planting and harvesting, operating without human drivers. Drones collect data and monitor crop health, relaying valuable information to farmers.

AI also plays a role in pest management by identifying central plants and determining the optimal pesticide types and dosages(Chen, Shih-Fang, 2022). Drones equipped with AI can spray pesticides more efficiently, saving both time and resources. Predictive analytics estimate precipitation and evapotranspiration, while ML models can be trained using soil samples and other data to provide insights into soil moisture, temperature, and condition. This allows for more efficient irrigation, benefiting both farmers' profits and the environment.

By precisely addressing the needs of farmland, AI helps ensure that crops receive the necessary resources, improving yields, fertilizer efficiency, and profitability. Precision agriculture promotes sustainability, environmental protection, and enhanced productivity. AI technology is being widely adopted across agricultural industries to increase efficiency.

With the shift from agrarian to urban lifestyles, the agricultural sector has faced a labour shortage. Traditional farming methods require significant manual labour, including tasks like planting, irrigating, harvesting, and weeding. AI offers automation solutions to address these challenges. Companies are developing autonomous agricultural bots to handle labour-intensive tasks, enhancing productivity, speed, cost-effectiveness, and quality.

AI applications in agriculture include predicting diseases, estimating soil retention, modelling crop development, evaluating pesticide and nutrient loss, and even assessing egg fertility. AI helps farmers make more informed decisions about crop and livestock production. The effective use of AI tools and reliable data collection are key indicators of growth in the agricultural sector.

Although AI has been tested across various industries over the past decade, its potential to improve agricultural decision-making has only recently become clear. By leveraging AI, farmers can make more efficient decisions that enhance crop and livestock output. AI-based data analytics are essential for mining the large datasets used in agriculture. This advanced technology uses machine learning algorithms to extract valuable insights from extensive datasets, helping to understand crop growth patterns, predict diseases, and provide tailored recommendations for fertilizers and pesticides. Sensors such as those that measure soil, weight, temperature, and light intensity are integrated with cameras and can be deployed via drones or robots to gather field data. Figure 2 describes flow of crop growth prediction as well as dieses detection(Chen, Shih-Fang, 2022).

Figure 2. AI in Agriculture

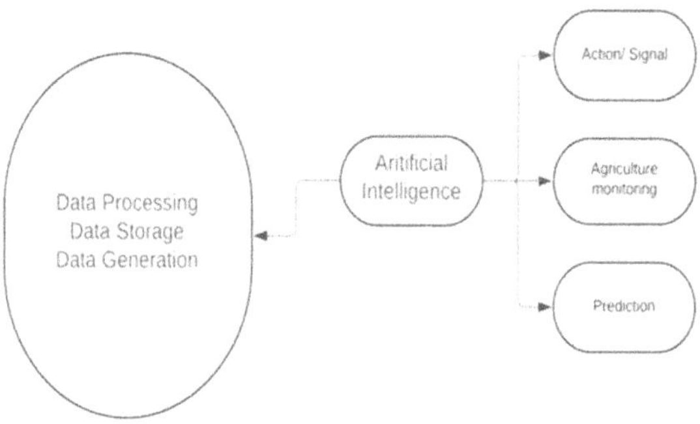

4.1.1. Agriculture parameters monitored by artificial intelligence

Agriculture is inherently labour-intensive, making labour shortages a common challenge. However, automation offers solutions, including self-driving tractors, smart irrigation systems, automated spraying, fertilizing systems, and AI-based harvesting robots. Despite these advancements, it can be difficult for software companies to effectively communicate the full potential of AI systems to farmers (Javaid, Mohd, et al, 2023). In agriculture, AI is employed for field harvesting, crop health monitoring, pest control, and detecting nutrient deficiencies. Machine learning (ML) and AI are replacing out-dated methods of forecasting and intelligence, making agriculture more adaptable through the integration of advanced technologies. For instance, biosensors now enable the monitoring of soil moisture and fertility. Instead of relying solely on simple linear regression models, AI gathers raw data and applies complex algorithms, such as neural networks, to predict weather trends based on non-linear dependencies. This allows AI to determine the optimal time for planting key crops like rice, wheat, and maize, which rely heavily on rainfall for growth.

Farms generate vast amounts of data daily on factors like temperature, soil conditions, water usage, and weather. AI and ML models analyse this data in real time, offering valuable insights on when to plant, which crops or hybrid seeds to select for higher yields, and more. These systems enhance precision agriculture by improving harvest accuracy and quality. AI technology plays a crucial role in the early detection of pests, diseases, and nutrient deficiencies on farms. AI sensors can also identify and target weeds before applying herbicides, thereby reducing chemical use and costs. Several tech companies have developed robots equipped with computer vision and AI to accurately monitor weeds and apply herbicide only where necessary.

AI has made significant progress (Wongchai A, Shukla SK,2022) in various industries, including sales, healthcare, architecture, marketing, and finance. In agriculture, devices based on AI, ML, and the Internet of Things (IoT) are now used to manage irrigation, secure fields, and fertilize crops. Proper irrigation is essential for achieving optimal yields, and large-scale watering was once a complex challenge for a small number of farmers. Smart irrigation systems, powered by ML models that calculate soil moisture and IoT devices, have emerged as a solution. These systems monitor crops and soil conditions using remote sensing, satellite imagery, drones, and precision technology.

Robots with autonomy are now being designed to handle labor-intensive tasks like harvesting more efficiently than human workers. Additionally, (Wongchai A, Shukla SK,2022) nano-based sensing mechanisms and smart sensors are being developed to detect infections, allergens, chemicals, and pollutants in food, water, and soil for plant and animal production systems. Figure 3 illustrate different aspects of AI based agricultural monitoring.

Agricultural land generates millions of data points daily, and AI enables the collection and analysis of this data, helping farmers make real-time decisions and addressing long-standing challenges. AI also assists in weather prediction, allowing farmers to plan the best times for sowing. In terms of water management, (Wongchai A, Shukla SK,2022) AI has improved efficiency by evaluating hybrid seeds and predicting yields before planting, reducing the risk of crop failure and boosting productivity. As fewer people choose farming as a profession, many farms face a shortage of skilled labor. Historically, farms required numerous workers, particularly for seasonal harvests, but labor shortages have increased as societies have become more urbanized. AI-powered bots have helped address these shortages.

Figure 3. AI potential for Agriculture Monitoring

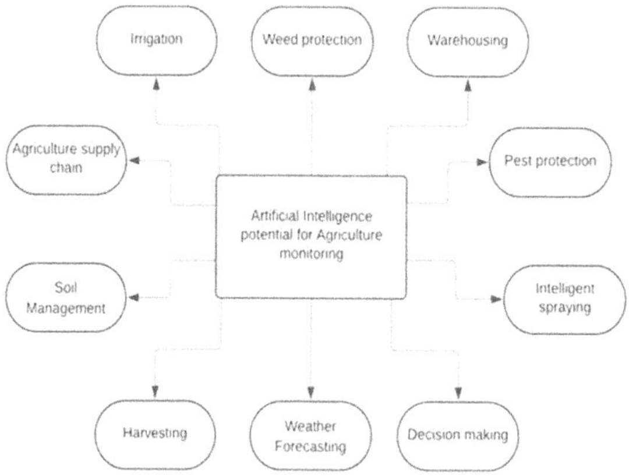

Emerging agricultural technologies (Chen, Shih-Fang, 2022) facilitate the collection of data that helps farmers monitor and optimize crop production. These technologies continuously adapt to changing environmental and ecosystem conditions, highlighting the importance of innovation in modern agriculture. By adopting new technology, the agricultural sector works toward sustainability goals, with AI playing a key role in improving both efficiency and productivity. AI is helping farmers reduce environmental impacts while enhancing operational processes. Consequently, the agricultural industry has widely embraced AI to optimize production, food systems, and bio system engineering, effectively addressing many of the challenges facing modern farming.

Data collected from sensors, drones, and satellites can play a crucial role in optimizing crop cultivation. AI analyses this information, enabling farmers to make well-informed decisions. For instance, in phenotyping, AI can evaluate a plant's biomass and characteristics, identifying patterns between diseased and healthy crops by analysing changes in biomass and environmental factors. This helps improve disease detection by pinpointing the specific causes. Identifying impurities and pests in crops through AI contributes to better yield management. Water management in agriculture has significant effects on hydrological, climatological, and agronomic systems. Advanced machine learning (ML) applications now support daily, weekly, or monthly evapotranspiration estimates, improving irrigation efficiency. Additionally, (Chen, Shih-Fang, 2022) ML models predict daily dew point temperatures, aiding in the assessment of weather conditions and evapotranspiration rates.

AI's vast potential will accelerate transformative changes, reshaping how food is produced and distributed from farm to table. Agriculture is currently in the midst of a digital transformation, with AI expanding the possibilities for cultivating new agricultural regions. This will also create job opportunities as new enterprises arise. AI makes controlled farming of innovative products, like insects, as well as traditional crops like leafy greens, more feasible and cost-effective. This technology assists farmers in tackling challenges such as climate change, water scarcity, and pest control.

With access to vast amounts of data, AI utilizes machine learning, deep learning, and other techniques to generate precise predictions that help farmers make smarter decisions about crops. AI's ability to process large datasets with unparalleled accuracy surpasses human capability. Many farmers are now employing AI-powered robots to take over tasks that once required manual labour. As urbanization grows, fewer people are choosing farming as a profession, leading to a decline in the availability of agricultural labour. AI helps bridge this gap by automating labour-intensive tasks.

4.1.2. Artificial Intelligence applications in agriculture

AI applications in agriculture are continually expanding, offering innovative solutions to long-standing challenges. One key area of AI application is the **prediction of weather patterns**, where machine learning models analyze historical weather data, satellite imagery, and real-time weather updates to forecast conditions such as rainfall, temperature variations, and droughts. This helps farmers optimize planting schedules and irrigation plans.

In addition, AI-driven **predictive models for plant diseases** can analyze environmental conditions, plant health data, and historical disease patterns to detect early signs of potential outbreaks. This proactive approach enables targeted interventions, reducing the need for widespread pesticide use. Furthermore, the rise of **self-driving tractors** powered by AI allows for autonomous, efficient field operations, reducing labor costs and improving precision in tasks like planting, tilling, and harvesting.

AI is also instrumental in **monitoring crops and soil health** through the use of remote sensing technologies and IoT devices, enabling real-time analysis of soil moisture, nutrient levels, and crop growth stages. AI algorithms can even **identify wasteful resource consumption patterns** by analyzing data from irrigation systems and fertilizer application, promoting more sustainable farming practices. **Anomaly detection** is another AI-driven innovation, where machine learning models identify unusual patterns in crop behavior, signaling issues such as pest infestations, water stress, or nutrient imbalances. Finally, AI systems are increasingly used to assess **farm-level financials**, helping farmers make informed decisions about resource allocation, operational costs, and investment planning, ultimately enhancing farm profitability and sustainability.

4.2. Artificial Intelligence in Healthcare

AI has revolutionized the healthcare industry, with applications spanning multiple areas (Shaheen, 2021). This study, based on a broad literature review, explores AI's role in healthcare by focusing on key aspects such as:

(i) medical imaging and diagnostics,
(ii) virtual patient care,
(iii) medical research and drug discovery,
(iv) patient engagement and adherence,
(v) rehabilitation,
(vi) administrative tasks.

AI has had a significant impact in detecting clinical conditions through medical imaging and diagnostics, aiding in the early detection of COVID-19, facilitating virtual patient care through AI-driven tools, and managing electronic health records. It also enhances patient engagement and adherence to treatment plans, reduces the administrative burden on healthcare professionals (HCPs), supports the discovery of new drugs and vaccines, identifies prescription errors, enables extensive data storage and analysis, and assists in technology-driven rehabilitation (Shaheen, 2021).

Figure 4. Role of AI in Healthcare

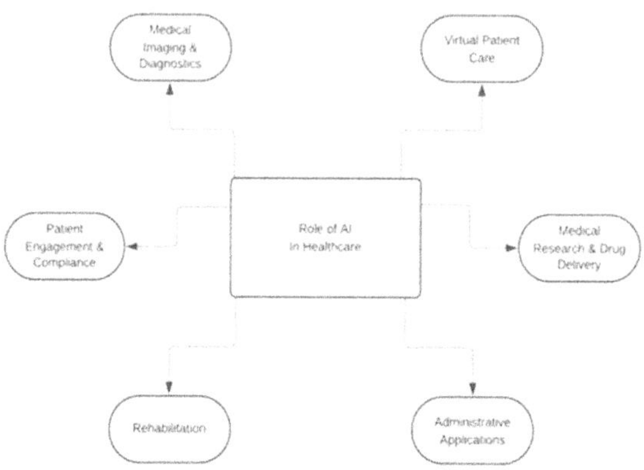

However, integrating AI into healthcare presents several technical, ethical, and societal challenges, including concerns about privacy, safety, autonomy, cost, informed consent, access, and effectiveness (Shaheen, 2021). TrinhHuy et al. (2025) explored event scale prediction in event-based social networks, which has potential applications in managing healthcare events and patient care. Similarly, Trinh et al. (2021) proposed a novel stratified block model for processing large-scale data, which can improve the management of healthcare datasets and optimize machine learning workflows. Proper governance of AI in healthcare is essential to ensure patient safety, establish accountability, and build healthcare professionals' confidence in its benefits. Effective governance is critical in addressing regulatory, ethical, and trust issues, which will facilitate broader acceptance and implementation of AI technologies. Since the onset of the COVID-19 pandemic, AI has spurred a transformation in healthcare, and this technological shift may mark a significant step toward meeting future healthcare demands. Figure 4 shows different healthcare operation where AI is role out.

4.2.1. Medical Imaging and Diagnostic Services

AI has become (Ghosh, P., 2021) a powerful tool for image analysis, increasingly adopted by radiologists for early disease diagnosis and reducing diagnostic errors, particularly in preventive care. Similarly, AI plays a vital role in analysing ECG and echocardiography charts, aiding cardiologists in making informed decisions. One example is the Ultromics platform, which was implemented at a hospital in Oxford. This platform uses AI to analyse echocardiography scans, identifying heartbeat patterns and detecting ischemic heart disease. AI has shown promising results in the early detection of diseases such as breast cancer, skin cancer, eye conditions, and pneumonia through various imaging techniques.

AI tool scan also analyse speech patterns to predict psychotic episodes and detect features of neurological diseases like Parkinson's. Recent studies have demonstrated AI's effectiveness in predicting diabetes onset using machine learning models. In one study, a two-class augmented decision tree model outperformed others in predicting diabetes-related variables. Additionally, a review highlighted AI's role in utilizing CT scans, X-rays, MRIs, and ultrasounds to diagnose COVID-19, showcasing AI's critical contribution in combating the virus.

A deep learning model (Bedi, G.;, 2015) called a transformer has been used in medical imaging tasks, including registration, detection, classification, image translation, segmentation, and video-based applications. Previous studies explored the use of transformers in distinguishing COVID-19 from pneumonia using X-ray and CT images, addressing the urgent need for efficient COVID-19 case management. One study employed the Image Net-pretrained vision transformer (ViT)-B/32 network to detect COVID-19 from chest X-ray patches.

Whig, Pawan et. al., (2024) proposed a new method based on chest CT scans for automatic COVID-19 detection, using a computer vision-based technique with wavelet Renyi entropy (WRE) and a three-segment biogeography-based optimization (3SBBO) algorithm. This approach, involving WRE for feature extraction, 3SBBO for optimizing neural network biases and weights, and a feed forward neural network (FNN) for image classification, outperformed methods like kernel-based extreme learning machine and radial basis function neural network.

Berg (2018) demonstrated the use of the ViT model to classify normal, malignant, and benign breast tissues based on ultrasound images. Their study showed that ViT achieved better accuracy in classifying ultrasound breast images compared to convolutional neural networks (CNNs).

4.2.2. Virtual Patient Care

Baig (2017) highlighted that advancements in wearable technology and the application of AI and machine learning (ML) in healthcare have already been explored extensively. As a result, patient monitoring and management through virtual care, supported by wearable technology, have become an integral part of standard healthcare practices. AI has also been utilized in managing chronic conditions like diabetes, hypertension, sleep apnea, and bronchial asthma through the use of non-invasive, wearable sensors. In one study, a smart sensor system was proposed to monitor individuals' home environments and track their health status and behaviours. This platform, which incorporates unobtrusive, biomedical, and wearable sensors, captures vital signs such as respiratory rate, heart rate, breathing patterns, blood pressure, and ECG. A smart device, like a tablet, is suggested to act as an interface between the sensors and the user, with data being sent to the cloud for analysis—particularly for elderly care.

A case report by Patel (2021) described a patient whose atrial fibrillation was detected after using a wearable device to monitor her ECG, following inconclusive traditional tests. The device's data helped physicians make an accurate diagnosis. In the area of mental health, Sukei et al. demonstrated that ML models could predict emotional states based on mobile sensor data, even when some data points are missing. This could become a valuable tool for physicians assessing patient moods, although more research is needed to address the challenges of incomplete data for further model development.

Since the onset of the COVID-19 pandemic, wearable devices have made progress in tracking physiological changes and enabling remote patient monitoring. Bogu and Snyder suggested that wearable sensors could help predict COVID-19 early on, and real-time data from wearable devices could enhance the understanding of clinical features that might otherwise be overlooked. AI, combined with ML and big data, could also predict disease progression, such as diabetic nephropathy, and diagnose COVID-19 in organ transplant patients. Yu et al. emphasized the importance of incorporating AI into bedside care during the pandemic, and the growing necessity for remote healthcare services due to COVID-19.

The development of the metaverse offers an even more advanced alternative to traditional telemedicine. A recent study noted that telemedicine services grew 38 times during COVID-19, partly because of reduced in-person consultations and efforts to prevent the virus's spread. The metaverse introduces new tools, such as virtual scanning and raw data sharing that are user-friendly, cost-effective, and efficient. In addition, AR glasses can enable real-time interactions with clinicians, providing live video and audio communication. AR solutions could be particularly useful in emergency situations, allowing remote physicians to offer timely, on-the-spot treatment.

Current technologies like AI, telepresence, blockchain, virtual reality (VR), AR, and digital twinning are transforming healthcare by providing low-cost solutions that improve patient outcomes. The met averse offers a virtual experience where human emotions and actions are simulated, blending social and financial elements from both the real and virtual worlds. AI can further enhance this immersive experience by strengthening the infrastructure of virtual environments and improving essential services.

4.2.3. Medical Research and Drug Discovery

AI is highly effective at analysing the large and intricate datasets involved in medical research. It can also be used to explore scientific literature, integrate diverse data types, and accelerate drug discovery. Pharmaceutical companies are increasingly relying on AI to streamline the drug development process. By using predictive analytics, scientists can identify suitable candidates for clinical trials and develop accurate models of biological processes. Machine learning (ML) plays a key role in the pre-trial phase of clinical trials, aiding in cohort selection, participant organization, and data collection and analysis. This can improve the patient-centric approach, generalizability, and overall success of clinical trials. However, there is a need for greater focus on addressing the functional and philosophical challenges of using ML in clinical trials.

Alongside ML, natural language processing (NLP) (Weissler, E.H et.al.,2021) has shown promise in improving participant management in clinical trials, although its impact on the quality of trials and participant experience remains uncertain. More research is needed to compare methods for optimizing participant management. In clinical research, generative AI can be employed to create synthetic data, enhancing datasets and increasing diversity. Researchers can also conduct trials in immersive, controlled environments using metaverse technologies, facilitating collaboration between geographically distant researchers. The met averse allows researchers to work together in a virtual space, mimicking the experience of being in the same room.

Additionally, AI tools like ChatGPT can assist in clinical trials by aiding data collection and providing information about trial processes. ChatGPT can help summarize relevant research and identify key findings, enabling medical researchers to efficiently navigate large volumes of online information. It can also function as a chatbot, translating complex medical terminology for researchers. However, using chatbots in medical research could raise additional ethical concerns that need to be addressed.

4.2.4. Patient Engagement and Compliance

Patient engagement and adherence are critical (Davenport, T et. al.,2019) factors that often determine the difference between positive and negative health outcomes. Non-compliance occurs when patients do not follow prescribed treatment plans or fail to take recommended medications. When patients are actively involved in their healthcare, they tend to experience improved health outcomes, including better healthcare utilization, lower costs, and enhanced patient satisfaction. A survey of healthcare leaders and executives revealed that fewer than 50% of patients were highly engaged in their treatment plans. Although healthcare providers create treatment plans aimed at improving acute or chronic conditions, the desired outcomes are often hindered when patients fail to make necessary behavioural changes, such as managing their weight, attending follow-up appointments, or adhering to prescribed treatments.

To address these challenges (Hukunda, B. et. al., 2023), AI has been increasingly implemented to boost patient engagement. Machine learning (ML) and workflow automation tools are being applied to deliver complex interventions across the care continuum. Real-time messaging alerts and personalized content designed to promote behavioural change at key moments represent a promising area of research aimed at improving patient compliance and engagement.

4.2.5. Rehabilitation

AI has introduced innovative applications in the realm of rehabilitation, spanning both physical aspects (such as robotics) and virtual components (such as informatics) (Anderson et. Al., 2019). A key branch of AI, machine learning (ML), focuses on developing algorithms that improve through experience. In rehabilitation, ML is applied in areas such as perioperative care, brain–computer interface technologies, myoelectric control systems, and symbiotic neuro prosthetics. Additionally, ML techniques have been utilized in musculoskeletal care for patient data analysis, clinical decision support, and diagnostic imaging. In therapeutic settings, AI-driven (Luxton, D. D., & Riek, L. D., 2019) cognitive tools have been used to evaluate rehabilitation exercises by interpreting signals from machines.

4.2.6. Administrative Applications

AI can help reduce administrative workloads by automatically filling in structured data fields from clinical notes, retrieving key information from previous medical records, and documenting patient interactions. For instance, nurses in the United States spend about 25% of their working hours on administrative and regulatory tasks. Implementing voice-to-text systems could save time for both nurses and physicians. While rule-based systems integrated with electronic health records (EHR) are common, they lack the precision that machine learning (ML)-based algorithms can provide (Wani, S.U.D et. al., 2022). According to Amazon is developing an advanced ML solution that extracts useful insights from unstructured EHR data and clinical notes. Similarly, Li et al. introduced BEHRT, a deep neural sequence transduction model designed for EHRs. BEHRT uses multiple embedding such as age, visit, and event details to represent a patient's clinical history and predict the likelihood of 301 conditions in future visits. This model significantly improves the accuracy of various tasks compared to current deep EHR models. Its flexible architecture allows it to incorporate diverse clinical data, such as diagnoses and medications, improving prediction precision. Additionally, a recent study introduced the hierarchical BEHRT (Hi-BEHRT) (Berg, S. Nudge., 2018) model, which uses a transformer-based approach for risk prediction. The study found Hi-BEHRT to significantly outperform other deep-learning models in predicting risks for patients with chronic conditions like diabetes, heart failure, stroke, and kidney disease.

4.3. AI in Robotics

The merging of robotics and AI is quickly becoming a driving force behind the creation of new industries, cutting-edge technologies, and improved productivity and efficiency in established sectors (Ness et. al., 2023). The ongoing advancements in AI within robotics are leading to increased recognition of its practical uses in various real-world applications. AI is playing a major role in reshaping multiple industries and improving daily life, with applications ranging from autonomous vehicles and customer service to healthcare and industrial robots. Despite concerns about AI and robotics potentially replacing human jobs, the World Economic Forum (WEF) predicts a net gain of 12 million jobs due to this technology by 2025. This growth presents a valuable opportunity for workforce retraining and skill development, aligned with the latest technological innovations.

The integration of AI and robotics offers significant potential to revolutionize job functions in various sectors. This includes automating repetitive tasks in manufacturing and adding adaptability and cognitive functions to monotonous processes. AI's possible uses in robotics are vast and diverse, making it an exciting area for exploration and understanding. Continue reading to gain deeper insights into the relationship between AI and robotics and learn how you can contribute to the future of this transformative field.

4.3.1. The Use of Artificial Intelligence Combined with Robotics Covers A Wide Range of Applications

The integration of AI in robotics has attracted considerable attention in recent years. This growing field of research and innovation examines how AI can be embedded into robotic systems to boost their functionality and efficiency. In robotics, AI has proven highly useful across a wide variety of applications. Its influence spans multiple industries, such as customer service and manufacturing, prompting a shift in how we perceive and interact with robotic technologies (Ness et. al., 2023). Figure 5 shows different robotics applications in AI.

Today, it's essential to explore the key areas where AI is being utilized alongside robotics. AI-driven chatbots, particularly in customer support, are becoming more common. These automated service agents can handle simple and repetitive queries without the need for human involvement. As interactions between AI systems and humans increase, these systems also enhance their learning capabilities. With the continual progress of AI technologies, It is expected that robots will increasingly be used in customer support across both online platforms and physical locations. AI has proven to be an invaluable tool in robotic assembly, especially in complex manufacturing industries like aerospace.

The integration of advanced visual systems with AI allows for real-time adjustments and corrections, (Ness et. al., 2023) enabling robots to autonomously learn optimal paths for specific tasks during operation. In the packaging industry, AI is utilized to boost efficiency, improve accuracy, and reduce costs. AI also streamlines the installation and relocation of robotic systems by refining and storing specific movements performed by the robots. Imaging technology is vital in industries like assembly and logistics, where accuracy is crucial. With AI, robots can achieve enhanced visual capabilities and image recognition, improving their ability to detect even the smallest details

Figure 5. AI in Robotics

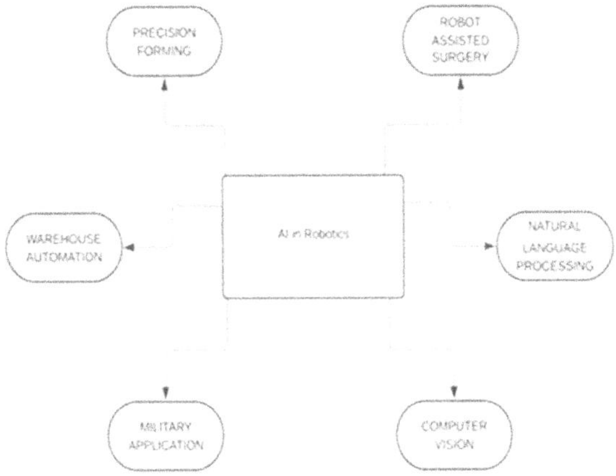

4.3.2. The Modern Era of Robotics and AI

In recent years (Ness, Stephanie et. al., 2023)the field of robotics and artificial intelligence has been marked by a notable trend toward miniaturizing electrical and mechatronic components, alongside a significant increase in computational power. These developments have led to the creation of more practical and efficient robotic systems. In 1973, Waseda University in Japan unveiled the WaBot, the first humanoid robot designed to mimic human movement. The WaBot had basic capabilities for moving, handling objects, and transporting items from one place to another. Five years later, in 1978, Unimation introduced an advanced version of the Unimate called the Programmable Universal Machine for Assembly (PUMA). PUMA quickly became a major reference point in both industry and academia for anthropomorphic robots and continues to be widely cited in current robotics research and literature. The field of reinforcement learning emerged in the 1980s, drawing from various academic disciplines and the concept of trial-and-error learning, which has its roots in early 18th-century psychological studies of animal behaviour. This form of learning, known as reinforcement, involves the development of specific behaviours through interaction with the environment, where stimuli are temporally linked to behaviour and reinforce particular patterns. Technically, this process is akin to solving an optimization problem with stochastic elements due to incomplete system knowledge. An extended version of the optimal control framework can be used to address this problem. Witten was among the first to apply adaptive optimum control methods in this context. Temporal-difference (TD) learning, inspired by animal learning psychology, plays a crucial role in modern reinforcement learning theory. TD learning differs from traditional reinforcement learning by updating the learner's behaviour or strategy after each action based on an estimate of the expected reward, rather than waiting for the actual reward. This approach is governed by the differences between successive estimates. Arthur Samuel first demonstrated this method in 1959 with his checkers-playing program. The actor-critic architecture, a refinement of reinforcement learning, was used in 1983 to solve the pole-balancing control problem. By 1989, the integration of optimal control methods with online learning marked a significant development in the field. Today, Chris Watkin's Q-Learning algorithm represents a breakthrough in combining time difference and optimal control methods. Additionally, the 1980s saw significant progress in robot manipulator control, with John J. Craig and Marc Raibert introducing a new hybrid control technique that allowed for simultaneous position and force control, enabling more compliant movements in robotic manipulators.

4.4. AI in Aviation Industry

Artificial Intelligence is having a profound impact on the aviation industry, driving significant advancements in safety, efficiency, and passenger satisfaction (Pawan Whig et. al.,2024). The integration of AI technologies into aviation operations marks a transformative shift, reshaping how airlines, airports, and air traffic control systems function which is basically shows in figure 6.

One of the most significant applications of AI in aviation is predictive maintenance. Modern aircraft are equipped with a multitude of sensors that generate vast amounts of data. AI algorithms analyse this data to identify patterns and anomalies that may indicate potential mechanical failures. By predicting these issues before they become critical, airlines can perform maintenance proactively, thereby reducing unexpected downtime, lowering repair costs, and enhancing overall flight safety. This predictive approach not only helps in extending the lifespan of aircraft components but also ensures a more reliable and efficient fleet operation.

In the realm of air traffic management, AI is revolutionizing flight scheduling and routing. Air traffic control systems are notoriously complex, dealing with numerous variables and constraints in real-time. AI-powered systems optimize flight paths, manage air traffic flow, and adjust schedules dynamically to minimize delays. This optimization is crucial for managing the crowded airspace and ensuring that flights are both timely and efficient. AI's ability to process large volumes of data quickly and accurately makes it an invaluable tool in enhancing the overall efficiency of air traffic management.

Passenger experience is another area where AI is making a significant difference. Intelligent virtual assistants and chatbots are increasingly being used to provide real-time support to travelers. These AI-driven tools can handle a wide range of customer interactions, from answering queries and providing flight information to assisting with booking and managing travel plans. By offering personalized recommendations and streamlined services, AI improves the overall passenger experience and reduces the workload on human customer service agents.

Machine learning algorithms also play a crucial role in optimizing fuel consumption and improving flight efficiency. By analysing data from various sources, including weather conditions, aircraft performance, and historical flight data, AI can help airlines make more informed decisions about fuel management and route planning. This not only leads to cost savings but also contributes to more sustainable aviation practices by reducing the carbon footprint of flights.

The development of autonomous aircraft represents one of the most exciting frontiers in aviation. AI technologies are being used to develop and refine autonomous systems that could eventually operate aircraft with minimal or no human intervention. While this has the potential to revolutionize the industry by enhancing operational efficiency and safety, it also presents challenges such as ensuring the reliability of these systems and addressing regulatory concerns.

Despite the numerous advantages AI brings to aviation, there are also significant challenges that need to be addressed. Data security is a major concern, as the integration of AI involves handling large volumes of sensitive information. Ensuring robust security measures and regulatory compliance is essential to protecting this data from cyber threats. Additionally, the rapid advancement of AI technology requires continuous updates to regulatory frameworks to ensure that new innovations are safe and reliable.

In summary, AI's role (Pawan Whig et. al.,2024) in the aviation industry is multifaceted and transformative. By enhancing predictive maintenance, optimizing air traffic management, improving passenger services, and contributing to more efficient and sustainable operations, AI is driving significant advancements in the field. As the technology continues to evolve, it will likely lead to even more innovations, reshaping the future of aviation while addressing the associated challenges and opportunities.

Figure 6. AI in Aviation

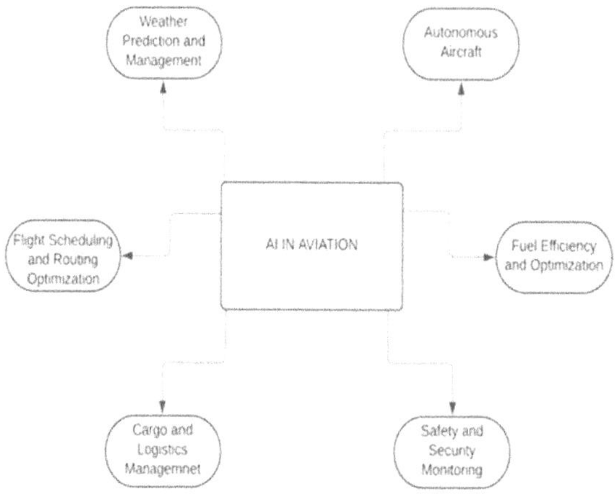

4.4.1. AI Applications in Aviation Industry

AI has a wide range of applications (Shmelova, Tetiana, et. al.,2023) in the aviation industry, each contributing to increased efficiency, safety, and passenger satisfaction. Here are some key applications:

4.4.2. Predictive Maintenance:

AI algorithms analyze data from aircraft sensors to predict potential mechanical failures before they occur. By identifying patterns and anomalies in real-time, airlines can perform maintenance proactively, reducing unexpected downtime and costly repairs.

4.4.3. Flight Scheduling and Routing Optimization:

AI optimizes (Shmelova, Tetiana, et al, 2023) flight schedules and routes by analyzing data from various sources, including weather conditions, air traffic, and historical flight patterns. This helps minimize delays, improve air traffic flow, and enhance operational efficiency.

4.4.3.1. Air Traffic Management:

AI systems assist air traffic controllers by providing real-time data analysis, predicting traffic congestion, and optimizing air traffic routes. This reduces the risk of collisions, manages airspace more effectively, and improves overall safety.

4.4.3.2. Passenger Experience Enhancement:

AI-driven virtual assistants and chatbots offer real-time support for passengers, handling inquiries, booking changes, and providing personalized recommendations. AI also helps in managing check-in processes, reducing wait times, and improving overall customer service.

4.4.3.3. Fuel Efficiency and Optimization:

(Shmelova, Tetiana, et al, 2023) Machine learning algorithms analyse flight data to optimize fuel consumption. By considering factors like weather, aircraft performance, and route efficiency, AI helps airlines reduce fuel costs and minimize their environmental impact.

4.4.3.4. Autonomous Aircraft:

AI is being used to develop autonomous or semi-autonomous aircraft systems. These systems aim to reduce the need for human pilots, enhance safety through advanced automation, and improve operational efficiency.

4.4.3.5. Safety and Security Monitoring:

AI systems are employed to enhance safety and security measures. For example, AI can analyse surveillance footage to detect suspicious behaviour or security breaches at airports. Additionally, AI is used in monitoring and detecting potential cyber security threats.

4.4.3.6. Flight Simulation and Training:

AI-powered flight simulators provide realistic training environments for pilots. These simulators can mimic various flight scenarios, including emergencies, and adapt to a pilot's learning pace, enhancing training effectiveness.

4.4.3.7. Cargo and Logistics Management

AI helps in optimizing cargo loading and logistics operations by analysing data related to cargo weight, distribution, and flight requirements. This improves the efficiency of cargo handling and ensures compliance with safety regulations.

4.4.3.8. Customer Insights and Personalization:

AI analyses customer data to provide personalized services and offers. By understanding passenger preferences and behaviour, airlines can tailor their services, enhance loyalty programs, and create targeted marketing strategies.

4.4.3.9. Weather Prediction and Management:

AI models predict weather patterns (Shmelova, Tetiana, et al, 2023) and turbulence, providing pilots and airlines with accurate and timely information. This helps in planning safer flight paths and mitigating the impact of adverse weather conditions.

4.5. AI in Finance

AI is at the forefront of a significant transformation in the finance industry, driving profound changes that enhance decision-making processes, streamline operations, and elevate customer experiences. The integration of AI technologies into finance leverages advanced algorithms, machine learning, and sophisticated data analytics to overhaul traditional practices and introduce innovative solutions to complex challenges (Cao, Longbing,2020).

At its core, AI's ability to process vast amounts of data at unprecedented speeds allows financial institutions to make more informed and accurate decisions. Machine learning algorithms analyse historical data and identify patterns that humans might overlook, providing deeper insights into market trends, consumer behaviour, and investment opportunities. This enhanced decision-making capability is crucial in a fast-paced financial environment where timely and precise actions are essential.

One of the most notable applications of AI in finance is the automation of routine tasks. By employing AI-driven systems for functions such as transaction processing, customer service, and compliance monitoring, financial institutions can significantly reduce operational costs and minimize human error. Robotic Process Automation (RPA) and AI chatbots, for instance, handle repetitive tasks and customer inquiries efficiently, freeing up human resources to focus on more strategic activities.

Fraud detection is another area where AI is making a substantial impact. Traditional methods of fraud detection often struggle to keep up with the sophistication of modern fraudsters. AI systems, however, analyse transaction patterns in real time, identifying anomalies and potential fraudulent activities with greater accuracy. Machine learning models continuously adapt and improve, enhancing the ability to detect and prevent fraudulent transactions before they cause significant harm.

Investment strategies are being revolutionized through AI-powered tools that analyse market data and predict future trends. Algorithms can evaluate large datasets to identify investment opportunities and risks, enabling more informed portfolio management. AI-driven trading systems also execute trades at optimal times, leveraging predictive analytics to maximize returns and minimize losses.

Personalization of financial services is another major benefit of AI. By analysing customer data and behaviour, AI systems can tailor financial products and services to individual needs and preferences. This level of personalization enhances customer satisfaction and loyalty, as clients receive recommendations and offers that are relevant to their financial goals and circumstances.

As AI technologies continue to evolve, (Cao, Longbing, 2020)they promise to redefine traditional financial practices. The integration of AI into the finance sector not only drives efficiency and accuracy but also introduces new opportunities for growth and innovation. Financial institutions that embrace AI can gain a competitive edge by leveraging these advanced technologies to better manage risks, optimize operations, and deliver superior customer experiences.

However, the adoption of AI in finance also brings challenges that must be addressed. Issues such as data privacy, ethical considerations, and the need for robust regulatory frameworks are crucial in ensuring the responsible use of AI technologies. Financial institutions must navigate these challenges while harnessing the potential of AI to achieve their strategic objectives.

In summary, AI (Cao, Longbing, 2020) is transforming the finance industry by enhancing decision-making, automating routine tasks, improving fraud detection, optimizing investment strategies, and personalizing financial services. As the technology continues to advance, its integration into finance will redefine traditional practices and unlock new opportunities for innovation and growth in an increasingly digital and data-driven world.

There are several examples such as, Lending, Fraud Detection, Insurance Sector, and Investing.

4.5.1. LENDING:

Lending in finance refers to the process by which a lender provides funds to a borrower with the expectation that the borrower will repay the loan over time, typically with interest. This fundamental financial activity is essential for facilitating personal, business, and economic growth. Here's a detailed overview of how lending works and its key components:

Lending can take various forms, including:

a) **Personal Loans:** Unsecured loans provided to individuals for personal use, such as consolidating debt or financing major purchases.

b) **Business Loans:** Loans given to businesses for purposes like expansion, equipment purchase, or working capital. These can be secured or unsecured (Garg, Vikas, et al, 2021).
c) **Mortgage Loans:** Loans specifically for purchasing real estate, where the property serves as collateral.
d) **Student Loans:** Loans designed to help students finance their education, often with deferred repayment options.
e) **Auto Loans:** Loans for purchasing vehicles, where the vehicle itself serves as collateral.

4.5.2. Fraud Detection:

Fraud detection involves identifying and preventing deceptive activities aimed at gaining an unfair advantage or financial benefit through illicit means. It is a critical component in various sectors, including finance, retail, and online services (Garg, Vikas, et al, 2021). Effective fraud detection combines technological tools, data analysis, and human oversight to safeguard against fraudulent actions.

Key Aspects of Fraud Detection:

a) **Data Analysis:** Advanced analytics play a crucial role in detecting fraud. By examining large volumes of transactions and user behaviours, sophisticated algorithms can identify unusual patterns or anomalies that may indicate fraudulent activity. This analysis often involves using statistical models, machine learning techniques, and artificial intelligence to flag suspicious behaviour.
b) **Pattern Recognition:** Fraud detection systems (Garg, Vikas, et al, 2021)are designed to recognize patterns that are indicative of fraudulent activities. These patterns can include irregular transaction sizes, unusual frequencies, or atypical user behaviours. For instance, a sudden spike in transactions from a single account or the use of multiple accounts from a single IP address can trigger alerts.
c) **Real-Time Monitoring:** To effectively prevent fraud, continuous real-time monitoring of transactions and activities is essential. This approach allows for immediate detection of suspicious actions and quick responses to mitigate potential damage. Real-time systems use algorithms that assess transactions as they occur, providing instant alerts to fraud investigators.
d) **Behavioral Analysis:** Understanding normal behaviour is key to identifying deviations that may signify fraud. Behavioural analysis involves tracking how users interact with systems and identifying deviations from their usual patterns. For example, if a user suddenly starts making large purchases or accessing sensitive information at odd hours, it may raise red flags.

e) **Machine Learning and AI:** Machine learning models and artificial intelligence enhance fraud detection by learning from historical data and improving their accuracy over time. These systems can adapt to new types of fraud as they emerge, making them more effective at catching sophisticated schemes that traditional methods might miss.
f) **Fraud Detection Frameworks:** Organizations often implement (Garg, Vikas, et al, 2021) comprehensive fraud detection frameworks that include various tools and processes. These frameworks typically involve risk assessments, fraud scoring, and multi-layered verification procedures to ensure that fraudulent activities are detected and addressed promptly.
g) **Human Oversight:** While technology plays a significant role, human judgment remains crucial in fraud detection. Experienced investigators analyse alerts and assess the context of flagged transactions to determine whether they are genuinely fraudulent or false positives. Human oversight ensures a balance between automation and practical decision-making.
h) **Regulatory Compliance:** Fraud detection systems must comply with regulatory standards and legal requirements to protect consumer data and privacy. Adhering to regulations helps build trust with customers and ensures that fraud detection practices are both effective and ethical.

4.5.3. Insurance Sector:

AI is increasingly shaping the insurance industry by enhancing various aspects of operations, customer service, and risk management. Here's how AI is transforming the insurance sector:

a) Claims Processing: AI streamlines the claims process by automating routine tasks, such as data entry and initial claim assessment. Machine learning algorithms can analyse claims data, assess damage through image recognition, and expedite approvals. This reduces processing times, minimizes human error, and improves customer satisfaction by providing faster claims resolutions.
b) Risk Assessment and Underwriting: AI enhances risk assessment and underwriting by analysing vast amounts of data from diverse sources, such as social media, public records, and IoT devices. Predictive analytics and machine learning models help insurers evaluate risks more accurately and set premiums based on individual risk profiles. This leads to more personalized insurance policies and better risk management.
c) Predictive Analytics: Predictive analytics powered by AI helps insurers anticipate future trends and customer needs. For example, AI can forecast claim frequencies, predict customer churn, and identify emerging risks. This proactive

approach allows insurers to make informed decisions, optimize their strategies, and allocate resources more effectively.

d) Customer Service and Engagement: AI-powered chatbots and virtual assistants improve customer service by providing instant responses to queries, handling routine tasks, and offering personalized recommendations. These AI tools enhance customer engagement, streamline policy management, and provide 24/7 support, leading to higher customer satisfaction and retention.
e) Enhanced Data Management: AI helps insurers manage and analyse large volumes of data more effectively. By employing advanced data analytics and machine learning techniques, insurers can gain deeper insights into customer behaviour, market trends, and operational performance. This data-driven approach supports better decision-making and strategic planning.

4.5.4. Investing:

AI is increasingly influencing the investing sector, transforming how investment decisions are made, portfolios are managed, and market trends are analysed. Here's a detailed look at the various roles AI plays in investing:

a) Algorithmic Trading: AI-driven algorithms are used for high-frequency trading, executing buy and sell orders at speeds and frequencies that are impossible for human traders. These algorithms analyse market data, identify trading opportunities, and execute trades based on predefined criteria. AI enhances the accuracy and efficiency of trading strategies, allowing for rapid response to market movements.
b) Portfolio Management: AI assists in optimizing portfolio management by analysing various factors such as risk tolerance, investment goals, and market conditions. Robo-advisors, powered by AI, offer personalized investment strategies and portfolio recommendations based on individual preferences and financial goals. These systems can automatically rebalance portfolios to maintain optimal asset allocation.
c) Sentiment Analysis: AI performs sentiment analysis by analysing news articles, social media posts, and other sources of public opinion to gauge market sentiment. Understanding market sentiment helps investors assess the potential impact of news events or public opinion on asset prices, enabling them to make more informed investment choices.
d) Enhanced Decision-Making: AI supports enhanced decision-making by providing advanced analytical tools and insights. Investors can leverage AI-powered models to evaluate complex data sets, simulate different scenarios, and assess

potential outcomes. This helps in making more informed and strategic investment decisions.

4.6. AI in Games

Modern computer games often use 3D (Kumar, Koushal et. al., 2012) animated graphics and increasingly sophisticated 3D sound effects to create a realistic experience. However, the so-called "AI" in most games doesn't reflect true artificial intelligence as understood in academic circles. Instead, it involves a variety of techniques aimed primarily at simulating believable intelligent behaviour. The term "game AI" encompasses a wide range of programming and design methods, such as path finding algorithms, neural networks, emotion and social models, finite state machines, rule-based systems, decision trees, and other related approaches. These techniques are employed to enhance the illusion of intelligence within the gaming environment.

4.6.1. Roles of AI in Games

AI has a significant impact on the gaming industry, enhancing gameplay, creating realistic environments, and improving player experiences (Millington, 2019). Here are some key applications of AI in games:

a) Non-Player Character (NPC) Behavior: AI is used to create intelligent and dynamic NPCs that react to player actions in realistic ways. Techniques such as finite state machines, behaviour trees, and utility-based AI help NPCs exhibit complex behaviours, including: **(1) Pathfinding:** NPCs navigate the game world using algorithms like A* or navigation meshes to find optimal routes. (2) **Decision Making:** NPCs make decisions based on the player's actions, environmental factors, and predefined rules, creating more immersive interactions.
b) Procedural Content Generation: AI can generate game content dynamically, including levels, maps, and quests. Procedural generation algorithms use AI to create diverse and unique game environments, such as: (1) **Terrain Generation:** Creating landscapes and terrains that are varied and natural-looking. (2) **Quest and Mission Generation:** Generating quests and missions based on player progress and preferences, ensuring a fresh experience for each play through.
c) Adaptive Difficulty: AI adjusts the difficulty level of a game in real-time based on player performance. Techniques include: (1) **Dynamic Difficulty Adjustment:** Modifying game parameters, such as enemy strength or puzzle complexity, to match the player's skill level. (2) **Personalized Challenges:** Tailoring challenges and obstacles to suit individual player preferences and abilities.

d) Behavioral AI: AI models simulate realistic behaviours and strategies for game characters and opponents. Examples include: (1) **Strategic Decision Making:** AI opponents use advanced strategies and tactics, such as planning and resource management, to provide challenging gameplay. (2) **Learning from Players:** AI systems analyse player behaviour to adapt and develop new strategies, creating a more engaging experience.

e) Natural Language Processing (NLP): NLP enables interactive dialogue systems and conversational agents within games. Applications include: (1) **Chatbots:** AI-driven chatbots engage players in meaningful conversations and provide information or assistance. (2) **Dynamic Dialogue:** Generating and managing dialogue based on player choices, enhancing narrative depth and immersion.

f) Animation and Motion: AI techniques are used to create realistic animations and movements for game characters. Examples include: (1) **Procedural Animation:** AI generates smooth and natural animations for character movements and interactions. (2) **Motion Synthesis:** Combining different motion capture data to create fluid and varied animations.

g) Player Behavior Analysis: AI analyses player behaviour to understand preferences and patterns, leading to personalized experiences. Applications include: (1) **Recommendation Systems:** Suggesting in-game content, such as items or missions, based on player interests and history. (2) **Behavior Prediction:** Anticipating player actions and adapting game elements accordingly.

h) Game Testing and Quality Assurance: AI assists in game testing by automating repetitive tasks and identifying potential issues (Melhart et al., 2023). Techniques include: (1) **Automated Testing:** AI-driven tools test various game scenarios, detecting bugs and inconsistencies more efficiently. (2) **Performance Monitoring:** Analysing game performance metrics to ensure smooth operation and optimal player experience.

i) Enhanced Storytelling and Narrative: AI enhances storytelling by creating adaptive narratives that respond to player choices and actions. Applications include: (1) **Branching Narratives:** Developing storylines that change based on player decisions, leading to multiple possible outcomes. (2) **Procedural Story Generation:** Creating unique and engaging story elements dynamically based on player interactions.

j) Real-Time Strategy (RTS) AI: In RTS games, AI manages complex tasks such as resource allocation, unit control, and strategic planning(Millington, Ian., 2019). Techniques include: (1) **Tactical Planning:** Developing strategies for unit deployment and combat based on the evolving game state. (2) **Resource Management:** Allocating resources and managing economy to optimize gameplay and maintain balance.

5. AI IN ADVANCING APPLIED LIFE SCIENCES

The role of AI in advancing applied life sciences is rapidly evolving, with AI transforming key areas like genomics, drug discovery, medical diagnostics, and personalized medicine. Exploring this topic requires a deep dive into AI techniques beyond machine learning, encompassing neural networks, natural language processing (NLP), computer vision, robotics, and evolutionary algorithms. Here's how different AI techniques contribute to various life sciences applications.

5.1. Understanding Neural Networks in Life Sciences

Neural networks, especially deep learning models (Wongchai et al., 2022), as shown in Figure 7, are inspired by the human brain's structure and function. Their application in life sciences is crucial for tasks like gene sequencing, disease prediction, and image analysis (e.g., MRI, CT scans).

Figure 7. Deep neural network

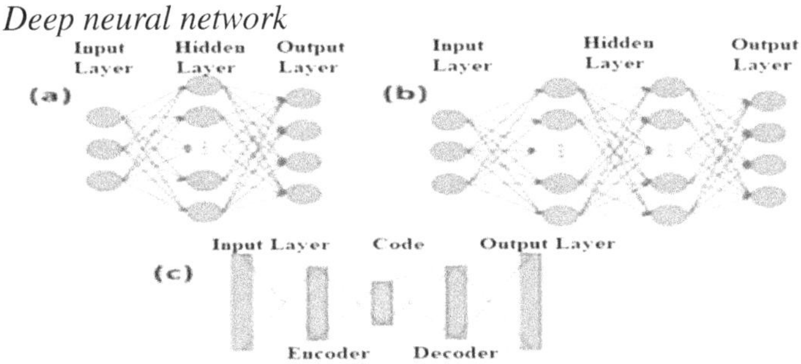

5.2. Broad AI Techniques in Life Sciences Beyond Machine Learning

AI applications in life sciences indeed extend beyond traditional machine learning, involving various methods tailored to specific challenges:

- **Natural Language Processing (NLP)**: Essential for interpreting unstructured clinical data, research papers, and electronic health records (EHR). NLP aids in identifying patterns in patient histories and synthesizing new insights (Weissler et al., 2021). Figure 8 illustrates the basic NLP processing flow.

Figure 8. NLP clinical data processing

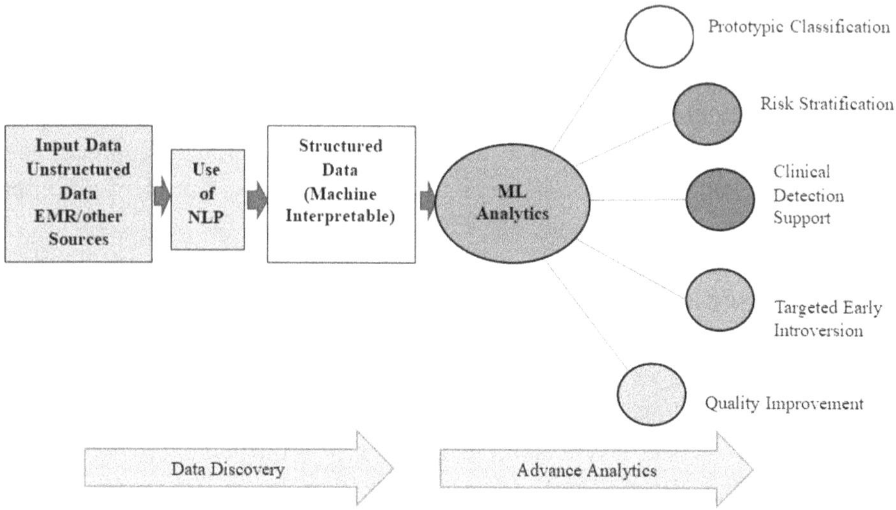

- **Computer Vision**: Utilized in analysing medical images, identifying microscopic organisms, and automating pathology slide analysis. Computer vision supports early disease detection and precise cell classification. Figure 9 describes image pre-processing stages to analysing medical images which most prominently affected on classification.

Figure 9. Image pre-processing

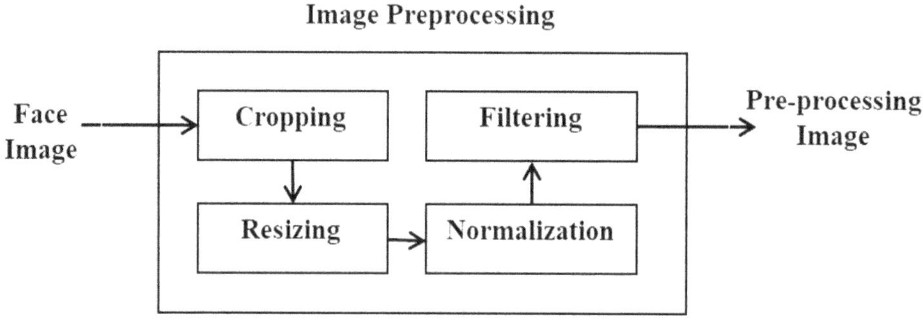

- **Evolutionary Algorithms**: These are employed to optimize bioinformatics workflows, predict protein folding structures, and simulate drug interaction effects, mimicking natural selection to refine solutions iteratively (Whig et al., 2024). Figure 10 illustrates the general optimization flow for framing the problem to find an optimized solution.

Figure 10. Optimized workflow

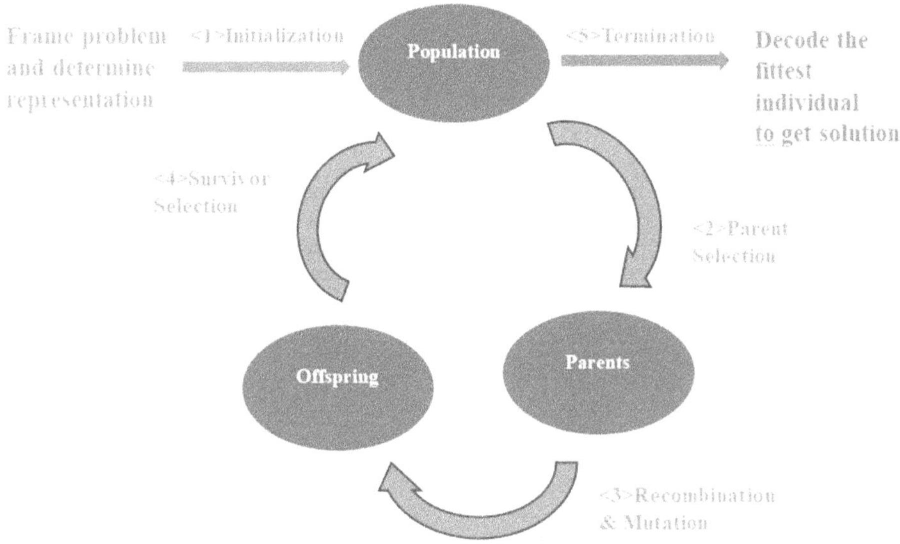

5.3. Broader Applications of AI in Life Sciences

While machine learning is prominent in life sciences, other AI techniques also play significant roles:

- **Robotics and Autonomous Systems**: Assist in laboratory automation, high-throughput screening, and surgical assistance, enhancing precision and reducing human error (Habuza et al., 2021).
- **Predictive and Prescriptive Analytics**: AI tools predict disease progression and suggest treatments based on historical data, particularly useful in managing chronic illnesses.
- **Reinforcement Learning (RL)**: Applied in adaptive drug dosing and personalized healthcare plans, figure 11 describes, reinforcement learning dynamically adjusts strategies to achieve optimal outcomes.

Figure 11. Reinforcement Learning

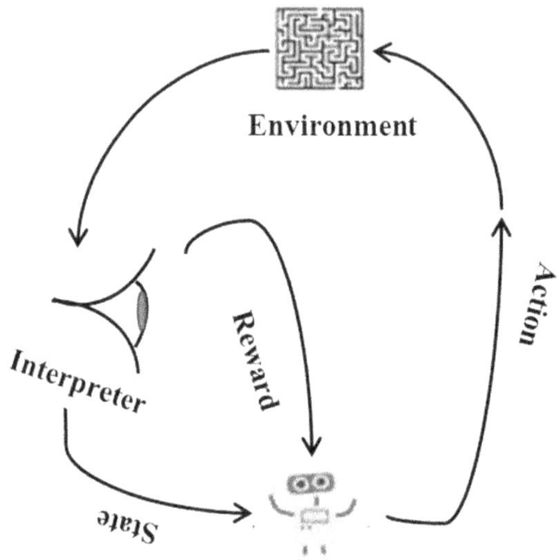

6. FUTURE SCOPE

The future scope of AI in advancing applied life sciences is promising, with transformative potential across various domains (Raj, Ravi, and Andrzej Kos., 2023). AI is set to revolutionize personalized medicine by tailoring treatments based on genetic and lifestyle data and expedite drug discovery through advanced predictive models. Enhanced genomics analysis and improved medical imaging will further enable precise diagnostics and targeted therapies. Predictive analytics and health monitoring will offer proactive disease management, while AI-driven robotics and automation will streamline healthcare delivery. As AI continues to evolve, it will drive innovations in research, patient care, and personalized medicine, ultimately leading to more effective treatments and improved health outcomes.

7. DRAWBACKS AND CHALLENGES

7.1. In Agriculture:

AI has the potential to revolutionize agriculture by improving efficiency, crop yields, and resource management. However, several drawbacks and challenges (Ben Ayed et. al.,2021) are associated with its implementation in this sector:

a) High Initial Costs

Expensive Technology: The upfront cost of implementing AI-powered tools like drones, sensors, and autonomous machinery is high, making it difficult for small-scale or low-income farmers to adopt these technologies.

Investment in Infrastructure: AI often requires substantial investments in infrastructure, such as high-speed internet, cloud services, and modern farming equipment, which may not be readily available in rural areas.

b) Data Dependency

Lack of Data: AI systems rely on large datasets to function effectively, but many farms, especially in developing regions, do not have access to sufficient historical data or real-time monitoring.

Data Privacy Concerns: As AI tools collect massive amounts of data from farms, there are concerns about who owns the data and how it is used, especially if shared with third-party companies.

c) Job Displacement

Automation of Labor: The use of AI-driven machinery and robots to automate tasks like planting, weeding, and harvesting may displace manual labor, particularly in regions where agriculture provides significant employment.

Loss of Traditional Knowledge: Automation and AI can lead to a reduction in the reliance on farmers' traditional knowledge and skills, which could erode local farming practices and expertise.

7.2. In Healthcare:

a) Algorithms Development Concerns

Biases in data collection (no date) processes can lead to potentially skewed outcomes in model development. For example, if datasets under-represent minority groups due to racial biases, it may result in inaccurate predictions. While techniques such as creating multi-ethnic training datasets are available to address these biases, AI models may also independently handle some biases, like the stereotype neural network designed to mitigate ambiguous effects. The effectiveness of these strategies in eliminating real-world bias remains to be seen.

Post-data collection, AI development faces additional challenges. Over fitting occurs when an algorithm identifies irrelevant associations between patient features and outcomes, leading to poor performance on new, unseen data. This happens when too many variables influence the results, causing the model to excel within the training dataset but fail to predict future events accurately. Another concern is data leakage, where an algorithm's high accuracy may be misleading if the dataset includes covariates that improperly correlate with the outcome. To address these issues, it is essential to validate results with new datasets.

b) Ethical concerns:

Ethical concerns have surrounded AI since its inception, with accountability emerging as a key issue, distinct from the data privacy and security challenges already recognized. The current system demands that someone be held accountable when poor decisions are made, especially in fields like medicine where the stakes are high. Many view AI as a “black box” because it can be difficult to understand how an algorithm arrives at a particular conclusion. While this “black-box” issue may be less concerning in lower-stakes applications, where the focus is on efficiency or operational improvements, the question of accountability becomes critical in healthcare settings. When AI is used to improve medical outcomes, the responsibility for errors becomes a significant concern. In cases of system failure, it can be unclear who should be held responsible. Blaming the doctor may not be appropriate if they did not develop or directly manage the algorithm, and holding the developer accountable might seem disconnected from the clinical environment

7.3. In Robotics:

AI in robotics brings (Habuza, Tetiana, et al., 2021) significant advancements, but it also presents several drawbacks and challenges across various aspects of technology and implementation. Here are some key drawbacks of AI in robotics:

a) High Development and Implementation Costs

Expensive to Develop: Developing AI-powered robots requires significant investment in research, hardware, and software, which makes it costly for many organizations to adopt.

Maintenance and Upgrades: AI-driven robots need regular maintenance, updates, and potential reprogramming, which adds to the operational costs.

b) Job Displacement

Automation of Jobs: AI-powered robots can perform tasks traditionally done by humans, leading to job displacement in industries such as manufacturing, logistics, and even services.

Skills Gap: As more tasks are automated, there is a growing demand for highly specialized skills to manage and maintain AI-driven systems, leaving behind workers without technical training.

c) Complexity and Technical Limitations

Lack of Flexibility: Many AI-powered robots are designed for specific tasks and have limited **adaptability**. When faced with unexpected situations or tasks outside their programming, they often fail to respond effectively.

Difficulty in Training: Training AI systems for complex decision-making tasks, such as navigating uncertain environments or interacting with humans in dynamic settings, is still a challenge.

d) Safety Concerns

Risk of Malfunctions: AI-driven robots may malfunction or make incorrect decisions, posing safety risks, especially in environments like healthcare, autonomous vehicles, and heavy industry.

Autonomous Decision Making: In certain applications, such as autonomous vehicles or military robotics, robots' autonomous decision-making capabilities raise concerns about their reliability and ethical implications.

7.4. In Aviation:

AI is transforming the aviation industry, offering numerous benefits such as enhanced safety, efficiency, and customer experience (Saraf et al., 2020). However, it also presents several challenges and drawbacks that need to be addressed:

a) High Initial Costs

Expensive Implementation: The integration of AI systems, such as predictive maintenance, autonomous operations, and air traffic management tools, requires substantial financial investment. This can be particularly challenging for smaller airlines or aviation companies with limited budgets.

Maintenance and Upgrades: AI systems require continuous updates, monitoring, and maintenance, which adds to operational costs over time.

b) Job Displacement

Automation Impact: AI's ability to automate tasks, from customer service chat-bots to autonomous aircraft and air traffic control, poses the risk of job displacement for pilots, air traffic controllers, and ground staff.

Changing Skill Requirements: The increasing use of AI in aviation demands a shift in skills. Employees need specialized knowledge to manage AI systems, creating a gap for those who lack the required technical expertise.

c) Safety Concerns

System Failures: AI systems, while sophisticated, can fail due to technical glitches or incorrect data inputs. In critical sectors like aviation, where safety is paramount, any malfunction could have catastrophic consequences.

Reliability of Autonomous Systems: The idea of autonomous planes or drones piloted entirely by AI raises concerns about the reliability and safety of these systems, especially in the face of unpredictable situations or emergencies.

7.5. In Games:

AI has brought significant advancements to the gaming industry, enhancing player experiences, creating more dynamic gameplay, and adding realism to game environments (Yin et al., 2023). However, the implementation of AI in games also comes with several drawbacks:

a) Lack of True Intelligence:

Predictable Behaviour: In many cases, game AI is programmed with a set of predefined behaviours, making it predictable over time. This can reduce the challenge and excitement for players who learn how the AI will react in certain situations.

Limited Adaptability: Most game AI lacks the ability to truly learn and adapt to different players or changing environments, limiting the depth of interaction and gameplay variety.

b) High Development Costs:

Complexity and Cost: Developing sophisticated AI for games can be expensive and time-consuming. Creating AI that can make real-time decisions, interact with complex environments, and provide engaging experiences requires significant resources.

Balancing AI Difficulty: Designers often struggle to find a balance between creating AI that is challenging but not frustrating for players, which adds to development challenges and costs.

c) Over-Reliance on AI:

Reduced Player Control: Some games rely too heavily on AI-driven elements, reducing the player's sense of control. In games with too much automation, players might feel like their actions have little impact on the outcome, reducing engagement.

Lack of Player Customization: AI-driven decisions can sometimes override player preferences or limit the ability to customize gameplay experiences, leading to dissatisfaction[36].

8. CONCLUSION

Artificial Intelligence (AI) is revolutionizing various industries, including agriculture, healthcare, robotics, aviation, and finance. In agriculture, AI supports better crop monitoring, yield forecasting, and resource management efficiency. In healthcare, AI improves diagnostic processes, tailors treatments, and streamlines administrative tasks, contributing to better patient care. The incorporation of AI in robotics has resulted in greater automation, precision, and flexibility across sectors like manufacturing and customer service. The aviation sector utilizes AI for predictive maintenance, flight planning, and improving safety measures. In finance, AI enhances fraud detection, risk management, and algorithmic trading, fostering a more secure and efficient financial landscape. Although each industry faces hurdles related to ethics, adoption, and trust, AI's rapid evolution promises improved productivity, informed decision-making, and greater innovation across these fields.

REFERENCES

Academy of Royal Medical Colleges. (2019) Artificial Intelligence in Healthcare. https://www.aomrc.org.uk/wpcontent/uploads/2019/01/Artificial_intelligence_in_healthcare_0119.pdf

Al Kuwaiti, A., Nazer, K., Al-Reedy, A., Al-Shehri, S., Al-Muhanna, A., Subbarayalu, A. V., Al Muhanna, D., & Al-Muhanna, F. A. (2023). A Review of the Role of Artificial Intelligence in Healthcare. *Journal of Personalized Medicine*, *13*(6), 951. DOI: 10.3390/jpm13060951 PMID: 37373940

Anderson, D. (2019). Artificial Intelligence and Applications in PM&R. *American Journal of Physical Medicine & Rehabilitation*, *98*(11), e128–e129. DOI: 10.1097/PHM.0000000000001171 PMID: 30839314

Baig, M. M., GholamHosseini, H., Moqeem, A. A., Mirza, F., & Lindén, M. (2017). A Systematic Review of Wearable Patient Monitoring Systems–Current Challenges and Opportunities for Clinical Adoption. *Journal of Medical Systems*, *41*(7), 115. DOI: 10.1007/s10916-017-0760-1 PMID: 28631139

Bedi, G., Carrillo, F., Cecchi. (2015). Automated analysis of free speech predicts psychosis onset in high-risk youths. *npj Schizophr 1, 15030. https://doi.org/.*DOI: 10.1038/npjschz.2015.30

Ben Ayed, R., & Hanana, M. (2021). Artificial intelligence to improve the food and agriculture sector. *Journal of Food Quality*, *2021*(1), 5584754.

Berg, S. (2018). Nudge Theory Explored to Boost Medication Adherence. American Medical Association., Available online www.ama-assn.org/delivering-care/patient-support-advocacy/nudge-theory-exploredboost-medication-adherence, Retrieved January 1, 2023, from.

Cao, L. AI in Finance: A Review (July 10, 2020). Available at *SSRN*: https://ssrn.com/abstract=3647625 or http://dx.doi.org/DOI: 10.2139/ssrn.3647625

Chen, S.-F., & Kuo, Y.-F. (2022). *Artificial Intelligence for Image Processing in Agriculture*. Sensing, Data Managing, and Control Technologies for Agricultural Systems., DOI: 10.1201/9781003214892

Davenport, T., & Kalakota, R. (2019). The potential for artificial intelligence in healthcare. *Future Healthcare Journal*, *6*(2), 94–98. DOI: 10.7861/futurehosp.6-2-94 PMID: 31363513

Emara, T. (2019). Adaptive Power-Saving Mechanism for VoIP Over WiMAX Based on Artificial Neural Network. In *Algorithms, Methods, and Applications in Mobile Computing and Communications* (pp. 158–177). IGI Global. https://doi.org/DOI: 10.4018/978-1-5225-5693-0.ch007

Emara, T. Z. (2016). Maximizing power saving for voip over wimax systems. [IJMC-MC]. *International Journal of Mobile Computing and Multimedia Communications*, *7*(1), 32–40. DOI: 10.4018/IJMCMC.2016010103

Emara, T. Z., Saleh, A. I., & Arafat, H. (2014). Power saving mechanism for VoIP services over WiMAX systems. *Wireless Networks*, *20*(5), 975–985. DOI: 10.1007/s11276-013-0650-5

Emara, T. Z., Trinh, T., & Huang, J. Z. (2023). Geographically distributed data management to support large-scale data analysis. *Scientific Reports*, *13*(1), 17783. DOI: 10.1038/s41598-023-44789-x PMID: 37853092

Garg, V., . . . (Eds.). Applications of Artificial Intelligence in Business and Finance: Modern Trends. CRC Press,(2021). ISBN 9781774639658,272 Pages 8 Color & 30 B/W Illustrations,Published July 8, 2024 by Apple Academic Press.

Ghosh, P. (2021). AI Early Diagnosis Could Save Heart and Cancer Patients. *Science Correspondent. BBC News. 2018.*

Girish Kumar jha (2007). Artificial Neural Networks and its applications. *ARI, New Delhi, 48416112*

Habuza, T., Navaz, A. N., Hashim, F., Alnajjar, F., Zaki, N., Serhani, M. A., & Statsenko, Y. (2021). AI applications in robotics, diagnostic image analysis and precision medicine: Current limitations, future trends, guidelines on CAD systems for medicine. *Informatics in Medicine Unlocked*, *24*, 100596. Advance online publication. DOI: 10.1016/j.imu.2021.100596

Hukunda, B., Rau, A., & Upadhyay, P. (2023). Reimaging Healthcare Opportunities with Artificial Intelligence. *Infosys Navigate Your Next. 2018. Available online:*https://www.infosys.com/industries/healthcare/featuresopinions/Documents/reimagining-healthcare-opportunities.pdf

Javaid, M., Haleem, A., Khan, I. H., & Suman, R. (2023). Understanding the potential applications of Artificial Intelligence in Agriculture Sector. *Advanced Agrochem*, *2*(1), 15–30.

Khorasani, E. S. (2008). Artificial intelligence: Structures and strategies for complex problem solving. Scalable Computing: Practice and Experience, 9(3).

Kumar, K., & Thakur, G. S. M. (2012). Advanced applications of neural networks and artificial intelligence: A review. International journal of information technology and computer science, 4(6), 57-68.

Luxton, D. D., & Riek, L. D. (2019). Artificial intelligence and robotics in rehabilitation. In Brenner, L. A., Reid-Arndt, S. A., Elliott, T. R., Frank, R. G., & Caplan, B. (Eds.), *Handbook of rehabilitation psychology* (3rd ed., pp. 507–520). American Psychological Association., DOI: 10.1037/0000129-031

Melhart, D., Togelius, J., Mikkelsen, B., Holmgård, C., & Yannakakis, G. N. (2023). The ethics of AI in games. *IEEE Transactions on Affective Computing*, *15*(1), 1–14. DOI: 10.1109/TAFFC.2023.3276425

Millington, I. (2019). *AI for Games*. CRC Press. DOI: 10.1201/9781351053303

Mohammed Yousef Shaheen. (2021). Applications of Artificial Intelligence (AI) in healthcare: A review. *Science OpenPreprints. DOI:* DOI: 10.14293/S2199-1006.1.SOR-.PPVRY8K.v1

Ness, S., Shepherd, N., & Xuan, T. (2023). Synergy Between AI and Robotics: A Comprehensive Integration. *Asian Journal of Research in Computer Science.*, *16*(4), 80–94. DOI: 10.9734/ajrcos/2023/v16i4372

Patel, D., & Tarakji, K. G. (2021). Smartwatch diagnosis of atrial fibrillation in patient with embolic stroke of unknown source: A case report. *Cardiovascular Digital Health Journal*, *2021*(2), 84–87. DOI: 10.1016/j.cvdhj.2021.01.001 PMID: 35265894

Raj, R., & Kos, A. (2023). Artificial Intelligence: Evolution, Developments, Applications, and Future Scope. *Przeglad Elektrotechniczny*, *99*(2).

Saraf, A. P.. (2020). *"Explainable artificial intelligence for aviation safety applications." AIAAAviation Forum*. Using Big Data to Increase Safety., DOI: 10.2514/6.2020-2881

Serey, J., Quezada, L., Alfaro, M., Fuertes, G., Vargas, M., Ternero, R., & Gutierrez, S. (2021). Artificial intelligence methodologies for data management. *Symmetry*, *13*(11), 2040. DOI: 10.3390/sym13112040

Shmelova, T.. (2023). Artificial intelligence methods and applications in aviation. In *Handbook of Research on AI Methods and Applications in Computer Engineering* (pp. 108–140). IGI Global., DOI: 10.4018/978-1-6684-6937-8.ch006

Trinh, T., Duc, L. P., Tran, C. T., Duy, T. T., & Emara, T. Z. (2021). A new stratified block model to process large-scale data for a small cluster. *International Conference on Artificial Intelligence and Big Data in Digital Era*, 253–263. Springer International Publishing Cham. https://doi.org/DOI: 10.1007/978-3-030-97610-1_21

TrinhHuy., T., Trinh, T., NguyenViet, H., Emara, T. Z., & VuongThi, N. (2025, February). *Event scale prediction in event-based social networks*. 11th EAI International Conference on Industrial Networks and Intelligent Systems, Da Nang, Vietnam.

Wani, S. U. D., Khan, N. A., Thakur, G., Gautam, S. P., Ali, M., Alam, P., Alshehri, S., Ghoneim, M. M., & Shakeel, F. (2022). Utilization of Artificial Intelligence in Disease Prevention: Diagnosis, Treatment, and Implications for the Healthcare Workforce. *Health Care*, *10*(4), 608. DOI: 10.3390/healthcare10040608 PMID: 35455786

Wei, C., Salloum, S., Emara, T. Z., Zhang, X., Huang, J. Z., & He, Y. (2018). A Two-Stage Data Processing Algorithm to Generate Random Sample Partitions for Big Data Analysis. In M. Luo & L.-J. Zhang (Eds.), *Cloud Computing -- CLOUD 2018* (Vol. 10967, Issue 2018060, pp. 347–364). Springer International Publishing. DOI: 10.1007/978-3-319-94295-7_24

Weissler, E. H., Naumann, T., Andersson, T., Ranganath, R., Elemento, O., Luo, Y., Freitag, D. F., Benoit, J., Hughes, M. C., Khan, F., Slater, P., Shameer, K., Roe, M., Hutchison, E., Kollins, S. H., Broedl, U., Meng, Z., Wong, J. L., Curtis, L., & Ghassemi, M. (2021). The role of machine learning in clinical research: Transforming the future of evidence generation. *Trials*, *22*(1), 537. DOI: 10.1186/s13063-021-05489-x PMID: 34399832

Whig, P., Kasula, B. Y., Yathiraju, N., Jain, A., & Sharma, S. (2024). *Transforming Aviation: The Role of Artificial Intelligence in Air Traffic Management. New Innovations in AI, Aviation, and Air Traffic Technology*. IGI Global., DOI: 10.4018/979-8-3693-1954-3.ch004

Wongchai, A., Shukla, S., Ahmed, A., Sakthi, U., Jagdish, M., & Kumar, R. (2022). Artificial intelligence - enabled soft sensor and internet of things for sustainable agriculture using ensemble deep learning architecture. *Computers & Electrical Engineering*, *102*, 108128. Advance online publication. DOI: 10.1016/j.compeleceng.2022.108128

Yin, Q. Y., Yang, J., Huang, K. Q., Zhao, M. J., Ni, W. C., Liang, B., ... & Wang, L. (2023). AI in human-computer gaming: Techniques, challenges and opportunities. Machine intelligence research, 20(3), 299-317.

Chapter 2

A Historical Perspective on Artificial Intelligence in Applied Life Sciences

Tamer Z. Emara
https://orcid.org/0000-0001-6730-9449
Damietta University, Egypt

Esraa Hassan
Kafrelsheikh University, Egypt

Samar Elbedwehy
Kafrelsheikh University, Egypt

Abeer Saber
Damietta University, Egypt

ABSTRACT

This chapter explores the transformative impact of Artificial Intelligence (AI) on applied life sciences, with a focus on environmental science. It begins by tracing the theoretical foundations of AI from early pioneers like Alan Turing and John McCarthy, and then discusses its evolution through the development of expert systems and machine learning techniques. The chapter highlights AI's significant contributions to environmental monitoring, climate change prediction, and biodiversity conservation, showcasing how AI enhances our understanding and management of complex systems. It also examines the integration of AI with satellite data and sensor networks to address environmental challenges. Concluding with future directions, the chapter addresses emerging trends and ethical considerations, emphasizing AI's role in supporting sustainable development goals. This overview provides a

DOI: 10.4018/979-8-3693-9208-9.ch002

foundational context for understanding AI's role in life sciences and sets the stage for exploring its broader applications in subsequent chapters.

1. INTRODUCTION

The convergence of artificial intelligence (AI) and life sciences marks one of the most transformative developments in modern science and healthcare. As a multidisciplinary field, life sciences encompass the study of living organisms, their interactions, and their environments, with the goal of understanding and improving health and well-being. AI, with its capability to analyze vast amounts of data, recognize patterns, and make predictions, offers unprecedented opportunities to advance this understanding and address complex biological and medical challenges.

AI's journey into life sciences began in the mid-20th century, with foundational theoretical work laying the groundwork for future applications. The evolution of AI from basic computational models to sophisticated machine learning algorithms has enabled researchers and practitioners to tackle problems that were previously intractable. Early efforts in AI focused on simulating human intelligence and decision-making processes, which laid the foundation for more specialized applications in the life sciences.

The integration of AI into life sciences has led to significant advancements in various domains, including genomics, drug discovery, personalized medicine, and agriculture. AI technologies have revolutionized our approach to analyzing genetic data, discovering novel therapeutic targets, and developing targeted treatments. In agriculture, AI-driven tools are enhancing productivity and addressing global food security challenges. These developments highlight AI's potential to drive innovation and improve outcomes across a wide range of applications.

This chapter provides a comprehensive overview of the role of AI in applied life sciences, with a focus on historical developments, key applications, and future directions. It begins by exploring the early theoretical foundations of AI, including seminal contributions by pioneers such as Alan Turing and John McCarthy. The chapter then examines the initial applications of AI in life sciences, highlighting early expert systems and their impact on biological research and medical practice.

Subsequent sections delve into the rise of machine learning and data-driven approaches, discussing how these advancements have transformed genomics, personalized medicine, and drug discovery. The chapter also explores AI's applications in agriculture, illustrating how AI technologies are reshaping farming practices and contributing to food security.

Finally, the chapter addresses the challenges and future directions for AI in life sciences, including data privacy and ethical considerations, the need for high-quality data, and emerging trends. By providing a thorough analysis of these aspects, the chapter aims to offer insights into the current state of AI in life sciences and its potential for future innovation.

The significance of AI in life sciences extends beyond its technical capabilities. AI represents a paradigm shift in how we approach complex biological and medical questions, offering new tools and methodologies for discovery and problem-solving. As AI technologies continue to evolve, they hold the promise of transforming research and practice in life sciences, leading to improved health outcomes, more effective treatments, and a deeper understanding of the living world.

In summary, this chapter sets the stage for a detailed exploration of AI's role in applied life sciences, emphasizing its transformative impact, current applications, and future potential. By examining the historical context, key developments, and ongoing challenges, the chapter provides a comprehensive overview of how AI is shaping the future of life sciences and healthcare.

2. EARLY THEORETICAL FOUNDATIONS

2.1 The Origins of AI

The origins of artificial intelligence (AI) are deeply rooted in the early 20th-century theories of computation and cognitive science, which set the stage for the development of intelligent machines. The idea of creating machines capable of intelligent behavior has been explored through various theoretical frameworks and technological advancements.

Alan Turing and the Concept of Machine Intelligence

Alan Turing's work in the 1950s was pivotal in shaping the theoretical foundations of AI. Turing's 1950 paper, "Computing Machinery and Intelligence," is considered a cornerstone of AI theory. In this influential work, Turing proposed the concept of machine intelligence and questioned whether machines could think. He introduced the Turing Test as a criterion for machine intelligence, suggesting that a machine could be considered intelligent if it could engage in a conversation indistinguishable from that of a human (Turing, 1950).

Turing's ideas were revolutionary in several respects. First, he proposed that intelligence could be evaluated based on a machine's ability to exhibit behavior equivalent to human intelligence, rather than by its internal workings. This approach emphasized the functional aspect of intelligence, focusing on outputs and interactions rather than the underlying mechanisms. Second, Turing's test introduced a pragmatic way to assess machine intelligence, which remains influential in contemporary discussions about AI capabilities.

Theoretical Contributions to Computation and Cognitive Science

Turing's work was informed by earlier developments in computation and cognitive science. The concept of a universal machine, which Turing described in his 1936 paper "On Computable Numbers," laid the theoretical groundwork for modern computers (Turing, 1936). The idea that a single machine could perform any computable task provided a foundational principle for the design of digital computers and the development of AI algorithms.

Additionally, Turing's exploration of cognitive processes and human intelligence influenced subsequent AI research. His work on pattern recognition and mathematical models of learning contributed to the development of early computational models and algorithms used in AI. Turing's emphasis on the mathematical nature of intelligence and learning processes helped bridge the gap between theoretical concepts and practical implementations.

2.2 The Dartmouth Conference

The Dartmouth Conference of 1956 is widely recognized as the formal inception of AI as an academic discipline. Organized by John McCarthy, this landmark event brought together a group of leading researchers who would become central figures in the field of AI (Bibel, 2014).

The Vision of the Dartmouth Conference

John McCarthy, along with Marvin Minsky, Nathaniel Rochester, and Claude Shannon, proposed that "every aspect of learning or any other feature of intelligence can in principle be so precisely described that a machine can be made to simulate it" (McCarthy et al., 1956). This ambitious vision outlined a broad and inclusive approach to AI research, suggesting that all aspects of intelligent behavior could be modeled and replicated by machines.

The conference aimed to explore the potential of artificial intelligence and establish it as a formal research area. The organizers anticipated that significant progress could be made in creating machines that could perform tasks requiring human-like intelligence. The Dartmouth Conference marked a crucial turning point in AI research, as it set the agenda for future investigations and established the foundational goals of the field.

Impact on AI Research and Development

The Dartmouth Conference had a profound impact on the direction of AI research. It established key research themes and methodologies that would shape the field for decades. The focus on simulating intelligent behavior and developing algorithms for learning and problem-solving became central to AI research. The conference also fostered a collaborative research environment, bringing together experts from diverse fields such as computer science, mathematics, and cognitive psychology.

Following the conference, researchers began to develop early AI systems and explore various approaches to achieving machine intelligence. The ideas and goals established at Dartmouth provided a framework for subsequent AI research, influencing the development of early AI technologies and applications.

2.3 Early Computational Models

The period following the Dartmouth Conference saw significant advancements in the development of computational models aimed at simulating intelligent behavior. These early models laid the groundwork for more sophisticated AI systems and contributed to our understanding of cognitive processes.

Early Neural Networks and Symbolic AI

Early efforts in AI research included the development of neural network models and symbolic AI approaches. Neural networks, inspired by the structure and function of biological neurons, were among the first computational models designed to replicate cognitive processes. One notable early neural network model was the Perceptron, developed by Frank Rosenblatt in the late 1950s. The Perceptron was designed to perform binary classification tasks by mimicking the way neurons process information (Rosenblatt, 1958). Although the Perceptron had limitations, it represented an early attempt to create machines capable of learning from data and making decisions.

Symbolic AI, also known as "good old-fashioned AI" (GOFAI), focused on representing knowledge and reasoning through formal symbols and rules. This approach sought to model cognitive processes by encoding human knowledge into algorithms and applying logical inference to solve problems. Early symbolic AI systems, such as those developed by Allen Newell and Herbert Simon, utilized rule-based approaches to simulate problem-solving strategies (Newell & Simon,

1972). Their work on the General Problem Solver (GPS) aimed to replicate human problem-solving techniques through formal algorithms, providing a framework for understanding how machines could perform complex cognitive tasks.

Development of Early AI Algorithms and Systems

In addition to neural networks and symbolic AI, early research in AI involved the development of algorithms and systems for solving specific problems. One of the significant contributions during this period was the development of search algorithms for problem-solving. Techniques such as breadth-first search and depth-first search were introduced to explore problem spaces and find solutions to complex tasks. These algorithms provided a foundation for developing AI systems capable of performing tasks such as automated theorem proving and game playing.

Another important development was the creation of early expert systems, which aimed to simulate human expertise in specific domains. Expert systems used rule-based approaches to encode knowledge and make decisions based on available information. These systems demonstrated the potential of AI to provide expert-level support in various applications, including medical diagnosis and engineering (Buchanan & Shortliffe, 1984; Wick & Thompson, 1992).

Impact on AI Research and Applications

The early computational models and algorithms developed during this period had a lasting impact on AI research and applications. The work on neural networks and symbolic AI provided foundational principles for understanding and developing intelligent systems. The early algorithms and systems demonstrated the feasibility of creating machines capable of performing complex cognitive tasks, paving the way for future advancements in AI technologies (Confalonieri et al., 2021).

The developments of this era also highlighted the interdisciplinary nature of AI research, as researchers from fields such as computer science, mathematics, and cognitive psychology collaborated to advance the field. The theoretical foundations and early models established during this period continue to influence contemporary AI research and applications, shaping the direction of the field and contributing to the ongoing exploration of machine intelligence.

3. INITIAL APPLICATIONS IN LIFE SCIENCES

3.1 Expert Systems in Life Sciences

In the 1960s and 1970s, the application of artificial intelligence (AI) to life sciences began to emerge through the development of expert systems. These systems were designed to emulate the decision-making capabilities of human experts by using knowledge-based approaches to solve complex problems. Expert systems leveraged heuristic rules and pattern recognition to provide solutions in various scientific domains.

Dendral: A Pioneer in Chemical Analysis

One of the earliest and most notable expert systems in the life sciences was Dendral, developed at Stanford University in the 1960s. Dendral was designed to assist chemists in determining the structure of organic molecules based on mass spectrometry data. The system employed heuristic rules and pattern recognition techniques to analyze chemical data and predict molecular structures (Buchanan & Shortliffe, 1984).

Dendral's development marked a significant milestone in the application of AI to chemical analysis. The system's ability to process complex mass spectrometry data and generate hypotheses about molecular structures demonstrated AI's potential to support scientific research. Dendral's success also highlighted the importance of incorporating domain-specific knowledge into AI systems, setting a precedent for future expert systems in various scientific fields.

Early Impact and Influence

The success of Dendral had a profound impact on the development of AI in life sciences. It demonstrated that AI systems could effectively emulate the decision-making capabilities of human experts and handle intricate scientific problems. The techniques and methodologies developed for Dendral influenced subsequent expert systems and contributed to the growth of AI applications in other areas of life sciences, including medicine and biological research.

3.2 The MYCIN System

Another pioneering AI system that made a significant impact in the life sciences was MYCIN, developed in the early 1970s at Stanford University. MYCIN was an expert system designed to assist in diagnosing and treating bacterial infections. It utilized a rule-based approach to simulate the reasoning of medical experts and provide recommendations based on patient symptoms and medical history (Shortliffe, 1976).

Design and Functionality of MYCIN

MYCIN was built upon a knowledge base of medical expertise and used a set of if-then rules to infer diagnoses and recommend treatments. The system's rule-based approach allowed it to process patient data, evaluate symptoms, and suggest appropriate antibiotic therapies. MYCIN's diagnostic capabilities were based on a combination of medical knowledge and heuristic rules, which enabled it to simulate the reasoning of experienced clinicians.

MYCIN's success was notable for several reasons. It demonstrated the potential of AI to enhance clinical decision-making and provide valuable support in medical diagnostics. The system's ability to generate accurate recommendations and assist clinicians in making informed decisions highlighted the practical applications of AI in healthcare.

Legacy and Influence

The development and success of MYCIN had a lasting impact on the field of medical AI. It laid the groundwork for future AI applications in healthcare and inspired the development of subsequent expert systems and diagnostic tools. MYCIN's approach to rule-based reasoning and knowledge representation became a foundational model for later AI systems in medicine and other life sciences.

3.3 Early AI in Biological Research

During the 1960s and 1970s, AI began to make its mark in biological research, with early applications utilizing AI technologies to model biological processes and analyze complex data. The development of programming languages and tools specifically designed for AI research played a crucial role in advancing these applications.

AI Programming Languages: LISP and Prolog

Two influential programming languages that emerged during this period were LISP and Prolog. LISP, developed by John McCarthy in the late 1950s, became a primary language for AI research due to its flexibility and suitability for symbolic processing. It was widely used in AI research to develop algorithms and systems for modeling biological processes (McCarthy, 1960).

Prolog, introduced in the early 1970s, was another significant programming language designed to support AI research. Prolog's focus on logic programming and symbolic reasoning made it well-suited for tasks involving complex data analysis and knowledge representation. Researchers used Prolog to develop systems for analyzing biological data and modeling biological processes, contributing to the advancement of AI in biological research (Colmerauer & Roussel, 1993).

Applications and Advances in Biological Research

Early AI applications in biological research included the development of models for understanding genetic processes, protein structures, and ecological systems. Researchers used AI techniques to analyze complex biological data, such as genetic sequences and protein interactions. These early applications demonstrated AI's potential to handle intricate scientific problems and provided insights into biological processes that were not easily obtainable through traditional methods.

For example, AI-based systems were used to model genetic inheritance patterns and predict the effects of genetic mutations on phenotypic traits. These models helped researchers gain a deeper understanding of genetic processes and contributed to the development of early bioinformatics tools.

Paving the Way for Future Methodologies

The initial applications of AI in biological research paved the way for more advanced methodologies and technologies in the future. The use of AI programming languages like LISP and Prolog facilitated the development of sophisticated models and systems for analyzing biological data. These early efforts laid the foundation for the integration of AI technologies into modern biological research and contributed to the growth of bioinformatics and computational biology.

3.4 Summary

The period from the 1960s to the 1980s saw significant advancements in the application of AI to life sciences. Expert systems like Dendral and MYCIN demonstrated AI's potential to support scientific research and clinical decision-making. Early AI applications in biological research, supported by programming languages such as LISP and Prolog, highlighted the ability of AI to handle complex data and model biological processes. These early developments laid the groundwork for future advancements in AI applications in life sciences, contributing to the evolution of bioinformatics, computational biology, and medical AI.

4. THE RISE OF MACHINE LEARNING AND DATA-DRIVEN APPROACHES

4.1 Introduction of Machine Learning

The 1980s and 1990s heralded a transformative shift in artificial intelligence (AI) research with the emergence of machine learning algorithms. Unlike traditional AI approaches that depended heavily on explicit programming and predefined rules, machine learning focused on enabling systems to learn from data and improve their performance autonomously. This shift marked a significant departure from the earlier AI models and was driven by several key factors.

Advancements in Computational Power

One of the primary drivers behind the rise of machine learning was the substantial increase in computational power. During the 1980s and 1990s, advancements in computer hardware, including the development of more powerful processors and increased memory capacity, allowed researchers to handle larger datasets and perform more complex calculations. This enhanced computational capability provided the foundation for training more sophisticated machine learning models and algorithms (Jiang et al., 2022).

Availability of Large Datasets

Another critical factor contributing to the rise of machine learning was the availability of large datasets. The growth of digital data and the development of data collection technologies enabled researchers to access vast amounts of information for training and evaluating machine learning models. This abundance of data allowed for more accurate and robust learning processes, as models could be trained on diverse and extensive datasets to identify patterns and make predictions (LeCun et al., 2015).

Paradigm Shift in AI Research

Machine learning introduced a paradigm shift in AI research by emphasizing the importance of data-driven approaches over rule-based systems. Instead of relying on predefined rules and algorithms, machine learning models learn from examples and adapt their behavior based on the patterns observed in the data. This approach allowed AI systems to handle more complex and dynamic problems, paving the way for advancements in various domains, including natural language processing, computer vision, and bioinformatics (Mitchell, 1997).

4.2 Neural Networks and Backpropagation

A major advancement during this era was the revival and refinement of neural network models, inspired by the structure and function of the human brain. The development of neural networks and the introduction of the backpropagation algorithm played a crucial role in the success of machine learning during the 1980s and 1990s.

The Revival of Neural Networks

Neural networks, which had their origins in the 1950s and 1960s, experienced a resurgence in the 1980s. These models, inspired by the interconnected neurons in the human brain, are designed to recognize patterns and learn from data through a network of nodes (neurons) organized in layers. The revival of neural networks was driven by advances in computational power and the development of more efficient training algorithms.

The Backpropagation Algorithm

The backpropagation algorithm, popularized by researchers such as Geoffrey Hinton and his colleagues, was a key development in the training of neural networks. Backpropagation is a supervised learning technique used to minimize the error in neural networks by adjusting the weights of the connections between neurons. This algorithm works by propagating the error gradient backward through the network, allowing the model to update its parameters and improve its performance over time (Rumelhart et al., 1986).

Impact on AI Applications

The introduction of backpropagation enabled the training of multi-layer neural networks, also known as deep neural networks. These networks, with their multiple layers of interconnected neurons, were capable of learning complex patterns and representations from data. The advancements in neural network training led to significant improvements in various AI applications, including image recognition, speech recognition, and natural language processing (Hinton et al., 2006) (T. Emara, 2018; T. Z. Emara, 2016; T. Z. Emara et al., 2014).

Applications and Breakthroughs

Neural networks and backpropagation led to several breakthroughs in AI applications. For example, convolutional neural networks (CNNs) became widely used for image recognition tasks, achieving remarkable accuracy in identifying objects and patterns in images. Similarly, recurrent neural networks (RNNs) and their variants, such as long short-term memory (LSTM) networks, were employed for natural language processing tasks, including language modeling and machine translation (Saber et al., 2021; Saber et al., 2021) (Ahmed et al., 2022) (Batra et al., 2024).

4.3 AI in Genomics and Bioinformatics

The rise of machine learning and the growth of genomic data marked a new era in the application of AI to life sciences. The advent of large-scale genomic projects, such as the Human Genome Project, generated an unprecedented volume of genetic data and opened new opportunities for applying AI technologies to genomics and bioinformatics (Collins and Fink, 1995) (Sakr et al., 2020).

The Human Genome Project and Data Explosion

The Human Genome Project, which began in the late 1990s and was completed in the early 2000s, was a landmark achievement in genomics. This international research initiative aimed to map and sequence the entire human genome, providing a comprehensive reference for understanding genetic variation and its role in health and disease. The project generated vast amounts of genomic data, which required advanced computational tools and techniques for analysis and interpretation.

AI-Driven Tools for Genomic Analysis

Machine learning and AI technologies played a crucial role in analyzing and interpreting the genomic data generated by the Human Genome Project. AI-driven tools were developed to identify genetic markers, predict the functional impact of genetic variations, and understand the molecular basis of diseases. These tools leveraged machine learning algorithms to process large datasets and extract meaningful insights from complex genetic information.

Applications in Personalized Medicine

The application of AI in genomics also contributed to the advancement of personalized medicine. By analyzing individual genetic profiles, AI-driven tools could provide insights into a person's susceptibility to certain diseases, predict treatment responses, and guide personalized therapeutic strategies. The integration of AI technologies into genomics and bioinformatics enabled researchers and clinicians to move towards more precise and individualized approaches to healthcare (Alnowaiser et al., 2024) (Saber et al., 2023) (Hassan et al., 2024).

Impact on Bioinformatics

AI technologies revolutionized the field of bioinformatics by providing new methods for analyzing biological data and uncovering patterns that were not easily detectable using traditional techniques. Machine learning algorithms facilitated the development of predictive models for gene function, protein interactions, and disease associations. These advancements accelerated research in molecular biology and contributed to the development of novel therapeutic targets and interventions (LeCun et al., 2015).

4.4 Summary

The rise of machine learning and data-driven approaches in the 1980s and 1990s marked a transformative period in AI research and its application to life sciences. The introduction of machine learning algorithms shifted the focus from explicit programming to data-driven learning, enabling AI systems to improve over time based on experience. The revival of neural networks and the development of the backpropagation algorithm led to significant advancements in various AI applications, including image recognition and natural language processing. The application of AI to genomics and bioinformatics, fueled by the explosion of genomic data, revolutionized the analysis of genetic information and paved the way for personalized medicine. These developments set the stage for continued progress in AI applications in life sciences and beyond.

5. AI IN GENOMICS AND PERSONALIZED MEDICINE

5.1 Advances in Genomic Data Analysis

The completion of the Human Genome Project (HGP) in 2003 was a landmark achievement in genomics, providing a comprehensive map of the human genome and opening new avenues for research and clinical applications. The sheer volume and complexity of genomic data generated by the HGP necessitated the development of advanced computational tools to analyze and interpret this information effectively.

The Challenge of Genomic Data Complexity

Genomic data is characterized by its high dimensionality and complexity, with millions of base pairs and numerous genetic variations to consider. Traditional data analysis methods were insufficient to handle the scale and intricacy of this data, leading to the need for more sophisticated computational approaches. The advent of AI technologies, particularly machine learning and deep learning, provided new capabilities for managing and extracting insights from genomic data (Mitra et al., 2008).

Machine Learning in Genomic Analysis

Machine learning algorithms have been instrumental in addressing the challenges of genomic data analysis. These algorithms are designed to learn from data and identify patterns that are not immediately apparent through traditional analysis methods. In genomics, machine learning techniques have been used to:

- **Predict Disease Susceptibility:** Algorithms can analyze genetic variations and identify associations with various diseases. For example, machine learning models have been employed to predict an individual's risk of developing complex diseases such as cancer, cardiovascular diseases, and diabetes based on their genetic profile.
- **Discover Novel Therapeutic Targets:** Machine learning has facilitated the identification of potential therapeutic targets by analyzing the interactions between genes, proteins, and other biological molecules. This approach has

led to the discovery of new drug targets and the development of targeted therapies.

- **Understand Gene Function and Regulation:** AI-driven tools have been used to model gene expression patterns and regulatory networks. These models help researchers understand the functional roles of genes and the regulatory mechanisms that control their expression, contributing to a deeper understanding of molecular biology.

Deep Learning and Genomic Data

Deep learning, a subset of machine learning involving neural networks with multiple layers, has also made significant contributions to genomic data analysis. Deep learning models are capable of handling large-scale data and capturing complex patterns and relationships. For instance, convolutional neural networks (CNNs) have been used to analyze DNA sequences and identify functional elements, while recurrent neural networks (RNNs) have been applied to study gene expression time series data (LeCun et al., 2015).

5.2 Personalized Medicine and AI

Personalized medicine, also known as precision medicine, aims to tailor medical treatments to individual patients based on their genetic, clinical, and lifestyle information. The integration of AI into personalized medicine has revolutionized the field by enabling more precise and effective treatments.

AI-Driven Patient Data Analysis

AI technologies, particularly machine learning algorithms, play a crucial role in analyzing patient data and providing insights for personalized treatment. Key applications include:

- **Predicting Treatment Outcomes:** Machine learning models can predict how patients will respond to specific treatments based on their genetic and clinical profiles. These predictions help clinicians select the most effective treatment strategies and avoid adverse reactions.
- **Identifying Optimal Therapeutic Strategies:** AI-driven approaches can analyze large datasets to identify patterns and correlations that guide the selection of optimal therapeutic strategies. For example, algorithms can recommend personalized drug regimens based on a patient's genetic makeup and previous treatment responses.

- **Developing Personalized Treatment Plans:** AI technologies facilitate the creation of personalized treatment plans by integrating data from various sources, including genetic information, medical history, and lifestyle factors. These plans are tailored to the individual patient's needs, improving treatment efficacy and patient outcomes (Kourou et al., 2015).

Advancements in Precision Medicine

The use of AI in personalized medicine has led to several advancements, including:

- **Targeted Therapies:** AI-driven research has led to the development of targeted therapies that specifically address the molecular mechanisms underlying diseases. For example, targeted therapies for cancer aim to inhibit specific genetic mutations or protein interactions that drive tumor growth.
- **Genomic Profiling:** Personalized medicine often involves genomic profiling, where AI tools analyze a patient's genetic data to identify mutations and variations associated with diseases. This information helps in diagnosing conditions and selecting appropriate treatments.
- **Predictive Modeling:** AI-based predictive models assist in forecasting disease progression and treatment outcomes. These models use historical data and machine learning algorithms to predict future health events, allowing for proactive management and intervention.

5.3 Case Studies in AI-Driven Genomics

Several case studies exemplify the transformative impact of AI on genomics and personalized medicine. These studies highlight the practical applications of AI in advancing medical research and clinical practice.

Case Study 1: AI in Cancer Genomics

AI algorithms have been used to identify genetic variants associated with cancer, leading to the development of targeted therapies. For example, the application of machine learning to genomic data from cancer patients has revealed mutations in genes such as BRCA1 and BRCA2, which are linked to increased risk of breast and ovarian cancer. AI-driven tools have enabled the identification of patients who may benefit from targeted treatments and preventive measures (LeCun et al., 2015).

Case Study 2: AI in Diabetes Research

In the field of diabetes research, AI technologies have been employed to analyze genetic and clinical data to identify risk factors and predict disease onset. Machine learning models have been used to analyze large datasets from electronic health records and genetic studies to uncover associations between genetic variants and diabetes risk. These insights have contributed to the development of personalized prevention and treatment strategies for diabetes (Mitra et al., 2008).

Case Study 3: AI in Drug Discovery

AI has also made significant contributions to drug discovery by analyzing genomic data to identify potential drug targets. For instance, AI-driven tools have been used to screen large chemical libraries and predict the interactions between drugs and target proteins. This approach has accelerated the drug discovery process and led to the identification of new therapeutic candidates (Kourou et al., 2015).

5.3 Summary

AI has played a pivotal role in advancing genomics and personalized medicine. The completion of the Human Genome Project highlighted the need for advanced computational tools to analyze complex genomic data. AI technologies, including machine learning and deep learning, have enabled researchers to predict disease susceptibility, discover novel therapeutic targets, and enhance personalized medicine. Case studies in cancer genomics, diabetes research, and drug discovery illustrate the transformative potential of AI in advancing medical research and improving patient outcomes.

6. AI IN DRUG DISCOVERY AND DEVELOPMENT

6.1 Early AI Applications in Drug Discovery

The application of AI in drug discovery has its origins in the early days of AI research, when researchers began exploring ways to use computational methods to support drug development. These early AI systems focused on optimizing chemical compounds, predicting drug interactions, and assisting in the design and evaluation of new drugs.

Heuristic Rules and Knowledge-Based Approaches

In the early stages, AI systems in drug discovery relied heavily on heuristic rules and knowledge-based approaches. These methods used predefined rules and expert knowledge to guide the drug discovery process. For example, early systems applied rule-based algorithms to predict the properties of chemical compounds and their interactions with biological targets. These systems aimed to assist chemists and pharmacologists in designing compounds with desirable characteristics and evaluating their potential as drug candidates (Sinha and Vohora, 2018).

Challenges and Limitations

While these early AI applications provided valuable insights, they were limited by the computational power and data availability of the time. The complexity of biological systems and the vast number of possible chemical interactions posed significant challenges. Nevertheless, these early efforts laid the groundwork for more sophisticated AI-driven approaches and demonstrated the potential of computational methods in drug discovery.

Foundations for Modern AI Techniques

The success of early AI applications in drug discovery established a foundation for the development of modern AI techniques. Researchers began to explore more advanced computational methods, including machine learning and data mining, to enhance the drug discovery process. These early efforts also highlighted the need for large-scale data and advanced algorithms, setting the stage for future advancements in AI-driven drug discovery.

6.2 Modern AI Techniques in Drug Discovery

The 21st century has seen a dramatic evolution in AI techniques, transforming drug discovery through the application of deep learning and advanced data analytics. Modern AI algorithms have revolutionized the field by providing new capabilities for analyzing complex biological data, predicting drug efficacy and safety, and streamlining the drug development process.

Deep Learning and Data Analytics

Deep learning, a subset of machine learning that involves neural networks with multiple layers, has become a cornerstone of modern AI in drug discovery. Deep learning algorithms are capable of learning complex patterns and representations from large datasets, making them well-suited for analyzing biological data. Key applications of deep learning in drug discovery include:

- **Predicting Drug Efficacy and Safety:** AI algorithms can analyze large datasets of drug-related information, including clinical trial results and patient records, to predict the efficacy and safety of drug candidates. These predictions help in selecting the most promising candidates for further development and reducing the risk of adverse effects.
- **Virtual Screening:** Virtual screening involves using computational methods to evaluate the potential of chemical compounds to interact with biological targets. AI-powered virtual screening techniques can analyze vast chemical libraries and predict the likelihood of successful drug-target interactions. This approach accelerates the identification of promising drug candidates and reduces the need for experimental screening.
- **Molecular Docking:** Molecular docking simulations are used to predict how a drug molecule binds to its target protein. AI techniques enhance molecular docking by improving the accuracy of binding affinity predictions and optimizing the docking process. This helps in designing drugs with better binding properties and increased efficacy.

Data Integration and Interpretation

Modern AI techniques also involve integrating and interpreting data from various sources, such as genomic, proteomic, and clinical data. AI-driven tools can combine data from different modalities to provide a comprehensive understanding of drug interactions and mechanisms of action. This integrated approach enables researchers to identify new drug targets, understand the underlying biology of diseases, and develop more effective therapies (Hasselgren and Oprea, 2024).

Accelerating Drug Development

AI has significantly accelerated the drug development process by automating various stages, from target identification to preclinical testing. By leveraging AI technologies, researchers can streamline the development pipeline, reduce the time and cost associated with drug discovery, and bring new therapies to market more efficiently.

6.3 Case Studies in AI-Enhanced Drug Development

Several high-profile case studies highlight the success of AI in drug discovery and development, demonstrating the transformative potential of AI-driven approaches.

Case Study 1: AI in Alzheimer's Disease Research

AI technologies have been applied to the discovery of novel drug candidates for Alzheimer's disease, a complex and challenging condition. Researchers have used machine learning algorithms to analyze large datasets of genetic, clinical, and imaging data to identify potential therapeutic targets and biomarkers. For example, AI-driven approaches have led to the identification of new drug candidates that target specific proteins involved in Alzheimer's pathology, such as amyloid-beta and tau (Mullowney et al., 2023).

Case Study 2: AI in COVID-19 Drug Discovery

The COVID-19 pandemic accelerated the use of AI in drug discovery and development. AI-driven approaches were employed to identify potential treatments and vaccines for COVID-19. Researchers used machine learning algorithms to analyze viral genomes, predict protein structures, and identify existing drugs that could be repurposed for COVID-19 treatment. These efforts led to the rapid development of several promising therapeutic candidates and vaccines, demonstrating AI's ability to respond to urgent public health challenges (Mullowney et al., 2023).

Case Study 3: AI in Oncology Drug Development

AI technologies have also made significant contributions to oncology drug development. For instance, AI-driven platforms have been used to analyze large-scale genomic data from cancer patients to identify genetic mutations and molecular signatures associated with different cancer types. These insights have led to the development of targeted therapies that specifically address the genetic drivers of cancer. AI has played a crucial role in optimizing clinical trial designs and identifying patient populations most likely to benefit from new treatments (Hasselgren and Oprea, 2024).

6.4 Summary

AI has profoundly impacted drug discovery and development, from its early applications using heuristic rules to modern techniques leveraging deep learning and advanced data analytics. The integration of AI technologies has enabled more efficient analysis of complex biological data, improved predictions of drug efficacy and safety, and accelerated the development of new therapies. Case studies in Alzheimer's disease, COVID-19, and oncology illustrate the transformative potential of AI in advancing drug discovery and bringing innovative treatments to market.

7. ARTIFICIAL INTELLIGENCE IN AGRICULTURE

7.1 Introduction to AI in Agriculture

Agriculture has long been the backbone of human civilization, but the integration of artificial intelligence (AI) has transformed modern agricultural practices. AI technologies offer innovative solutions to address challenges such as crop yield optimization, pest and disease management, and resource efficiency. By leveraging data-driven approaches, AI enhances productivity and sustainability in agriculture.

The adoption of AI in agriculture can be traced back to the early 2000s when advancements in computational power and data collection technologies began to influence farming practices. Initially, AI applications were limited to basic data analysis, but as technology evolved, more sophisticated AI systems emerged, revolutionizing the agricultural industry (Wolfert et al., 2017).

7.2 Applications of AI in Agriculture

Precision Agriculture

One of the most impactful applications of AI in agriculture is precision farming. Precision agriculture involves using AI-driven technologies to monitor and manage crop production with high accuracy. Key components include:

- **Remote Sensing:** AI algorithms analyze data from satellites, drones, and other remote sensing technologies to monitor crop health, soil conditions, and weather patterns. This information enables farmers to make data-driven decisions about irrigation, fertilization, and pest control.
- **Predictive Analytics:** AI models predict crop yields and optimize resource use based on historical data and real-time inputs. For instance, predictive

analytics can forecast crop performance under different weather conditions, helping farmers plan planting and harvesting schedules more effectively (Kamilaris & Prenafeta-Boldú, 2018).

Pest and Disease Management

AI technologies play a crucial role in managing pests and diseases, which are significant threats to crop production. AI-driven solutions include:

- **Image Recognition:** Machine learning algorithms analyze images of crops to detect signs of pest infestations or diseases. This approach allows for early detection and targeted intervention, reducing the reliance on broad-spectrum pesticides (Mohanty et al., 2016).
- **Smart Spraying:** AI systems control and optimize the application of pesticides and herbicides. By using data from sensors and imaging technologies, AI can precisely target affected areas, minimizing chemical use and environmental impact (Shaikh et al., 2022).

Automated Farming Equipment

Automation in farming is significantly enhanced by AI. Modern agricultural machinery equipped with AI technologies includes:

- **Autonomous Tractors:** AI-powered autonomous tractors perform tasks such as planting, tilling, and harvesting with minimal human intervention. These tractors use sensors and AI algorithms to navigate fields, avoid obstacles, and optimize field operations (Torii, 2000).
- **Robotic Harvesters:** AI-driven robotic systems are designed to pick fruits and vegetables with high efficiency and precision. These robots use computer vision and machine learning to identify ripe produce and harvest it without damaging the plants (Yépez-Ponce et al., 2023).

Supply Chain Optimization

AI also contributes to optimizing agricultural supply chains. Applications include:

- **Demand Forecasting:** AI algorithms predict consumer demand for agricultural products, helping farmers and distributors manage inventory and reduce waste. By analyzing market trends and historical sales data, AI provides insights into future demand patterns (Ganeshkumar et al., 2023).

- **Logistics Management:** AI systems optimize transportation routes and schedules for agricultural products, reducing costs and improving delivery efficiency. This application ensures that fresh produce reaches markets promptly and with minimal spoilage (Torii, 2000).

7.3 Case Studies in AI-Enhanced Agriculture

Case Study 1: AI in Precision Irrigation

A notable example of AI in agriculture is the use of precision irrigation systems. For instance, the company CropX has developed a soil sensing technology that uses AI to analyze soil moisture levels and weather data. This system provides real-time recommendations for irrigation, resulting in significant water savings and improved crop yields ("Irrigation planning - Use every drop wisely," 2024).

Case Study 2: AI for Pest Management

Another example is the use of AI for pest management by the startup Prospera Technologies. Prospera's AI-driven platform uses computer vision to monitor crops and identify pest infestations. The system provides actionable insights to farmers, enabling timely and targeted pest control measures. This approach has been shown to reduce pesticide use and increase crop productivity (Prospera, 2020).

Case Study 3: AI in Crop Monitoring

The startup Taranis employs AI to offer high-resolution crop monitoring using drone imagery and machine learning. Taranis' platform detects and classifies crop diseases, nutrient deficiencies, and pest infestations. The insights provided by the platform help farmers make informed decisions about crop management, leading to improved yields and reduced losses (*Taranis: Leading Crop Management Software*, 2023).

7.4 Summary

AI is transforming agriculture through precision farming, pest and disease management, automated equipment, and supply chain optimization. By integrating AI technologies, the agricultural sector can address challenges related to productivity, resource efficiency, and environmental impact. Case studies demonstrate the practical applications of AI in enhancing agricultural practices, highlighting the technology's potential to revolutionize the industry and contribute to global food security.

8. ARTIFICIAL INTELLIGENCE IN ENVIRONMENTAL SCIENCE

Artificial Intelligence has emerged as a powerful tool in environmental science, offering innovative solutions to complex challenges related to the environment and sustainability. As concerns about climate change, resource depletion, and biodiversity loss grow, AI technologies provide new opportunities for monitoring, analyzing, and managing environmental systems.

8.1 Enhancing Environmental Monitoring

Remote Sensing and Satellite Data Analysis

AI-driven techniques have revolutionized environmental monitoring through the analysis of remote sensing and satellite data. Machine learning algorithms can process vast amounts of satellite imagery to track changes in land use, deforestation, and urbanization. For example, AI models are used to monitor forest cover, detect illegal logging activities, and assess the impact of natural disasters. The integration of AI with satellite data allows for real-time monitoring and more accurate assessments of environmental changes (Zhu et al., 2017).

Air and Water Quality Monitoring

AI technologies also play a crucial role in monitoring air and water quality. Sensor networks combined with AI algorithms can detect and predict pollution levels, identify sources of contamination, and assess the impact on human health and ecosystems. For instance, AI models can analyze data from air quality sensors to predict pollution events and provide actionable insights for public health interventions (Bainomugisha et al., 2024).

8.2 Climate Change Prediction and Mitigation

Climate Modeling and Simulation

AI enhances climate modeling and simulation by improving the accuracy of predictions and understanding of complex climate systems. Machine learning techniques can analyze historical climate data to identify patterns and trends, refine climate models, and predict future climate scenarios. AI-driven models are used to simulate the effects of different climate policies and interventions, helping policymakers make informed decisions (Bauer et al., 2023).

Carbon Footprint Analysis

AI is also employed to assess and reduce carbon footprints. By analyzing data from various sources, including industrial processes and transportation systems, AI algorithms can identify opportunities for reducing greenhouse gas emissions. For example, AI models can optimize energy consumption in buildings and transportation networks, leading to more efficient and sustainable practices (Jin et al., 2024).

8.3 Biodiversity Conservation and Ecosystem Management

Species Identification and Monitoring

AI technologies facilitate the identification and monitoring of species, which is essential for biodiversity conservation. Machine learning algorithms can analyze audio and visual data to identify animal species, track their populations, and monitor their behaviors. For instance, AI-powered acoustic monitoring systems can detect and classify bird calls, providing valuable data for conservation efforts (Sharma et al., 2023).

Habitat Modeling and Restoration

AI is used to model and restore habitats by analyzing environmental data and predicting the impacts of different restoration strategies. Machine learning algorithms can identify areas of ecological importance, assess habitat degradation, and guide restoration efforts. AI-driven approaches enable more effective and targeted habitat restoration, contributing to the preservation of ecosystems and species (Urli et al., 2016).

8.4 Case Studies in Environmental AI

Case Study 1: AI for Predicting Wildfire Risk

AI has been successfully applied to predicting wildfire risk by analyzing meteorological data, vegetation patterns, and historical fire occurrences. For example, AI models can forecast fire behavior and assess the likelihood of ignition, enabling better preparedness and response strategies. These models have proven effective in reducing the impact of wildfires and improving fire management practices (Sayad et al., 2019).

Case Study 2: AI in Marine Conservation

In marine conservation, AI technologies are used to monitor ocean health and protect marine biodiversity. Machine learning algorithms analyze data from underwater sensors and drones to track marine species, detect illegal fishing activities, and assess the health of coral reefs. These applications contribute to the conservation of marine ecosystems and the sustainable management of ocean resources (Wang et al., 2021).

Case Study 3: AI for Precision Agriculture

AI-driven precision agriculture techniques are used to optimize the use of resources in farming and minimize environmental impact. By analyzing data from sensors, drones, and satellite imagery, AI models can provide insights into soil health, crop conditions, and pest management. These techniques lead to more sustainable agricultural practices and reduce the environmental footprint of farming (Erickson and Fausti, 2021).

8.5 Future Directions in Environmental AI

Integration with Emerging Technologies

The future of AI in environmental science will likely involve the integration of AI with emerging technologies such as the Internet of Things (IoT), blockchain, and advanced robotics. IoT devices can provide real-time environmental data, while blockchain can enhance data transparency and traceability. Advanced robotics may assist in environmental monitoring and conservation efforts, leading to more comprehensive and effective solutions (Bhati et al., 2024).

AI for Sustainable Development Goals

AI will play a crucial role in achieving the United Nations Sustainable Development Goals (SDGs), particularly those related to climate action, life below water, and life on land. By providing innovative solutions for environmental monitoring, resource management, and conservation, AI technologies will support global efforts to create a more sustainable and resilient future (Garrido et al, 2015).

Ethical and Social Considerations

As AI technologies advance, it is essential to address ethical and social considerations related to their deployment. Ensuring equitable access to AI tools, addressing biases in AI algorithms, and promoting transparency in environmental decision-making will be crucial for the responsible use of AI in environmental science (Floridi, 2023; TrinhHuyet al, 2025).

9. CHALLENGES AND FUTURE DIRECTIONS

9.1 Data Privacy and Ethical Considerations

The integration of AI in life sciences presents significant ethical and privacy challenges that must be addressed to ensure responsible use and public trust.

Data Privacy Concerns

The use of personal health data for AI-driven research and decision-making introduces concerns about data privacy. Health records, genetic information, and other sensitive data are critical for training AI models but must be protected against unauthorized access and misuse. Regulatory frameworks such as the Health Insurance Portability and Accountability Act (HIPAA) in the United States and the General Data Protection Regulation (GDPR) in the European Union provide guidelines for data protection but need to be continually updated to address emerging AI technologies (Obermeyer et al., 2019).

Ethical Considerations

Ethical considerations in AI research and application involve ensuring that AI systems are used fairly and without bias. AI algorithms can inadvertently perpetuate existing biases in data, leading to unequal treatment across different demographic groups. Researchers and developers must work to identify and mitigate biases in AI systems to promote equity and fairness (Obermeyer et al., 2019).

Informed Consent and Transparency

Informed consent is crucial when using personal data for AI research. Participants should be aware of how their data will be used and have the option to withdraw consent at any time. Additionally, transparency in AI algorithms and decision-making processes is essential for accountability. Stakeholders must understand how AI systems arrive at conclusions and recommendations to ensure they can be effectively scrutinized and validated (Obermeyer et al., 2019).

Ongoing Dialogue

Ensuring the responsible use of AI in life sciences requires ongoing dialogue between researchers, policymakers, and the public. Establishing clear ethical guidelines and regulations, promoting transparency, and addressing privacy concerns are critical steps toward fostering trust and ensuring that AI technologies are used ethically and responsibly (Obermeyer et al., 2019).

9.2 The Need for High-Quality Data

AI systems in life sciences depend heavily on high-quality data to produce accurate and reliable results. The challenges related to data quality, sharing, and integration are critical to overcoming barriers in AI research and application.

Challenges in Data Quality

High-quality data is essential for training and validating AI models. However, many datasets in the life sciences are incomplete, inconsistent, or contain errors. Ensuring data accuracy and consistency is crucial for building effective AI models. Researchers must implement rigorous data cleaning and validation procedures to improve data quality (Fillinger et al., 2019) (Trinh et al., 2022; VuThi et al., 2024).

Data Sharing and Integration

Effective AI research and application require access to diverse datasets from various sources. Data sharing and integration pose significant challenges due to privacy concerns, data ownership issues, and technical difficulties in harmonizing different data formats. Collaborations between institutions and the development of standardized data-sharing protocols are necessary to facilitate data integration and enhance the utility of AI in life sciences (Fillinger et al., 2019)(T. Z. Emara et al., 2023; T. Z. Emara & Huang, 2019a, 2020).

Annotated Datasets

Annotated datasets, which include labeled data that provides context and meaning, are vital for training supervised AI models. The creation of high-quality annotated datasets is resource-intensive and requires domain expertise. Addressing the need for more comprehensive and accurately labeled datasets will support the development of more effective AI models and applications in life sciences (Fillinger et al., 2019) (T. Z. Emara & Huang, 2019b).

9.3 Future Trends in AI and Life Sciences

Several emerging trends are expected to shape the future of AI in life sciences, driving innovation and expanding the potential applications of AI technologies.

Advancements in AI Technologies

The continued advancement of AI technologies, including deep learning and reinforcement learning, will drive progress in genomics, drug discovery, and personalized medicine. Deep learning algorithms, with their ability to model complex patterns and representations, will enhance our understanding of biological systems and improve predictive accuracy. Reinforcement learning, which focuses on training models through trial and error, has the potential to optimize treatment strategies and drug development processes (Esteva et al., 2019).

Integration with Emerging Fields

AI's integration with emerging fields such as synthetic biology and systems medicine holds the potential for groundbreaking discoveries and applications. Synthetic biology, which involves designing and constructing new biological parts and systems, can benefit from AI-driven approaches for optimizing gene editing and metabolic engineering. Systems medicine, which aims to understand diseases as complex systems involving multiple biological processes, can leverage AI to integrate and analyze multi-omic data, leading to more personalized and effective treatments (Esteva et al., 2019).

AI for Global Health Challenges

AI has the potential to address global health challenges by improving disease surveillance, predicting outbreaks, and optimizing healthcare delivery. For instance, AI models can analyze global health data to identify patterns and predict the spread of infectious diseases, supporting timely and targeted public health interventions. Additionally, AI can enhance the accessibility and quality of healthcare in underserved regions by providing remote diagnostic and treatment support (Esteva et al., 2019).

Ethical and Regulatory Developments

As AI technologies continue to evolve, ethical and regulatory frameworks will need to adapt to address new challenges. Developing robust guidelines for the ethical use of AI, ensuring data privacy, and promoting transparency will be critical for maintaining public trust and ensuring that AI benefits are realized responsibly (Obermeyer et al., 2019; Fillinger et al., 2019).

9.4 Summary

The integration of AI in life sciences presents both opportunities and challenges. Addressing data privacy and ethical considerations, ensuring high-quality data, and exploring future trends will be crucial for advancing AI technologies and maximizing their impact. By fostering collaboration and innovation, the life sciences community can harness the power of AI to drive progress and address pressing global health challenges.

10. CONCLUSION

Artificial Intelligence (AI) has profoundly transformed the landscape of applied life sciences, marking a new era of innovation and discovery. From its early theoretical foundations to its current applications and future potential, AI has demonstrated its capability to enhance our understanding of biological systems, advance medical research, and improve healthcare outcomes.

Reflecting on Early Developments

The journey of AI in life sciences began with foundational concepts in the mid-20th century, as pioneers like Alan Turing and John McCarthy laid the groundwork for computational models and expert systems. The Dartmouth Conference of 1956 set the stage for AI as a formal academic discipline, leading to the development of early AI systems such as Dendral and MYCIN. These systems provided initial insights into the potential of AI to assist in chemical analysis and medical diagnosis, laying the groundwork for future advancements.

Advancements in Machine Learning and Data-Driven Approaches

The 1980s and 1990s marked a significant shift with the rise of machine learning, driven by advancements in computational power and the availability of large datasets. The revival of neural networks and the development of algorithms such as backpropagation paved the way for sophisticated AI models capable of handling complex patterns and representations. AI's integration into genomics and bioinformatics further demonstrated its potential to analyze vast amounts of genetic data, leading to breakthroughs in personalized medicine and drug discovery.

AI's Impact on Genomics and Drug Development

The completion of the Human Genome Project in 2003 catalyzed the application of AI in genomics and personalized medicine. AI technologies enabled researchers to analyze genetic variations, predict disease susceptibility, and develop targeted therapies. Similarly, AI's role in drug discovery has evolved from heuristic methods to advanced techniques such as deep learning and molecular docking, accelerating the identification of promising drug candidates and reducing development timelines.

Emergence of AI in Agriculture

AI's applications extend beyond traditional life sciences into agriculture, where technologies such as precision farming, pest and disease management, and automated equipment are revolutionizing farming practices. AI-driven solutions are enhancing productivity, optimizing resource use, and addressing global food security challenges, demonstrating the broad impact of AI across diverse domains.

Addressing Challenges and Future Directions

As AI continues to advance, several challenges must be addressed to ensure its responsible and effective use. Data privacy and ethical considerations remain paramount, requiring robust frameworks and ongoing dialogue among researchers, policymakers, and the public. The need for high-quality data and effective data-sharing protocols is essential for maximizing AI's potential in life sciences. Looking ahead, trends such as advancements in deep learning, integration with emerging fields, and AI's role in global health challenges will shape the future of AI in life sciences.

Final Thoughts

AI has proven to be a transformative force in applied life sciences, offering new opportunities for discovery and innovation. By addressing existing challenges and embracing future trends, the life sciences community can continue to harness the power of AI to drive progress, improve health outcomes, and contribute to a better understanding of the complex biological processes that underpin human health and disease. The ongoing evolution of AI technologies promises to bring even greater advancements, further enhancing our ability to tackle some of the most pressing challenges in life sciences and beyond.

REFERENCES

Ahmed, A. S., Keshk, A. E. M., Abo-Seida, O., & Sakr, M. (2022). Tumor detection and classification in breast mammography based on fine-tuned convolutional neural networks. *IJCI.International Journal of Computers and Information*, *9*(1), 74–84.

Alnowaiser, K., Saber, A., Hassan, E., & Awad, W. A. (2024). An optimized model based on adaptive convolutional neural network and grey wolf algorithm for breast cancer diagnosis. *PLoS One*, *19*(8), e0304868. DOI: 10.1371/journal.pone.0304868 PMID: 39159151

Anjum, M. N., Cheema, M. J. M., Hussain, F., & Wu, R. S. (2023). Precision irrigation: Challenges and opportunities. *Precision Agriculture*, •••, 85–101.

Bainomugisha, E., Warigo, P. A., Daka, F. B., Nshimye, A., Birungi, M., & Okure, D. (2024). AI-driven environmental sensor networks and digital platforms for urban air pollution monitoring and modelling. *Societal Impacts*, *3*, 100044. DOI: 10.1016/j.socimp.2024.100044

Batra, R., Arora, S., Sharma, M. M., Rana, S., Raheja, K., Saber, A., & Shah, M. A. (2024). Integration of LSTM networks with gradient boosting machines (GBM) for assessing heating and cooling load requirements in building energy efficiency. *Energy Exploration & Exploitation*, *42*(6), 01445987241268075. DOI: 10.1177/01445987241268075

Bauer, P., Dueben, P., Chantry, M., Doblas-Reyes, F., Hoefler, T., McGovern, A., & Stevens, B. (2023). Deep learning and a changing economy in weather and climate prediction. *Nature Reviews. Earth & Environment*, *4*(8), 507–509. DOI: 10.1038/s43017-023-00468-z

Bhati, N., Duggar, R., & Saber, A. (2024). Empowering Safety by Embracing IoT for Leak Detection Excellence. In Innovations in Machine Learning and IoT for Water Management (pp. 231-251). IGI Global.

Bibel, W. (2014). Artificial Intelligence in a historical perspective. *AI Communications*, *27*(1), 87–102. DOI: 10.3233/AIC-130576

Buchanan, B. G., & Shortliffe, E. H. (1984). *Rule-Based Expert Systems: The MYCIN Experiments of the Stanford Heuristic Programming Project*. Addison-Wesley.

Collins, F. S., & Fink, L. (1995). The Human Genome Project. *Alcohol Health and Research World*, *19*(3), 190–195. PMID: 31798046

Colmerauer, A., & Roussel, P. (1993). *The birth of Prolog*. The Second ACM SIGPLAN Conference on History of Programming Languages. Presented at the HOPL93: History of Programing Languages 1993, Cambridge Massachusetts USA. DOI: 10.1145/154766.155362

Confalonieri, R., Coba, L., Wagner, B., & Besold, T. R. (2021). A historical perspective of explainable Artificial Intelligence. *Wiley Interdisciplinary Reviews. Data Mining and Knowledge Discovery*, *11*(1), e1391. Advance online publication. DOI: 10.1002/widm.1391

Emara, T. (2018). Adaptive Power-Saving Mechanism for VoIP Over WiMAX Based on Artificial Neural Network. In Waluyo, A. B. (Ed.), *Algorithms, Methods, and Applications in Mobile Computing and Communications* (pp. 158–177). IGI Global., DOI: 10.4018/978-1-5225-5693-0.ch007

Emara, T. Z. (2016). Maximizing Power Saving for VoIP over WiMAX Systems. *International Journal of Mobile Computing and Multimedia Communications*, *7*(1), 32–40. DOI: 10.4018/IJMCMC.2016010103

Emara, T. Z., & Huang, J. Z. (2019a). A distributed data management system to support large-scale data analysis. *Journal of Systems and Software*, *148*, 105–115. DOI: 10.1016/j.jss.2018.11.007

Emara, T. Z., & Huang, J. Z. (2019b). RRPlib: A spark library for representing HDFS blocks as a set of random sample data blocks. *Science of Computer Programming*, *184*, 102301. DOI: 10.1016/j.scico.2019.102301

Emara, T. Z., & Huang, J. Z. (2020). Distributed Data Strategies to Support Large-Scale Data Analysis Across Geo-Distributed Data Centers. *IEEE Access : Practical Innovations, Open Solutions*, *8*, 178526–178538. DOI: 10.1109/ACCESS.2020.3027675

Emara, T. Z., Saleh, A. I., & Arafat, H. (2014). Power saving mechanism for VoIP services over WiMAX systems. *Wireless Networks*, *20*(5), 975–985. DOI: 10.1007/s11276-013-0650-5

Emara, T. Z., Trinh, T., & Huang, J. Z. (2023). Geographically distributed data management to support large-scale data analysis. *Scientific Reports*, *13*(1), 17783. Advance online publication. DOI: 10.1038/s41598-023-44789-x PMID: 37853092

Erickson, B., & Fausti, S. W. (2021). The role of precision agriculture in food security. *Agronomy Journal*, *113*(6), 4455–4462. DOI: 10.1002/agj2.20919

Esteva, A., Robicquet, A., Ramsundar, B., Kuleshov, V., DePristo, M. A., Chou, K., Cui, C., Corrado, G. S., Thrun, S., & Dean, J. (2019). A guide to deep learning in healthcare. *Nature Medicine*, *25*(1), 24–29. DOI: 10.1038/s41591-018-0316-z PMID: 30617335

Fillinger, S., de la Garza, L., Peltzer, A., Kohlbacher, O., & Nahnsen, S. (2019). Challenges of big data integration in the life sciences. *Analytical and Bioanalytical Chemistry*, *411*(26), 6791–6800. DOI: 10.1007/s00216-019-02074-9 PMID: 31463515

Floridi, L. (2023). The ethics of artificial intelligence: Principles, challenges, and opportunities.

Ganeshkumar, C., Jena, S. K., Sivakumar, A., & Nambirajan, T. (2023). Artificial intelligence in agricultural value chain: Review and future directions. *Journal of Agribusiness in Developing and Emerging Economies*, *13*(3), 379–398. DOI: 10.1108/JADEE-07-2020-0140

Garrido, M.J., Villareal, M.G., Calderon, D., Muñoz, A., Macintyre, A., Mulliez, A., Goldberg, B., Felice, C.D., Salamanca, D., Martucci, D., Ubico, D., Gutnisky, G., Sipalla, H., Cornelio, I.D., Lozanoska, J., Amaya, J., Garcia, J., Araque, L., Estrada, M.S., Garita, M., Tarre, P., Gomez, R., Cornejo-Hösl, R., Ryan, R., Salojärvi, V., & Gonzalez, Z. (2015). The 2030 Agenda for Sustainable Development. *An Insider's Guide to the UN.*

Hassan, E., Saber, A., & Elbedwehy, S. (2024). Knowledge distillation model for Acute Lymphoblastic Leukemia Detection: Exploring the impact of nesterov-accelerated adaptive moment estimation optimizer. *Biomedical Signal Processing and Control*, *94*, 106246. DOI: 10.1016/j.bspc.2024.106246

Hasselgren, C., & Oprea, T. I. (2024). Artificial intelligence for drug discovery: Are we there yet? *Annual Review of Pharmacology and Toxicology*, *64*(1), 527–550. DOI: 10.1146/annurev-pharmtox-040323-040828 PMID: 37738505

Hinton, G. E., Osindero, S., & Teh, Y. W. (2006). A fast learning algorithm for deep belief nets. *Neural Computation*, *18*(7), 1527–1554. DOI: 10.1162/neco.2006.18.7.1527 PMID: 16764513

Irrigation planning - Use every drop wisely. (2024, April 23). CropX. https://cropx.com/cropx-system/irrigation-planning/

Jiang, Y., Li, X., Luo, H., Yin, S., & Kaynak, O. (2022). Quo vadis artificial intelligence? *Discover Artificial Intelligence*, *2*(1), 4. Advance online publication. DOI: 10.1007/s44163-022-00022-8

Jin, Y., Sharifi, A., Li, Z., Chen, S., Zeng, S., & Zhao, S. (2024). Carbon emission prediction models: A review. *The Science of the Total Environment, 927*, 172319. DOI: 10.1016/j.scitotenv.2024.172319 PMID: 38599410

Kamilaris, A., & Prenafeta-Boldú, F. X. (2018). Deep learning in agriculture: A survey. *Computers and Electronics in Agriculture, 147*, 70–90. DOI: 10.1016/j.compag.2018.02.016

Khoury, M. J., & McCabe, L. L. (2015). Personalized medicine: From aspiration to reality. *Journal of the American Medical Association, 314*(14), 1451–1452. PMID: 26461994

Kourou, K., Exarchos, T. P., Karamouzis, M. V., & Papaloukas, C. (2015). Machine learning applications in cancer prognosis and prediction. *Computational and Structural Biotechnology Journal, 13*, 8–17. DOI: 10.1016/j.csbj.2014.11.005 PMID: 25750696

LeCun, Y., Bengio, Y., & Hinton, G. (2015). Deep learning. *Nature, 521*(7553), 436–444. DOI: 10.1038/nature14539 PMID: 26017442

McCarthy, J. (1960). Recursive Functions of Symbolic Expressions and Their Computation by Machine, Part I. *Communications of the ACM, 3*(4), 184–195. DOI: 10.1145/367177.367199

McCarthy, J., Minsky, M. L., Rochester, A., & Shannon, C. E. (1956). *A Proposal for the Dartmouth Summer Research Project on Artificial Intelligence*. Dartmouth College.

Mitchell, T. M. (1997). *Machine Learning*. McGraw-Hill.

Mitra, S., Datta, S., Perkins, T., & Michailidis, G. (2008). *Introduction to machine learning and bioinformatics*. CRC Press. DOI: 10.1201/b17186

Mohanty, S. P., Hughes, D. P., & Salathé, M. (2016). Using deep learning for image-based plant disease detection. *Frontiers in Plant Science, 7*, 1419. DOI: 10.3389/fpls.2016.01419 PMID: 27713752

Mullowney, M. W., Duncan, K. R., Elsayed, S. S., Garg, N., van der Hooft, J. J., Martin, N. I., Meijer, D., Terlouw, B. R., Biermann, F., Blin, K., Durairaj, J., Gorostiola González, M., Helfrich, E. J. N., Huber, F., Leopold-Messer, S., Rajan, K., de Rond, T., van Santen, J. A., Sorokina, M., & Medema, M. H. (2023). Artificial intelligence for natural product drug discovery. *Nature Reviews. Drug Discovery, 22*(11), 895–916. DOI: 10.1038/s41573-023-00774-7 PMID: 37697042

Newell, A., & Simon, H. A. (1972). *Human Problem Solving*. Prentice-Hall.

Obermeyer, Z., Powers, B. W., Vogeli, C., & Mullainathan, S. (2019). Dissecting racial bias in an algorithm used to manage the health of populations. *Science, 366*(6464), 447–453. DOI: 10.1126/science.aax2342 PMID: 31649194

Prospera. (2020). *AI for precision agriculture*. Retrieved from https://www.prospera.ag

Rosenblatt, F. (1958). The Perceptron: A Probabilistic Model for Information Storage and Organization in the Brain. *Psychological Review, 65*(6), 386–408. DOI: 10.1037/h0042519 PMID: 13602029

Rumelhart, D. E., Hinton, G. E., & Williams, R. J. (1986). Learning representations by backpropagating errors. *Nature, 323*(6088), 533–536. DOI: 10.1038/323533a0

Saber, A., Hussien, A. G., Awad, W. A., Mahmoud, A., & Allakany, A. (2023). Adapting the pre-trained convolutional neural networks to improve the anomaly detection and classification in mammographic images. *Scientific Reports, 13*(1), 14877. DOI: 10.1038/s41598-023-41633-0 PMID: 37689757

Saber, A., Sakr, M., Abo-Seida, O. M., & Keshk, A. (2021b, May). Automated Breast Cancer Detection and Classification Techniques–A survey. In 2021 International Mobile, Intelligent, and Ubiquitous Computing Conference (MIUCC) (pp. 200-207). IEEE.

Saber, A., Sakr, M., Abou-Seida, O., & Keshk, A. (2021a). A novel transfer-learning model for automatic detection and classification of breast cancer based deep CNN. *Kafrelsheikh Journal of Information Sciences, 2*(1), 1–9. DOI: 10.21608/kjis.2021.192207

Sakr, M., Saber, A. M., Abo-Seida, O., & Keshk, A. (2020). Machine learning for breast cancer classification using k-star algorithm. *Applied Mathematics & Information Sciences, 14*(5), 855–863. DOI: 10.18576/amis/140513

Sayad, Y. O., Mousannif, H., & Al Moatassime, H. (2019). Predictive modeling of wildfires: A new dataset and machine learning approach. *Fire Safety Journal, 104*, 130–146. DOI: 10.1016/j.firesaf.2019.01.006

Shaikh, T. A., Rasool, T., & Lone, F. R. (2022). Towards leveraging the role of machine learning and artificial intelligence in precision agriculture and smart farming. *Computers and Electronics in Agriculture, 198*, 107119. DOI: 10.1016/j.compag.2022.107119

Sharma, S., Sato, K., & Gautam, B. P. (2023). A Methodological Literature Review of Acoustic Wildlife Monitoring Using Artificial Intelligence Tools and Techniques. *Sustainability (Basel), 15*(9), 7128. DOI: 10.3390/su15097128

Shortliffe, E. H. (1976). *Computer-Based Medical Consultations: MYCIN*. Elsevier.

Sinha, S., & Vohora, D. (2018). Drug discovery and development: An overview. *Pharmaceutical medicine and translational clinical research*, 19-32.

Taranis: Leading Crop Management Software. (2023, October 30). Taranis. https://www.taranis.com/

Torii, T. (2000). Research in autonomous agriculture vehicles in Japan. *Computers and Electronics in Agriculture*, *25*(1-2), 133–153. DOI: 10.1016/S0168-1699(99)00060-5

Trinh, T., Duc, L. P., Tran, C. T., Duy, T. T., & Emara, T. Z. (2022). A New Stratified Block Model to Process Large-Scale Data for a Small Cluster. In *Lecture Notes on Data Engineering and Communications Technologies* (Vol. 124, pp. 253–263). Springer Science and Business Media Deutschland GmbH., DOI: 10.1007/978-3-030-97610-1_21

TrinhHuy., T., Trinh, T., NguyenViet, H., Emara, T. Z., & VuongThi, N. (2025, February). *Event scale prediction in event-based social networks*. 11th EAI International Conference on Industrial Networks and Intelligent Systems, Da Nang, Vietnam.

Turing, A. M. (1936). On Computable Numbers, with an Application to the Entscheidungsproblem. *Proceedings of the London Mathematical Society*, *42*(1), 230–265. DOI: 10.1112/plms/s2-42.1.230

Turing, A. M. (1950). Computing Machinery and Intelligence. *Mind*, *59*(236), 433–460. DOI: 10.1093/mind/LIX.236.433

Urli, T., Brotánková, J., Kilby, P., & Hentenryck, P. V. (2016). Intelligent Habitat Restoration Under Uncertainty. *AAAI Conference on Artificial Intelligence*.

Wick, M. R., & Thompson, W. B. (1992, March). Reconstructive expert system explanation. *Artificial Intelligence*, *54*(1–2), 33–70. DOI: 10.1016/0004-3702(92)90087-E

Wolfert, S., Ge, L., Verdouw, C., & Bogaardt, M. J. (2017). Big data in smart farming – A review. *Agricultural Systems*, *153*, 69–80. DOI: 10.1016/j.agsy.2017.01.023

Yépez-Ponce, D. F., Salcedo, J. V., Rosero-Montalvo, P. D., & Sanchis, J. (2023). Mobile robotics in smart farming: Current trends and applications. *Frontiers in Artificial Intelligence*, *6*, 1213330. DOI: 10.3389/frai.2023.1213330 PMID: 37719082

Chapter 3
AI-Driven Innovations: Transforming Healthcare, Agriculture, and Environmental Science

Muhammad Usman Tariq
https://orcid.org/0000-0002-7605-3040
Abu Dhabi University, UAE & University College Cork, Ireland

ABSTRACT

This chapter explores the integration of Artificial Intelligence (AI) in the life sciences, highlighting its impact on research, medical care, and understanding complex biological systems. AI's capacity to process vast datasets and identify hidden patterns is revolutionising fields such as drug discovery, genomics, clinical diagnostics, and healthcare administration. It examines the historical development of AI in life sciences, from rule-based systems like MYCIN to modern data-driven approaches enhanced by deep learning, machine learning, and natural language processing. Notable applications include improved patient outcomes, accelerated drug discovery, enhanced diagnostic accuracy, and personalised medicine. The chapter also discusses AI's transformative potential across various life science sectors, especially drug discovery, where it optimises candidate selection, reduces costs, and repurposes existing drugs. Overall, the integration of AI signifies a paradigm shift in life sciences, promising greater efficiency, innovation, and responsiveness to global health challenges.

DOI: 10.4018/979-8-3693-9208-9.ch003

INTRODUCTION

Incorporating Artificial Intelligence (AI) in life sciences marks a substantial turning point in how we commence research, offer medical care, and comprehend complicated biological systems. Artificial Intelligence's capability to assess immense amounts of data and disclose patterns that would otherwise remain hidden is altering several sectors in life sciences, from drug discovery to clinical diagnostics. This introduction investigates the overview of Artificial Intelligence's incorporation, historical evolution, and transformative potential across several life science domains (Altayeb et al., 2024). Artificial intelligence includes technologies involving "deep learning, machine learning, and NLP." It can evaluate data at unparalleled accuracy and speed. Artificial Intelligence is increasingly used in life sciences to improve decision-making procedures, enhance patient outcomes, and streamline operations. Artificial Intelligence utilizes experiment data in research labs, recognizes correlations, and generates hypotheses. For instance, in genomics, Artificial Intelligence can process sequencing data to explore genetic variants linked with diseases, paving the way for customized medicine. In clinical settings, AI-driven diagnostic tools can interpret medical images, like MRIs and X-rays, generally more accurately than human radiologists. Artificial intelligence incorporation extends beyond clinical and research applications to involve administrative procedures in medical care. Natural language processing (NLP) extracts related information from research articles and clinical notes, thus supporting better clinical decision-making and rapidly broadcasting knowledge (Akter, 2024).

BACKGROUND

The theoretical foundations of rural social entrepreneurship highlight the relationship between economic activity and social effect. Fundamentally, social entrepreneurship is starting companies that put social change ahead of financial viability and concentrate on developing creative, market-driven solutions to address societal issues. In rural areas, these issues frequently include poverty, lack of access to education and healthcare, and environmental deterioration. Knowledge rural social entrepreneurship requires an understanding of "embeddedness," as these businesses are firmly established in the communities in which they operate, utilizing local networks, resources, and customs to provide sustainable and timely solutions.

The history of artificial intelligence in life sciences can tracked back to the 1960s, when early computing technology started to impact different domains. Initial applications of artificial intelligence were mainly rule-based MYCIN, designed to diagnose bacterial infections and suggest antibiotics depending on patient information. However, these systems were restricted in scope and heavily dependent on expert knowledge. As computing power enhanced in the 1980s and 1990s, the emphasis switched to data-driven approaches. Machine learning started gaining traction, with scholars implementing algorithms that could learn from data instead of depending solely on predefined rules. The emergence of genomic data in the last decade of the 1990s attained a significant milestone, leading to artificial intelligence applications that could assess complicated biological information. In recent years, the COVID-19 pandemic has augmented the acceptance of artificial intelligence in life sciences. Scholars implemented artificial intelligence to model disease spread, recognize possible treatments, and assess vaccine efficacy, showcasing the technology's capability to respond rapidly to international health challenges (Tariq, 2025a).

The Transformative Potential Of AI Across Various Life Science Sectors

The transformative potential of artificial intelligence across life sciences is deep, affecting different key sectors. AI is transforming the drug detection procedure by significantly decreasing the expenses and time linked with introducing new therapies to the market. Conventional drug development is a long procedure that can take over a decade and generally fails because of safety profiles or poor efficacy. Artificial intelligence algorithms can forecast molecular interactions, screen vast chemical libraries, and enhance chemical structures, thereby recognizing promising drug candidates more effectively. For instance, firms like Atomwise use artificial intelligence to forecast how various compounds interact with particular targets, letting scholars emphasize the most promising candidates. Additionally, artificial intelligence can assist in repurposing current drugs for new therapeutic uses by assessing previous data for indications not previously considered. Artificial intelligence's capability to assess genomic data revolutionizes custom-made medicine, where treatments are altered to distinct genetic profiles. By processing sequencing data, artificial intelligence algorithms can recognize genetic variants contributing to diseases, assessing the development of targeted therapies. For example, mediums such as 23andME use artificial intelligence to offer insights into possible health risks and genetic predispositions. In oncology, AI-powered assessment can support recognizing the most efficient treatments dependent on a patient's unique tumor profile, leading to enhanced outcomes and decreased trial-and-error approaches in therapy selection. For instance, platforms like 23andMe utilize AI to provide insights into genetic predispositions and potential health risks. In oncology, AI-driven analysis can help identify the most effective treatments based on a patient's unique tumor profile. This leads to enhanced results and condensed trial-and-error approaches in therapy selection (Veeramanju, 2023).

In clinical diagnostics, artificial intelligence models improve diagnostic speed and accuracy. Machine learning algorithms can assess medical images, like MRIs and CT scans, detecting anomalies with high accuracy. The capability enhances diagnostic precision and decreases the workload for radiologists, letting them concentrate on complicated cases that need human expertise. Firms such as Zebra Medical Vision and AIdoc are leading the path in developing artificial intelligence solutions that can detect different medical conditions, from tumors to fractures, with remarkable precision. By incorporating artificial intelligence into the diagnostic workflow, medical care providers can confirm accurate and timely diagnoses, ultimately enhancing patient care. Artificial intelligence plays a dynamic role in public health by modeling virus eruptions, forecasting health trends, and optimizing resource allocation. During the global COVID-19 pandemic, artificial intelligence was instrumental in tracing the spread of the virus, forecasting case numbers, and informing public health responses. Epidemiological models driven by artificial intelligence can assess data from diverse sources, including social media, environmental factors, and mobility patterns, to predict possible outbreaks. This predictive ability allows public health officials to adopt timely interventions, ultimately saving lives (Tariq, 2025b).

Artificial intelligence-driven tools improve care by offering customized health suggestions and monitoring capabilities. Mobile applications and wearable devices fortified with artificial intelligence can trace significant signs, detect anomalies, and prepare medical care providers when interventions are required. Furthermore, artificial intelligence chatbots and virtual assistants enhance patient engagement by answering queries and providing reminders for medical observance. These tools improve the patient experience and allow individuals to take a dynamic role in handling their health. The assimilation of artificial intelligence in life sciences depicts a transformative switch, providing unprecedented prospects for research, patient care, treatment, and diagnosis. From accelerating drug discovery to allowing customized medicine and improving clinical diagnostics, artificial intelligence is composed to reshape the landscape of biological and healthcare research. As technology evolves, the collaboration between artificial intelligence and life sciences will undoubtedly lead to novel breakthroughs, enhancing health outcomes and elevating the comprehension of biological systems. The future of life sciences, amplified by artificial intelligence, promises to be both impactful and exciting, paving the path for innovations that can intensely enhance human well-being and health (Tariq, 2025c).

Harnessing Data Science in Life Sciences

Incorporating data science into the life sciences domain has evolved as a transformative force, redesigning research, treatment methodologies, and diagnostics. By implementing advanced analytical techniques and harnessing big data, life sciences are evolving to provide precision medicine, enhance medical care outcomes, and accelerate drug discovery. This detailed assessment investigates the critical role of big data in driving artificial intelligence advancements, methods for extracting actionable insights from case studies, and complex biological data that pinpoint successful data-driven applications. In life sciences, the amount of data generated is astounding. With advancements in sequencing technologies, massive datasets such as those from the Human Genome Project have become usual. These data offer insights into genetic predispositions and possible therapeutic targets. EHRs contain detailed patient histories, demographics, and treatment results, creating a rich dataset for assessment. Metabolomic and proteomic data types provide insights into metabolic pathways and protein expressions, which are required for comprehending diseases at a molecular level.

Artificial intelligence can optimize resource allocation in hospital settings and clinical trials, enhancing efficiency and decreasing costs. Techniques for extracting actionable insights from complex biological data. There is no question that Artificial Intelligence (AI) is going to advance life sciences, and data integration plays a pivotal role in exploiting the AI in life sciences: bringing together many disparate data sources (clinical, social, proteomic determinants) to provide a complete picture of a patient's health. This integration allows us to analyze detail and explain more and make better decisions. Data warehousing is a method used for achieving an effective data integration by centralizing data from several sources to achieve broader analysis. Centralised databases make it easy for scholars and health care professionals to get the insights they need for more informed research outcomes and better care for patients. With interoperability standards these are also very important, as they establish a common framework to ensure that many different systems and databases communicate with each other without any interference. Such standardisation is crucial to ensure easiness and accuracy of dealing with data, and its consistency. Machine learning (ML) techniques play a key role in analysing biological and medical data, being very different techniques to tackle the complexity and volume of biological and medical information. However, supervised learning — a popular ML approach — with an ability to learn from labelled datasets is ideal for tasks such as disease classification. For instance, genes can be used to assign tumour types and subsequently more specific diagnosis, and therapy options can be developed. In contrast with supervised learning, unsupervised learning performs well in exploring data raw data that has no label. Targeted therapies are particularly valuable because they allow us to identify subgroups within patient populations, and leverage these to develop therapies specifically tailored to that subgroup. For example, clustering approaches can discover new types of a disease, resulting in a more personalised treatment plan. Medical image analysis is exceptionally well suited to analysis through deep learning, especially through neural networks, such as convolutional neural networks (CNNs). These networks can recognize complex medical images, such as MRI scans and X-rays, with almost perfect accuracy and make them indispensable research diagnostic tools of our era.

Just as much as for unstructured data, natural Language Processing (NLP) methods are as vital for extracting insights from natural data, which is a typical kind of data in medical research and clinical practice. Research with NLP techniques like text mining enables researchers to comb through enormous amounts of scientific literature, patient narratives, and clinical notes to find key information to support understanding of treatment outcomes and understanding of the underlying disease mechanisms. And another NLP application used for the analysis of the patient feedback and social media data in order to understand public perception of diseases, treatments or outbreaks is known as Sentiment analysis. By analyzing experiences and public sentiment at real time, this analysis provides real time insights to patient experiences and public sentiment, which could help healthcare providers make data driven adjustments to care strategies and public health initiatives. A foundation tool in life sciences statistical analysis remains a cornerstone for the way researchers interpret complex biological data. Regression analysis is commonly utilized for relationship discovery between variables and to discover cause and effect between disease phenotypes and genetic mutations. For instance, regression models can determine how certain genetic mutations lead to the development of some diseases for later determination of potential therapeutic targets. When it comes to clinical trials, survival analysis is used to evaluate whether treatment is effective, based on time to event data (i.e., time from when a patient enters the trial until that patient experiences a particular clinical outcome). This technique allows for evaluating new treatments and determining optimized treatments for patients. Another crucial approach is network analysis which models biological systems as networks and interpret interactions driving cellular pathways or disease processes. By essentially using math to analyze this kind of data, researchers can find key regulatory nodes and regulatory interactions that could be targets for intervention. For example, knowing how particular proteins work together within a network can help us discover drugs that kill off disease causing pathways. These further complements pathway analysis which map out the biological pathways in diseases to identify potential targets of drug and biomarkers (Tariq, 2024a).

AI APPLICATIONS FOR LIFE SCIENCES

While a large part of life sciences involves searching for new molecules that could potentially be used as therapeutics, the search for new and improvement treatment has been overwhelmed by the explosion of its size and complexity. Such technologies are now driving large improvements in healthcare diagnostics, drug discovery and the like in bioinformatics. Having an understanding of the underlying AI algorithms gives you a clear idea of how the advances are revolutionizing the medical and biological sciences.

AI Algorithms in Healthcare Diagnostics

On the other hand, AI tools based on introducing AI to the medical image analysis, patient history analysis, and laboratory results have revolutionized medical diagnostics. Convolutional Neural Networks (CNNs) are one of the most impactful techniques in this field and has reshaped radiology and pathology. In fact, CNNs can process complex medical images at multiple layers, and thereby are able to identify fine patterns that even human radiologists might struggle to identify. By the way, ResNet, VGG, and Efficient have performed miracles in diagnosing diseases such as pneumonia, tuberculosis, and cancer by using X-rays, MRI and CT scans imaging modalities. Other than imaging analysis, Recurrent Neural Networks (RNNs), Long Short-Term Memory Networks (LSTMs), and other algorithms are used to process sequential medical data such as ECGs and patient histories. However, these models account for temporal dependencies present within the medical datasets and thus improve predictions of disease progression and patient deterioration (Tariq, 2024b).

AI in Drug Discovery

AI driven approaches have greatly optimized the drug discovery process, something heritage which was normally time consuming and costly. However, predicting interactions of molecular structures and biological targets is an important task performed by Deep Neural Networks (DNNs). These networks enable researchers to drug design with very high precision; accelerating early stage discovery.

CASE STUDIES ON SUCCESSFUL DATA-DRIVEN AI APPLICATIONS

Case Study 1: IBM Watson for Oncology

IBM Watson for Oncology is a significant instance of artificial intelligence's impact on cancer care. Watson assists oncologists in making customized treatment suggestions by assessing huge amounts of patient data, medical literature, and clinical guidelines. Watson can recommend treatment options personalized to the genetic makeup of a patient's tumor, improving the accuracy of cancer therapies. In tests, Watson has verified the capability of providing treatment suggestions faster than human oncologists, allowing quicker decision-making in crucial situations. Prior research shows that Watson's suggestions often align with expert oncologists' decisions, depicting its potential to improve patient care (Asvial et al., 2023).

Case Study 2: DeepMind's AlphaFold

DeepMind's AlphaFold represents a revolutionary achievement in protein structure prediction. Accurate protein folding is critical for comprehension of biological functions and drug design. Revolutionizing Structural Biology: AlphaFold can predict protein structures with unprecedented precision, addressing a long-standing challenge in biology. Scholars can recognize possible drug targets by accurately modeling protein interactions, streamlining the drug discovery procedure. DeepMind has made AlphaFold's forecasts publicly accessible, fostering collaboration and accelerating research worldwide (Liyakat, 2024).

Case Study 3: Tempus

Tempus leverages data science to offer insights for cancer treatment. By combining clinical and molecular data, Tempus supports oncologists in making data-driven decisions. Tempus incorporates genomic data, clinical results, and treatment histories, leading to a holistic comprehension of cancer. Oncologists can access updated information about treatment options depending on the latest research, enhancing patient care. Tempus implements machine learning to forecast patient responses to treatments, allowing customized therapeutic strategies.

Case Study 4: Zebra Medical Vision

Zebra Medical Vision utilizes deep learning algorithms to assess medical imaging data, improving diagnostic precision for radiologists. Automated image analysis: Zebra's algorithms can detect different conditions, like cardiovascular cancers and diseases, from imaging data, decreasing the burden on radiologists. By providing quick analysis of imaging data, Zebra improves workflow efficiency in clinical settings, letting radiologists emphasize complex cases. Zebra collaborates with healthcare providers to incorporate its solutions into previous workflows, demonstrating the practical applicability of Artificial Intelligence in medicine.

Incorporating data science into life sciences is redesigning the healthcare landscape. Big data drives Artificial Intelligence improvements that improve the comprehension of diseases, customize treatment, and optimize resource allocation. Scholars can extract actionable insights from complex biological data by implementing data integration, natural learning processing, and machine learning. Successful case studies such as IBM Watson for Oncology, Tempus, Zebra Medical Vision, and DeepMind's Alphafold exemplify the transformative potential of data-driven applications in augmenting patient care and progressing medical research. As technology advances, life science's future is promising for more efficient and personalized healthcare solutions.

ETHICAL FRONTIERS: NAVIGATING AI IN LIFE SCIENCES

The rapid progression of artificial intelligence in life sciences brings unique prospects for innovation, enhanced patient outcomes, and improved research capabilities. However, as we accept these technologies, we must also oppose particular ethical challenges. Problems surrounding Bias, privacy, and regulatory compliance need vigilant navigation to confirm that artificial intelligence in healthcare is equitable and efficient. This detailed assessment addresses the ethical frontiers in artificial intelligence-powered life sciences, emphasizing privacy concerns, regulatory considerations, and bias mitigation (Neethirajan, 2024). Incorporating artificial intelligence in healthcare depends heavily on using large datasets, generally consisting of sensitive patient information. The significance of maintaining patient privacy cannot be exaggerated. Patients trust healthcare providers to safeguard their data, and breaches can lead to severe outcomes involving identity theft and loss of patient trust (Kalyanaraman et al., 2022). Artificial intelligence systems generally need extensive data collection, raising concerns about how this data is used and obtained. Informed consent is the foundation of ethical medical practice. Patients should understand:

- "What data is being collected?" It includes personal health information, genomic data, and other related details.
- "How will the data be used?" Patients should be informed whether their data will be used for research, commercial purposes, or artificial intelligence training.
- "Who will have access to their data?" Clarity about sharing with third parties, involving scholars, regulatory bodies, and tech firms, is essential.

Data Security Measures

The subtle nature of healthcare data entails severe security procedures to defend against breaches. It involves:

- Encrypting data stationary and in transit helps defend delicate information from illegal accessibility.
- Applying firm access controls confirms that only official personnel can access patient information, decreasing the threat of internal breaches.
- Conducting routine security audits supports recognizing susceptibilities and confirms agreement with data protection principles.
- Firms must have quick incident response plans to address any data breaches quickly, decreasing possible patient harm.

Regulatory Frameworks

Different regulations govern data security and confidentiality in medical care, involving:

- "Health Insurance Portability and Accountability Act (HIPAA)": In the United States, HIPAA sets national standards for preserving patient health data.
- "General Data Protection Regulation (GDPR)": In the EU, GDPR sets firm data privacy and protection rules, providing people greater control over their data.

Mitigating Bias in AI Algorithms for Fair and Transparent Life Science Solutions

Bias in artificial intelligence algorithms can have profound reflective implications, specifically in life sciences, where decisions can directly impact patient care. Biase can result from:

- The algorithms may generate skewed outcomes if the data used to train artificial intelligence frameworks does not represent the wider population. For example, artificial intelligence may perform poorly when implemented in others if a dataset mainly features one demographic group.
- Bias can also be introduced during the algorithm development procedure, whether via the design selection or features of the model itself.

The consequences of biased artificial intelligence algorithms in medical care can be severe, involving:

- Patients from understated groups may get suboptimal care if artificial intelligence systems fail to assess their particular health requirements.
- Artificial intelligence systems trained on biased data may continue previous health disparities, further relegating vulnerable populations.

To create transparent and fair artificial intelligence solutions in life sciences, firms can implement different strategies:

- Efforts should be made to involve diverse patient populations in training datasets. This involves considering gender, age, socioeconomic status, and gender to confirm broad applicability.
- Regularly auditing artificial intelligence algorithms for Bias can support recognizing and correcting problems. Methods like fairness metrics can evaluate whether algorithms offer equitable results across different demographic groups.
- Promoting transparency in creating artificial intelligence models lets stakeholders comprehend how decisions are made. This can involve documenting the rationale for feature selection and model design.
- Including diverse stakeholders, including patients, ethicists, and healthcare providers, in the progress procedure can provide valued understanding and help recognize possible biases.

Education And Training

Educating data scientists, healthcare providers, and policymakers about the significance of bias mitigation is vital. Training programs should focus on the following:

- Recognizing that Bias can exist in algorithms and data is the first step towards addressing it.

- Recognizing that Bias can exist in algorithms and data is the first step towards addressing it.
- Training should involve ethical considerations in artificial intelligence development, confirming that equity and fairness are prioritized.

Regulatory And Ethical Considerations For Ai Implementation In Sensitive Fields

As artificial intelligence technologies commence to evolve, so too must the regulatory models that govern their use in medical care. Fundamental regulatory considerations involve:

- Regulatory bodies must assess artificial intelligence applications for effectiveness and safety before they can be applied in clinical settings. It is specifically vital to diagnostic treatment and tool suggestions.
- Continous monitoring of artificial intelligence systems after implementation is vital to recognize any unforeseen problems, specifically as algorithms may evolve with new data (Biswas et al., 2023).

Ethical Models

Models of what is wrong and what is right serve as a compass for the ethical use of Artificial Intelligence in life science. In fact, these models are based on core ethical principles that assure AI implementations are based on medical ethics and patient welfare. Beneficence is one of the fundamental principles, which requires that AI applications seek to improve patients' health and wellbeing. Before AI tools are used in medical practice, they must go through rigorous evaluation in order for benefits clearly and tangibly to emerge. This means that we should ensure that in the healthcare setting, AI algorithms will improve the diagnostic accuracy, treatment efficacy or operational efficiency. To ensure that AI systems always serve patients as intended, and also play a role in improving meaningful healthcare, regular monitoring and validation are necessary. Nonmaleficence, doing no harm, is another core ethical principle. Proactive efforts are absolutely needed to address the risk concerns arising from AI technologies. But AI algorithms can perpetuate AI's own biases of the data it's been trained from – which can in turn result in outcomes that harm select patient groups. For this, we need all developers to put in place sound measures to ensure that, when AI outputs are not biased, they are fair across different groups of people. Secondarily, the non-maleficence presupposes protection of the data security, since a breach allows not only losing the data confidentiality but can also undermine the patient's trust. In order to minimise the harm arising from data breaches or misuse of sensitive patient data processed by AI systems, healthcare institutions need to invest in advanced cybersecurity measures. Access and fair distribution of benefits provided through the use of AI is another principle that guide the ethical use of AI in life science, where justice is another important principle to guide its use in life science. While AI technologies can be used to change the way healthcare is delivered, this can only be done as long as these advantages are available to everyone, including marginalised or underserved populations. It is my suggestion that regulatory framework should push the adoption of cheaper and more accessible AI which could reduce health care disparities. For example, AI driven diagnostic tools should not be limited to well-resourced hospitals but should adapt for use in remote or low income areas. In addition to fair implementation of the AI, the justice of the implementation of it also requires that this is done with the minimization of systemic biases which could affect thereby some groups such as racial minorities or people economically disadvantaged. To make healthcare innovation inclusive, the models need to be trained on different types of datasets representative of a broader population (Nwanko et al., 2022).

As much as the ethical use of AI in areas of life sciences, the principle of autonomy is equally important. This means patients have the right to make informed decisions about their care and to that, AI-driven treatments or diagnostic tools are included. Preserving patient autonomy requires that AI decisions in medical decision making remain transparent. The benefits and the risks that come with AI use should be communicated to patients in the same context they would be talking to their doctor or nurse. For example, when AI is making diagnostic decisions for providers, they have to explain the role for patients to know how these tools factor into their care. Thus, the process of informed consent facilitates active participation of patients in the decisions on their treatment and putting an emphasis on trust relations between patients and healthcare providers. To be implemented ethically in life sciences, AI implementation also requires collaborative governance where many stakeholders collaborate to develop, regulate, and deploy AI technologies. Regulatory agencies are one major stakeholder group which holds the key role in drawing up ethical standards and guidelines for use of AI in healthcare. For instance, these agencies must impose that the AI tools follow the safety regulations and data privacy, and ethical standards for medical use. It is important that regulatory oversight has enough flexibility but is also comprehensive to shelter public goods and support innovation. Responsible use of regulatory efforts can and do take place with AI developers and regulators working together to foster the development of adaptive guidelines that can respond to the rapid technological advancements being made (Raj et al., 2024).

Another key group of the collaborative governance of AI in life sciences is also the healthcare providers. AI should be developed and implemented by clinicians facing real clinical needs and promising the achievement of better patient outcomes.

Because they understand the realities of patient care, these providers' knowledge is essential to engineering AI applications as functional and intuitive as they are in the real world. Through an integration of AI tools within clinical workflow, healthcare professionals can give feedback about AI and shape AI systems, so they become better, more reliable, and more beneficial for the provision of patient care. Moreover, training programs ought to educate healthcare providers on how to harness AI technologies appropriately and induce the capability to make well informed decisions in an environment with AI technology intervention. In addition, both ethicists and scholars' matter in their contributions to AI governance when it comes to asking what the ethical responsibilities are of AI use in the life sciences. Ethicists can contribute to the process by identifying and discussing potential ethical issues raised by AI technologies, like bias, privacy concerns, or unintended consequences. Ethicists can capture the moment to not only inform decisions about ethical considerations from beginning of stages, but also to incorporate ethical norms into AI models from development to operationalization. However, scholars provide when they do research to determine how AI affects healthcare outcomes, patient satisfaction, and the ethical model, giving evidence based advice to policymakers and AI designers. Making AI solutions patient centred requires the presence of patient advocacy groups. These groups are patient voices in AI discussions that help identify what patient needs, preferences and concerns are. By interacting with patient advocacy organisations, AI developer, and healthcare providers, together they can build technologies that are more in line with patient realities and respond to patient need in the real world. Feedback from patient groups, for instance, can help shape the development of AI tools that are easier to use, more user friendly and more customised to various patient groups. Patient advocacy groups involved in AI policy conversations also build trust and accountability by giving patients a concrete way to see their voices considered and woven into AI based healthcare solutions developing power (Salehi, 2024).

International Considerations

As artificial intelligence technologies are applied globally, international cooperation is vital to address ethical concerns. Different countries have diversified regulations and ethical standards, which can complicate implementing and developing artificial intelligence solutions. Collaborative efforts can help. Developing universal standards for data protection, ethical considerations, and algorithm transparency can facilitate the responsible use of artificial intelligence across borders. Countries can learn from one another's experiences in adopting artificial intelligence technologies, supporting the recognition of efficient strategies for addressing ethical challenges.

Artificial intelligence in life sciences has vast prospects of enhancing patient care and driving innovation. Furthermore, ethical considerations must remain at the lead of this technological enhancement. Addressing privacy concerns and confirming data security is paramount to maintaining patient trust. Additionally, mitigating Bias in artificial intelligence algorithms is vital to promoting equity and fairness in medical care solutions. Finally, ethical and regulatory frameworks must evolve with artificial intelligence technologies to confirm their responsible and safe implementation.

By navigating these ethical frontiers, stakeholders can harness the power of artificial intelligence in life sciences while protecting the health and rights of patients. This balanced approach will foster innovation and promote a more transparent and equitable healthcare system.

REVOLUTIONIZING DRUG DISCOVERY WITH AI

The drug discovery landscape is experiencing a significant transformation powered by artificial intelligence. Conventionally, the drug development procedure has been expensive, lengthy, and troubled, with high rates of failure. Moreover, artificial intelligence technologies are streamlining this procedure by improving target identification, accelerating virtual screening, and optimizing drug design. This detailed analysis investigates how artificial intelligence-powered methodologies are revolutionizing drug discovery, pinpointing fundamental methodologies, virtual screening acceleration, and impactful case studies that depict artificial intelligence's role in bringing new therapies to market (Dhanwe et al., 2024).

Understanding Target Identification

Target identification is vital in drug discovery, including identifying biological molecules. These general proteins play a dynamic role in disease procedures—the progress of subsequent drug design centers on precisely identifying these targets (Aldoseri et al., 2024). Artificial intelligence methodologies are refining the precision and productivity of target identification via different approaches. Machine learning algorithms can assess biological data involving genomic, metabolomic, and proteomic information to recognize possible drug targets. These algorithms can detect correlations and patterns that human scholars may overlook. By assessing gene expression data, artificial intelligence models can recognize downregulated and upregulated genes in particular diseases, suggesting possible targets for drug intervention (Bhardwaj et al., 2022). Artificial intelligence can help map biological networks, uncovering interactions between pathways and proteins, including in disease. This systems biology approach lets scholars recognize crucial nodes that may serve as efficient drug targets. NLP methods allow for the extraction of valuable information from scientific works, patents, and clinical trial reports. It can help scholars recognize new targets and comprehend previous ones better (Khang, 2024). Once possible targets have been recognized, the subsequent step is drug design, where artificial intelligence methodologies can significantly improve the procedure. Artificial intelligence tools can forecast the three-dimensional structures of target proteins, allowing the design of small molecules that fit into their active sites. Deep learning frameworks, like convolutional neural networks (CNNs), can assess protein structures and forecast how different compounds will interact with them. Machine learning models can guess the pharmacodynamics and pharmacokinetics of compounds, offering insights into how drugs will behave in the body. It helps supporters prioritize which compounds to advance into clinical testing (Tariq, 2024c).

Accelerating Virtual Screening Processes Through Machine Learning

Virtual screening is a technological method utilized to recognize possible drug entrants from large libraries of compounds. Conventional techniques can be costly and time-consuming, making them less viable for modern drug discovery demands (Salehi, 2024). Machine learning and artificial intelligence are revolutionizing virtual screening in several key ways. Machine learning algorithms can assess large compound libraries faster than conventional methods. This quickens the recognition of capable candidates and decreases the time spent in the early stages of drug discovery. Artificial intelligence models can be trained to forecast the binding affinity of compounds to target proteins. By utilizing historical data from previous drug discovery projects, these models can score compounds depending on their likelihood of success, letting scholars emphasize the most promising candidates. Artificial intelligence can optimize multiple properties of compounds concurrently, like toxicity, potency, and selectivity. This holistic approach enhances the prospects of recognizing viable drug candidates. Integrating artificial intelligence with high-throughput screening methods lets scholars simulate how thousands of compounds connect with target proteins in silico. This significantly improves the competence of the screening procedure (Singh et al., 2024).

Real-World Applications Of AI In Virtual Screening

Different firms have successfully applied artificial intelligence-powered virtual screening procedures, leading to breakthroughs in drug discovery. Artificial intelligence can assess compound diversity in libraries to confirm that a wide range of structures is being assessed. It decreases the risk of overlooking possible drug candidates. Conventional docking techniques can be enhanced with machine learning by training algorithms to identify successful binding modes depending on historical data, leading to more precise predictions of compound-target interactions (Akintuyi, 2024). Case studies on AI's role in bringing new therapies to market

Case Study 1: BenevolentAI and ALS Drug Discovery

BenevolentAI is a revolutionary artificial intelligence firm focused on drug discovery. One of its prominent achievements is recognizing a prospective treatment for amyotrophic lateral sclerosis (ALS). BenevolentAI used its artificial intelligence platform to assess vast datasets involving genomic and clinical data and recognize the target for ALS treatment. By incorporating multiple data types, the artificial intelligence system pinpointed a new target linked with disease progression. BenevolentAI implemented generative artificial intelligence methods to design new compounds after recognizing the target. This resulted in a candidate that depicted promise in preclinical frameworks. The firm advanced its candidate into clinical trials, demonstrating how artificial intelligence can accelerate drug development and bring new therapies to needy patients (Gul & Banday, 2024).

Case Study 2: Insilico Medicine and COVID-19

Insilico Medicine leveraged its AI-powered drug discovery medium to recognize possible treatments for COVID-19 quickly. Rapid Target Identification: Insilico's AI algorithms assessed protein structures of the SARS-CoV-2 virus and speedily recognized promising targets for drug intercession. This rapidity was critical in the face of the global pandemic. The firm used machine learning to screen large libraries of prevailing compounds against the recognized targets, suggestively quickly identifying possible candidates. Insilico efficaciously recognized a candidate drug within weeks, showcasing the influence of AI in speedily responding to developing health emergencies.

Case Study 3: Atomwise and Alzheimer's Disease

Atomwise, a leader in AI-driven drug discovery, implemented its technology to Alzheimer's disease, a condition with few efficient treatments (Owen, 2024). Atomwise used its AI platform to execute virtual high-throughput screening of millions of compounds, speedily recognizing those that could hinder the target proteins related to Alzheimer's. The firm's machine learning algorithms were trained on wide-ranging datasets, leading to the precise forecast of compound efficiency and security. Atomwise partnered with numerous research organizations to further authenticate its conclusions, signifying the significance of collaboration in progressing drug discovery.

Case Study 4: Moderna and mRNA Therapeutics

While Moderna is mainly known for its COVID-19 vaccine, the firm has leveraged AI in its broader drug discovery efforts, mainly in mRNA therapeutics. Moderna used AI to enhance the design of its mRNA constructs, refining their constancy and efficiency. This computational method permitted quick repetitions and optimizations. Incorporating AI into Moderna's pipeline has simplified a more agile approach to drug development, allowing the firm to hinge speedily in response to evolving health threats. Moderna is discovering numerous mRNA-based therapies for conditions beyond infectious diseases, emphasizing the flexibility of AI in controlling drug discovery across multiple therapeutic domains.

Incorporating AI in drug discovery transforms how new therapies are recognized and established. From improving target identification to quickening virtual screening procedures, AI methodologies are restructuring workflows and enhancing the success rates of drug candidates. Case studies from organizations like BenevolentAI, Insilico Medicine, Atomwise, and Moderna exemplify the reflective impact of AI on bringing advanced therapies to market. As the pharmaceutical landscape advances, the collective efforts between AI technologies and conventional drug discovery procedures will be essential in addressing unmet medical requirements and quickening the expansion of effective treatments. The future of drug discovery is not just about technology; it's about connecting the power of AI to change healthcare and advance patient outcomes worldwide (Brian & Bommu, 2022).

PRECISION MEDICINE: TAILORING TREATMENT WITH AI

Precision medicine depicts a transformative method of healthcare, stressing the customization of medical treatment to individual characteristics, requirements, and likings of patients. Artificial intelligence is indispensable in this standard shift, allowing the progress of personalized treatment plans, enhancing diagnoses, and improving patient results. AI algorithms assess data to recognize patterns and insights that can advise personalized treatment plans. Machine learning procedures, like supervised and unsupervised learning, process diverse datasets comprising genomic information, clinical histories, and lifestyle factors. For example, algorithms can forecast how a patient's exclusive genetic makeup impacts their response to precise medications, letting healthcare providers opt for the most efficient therapies while lessening adverse impacts. Furthermore, AI can support recognizing appropriate clinical trials for patients depending on their particular health profiles, improving access to cutting-edge treatments that may be more suitable than standard options (Ahmadi, 2023). AI-driven predictive analytics can support clinicians in recognizing possible health risks and suggesting apt interventions, confirming that treatment plans are recurrently updated depending on existing data. AI's impact on precision medicine is progressively evident in numerous real-world applications.

AI algorithms assess genomic data from tumor samples to recognize mutations that particular therapies may target. For example, IBM Watson for Oncology implements AI to support oncologists in choosing tailored treatment options depending on the molecular profile of tumors. Machine learning models can forecast cardiovascular events by assessing EHRs, patient demographics, and lifestyle factors. Firms like Tempus use AI to adapt treatment plans for patients with heart disease, enhancing outcomes via personalized intrusions (Kazi, 2024). AI-powered mediums can help recognize and diagnose rare diseases by assessing genetic and clinical data, simplifying diagnoses quicker and more precisely. This method allows personalized treatment strategies that address the exceptional requirements of patients with rare conditions.

AI transforms precision medicine by evolving personalized treatment plans, incorporating patient data for improved diagnosis, and signifying real-world applications across various medical domains (Jeffrey & Bommu, 2024).

AI in Medical Imaging: A New Diagnostic Era

Integrating artificial intelligence (AI) into medical imaging has transformed the field, offering advanced tools that improve diagnostic precision, streamline workflows, and advance patient outcomes. AI technologies, mainly in image dissection and sorting, are making substantial strides, steering in a new era of diagnostics. Medical imaging includes numerous modalities, including X-rays, MRI, CT scans, and ultrasound, each needing accurate analysis for effective diagnosis. AI algorithms, mainly those leveraging deep learning methods, have shown incomparable competencies in image segmentation—recognizing and separating related structures within images. For example, convolutional neural networks (CNNs) can competently discriminate between healthy and diseased tissues, allowing more precise identification of tumors and lesions. These progresses range to classification tasks, where AI can classify images depending on precise standards, such as the presence of cancer or other pathologies. By training on vast datasets, AI models learn to identify patterns and anomalies that human radiologists may find elusive or easily overlooked. This ability improves the diagnostic process and lessens the cognitive load on healthcare professionals, letting them emphasize complex cases that need human expertise. AI-powered imaging tools have been exposed to meaningfully improve diagnostic precision. A projecting instance is using AI to detect breast cancer through mammography. Research has verified that AI algorithms can beat human radiologists in terms of understanding and specificity, leading to earlier and more precise detections of growths. This development is crucial, as early detection is often related to better predictions and survival rates.

Furthermore, AI tools assist in lessening false positives and negatives, which are common issues in radiology. By providing a second opinion, these systems can flag possible problems that need further investigation, reducing the risk of misdiagnosis. As AI progresses, its incorporation into routine clinical practice promises to regulate diagnostic procedures, leading to more reliable and consistent results across different healthcare settings. Many case studies exemplify the effective implementation of AI in medical imaging. One prominent instance is deploying an AI algorithm for detecting lung cancer in CT scans. Scholars trained a model on thousands of scans, attaining remarkable results that verified the algorithm's capability to recognize nodules with high precision. Following its authentication, this technology has begun to be executed in clinical settings, suggestively supporting radiologists in their diagnostic efforts. In pathology, AI has been implemented to assess histopathological images. One study depicted an AI system that precisely categorized numerous types of tumors from biopsy samples. The algorithm matched and, in some cases, exceeded the performance of expert pathologists. This application is mainly promising given the mounting demand for effective and precise pathology services, specifically in resource-limited settings. The infusion of AI into medical imaging signifies a paradigm shift in diagnostics. By improving image segmentation and classification, AI tools enhance diagnostic precision and nurture a more effective healthcare system. With constant research and improvements, the future of medical imaging looks promising. As these technologies mature, they will undoubtedly play a vital part in determining the landscape of contemporary medicine.

NLP in Healthcare: Unleashing the Power of Text Data

Natural Language Processing (NLP) advances healthcare by revealing insights from the huge amounts of unstructured text data in electronic health records (EHRs). This technology allows healthcare providers to excerpt valuable information, improve clinical decision-making, and enhance patient outcomes. EHRs comprise a wealth of data, including patient histories, clinical notes, and treatment plans. NLP algorithms can assess this text to recognize critical data points, such as symptoms, diagnoses, and medication lists. By powering the extraction of structured data from unstructured notes, NLP lessens the burden on healthcare experts and allows for more effective data management (Farahani & Ghasemi, 2024). AI-driven text analysis improves clinical decision-making by offering insights that notify treatment options and enhance patient care. For instance, NLP can assess trends in patient data to recognize potential risks or suggest evidence-based intrusions. By synthesizing information from multiple sources, NLP systems can support clinicians in making informed, apt choices. Numerous case studies focus on the effective application of NLP in healthcare. One prominent instance is the use of NLP to forecast patient readmissions. By assessing clinical notes, scholars developed a model that recognizes high-risk patients based on historical data, allowing for targeted intrusions. Another case included using NLP to rationalize clinical trial recruitment. By mining EHRs for related patient characteristics, scholars suggestively condensed the time required to find appropriate trial candidates, thus quickening the drug development procedure. NLP is harnessing the power of text data in healthcare, offering actionable insights that improve clinical practices and enhance patient care. As this technology advances, its effect on the healthcare landscape will likely progress, leading to more effective and active healthcare delivery (Agrawal et al., 2024).

Precision Agriculture: AI for Sustainable Food Production

Precision agriculture utilizes AI to enhance crop yields and resource use, allowing agriculturalists to make data-based decisions. It implements autonomous robotics for effective farming practices, endorsing sustainability while meeting the increasing global food demand (Akter, 2024). AI-driven analytics optimize crop yields and resource use by analyzing vast weather patterns, soil health, and crop performance datasets. By modifying practices to precise conditions, AI helps maximize efficiency while lessening waste-reducing inputs like water and fertilizers. This data-driven method improves efficacy and supports sustainable agriculture by endorsing liable resource management. AI monitoring systems allow early recognition of crop diseases and pests by utilizing advanced image recognition and machine learning algorithms to assess visual information from fields. These systems can recognize symptoms, such as discoloration or abnormal growth before they become noticeable to the naked eye. Farmers can adopt targeted interventions by quickly providing real-time alerts and dropping the spread of bugs and pest infestations. This active approach minimizes crop loss, declines dependence on chemical treatments, and endorses healthier, more supportable farming practices. Autonomous agricultural robotics alter farming practices by performing vital tasks accurately and competently. These robots can handle numerous operations, comprising planting, weeding, and harvesting, using advanced sensors and AI algorithms to steer fields and evaluate crop health. By systematizing labor-intensive procedures, they suggestively decrease the requirement for manual labor, lessen operational costs, and minimize human error.

AI-Powered Environmental Science: Protecting Our Planet

Artificial intelligence is vital in environmental science, providing innovative solutions to some of the planet's most persistent issues. From climate modeling to biodiversity maintenance and pollution detection, AI technologies are improving our ability to comprehend and protect the environment (Glady et al., 2024). AI is transforming climate modeling by administering data from numerous sources, comprising satellite imagery and historical climate records. Machine learning assesses this data to recognize patterns and make precise predictions about future climate situations. These models help scientists comprehend possible impacts on ecosystems, weather patterns, and sea-level rise, allowing policymakers to develop operative climate adaptation and mitigation strategies (Aagaard & Tucci, 2024). AI technologies are progressively used in biodiversity conservation efforts. AI-power-driven monitoring systems, such as remote sensing and acoustic analysis, can more proficiently trace wildlife populations and their habitats. For instance, machine learning algorithms can assess camera trap images to recognize species and monitor their movements. This data helps conservationists evaluate ecosystem health, detect threats to biodiversity, and apply targeted conservation measures (Javaid et al., 2023). Advanced algorithms can assess sensor data to observe air and water quality, recognizing real-time pollution hotspots. This information allows for timely interventions to alleviate environmental hazards. Furthermore, AI is used in disaster management, forecasting the incidence of natural disasters like floods and wildfires. By evaluating historical data and environmental conditions, AI models can forecast probable disasters, aiding communities in preparing and responding proficiently. AI is a vital tool in environmental science, driving enhancements in climate modeling, biodiversity conservation, pollution detection, and disaster management (Talebi et al., 2024).

Synthetic Biology and Biomanufacturing: Engineering the Future with AI

AI algorithms can simulate and forecast how biological components will interrelate, restructuring the design of synthetic organisms for tasks like drug production or biofuel generation. This lessens the trial-and-error phase in traditional biology, fast-tracking innovation. In biomanufacturing, AI optimizes procedures by evaluating vast datasets from production systems. Machine learning can recognize key parameters that improve yield, decrease costs, and minimize waste, contributing to more sustainable practices. For example, AI can fine-tune fermentation conditions to increase output using fewer resources. Several case studies highlight practical AI applications. For instance, scholars used AI to engineer yeast strains for bioethanol production, suggestively increasing yield. Another research showcased AI's role in improving enzyme production for industrial applications, leading to more effective manufacturing procedures. AI transforms synthetic biology and biomanufacturing, driving industrial biotechnology efficacy, sustainability, and innovation (Holzinger et al., 2023).

Bioinformatics: AI as a Tool for Uncovering Biological Insights

AI is becoming a crucial tool in bioinformatics, altering genomic and proteomic data analysis to reveal vital biological insights. Machine learning algorithms surpass in administering vast amounts of genomic and proteomic data, allowing researchers to manually recognize patterns and correlations that would be tough to determine. These algorithms can analyze DNA sequences, protein structures, and expression profiles, offering insights into gene functions and interactions (Adewusi et al., 2024). By incorporating diverse data sources, such as patient clinical data and molecular profiles, AI models can forecast which biological molecules are linked with particular conditions. This fast-tracks the discovery of possible targets for new treatments, suggestively improving precision medicine. In real-world applications, AI has been used in cancer research to recognize unique genetic mutations related to tumor types, aiding in developing targeted therapies. Another application is drug discovery, where AI models rationalize identifying compounds with therapeutic possible by assessing biological data and forecasting their efficiency. AI is transforming bioinformatics, driving developments in data analysis, biomarker discovery, and therapeutic innovation (Samuel, 2024; Kazi, 2024)).

AI in Biomedical Engineering: Innovating Medical Devices

AI algorithms rationalize the design procedure of medical devices by analyzing vast datasets, forecasting device performance, and optimizing designs for precise clinical applications. This lets engineers create smarter devices that can accurately accommodate patient requirements. Intelligent device design integrates AI to monitor patient health uninterruptedly and provide personalized feedback. (Nwankwo et al., 2022). Notable case studies exemplify AI's impact. For instance, AI-improved prosthetics can learn and adjust user movements, refining functionality and comfort. Another case included AI in developing innovative insulin delivery systems, which automatically adjust dosages based on real-time glucose monitoring, suggestively helping diabetic patients. AI is essential in public health because it improves disease scrutiny, refines resource allocation, and informs health planning (Chisom et al., 2024). AI algorithms assess information from numerous sources—such as social media, healthcare reports, and environmental factors—to detect patterns that specify possible disease outbreaks. By forecasting outbreaks before they occur, public health officials can implement timely intrusions, reducing the spread of diseases and shielding communities. AI supports optimizing resource allocation by analyzing demographic data, disease occurrence, and healthcare infrastructure. This confirms that resources, such as vaccines or medical personnel, are efficiently deployed during crises. AI-driven models can simulate different scenarios to aid policymakers in making informed decisions about public health plans. Case studies exemplify efficacious applications of AI in global public health. During the COVID-19 outbreak, AI tools were used to trace virus outbreaks and inform vaccination strategies, demonstrating their efficiency in real-time crisis management. Another case included AI in malaria control, where predictive models helped target interventions in high-risk areas. AI is crucial for protecting communities via enhanced disease surveillance, efficient resource allocation, and strategic public health planning (Akintuyi, 2024). AI makes substantial paces across various life sciences domains, with impactful applications in healthcare, agriculture, and environmental science. Detailed analysis of successful AI implementations across healthcare, agriculture, and environmental science. One protruding case is the use of AI in radiology, where algorithms like Google's DeepMind have confirmed superior precision in detecting breast cancer from mammograms compared to human radiologists. This achievement augments diagnostic precision and decreases load, letting clinicians focus on complex cases (Raj et al., 2024). Firms like Climate Corporation use AI to evaluate weather patterns and soil conditions, providing farmers with actionable insights to optimize crop yields. This method has led to more sustainable farming practices and augmented productivity by tailoring interventions to precise environmental conditions (Anwar et al., 2023). AI has also shown promise in environmental science. For instance,

IBM's Green Horizon project implements AI to forecast pollution levels and enhance city energy consumption. This initiative helps in urban planning and contributes to cleaner air quality.

LESSONS LEARNED FROM INDUSTRY LEADERS AND INNOVATORS

From these cases, industry leaders focus on the significance of interdisciplinary collaboration and data quality. Ensuring accessibility to high-quality data and fostering stakeholder associations are crucial for efficacious AI implementation. Additionally, incessant iteration and validation of AI models are vital for building trust and attaining meaningful results in real-world applications. (Pachiappan et al., 2024). As AI progresses, numerous emerging technologies promise to impact life sciences suggestively. Natural language processing (NLP) innovations will improve data analysis capabilities, allowing more nuanced insights from unstructured medical records. Moreover, progressions in generative models, such as those creating synthetic biological data, could fast-track drug discovery and personalized medicine. However, widespread AI adoption faces issues. Incorporating AI into prevailing systems can be complex and expensive, impeding smaller organizations from advancing from these advancements. There's also a requirement for robust regulatory frameworks to confirm that AI applications are safe and effective. To advance AI-driven innovation in life sciences, strategic suggestions involve fostering interdisciplinary collaboration between data scientists and domain specialists to confirm that AI models are related and precise. Investment in training programs for healthcare professionals on AI technologies is vital for building trust and capability. Also, promoting open data initiatives can improve the quality and availability of datasets, driving innovation while addressing ethical considerations. In summary, while AI holds massive potential for life sciences, addressing issues and executing strategic initiatives will be critical for realizing its benefits (Popescu et al., 2024).

CONCLUSION

Artificial Intelligence integration in life sciences is a major turning point for research, medicine and understanding biological systems. By processing huge datasets, uncovering hidden patterns and helping with decision-making, AI has provided major improvements in drug discovery, genomics, clinical diagnostics and healthcare management. AI has been instrumental in accelerating drug discovery, improving personalised medicine to improve patient outcomes, streamlining operations and lower the time and cost associated with traditional methods. The historical evolution of AI -- from rule based systems to advanced machine learning -- represents how AI has grown in importance and how it can help meet the challenges of addressing global health. But AI's responsible and broad application in life sciences requires addressing ethical considerations, data privacy and regulatory challenges. Indeed, the future of healthcare and biological research will increasingly rely on AI, and its ability to drive innovation, efficiency and precision across a variety of life science domains underscores the critical importance of AI in future healthcare and biological research.

REFERENCES

Aagaard, A., & Tucci, C. (2024). AI-Driven Business Model Innovation: Pioneering New Frontiers in Value Creation. In *Business Model Innovation: Game Changers and Contemporary Issues* (pp. 295–328). Springer International Publishing. DOI: 10.1007/978-3-031-57511-2_10

Adewusi, A. O., Asuzu, O. F., Olorunsogo, T., Iwuanyanwu, C., Adaga, E., & Daraojimba, D. O. (2024). AI in precision agriculture: A review of technologies for sustainable farming practices. *World Journal of Advanced Research and Reviews, 21*(1), 2276–2285. DOI: 10.30574/wjarr.2024.21.1.0314

Agrawal, K., & Nargund, N. (2024, January). Deep Learning in Industry 4.0: Transforming Manufacturing Through Data-Driven Innovation. In *International Conference on Distributed Computing and Intelligent Technology* (pp. 222-236). Cham: Springer Nature Switzerland. DOI: 10.1007/978-3-031-50583-6_15

Ahmadi, A. (2023). The AI Food Revolution: Reshaping Food Sciences through Artificial Intelligence. [IJBLS]. *International Journal of BioLife Sciences, 2*(1), 62–71.

Akintuyi, O. B. (2024). Adaptive AI in precision agriculture: a review: investigating the use of self-learning algorithms in optimizing farm operations based on real-time data. *Research Journal of Multidisciplinary Studies, 7*(02), 016-030.

Akintuyi, O. B. (2024). AI in agriculture: A comparative review of developments in the USA and Africa. *Research Journal of Science and Engineering, 10*(02), 060-070.

Akter, M. S. (2024). AI for sustainability: leveraging technology to address global environmental. *Journal of Artificial Intelligence General science (JAIGS) ISSN: 3006-4023, 3*(1), 40-48.

Akter, M. S. (2024). Harnessing technology for environmental sustainability: utilizing AI to tackle global ecological challenge. *Journal of Artificial Intelligence General science (JAIGS) ISSN: 3006-4023, 2*(1), 61-70.

Aldoseri, A., Al-Khalifa, K. N., & Hamouda, A. M. (2024). AI-Powered Innovation in Digital Transformation: Key Pillars and Industry Impact. *Sustainability (Basel), 16*(5), 1790. DOI: 10.3390/su16051790

Altayeb, J. M., Eleyan, H., Wishah, N. D., Elmahmoum, A. E., Khalil, A. J., Abu-Nasser, B. S., & Abu-Naser, S. S. (2024). AI-Driven Innovations in Agriculture: Transforming Farming Practices and Outcomes.

Alzamily, J. Y., Bakeer, H., Almadhoun, H., Abunasser, B. S., & Abu-Naser, S. S. (2024). Artificial Intelligence in Healthcare: Transforming Patient Care and Medical Practices.

Anwar, H., Anwar, T., & Mahmood, G. (2023). Nourishing the Future: AI-Driven Optimization of Farm-to-Consumer Food Supply Chain for Enhanced Business Performance. *Innovative Computing Review, 3*(2).

Asvial, M., Zagloel, T. Y. M., Fitri, I. R., Kusrini, E., & Whulanza, Y. (2023). Resolving Engineering, Industrial and Healthcare Challenges through AI-Driven Applications. *International Journal of Technology*, *14*(6), 1177. DOI: 10.14716/ijtech.v14i6.6767

Bhardwaj, A., Kishore, S., & Pandey, D. K. (2022). Artificial intelligence in biological sciences. *Life (Chicago, Ill.)*, *12*(9), 1430. PMID: 36143468

Biswas, A., Kumari, A., Gaikwad, D. S., & Pandey, D. K. (2023). Revolutionizing biological science: The synergy of genomics in health, bioinformatics, agriculture, and artificial intelligence. *OMICS: A Journal of Integrative Biology*, *27*(12), 550–569. DOI: 10.1089/omi.2023.0197 PMID: 38100404

Brian, K., & Bommu, R. (2022). Revolutionizing Healthcare IT through AI and Microfluidics: From Drug Screening to Precision Livestock Farming. *Unique Endeavor in Business & Social Sciences*, *1*(1), 84–99.

Chisom, O. N., Biu, P. W., Umoh, A. A., Obaedo, B. O., Adegbite, A. O., & Abatan, A. (2024). Reviewing the role of AI in environmental monitoring and conservation: A data-driven revolution for our planet. *World Journal of Advanced Research and Reviews*, *21*(1), 161–171. DOI: 10.30574/wjarr.2024.21.1.2720

Dhanwe, S. S., Abhangrao, C. M., & Liyakat, K. K. S. (2024). AI-driven IoT in Robotics: A Review. *Journal of Mechanisms and Robotics*, *9*(1), 41–48.

Farahani, M. S., & Ghasemi, G. (2024). How artificial intelligence plays a role in achieving sustainable development goals? *Sustainable Economies*, *2*(3), 66–66. DOI: 10.62617/se.v2i3.66

Glady, J. B. P., D'Souza, S. M., Priya, A. P., Amuthachenthiru, K., Vikram, G., & Boopathi, S. (2024). A Study on AI-ML-Driven Optimizing Energy Distribution and Sustainable Agriculture for Environmental Conservation. In *Harnessing High-Performance Computing and AI for Environmental Sustainability* (pp. 1–27). IGI Global. DOI: 10.4018/979-8-3693-1794-5.ch001

Gul, D., & Banday, R. U. Z. (2024). *Transforming Crop Management Through Advanced AI and Machine Learning: Insights into Innovative Strategies for Sustainable Agriculture*. AI, Computer Science and Robotics Technology.

Holzinger, A., Keiblinger, K., Holub, P., Zatloukal, K., & Müller, H. (2023). AI for life: Trends in artificial intelligence for biotechnology. *New Biotechnology*, *74*, 16–24. DOI: 10.1016/j.nbt.2023.02.001 PMID: 36754147

Javaid, M., Haleem, A., Khan, I. H., & Suman, R. (2023). Understanding the potential applications of Artificial Intelligence in Agriculture Sector. *Advanced Agrochem*, *2*(1), 15–30. DOI: 10.1016/j.aac.2022.10.001

Jeffrey, L., & Bommu, R. (2024). Innovative AI Solutions for Agriculture: Enhancing CropManagement and Yield. *International Journal of Advanced Engineering Technologies and Innovations*, *1*(3), 203–221.

Kalyanaraman, A., Burnett, M., Fern, A., Khot, L., & Viers, J. (2022). Special report: The AgAID AI institute for transforming workforce and decision support in agriculture. *Computers and Electronics in Agriculture*, *197*, 106944. DOI: 10.1016/j.compag.2022.106944

Kazi, K. (2024). AI-Driven IoT (AIIoT) in Healthcare Monitoring. In *Using Traditional Design Methods to Enhance AI-Driven Decision Making* (pp. 77-101). IGI Global.

Khang, A. (Ed.). (2024). *AI-Driven Innovations in Digital Healthcare: Emerging Trends, Challenges, and Applications: Emerging Trends*. Challenges, and Applications.

Liyakat, K. K. S. (2024). Smart Agriculture based on AI-Driven-IoT (AIIoT): A KSK Approach. *labour, 18*, 19.

Neethirajan, S. (2024). Artificial intelligence and sensor innovations: Enhancing livestock welfare with a human-centric approach. *Human-Centric Intelligent Systems*, *4*(1), 77–92. DOI: 10.1007/s44230-023-00050-2

Nwankwo, W., Nwankwo, C. P., & Wilfred, A. (2022). Leveraging on Artificial Intelligence to Accelerate Sustainable Bioeconomy. *IUP Journal of Knowledge Management, 20*(2).

Owen, J. (2024). The Pinnacle of AI-Driven Robotics: Achieving New Heights in Technology.

Pachiappan, K., Anitha, K., Pitchai, R., Sangeetha, S., Satyanarayana, T. V. V., & Boopathi, S. (2024). Intelligent Machines, IoT, and AI in Revolutionizing Agriculture for Water Processing. In *Handbook of Research on AI and ML for Intelligent Machines and Systems* (pp. 374–399). IGI Global.

Popescu, S. M., Mansoor, S., Wani, O. A., Kumar, S. S., Sharma, V., Sharma, A., Arya, V. M., Kirkham, M. B., Hou, D., Bolan, N., & Chung, Y. S. (2024). Artificial intelligence and IoT driven technologies for environmental pollution monitoring and management. *Frontiers in Environmental Science*, *12*, 1336088. DOI: 10.3389/fenvs.2024.1336088

Raj, N., Masih, J., Selvaraj, K., Pavithra, G., & Ramkumar, E. V. (2024, April). IoT & AI Enabled Information and Communication Technologies in Transforming Rural Communities via Digital Agriculture. In *2024 Ninth International Conference on Science Technology Engineering and Mathematics (ICONSTEM)* (pp. 1-6). IEEE. DOI: 10.1109/ICONSTEM60960.2024.10568764

Raliya, R. (2024). Artificial Intelligence for Precision and Sustainable Agricultural. *ACS Agricultural Science & Technology*, *4*(6), 628–630. DOI: 10.1021/acsagscitech.4c00290

Salehi, F. (2024). The role of artificial intelligence in revolutionizing the agriculture industry in Canada. *Asian Journal of Research and Review in Agriculture*, 70-78.

Samuel, P. K. (2024). AI for addressing challenges in agriculture. *Agricultural Engineering Today*, *48*(1), 5–8.

Singh, M., Joshi, M., Tyagi, K. D., & Tyagi, V. B. (2024). Future Professions in Agriculture, Medicine, Education, Fitness, Research and Development, Transport, and Communication. *Topics in Artificial Intelligence Applied to Industry*, *4*(0), 181–202. DOI: 10.1002/9781394216147.ch10

Talebi, E., & Nezhad, M. K. (2024). Revolutionizing animal sciences: Multifaceted solutions and transformative impact of AI technologies. *CABI Reviews, 19*(1).

Tariq, M. U. (2024a). Multi-Agent Models in Healthcare System Design. In Dall'Acqua, L. (Ed.), *Bioethics of Cognitive Ergonomics and Digital Transition* (pp. 143–170). IGI Global., DOI: 10.4018/979-8-3693-2667-1.ch008

Tariq, M. U. (2024b). Social Innovations for Improving Healthcare. In Chandan, H. (Ed.), *Social Innovations in Education, Environment, and Healthcare* (pp. 302–317). IGI Global., DOI: 10.4018/979-8-3693-2569-8.ch015

Tariq, M. U. (2024c). Leveraging AI for Entrepreneurial Innovation in Healthcare. In Özsungur, F. (Ed.), *Generating Entrepreneurial Ideas With AI* (pp. 192–216). IGI Global., DOI: 10.4018/979-8-3693-3498-0.ch009

Tariq, M. U. (2025a). Navigating the Convergence of IoMT and Generative AI in Global Healthcare: Opportunities, Challenges, and Ethical Considerations. In Kumar, V., Katina, P., & Zhao, J. (Eds.), *Convergence of Internet of Medical Things (IoMT) and Generative AI* (pp. 513–534). IGI Global Scientific Publishing., DOI: 10.4018/979-8-3693-6180-1.ch022

Tariq, M. U. (2025b). AI-Powered Breakthroughs: Revolutionizing Cognitive Psychology and Neuropsychology With Machine Learning. In Bansal, R., Maqableh, T., Shuklaa, G., Rabby, F., & Lathabhavan, R. (Eds.), *Transforming Neuropsychology and Cognitive Psychology With AI and Machine Learning* (pp. 65–92). IGI Global Scientific Publishing., DOI: 10.4018/979-8-3693-9341-3.ch003

Tariq, M. U. (2025c). Ethics and Conduct: Navigating the Moral Landscape of Virtual. In Darwish, D. (Ed.), *Building Power, Safety, and Trust in Virtual Communities* (pp. 297–332). IGI Global., DOI: 10.4018/979-8-3693-3868-1.ch014

Veeramanju, K. T. (2023). Revolutionizing agriculture: A case study of IBM's AI innovations. [IJAEML]. *International Journal of Applied Engineering and Management Letters*, *7*(4), 95–114.

KEY TERMS AND DEFINITIONS

Deep Learning: A subset of machine learning using neural networks to model complex data patterns and make decisions.

Genomics: The study of an organism's complete set of DNA, including genes and their functions.

Natural Language Processing (NLP): A technology enabling machines to understand, interpret, and respond to human language.

Personalised Medicine: Tailoring medical treatments based on individual genetic, environmental, and lifestyle factors.

Chapter 4
AI in Bioinformatics Unlocking the Secrets of Life

P. Venkadesh
https://orcid.org/0000-0001-6582-3153
V.S.B. College of Engineering Technical Campus, Coimbatore, India

S. V. Divya
https://orcid.org/0009-0004-4321-8066
V.S.B. College of Engineering Technical Campus, Coimbatore, India

V. Mathumitha
https://orcid.org/0009-0003-5749-2659
Bharath Institute of Higher Education and Research, Chennai, India

S. Ninisha Nels
Noorul Islam Centre for Higher Education, Kumaracoil, India

J. Balaji
Nandha College of Technology, Erode, India

Aiswariya Ramesh
https://orcid.org/0009-0003-1611-757X
V.S.B. College of Engineering Technical Campus, Coimbatore, India

ABSTRACT

Bioinformatics, where Artificial Intelligence (AI) meets biology, as transforming our understanding of life. By combining machine learning algorithm with tools like blast and cluster Omega, scientists are deciphering DNA and predicting protein structure. AI models such as CNN's and RNN's, help analyze DNA sequence, identify genetic variation and predict mutation effects. Alpha fold, developed by deep mind has made groundbreaking strides and protein folding, essential for drug discovery and under disease research. AI platforms like atom wise and benevolent AI accelerate drug discovery by analyzing massive datasets. AI even uses Natural Language Processing (NLP) to process scientific literature, uncovering insights and suggesting result direction. Looking ahead, AI ensures transparency, making advancement more

DOI: 10.4018/979-8-3693-9208-9.ch004

trustworthy. Together, bioinformatics and AI are reshaping feels like personalized medicine and synthetic biology. With endless possibilities the next breakthrough in biology may be just algorithm away, our journey into frontier has only begun.

1. INTRODUCTION TO BIOINFORMATICS AND AI

1.1. Overview Of Bioinformatics:

Alright, buckle up folks! You're about to enter the world of bioinformatics---yes, that's right, bioinformatics. Sounds fancy right? But really, it's just a cool mash up of biology, computer science and statistics. Think of it as the Avengers of science, except instead of Figurehting Thanos, bioinformatics Figurehts data overload. And trust me there's a LOT of data. Every second, researchers are cranking out more biological data than a Starbucks on Monday morning churns out lattes. In simple terms, bioinformatics is about using computers to make sense of all that crazy biological data. It's like when you open 50 browser tabs at once and your brain goes like, "Wait, what? ---except bioinformatics helps scientists find what they're looking for without their brains melting down.

At its core, bioinformatics helps with things like:

- **Analyzing Genomic Data:** Ever wondered how scientists Figureure out what each gene does? bioinformatics is like google maps for genomes.
- **Studying Evolutionary Relationships:** wondering if humans are more closely related to chimpanzees or they more closely related to eating bananas? Bioinformatics traces the genetic family tree!
- **Understanding Protein Structures:** Proteins are like the Lego blocks of life. If you want to design a cool new drug, you better know how those blocks fit together.
- **Predicting Drug Interactions:** Bioinformatics can run simulations of how drugs might behave before you start experimenting. That's like prediction how much you'll regret spicy tacos---before actually eating them!

1.2. Importance in Modern Biology:

Now, imagine trying to read a 3-billion-letter-long instruction manual. That's the human genome, folks without bioinformatics, you'd spend your entire life flipping through Pages, trying to make sense of it all. So, bioinformatics is basically like

Ctrl+F for Scientists. It's fast, efficient and it saves time---no more head-banging against textbooks.

Bioinformatics has brought us some of the coolest discoveries, like:

- **Personalized Medicine:** Because "one-size-fits-all" is for socks, not medical treatments. Now, doctors can tailor treatments to your genetic makeup.
- **Agricultural Innovations:** Think drought-resistant crops and bigger, better plants that feed the world.
- **Drug Discovery:** It's speeding up how we find new medicines by simulating molecular interactions like a video game---except this game could save lives!

Without bioinformatics, modern biology would be stuck in stone age, flipping through endless pages of data, trying to find the one page that matters. It's the glue that holds all the mind-boggling discoveries of today's biology together.

1.3. Brief History and Evolution of Bioinformatics

Flashbacks to the 60s and 70s: computers were like dinosaurs and bioinformatics was just a baby. Back then, a few nerdy scientists (the real MVPs) were playing around with early computers, sequencing tiny bits of DNA. It was like trying to solve a jigsaw puzzle---without knowing what the final picture looked like. But the real game-changer came with the Human Genome Project (1990-2003) ---the mega blockbuster of science, with an all-star international cast. They managed to sequence the entire human genome. That's 3 billion base pairs, by the way. Imagine doing that without computers. It's like trying to count all the grains of sand on beach with a magnifying glass, not easy.

Since then, bioinformatics has grown up fast:

- **1988:** GenBank was born---the first public database for DNA sequences. It's like the Wikipedia of genetics.
- **1990s:** The invention of bioinformatics algorithm like BLAST (Basic Local Alignment Search Tool) was introduced by Stephen Altschul et al. let's just say it's an awesome tool for comparing genetic sequence.
- **2000s:** Next-Gen sequencing made it faster and cheaper to sequence genomes. More data means more problems--- but also it means more solutions, thanks to bioinformatics.
- **Today:** AI and Machine Learning for taking bioinformatics to new heights. We're predicting protein structures, stimulating biological system and even working on designing synthetic life forms. let's all hail to the future with bioinformatics!

1.4. Role of AI in Bioinformatics

Now it's time to talk about our new superhero sidekick: AI. If bioinformatics is Iron Man, then AI is a sleek, shiny suit of armor worn by Iron Man, that turns it from awesome to unstoppable. Machine Learning (ML) and Deep Learning are like bioinformatics' secret weapons---they help make sense of data that's too complex or messy for traditional methods.

Here's how AI is shaking things up:

- **Predicting Protein Structure:** Remember that jigsaw puzzle from earlier? Using AI is like having a cheat sheet, for example DeepMind's AlphaFold literally solved one of the biggest challenges in biology ---predicting how proteins fold.
- **Genome Annotation:** AI speeds up genome annotation, which is used for Figureuring out what all the genes actually do. It's like having a super-fast personal assistant who can sort through an entire library in seconds.
- **Drug Discovery:** AI can stimulate how drugs interact with biological molecules, which help us avoid research and also reduce costs and time. It's like taste-testing every possible ice cream flavor without even leaving your house.
- **Predicting Diseases**: AI builds predictive models that estimate your risk for diseases like cancer or diabetes based on your genetic data. It's like having a crystal ball for your health but in a scientific way.

1.5 Literature review

Mount (2004) offers a foundational perspective on bioinformatics methods, including sequence alignment, genome annotation, and tools critical for data analysis in biology. Baxevanis (2004) covers practical bioinformatics applications in genomics and proteomics, essential for understanding gene functions and evolutionary relationships. Camacho (2018) examines how machine learning assists in interpreting complex biological networks, especially in predicting interactions and responses to various stimuli. (Ching, et al., 2018) discusses deep learning's transformative role in healthcare, including challenges in data quality, interpretability, and scalability for bioinformatics task. Marx (2013) reviews the challenges of big data in genomics, focusing on computational tools for storage, analysis, and retrieval in bioinformatics. (Stephens et al., 2015) provides an overview of the big data explosion in genomics, analyzing techniques for managing vast genomic datasets effectively. (Jumper et al., 2021) covers AlphaFold's contribution to accurately predicting protein structures, which significantly advances personalized medicine and protein-related research. Mardis (2017) explores DNA sequencing advances, discussing technologies like

Illumina and PacBio, essential for generating high-throughput data analyzed by AI. (Zhavoronkov et al., 2019) reviews AI methodologies that identify drug candidates and biomarkers, which have significantly reduced the cost and time of drug development. (Vamathevan et al., 2019) examines machine learning techniques for drug discovery, focusing on target identification, molecule generation, and candidate screening. (Litjens et al. 2017) ensures a comprehensive survey of AI applications in image analysis, highlighting AI's capability to automate diagnosis in areas like MRI and CT scans. (Esteva et al., 2019) discusses how deep learning assists healthcare professionals by analyzing large volumes of medical images and making accurate diagnoses. (Floridi et al., 2019) establishes ethical principles for AI in healthcare, emphasizing privacy, accountability, and fairness in algorithmic decision-making. (Mittelstadt et al., 2016) analyzes ethical implications and concerns in data-intensive fields, focusing on transparency, fairness, and consent in AI-powered healthcare. A study by (Torkzadehmahani et al. 2020) provides a structured overview of privacy-preserving AI techniques in biomedicine, highlighting federated learning as a promising method. A study by (Jiang et al. 2022) introduces a differential privacy-based scheme for sharing genomic datasets to enhance the reproducibility of genome-wide association studies (GWAS) outcomes, balancing data utility with privacy protection. A study by (Li et al. 2024) evaluated FL models for Parkinson's disease prediction using genomic and transcriptomic data, demonstrating that FL models achieved performance comparable to traditional centralized models. The study also highlighted the importance of data distribution among participating institutions in influencing model performance.

2. THE DATA DELUGE: GENOMICS AND BIG DATA

2.1. Understanding Genomic Data

Alright, let's dive into the deep end of the gene pool. Genomic data--- what is it, why does everyone geek out over it and why should you care about all this? Well, think of genomic data as life's ultimate cheat code. It's the blueprint of every living Organism, written in the language of DNA: A, T, C, G. These 4 little letters carry all the instruction needed to build you, me, your dog, that Rose in your garden and even the vegetables and fruits in your fridge.

There are a few different types of data we can pull from life's blueprint:

- **Genomic Data:** This data gives detail about the whole DNA sequence of an Organism. This is a big boss of biological data, telling us what's possible in terms of development, health and disease.

- **Proteomic Data:** This data gives detail about all the proteins expressed by an Organism. If the genome is the instruction manual, proteins are the actual products, that is they are the real workers making things happen.
- **Transcriptomic Data:** This data gives detail about all the RNA produced by the genome, showing us, which genes are being turned 'on' at any given time, like a biological light switch.
- **Metabolomic Data:** This data tracks the chemical reaction happening inside the cells--- essentially, who's doing what? when? And how? in your biological machinery.
- **Epigenomic Data:** These are the chemical modification on the top of your DNA that influence how genes get expressed. It's like tweaking your car's engine without actually changing the model.

2.2. AI's Role in Managing Big Data

Now, let's get to the juicy part of how on earth do we make sense of all this data? Imagine trying to organize 3 billion Lego pieces, and that's just for one human. We have sequence genomes for plants, animals, bacteria--- everything! We are talking petabytes [millions of gigabytes] of data, and it's coming at us faster than a tsunami. Here's how AI helps us keep the chaos in check:

- **Handling Big Data Volumes:** Genomic data is ridiculously huge. It's like trying to find a specific YouTube video in an ocean of video clips. AI--- especially machine learning (ML) and deep learning (DL)--- can shift through these massive data sets faster than any human ever could. For example, AI algorithms can analyze DNA sequences from millions of people to find genetic markers for disease like Alzheimer's. Finding that one mutation it's like spotting a needle in a Mount Everest -sized haystack but AI gets it done and it does it fast.
- **Dealing With Unstructured Data:** In biology, data isn't always nice and neat. Often, it's incomplete, noisy and messy. Think of your inbox after ignoring it for a month, it will be chaos. AI doesn't care, It can handle unstructured data, clean it up, and find the patterns that matter. For example, in genomics, sequencing errors happen. Gaps exit, I can predict what should be there with a 100% accuracy, patching up those holes like a data wizard.
- **Extracting Meaning From Complexity:** so, we have got all this data. But what does it mean? AI is like a Sherlock Holmes of bioinformatics, finding connections and patterns in places human would never think to look for. It can reveal how certain gene mutation led to disease or how different factors

like genes, proteins and environmental stuff interact to increase cancer risk .AI doesn't just crunch number---it solves mystery!

- **Predictive Analytics:** AI can predict the future. Seriously, it can look at someone's DNA and predict their risk of developing a disease, like diabetes or cancer, long before they show symptoms. It's like having your own futuristic health predictor, minus crystal ball and vague fortunes.

2.3. Tools of the AI Trade

So, what magical tools are scientists using to make this all happen? Let's meet a few AI heavy- hitters:

- **TensorFlow:** Developed by Google, TensorFlow is an open-source platform used for machine learning. It can process massive data sets and make prediction, like which genes are going to get express or however disease might develop based on your genetic code.
- **Deep Variant:** It is also from Google this tool focuses on making DNA sequencing more accurate. It uses deep learning to identify errors and genomic data and fix them, so we are not basing our research on a typo.
- **AlphaFold:** This AI made headlines by solving one of biology's biggest challenges--- predicting how proteins fold. Why does this matter? because knowing a protein's structure helps us design better drugs. It's like finally getting the instruction manual for how life's molecular machines work.
- **IBM Watson:** This bioinformatics genius can read genomic data and cross references it with scientific literature, helping doctors make sense of complex genetic profiles and tailored treatments to individual patients.

2.4. Genomic Data as a Messy Closet

Alright, let's break this down in a way that makes sense. Imagine your genome is a closet. Not just any closet--- a closet that hasn't been cleaned out in years. There are random socks, couple of winter coats, old T-shirts and shoes you thought you lost and your favorite pair of jeans is buried somewhere in that mess. Now imagine the above situation in terms of biology and multiply that by millions of closets, each one packed to the brim with 3 billion bits of information, that's what you call a genomic data.

3. AI TECHNIQUES IN BIOINFORMATICS: WHERE MAGIC MEETS SCIENCE

Welcome to the wild, wonderful, and sometimes Mind-bending world of AI In bioinformatics! If you have ever thought," AI sounds cool, but what exactly is it doing in bioinformatics?" --- you are in for a treat. Hold tight because we are driving into machine learning (ML), deep learning (DL) and natural language processing (NLP), all in fun and casual way!

3.1. Machine Learning and Deep Learning: From Curious Kid to Genius Researcher

3.1.1. Machine Learning (ML)

Think of ML as teaching a computer to be a curious little scientist. You don't have to spell out every single instruction. The types of ML are represented by Figure.1.

Figure 1. Types of ML

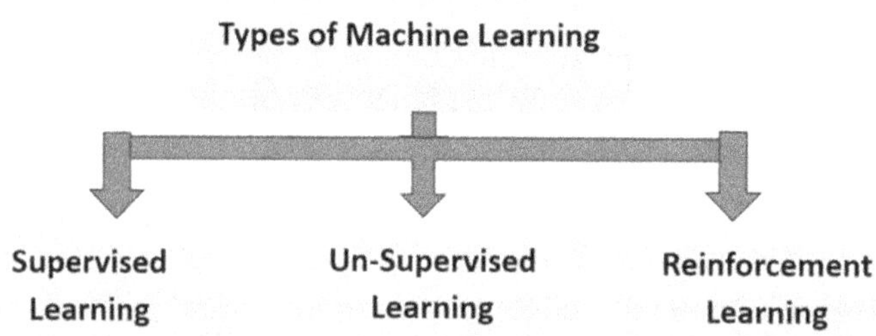

There are 3 types of ML;

- **Supervised learning:** Picture this as giving the computer a CHEAT SHEET. You provide it with the right answers during training, so it knows what to look for. Imagine feeding it a bunch of genetic profiles of patients labeled with who developed a certain disease. Now, the computer is like, "Got it! I will learn from these examples and predict if someone new gets the same disease!"
- **Unsupervised learning:** It's like not giving a cheat sheet and handing your computer a big jigsaw puzzle without showing it the final picture. In bioinfor-

matics, this helps cluster genes or proteins with similar functions, but without any prior clue of what those functions might be.

- **Reinforcement learning:** Have you ever rewarded yourself with pizza for completing a big task? Reinforcement learning works similarly. The algorithm learns by getting "**rewards**" or "**penalties**" based on its action. It's still an underdog in bioinformatics but could be a game changer for optimizing things like protein production.

3.1.2. Deep Learning (DL):

If AI is like a curious kid, then DL is like letting that kid grow into a genius level researcher with the photographic memory. DL models are built on artificial neural networks--- mimicking the human brain's way of processing information. And just like our brains, DL can tackle some really complex problems. Imagine layers in a DL model like layers on an onion the outer layer might recognize simple stuff like lines or edges while, the deeper layers recognize more complicated things like protein structure.

Deep learning really shines in task like:

- **Genomic Sequencing:** tools like deep variant uses DL to read DNA sequences more accurately than ever before. It's like improving your reading glasses but for genomes!
- **Protein Structure Prediction:** alpha fold predicts how protein fold based on the amino acid sequence, something scientists have been scratching their heads over for decades.
- **Drug Discovery:** DL is there a new hero in the town for predicting how different molecules will interact with proteins. This helps find potential new drugs faster--- and possibly the next miracle cure for... Well, who knows what!

3.1.3. Real- Life Bioinformatics Applications of ML and DL

Now that you have got a basic idea of ML and DL, let's talk about how they are used in bioinformatics.

- **Gene Expression Analysis:** ML goes through thousands of genes to Figureure out which ones are being expressed in certain conditions like disease. It's like trying to Figureure out which songs are on your playlist by just listening a few seconds instead of listening thousands of songs.

- **Disease Risk Prediction:** have you ever wonder if your genes hold the key to predicting your future health? ML models analyze genetic variants and lifestyle data to predict the likelihood of developing certain diseases from cancer to heart disease, it's like getting a sneak peek at your DNA based fortune!
- **Personalized Medicine:** ML helps in Giving tailor treatment according to your genetic makeup, lifestyle and even microbiome. Think of it as having a Doctor Who knows exactly what works for your body--- like customized medicine and treatment to Your body.
- **Image Based Diagnosis:** DL is a rock star when it comes to analyzing medical images. Whether it's MRI scans, CT scans or microscopic images, DL algorithms can identify diseases like cancer faster than a human eye ever could.
- **Drug Discovery and Repurposing:** DL models predict how molecules will interact with biological targets, meaning we can find new drugs quicker. It's like Netflix suggesting what to watch next, except its suggesting what medicine might save lives.

3.2. Natural Language Processing (NLP) in Bioinformatics: The Next Text Whisperer

Let's talk about another AI superhero in bioinformatics that is NLP. Unlike machine learning and deep learning, which deal with structured data [numbers, images, etc]. And in science there is a lot of words and these words worth millions of papers. No human can keep up with that word flood, but NLP can.

Here's how NLP is changing the game:

- **Literature Mining:** NLP can read thousands of scientific papers, pulling out key information about genes, diseases or drugs. It's like having a super speed reader who never gets tired. Forget cramming for exam-- -NLP could ace all your tests.
- **Text Classification:** NLP algorithms can sort papers based on topics, so bioinformaticians can find the right research in a sea of studies. It's like having a librarian who already knows exactly what you're looking for before you ask.
- **Entity Recognition**: NLP can automatically pick up important biological entities, like gene names or diseases in scientific papers. It's like having a magical highlighter that knows exactly what's important in your textbook.
- **Question- Answering System:** fancy asking an AI, "what genes are associated with Alzheimer's?" And having it come through all the relevant research to answer. With advanced NLP systems like Bio BERT, this is becoming a reality. Your next research assistant might just be an algorithm.

4. AI GENOMIC SEQUENCING: DECODING THE BLUEPRINT OF LIFE

All right, buckle up, because we are about to drive into the world of genomics sequencing--- a topic that's fascinating as it is complex. Imagine cracking the code of life itself, from the tiniest bacteria to most intricate of Organism. Pretty wild, right? But, as cool as that sounds, DNA sequencing is no walk in a park. It's more like trying to read the world's longest novel, except half the chapters are missing, the plot is scattered, and there are some serious Typos.

4.1. How AI Improves Sequencing Technology

Sequencing has come a long way since the first human genome was cracked open in 2003 to stop back then, it took over a decade and $ 2.7 billion to sequence it. But today you can get your genome sequence in just a few days for under $10,000. It's progress right but it's still not perfect. Sequencing DNA can be noisy, messy, and full of errors. It's like trying to put together a jigsaw puzzle that's been chewed up by your dog. Here's how AI helps the sequencing wizards over at Illumina and PacBio.

4.2. Illumina Sequencing and AI Integration

Illumina is the king of next generation sequencing (NGS). They take your DNA, chop it up into smaller bits, sequence those bits, and then piece them back together. Sounds easy right? But it's not. There's always room for errors, especially in tricky parts of the genome. That's where AI steps in like a super-smart editor. Illumina's platforms are starting to integrate AI- driven tools like Deep Variant. It is like the meticulous friend who proofread your work and catches all your embarrassing typos

4.3. PacBio and Long-Read Sequencing

If Illumina's method is like reading tiny fragments of a book, PacBio's long read sequencing is more like reading whole chapter at once. It's especially handy for tackling the more complicated parts of the genome, where short reads just don't cut it. AI make sure even the messy part of the genome gets read like a best seller. Here PacBio is represented by Figure.2.

Figure 2. PacBio

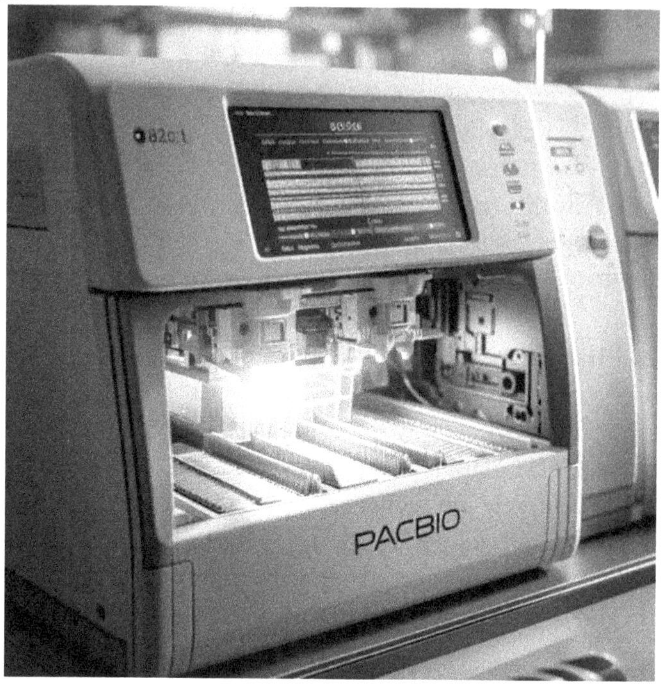

4.4. AI's Role in Analyzing Sequencing Data: Real- Life Magic

4.4.1. Tailoring Treatments With AI

Imagine going to the doctor and instead of getting the same old prescription everyone else gets, you walk out with the treatment tailored specifically for your genes. That's the power of AI in personalized medicine. For example, take cancer treatment. by sequencing a patient's tumor DNA, doctor can Figureure out which Genetic mutations are causing a problem. AI tools like IBM Watson for genomics is then used to cross reference this data with thousands of scientific papers and suggest a personalized treatment plan. IBM Watson is like the doctor's nerdy, well-read assistant, working behind the scenes to find the perfect solution--- sometimes even suggesting drugs that doctors hadn't considered.

4.4.2. AI in Population Genomics:

AI isn't just a one-on-one kind of deal; It's great for large- scale projects too. In 2020 researchers used AI to analyze genomic data from over 100,000 people as a part of the UK Biobank project. By using machine learning algorithms, they uncovered genetic markers linked to diseases like diabetes and heart disease. The AI sifted through mountains of data, spotting patterns that even the most caffeinated human researcher would have missed. Thanks to AI, we are getting better at predicting genetic risk and finding new ways to treat complex diseases.

4.4.3. AI in Evolutionary Biology

AI isn't just useful for modern medicine--- it's also helping us get a peek into the past. In 2018, scientists used AI to analyze the genomes of 200+ mammal species, uncovering new insights into how certain traits evolved. AI crunched through all the data, revealing trends in genetic evolution that have been overlooked for years. It's like having a time machine that helps us understand how different species, including humans, got to where we are today.

4.5. AI as a Super Librarian: A Humorous Metaphor

As mentioned earlier that DNA sequencing is like trying to read a novel with missing pages. Here's a full picture imagine you're reading a thriller come on just when you are about to find out who the culprit is, Suddenly the page next went missing. The story skips ahead, but you have no idea what's going on. You start gathering all the clues together, but the suspense is making you go out of your mind. This is the exact scenario that scientists deal with when sequencing genomes. DNA is long and it is full of errors and a missing bit. Sequencing technologies [like Illumina and PacBio].

5. AI FOR PREDICTIVE ANALYTICS IN HEALTH AND DISEASE: YOUR DATA-DRIVEN CRYSTAL BALL

AI is a new age fortune teller that doesn't rely on vague guesses but on your genetic makeup, lifestyle choices and environmental factor. It's like having high tech health horoscope, except this one can actually help you live longer !

5.1. Using AI to Predict Disease Risk: From Blurry Guesswork to 20/20 Vision

Predicting health risk isn't a new thing. Doctors have been doing it for years by looking at factors like family history, diet, and age. But let's be honest- --that's like trying to Figureure out who stole the cookies from the jar by squinting through a foggy window. It's not wrong, just not super accurate. AI, on the other hand gives us sharp high-definition clarity by analyzing a much broader and more complex array of data points, especially genetic information. Think of it as upgrading from a VHS tape to 4K streaming!

5.2. Genetic Factors: DNA's Sneaky Clues

DNA is like an ultimate instruction manual for your body. But here's the kicker-- it's also holding a ton of sneaky clues about what could go wrong. For instance, certain tiny changes in your DNA, called single nucleotide polymorphism (SLP), can either boost or lower your risk for developing specific diseases. For example, take breast cancer. Mutation in the BRCA1 and BRCA2 genes can make your risk skyrocket. But breast cancer's not the only disease lurking in your genetic code.

5.3. Lifestyle And Environmental Factors: It's Not All in the Genes

Of course, your DNA isn't calling all the shots. What you eat, how much you exercise, whether you smoke all those choices fit into the equation too. And let's not forget the stuff you can't control, like air quality or the fact that your upstairs neighbors throw late night parties.AI is smart enough to bring all of this into the equation say you have got a genetic predisposition for type 2 diabetes. If you are bingeing on sugary snacks and avoiding the gym like the Plague, AI can combine those lifestyle habits with your genetic risk and give you a more accurate picture.

5.4. Real-World Examples: AI in Action

5.4.1. Predicting Heart Disease Risk:

AI is stepping in to help predict who's at risk of having heart disease. Enter the Framingham Heart Study, which has collected data for over 70 years from thousands of participants. AI models trained on this data sets can predict your risk of heart disease by analyzing not just the usual suspect [like age, blood pressure, cholesterol] but also genetic markers and lifestyle habits. Tools like Cardio Risk are now being used in real time to help doctors make more informed decision. It's like having a data-driven health coach that knows exactly when to tell you to put down the double cheeseburger.

5.4.2. AI and Cancer Risk Prediction:

cancer prediction is another area where AI is making huge strides. IBM Watson for genomics analyzes genetic data from cancer patient to predict their risk of developing certain types of cancer IBM Watson doesn't just crunch numbers; It cross-reference patients' genome with years of research to make highly personalized treatment recommendations. Also take lung cancer, for example. Watson AI algorithm help doctors predict which treatment are most likely to work based on a patient's genetic data. It's like having a super-nerdy, ultra- dedicated medical intent that never sleeps and knows everything about cancer.

5.4.3. Preventing Type 2 Diabetes:

Type 2 diabetes is the preventable disease. AI driven models, like the ones from DeepMind, are being used to predict diabetes risk long before symptoms appear. These models combine genetic, lifestyle, and clinical data to predict whether someone is likely to develop type 2 diabetes in the next decade. And the best part is, if AI tells you that you are at high risk, there's still plenty of time to make changes to your diet and exercise habits to keep the disease at Bay. It's like getting an early warning before the ice bug hits the Titanic---so you can steer clear and live your best life.

5.5. AI, The Game Changing Element

Whether it's heart disease, cancer or diabetes, AI is changing the game by giving doctors and patients a data- Powered crystal ball to predict, prevent and manage disease with more accuracy than ever before. Keep those healthy habits going and AI might just give you a heads up when you need it the most!

6. AI POWERED PERSONALIZED MEDICINE:

In this section, we will break down what personalized medicine is, how AI is revolutionizing it, and take a deep dive into a real-world case where AI help create personalized cancer treatment. Who says healthcare can't be fun, we will end with a story about ordering personalized medicine like a pizza. Grab a slice of knowledge and dig in!

6.1. What is Personalized Medicine?

Personalized medicine is healthcare's responsibility to the one-size-fits-all approach. Instead of giving everyone the same treatment--- like those bland, one -favor -fits -all pizzas--- personalized medicine creates a plan that's as unique as you are. Here's how it works: personalized medicine use uses your genetic data, lifestyle and environmental factors to craft a treatment plan that fits you. Days of treatments that work for some and miss the mark for others are long gone. With personalized medicine, your therapy is custom-made for your biological footprint. Think of it as getting custom fitted shoes that doesn't pinch your toes but here it's your healthcare, we are fine tuning! Instead of shoes.

6.2. How AI is Tailoring Treatment Like a Pro

AI can crunch vast amount of data faster than you can say “extra pepperoni”. From genetic sequence to your medical history, AI analyzes it all to find patterns that humans might miss. Let’s explore it in detail

6.2.1. The Role of AI in Analyzing Microbial Communities:

Artificial intelligence AI has become a pivotal tool in analyzing microbial communities particularly through the lens of genomic data.

- **Taxonomic Profiling and Classification:** AI algorithms especially machine learning models have been employed to classify microorganism with high accuracy to enhance microbial identification. A 2024 study utilized machine learning techniques to analyze genomic sequence, leading to improved identification and understanding of microbial diversity. AI tools like BERTax and TEPI (Taxonomy-aware Embedding and Pseudo- Imaging) are used for this purpose. BERTax applies transformer-based models to taxonomic classification, leveraging natural language processing techniques for genomic data analysis. TERPI represents genomes as pseudo-images and maps them to taxonomy-aware embedding space, facilitating zero-shot genome classification.
- **Functional Annotation and Prediction: AI** has been instrumental in predicting the functions of genes within microbial genomes. A comprehensive review in 2024 highlighted AI driven approaches that enhance the annotation of microbial genes, fertilizing a deeper understanding of microbial loads in various environments for predicting gene functions. Some AI driven tools such as ‘SAFPred’ which employs protein embedding and incorporates con-

served synteny across bacterial species to predict gene functions, 'GeNLP' which utilize natural language processing to explore and predict microbial gene functions by analyzing genomic and protein sequence and 'DeepFRI' which is integrated into workflows for metagenomic data analysis and uses deep learning to improve the annotation coverage of microbial genes, significantly enhancing functional annotation.

- **Microbe Environment Interaction:** To predict microbial dynamics AI model have been developed to infer interaction between microbes and their environments. A study in 2023 introduces a tool called MetConSIN that uses AI to predict microbial community dynamics based on genomic and metagenomic data aiding in the understanding of microbial ecology. MetConSIN stands for "metabolically contextualized species interaction networks". It aids in understanding microbial ecology by inferring networks of interaction between microbes and environmental metabolites over time.
- **Anti-Microbial Resistance Prediction:** AI has been applied to predict antimicrobial resistance (AMR) by analyzing genomic data. A 2024 review discussed the use of machine learning models in identifying assistance pattern, which is crucial for developing effective treatment strategies. DnnARs, an artificial intelligent technique designed for prediction off antimicrobial resistance in bacteria. This tool employs deep learning models to analyze patterns of antibiotic resistance, aiding in effective management and response strategies.

6.2.2. How does AI Analyze All that Data?

- **Genomic Data:** AI looks at your DNA to spot Genetic mutation or quirks that could affect how you respond to treatments. Let's say you have a mutation in the TP53--- this could mean certain chemotherapy options won't work as well for you. AI spots that mutation and helps your doctor fix something that will get the job done.
- **Medical Records:** imagine AI going through your entire medical history like a detective, spotting patterns in your allergies, family history or past treatments and then AI will be like, "you have had a reaction to that drug before? Noted. Let's avoid that and find something better".
- **Scientific Literature:** AI doesn't just rely on old knowledge; It's constantly Keep up with the latest research paper, clinical trials and drug studies. It's like having a Doctor Who's up to date on all the latest science while you barely to keep up with your favorite Netflix shows.

6.3. AI Powered Drug Recommendation: Your Personal Pharmacist

Once AI has all your genetic info, medical history, and lifestyle factors, it can recommend treatments that fit you like a glove. For example, in cancer treatment, AI can analyze the genetic profile of tumor and recommend a drug that's likely to work best for this specific type of cancer. It's precision at molecular level--- like getting your coffee order exactly the way you want it, down to the perfect foam-to-milk ratio.

6.4. Personalized Cancer Treatment with AI

One of the most jaw-dropping uses of AI in personal medicine is in cancer treatment. Cancer is a Notoriously tricky disease--- No cases are exactly alike, which is why traditional treatment like chemotherapy often work for some and fail for others. But with AI in the mix, we can fine-tune cancer therapies like never before.

7. AI IN DRUG DISCOVERY: SPEEDING UP THE PROCESS

Let's be real: developing new drugs like dating--- Long, expensive and filled with bad matches before you finally find 'the one'. Traditional drug discovery is no different. It's a laborious process that can take years or even decades, costs billions of dollars and lead to countless failed trials. In this section, we will dive into the slow, exhausting and time-consuming world of traditional drug discovery and show you how AI is revolutionizing the process of drug discovery. The Steps in Drug Development are represented by Figure.3.

Figure 3. Steps in Drug Development

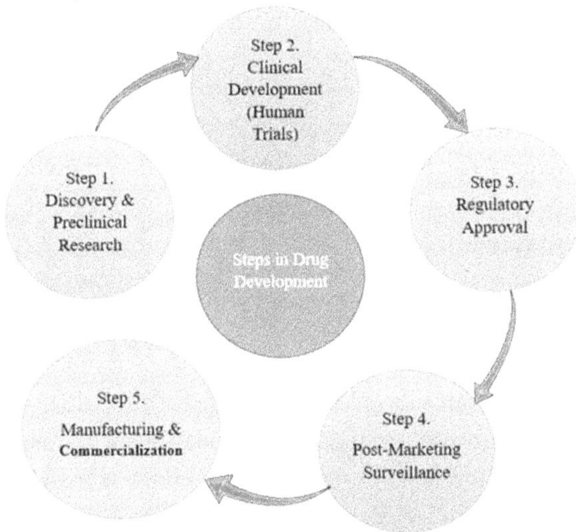

7.1. Challenges in Traditional Drug Discovery

Traditional drug discovery is notoriously slow and expensive and not exactly a walk in the park. Let's break it down:

- **Target Identification:** scientists first need to identify a biological target, like a protein or a gene, that's involved in a disease. Think of it like finding the one person in a crowd who might be a match, it can take years of trial and error.
- **Compound Screening:** once they have got the target, researchers will screen thousands sometimes even millions of compounds to see if any can interact with the target in a meaningful way. Imagine going on 100,000 first dates trying to Figureure out if anyone's a good fit. It will be exhausting right.
- **Preclinical Testing:** After a promising compound is found, it goes through rigorous testing in the lab and on animals to ensure it's safe and effective, most of them fail here and researchers end up back at square one all over again.
- **Clinical Trials:** if the compound survives the earlier stage, it moves to clinical trials with humans. These trials are divided into 3 phases, each one more expensive and time consuming than the last and guess what? The majority of drugs fail here too.

- **Approval:** finally, if a drug makes it through the gauntlet of trials, it still needs approval from regulatory agencies, a process that can take another year or more

In total, it takes 10 to 15 years and over $2.6 billion to bring a new drug to the market and more than 90% of drug candidates failed somewhere along the way. Here is where AI swoops in as a matchmaker, ready to speed things up and lower the cost of drug discovery.

7.2. AI-Driven Drug Discovery

Here enters AI the ultimate wingman. AI is transforming drug discovery by acting as a matchmaker between disease and compounds, but way faster than traditional methods. By using machine learning and massive data sets, AI can predict which compound are more likely to work, cutting down on time, cost and frustration.

7.2.1. How AI Predicts Drug Interaction and Success

AI's secret weapon is pattern recognition. While traditional drug discovery involves a lot of trial and error, ÁI can process huge amounts of data to spot patterns that would take human years to find. Here's how AI makes it all happen:

- **Virtual Screening:** instead of physically testing thousands of compounds in a lab, AI can stimulate how different compounds might interact with a disease target. This allows researchers to quickly narrow down the most promising candidates, saving a lot of time and lab coats
- **Predicting Drug- Target Interaction:** AI algorithms can predict how well a compound will bind to a biological target by analyzing the structures of Both. It's like AI is the ultimate dating app, which swipes right on, compounds that shows the most potential.
- **Optimizing Drug Candidates:** if a compound has potential but isn't quite "the one", AI can suggest tweaks to improve its efficiency and safety. It's like refining your dating profile, small changes can lead to a way better result.
- **Repurposing Existing Drugs:** sometimes, drugs developed for one disease turn out to work for another. AI is great at spotting these hidden gems, saving time and money because these drugs have already passed safety testing. It's like realizing your best friend is actually your soul mate after all those years.

7.3. AI In Action: Real World Success Stories

AI isn't just a cool futuristic concept; it's already making a difference in a real world. Here are some cases where AI played Cupid and Help discovering new drugs way faster than usual.

- **Insilico Medicine and Drug Discovery for Fibrosis:** Silico medicine, biotech company, used AI to discover a new drug for fibrosis, a condition that causes scarring in tissues. Traditional methods would have taken years, but Insilico's AI found a promising candidate in just 46 days. you read that right. Less than two months. The drug moved into Preclinical testing faster than you can say "clinical trial".
- **Benevolent AI and ALS Treatment:** Benevolent AI applied its AI-powered algorithms to amyotrophic lateral sclerosis (ALS), a devastating neurodegenerative disease. Their AI found a potential treatment by analyzing existing drugs and genetic data. This AI driven process saved years off the time and the treatment is already in clinical trials.
- **Atom Wise and Drug Repurposing for Zika Virus:** When the Zika Virus outbreak hit in 2016, atom wise jumped in with their AI system to screen existing drugs and see if any could be repurposed. In just a few weeks, their AI identified two drugs that could potentially treat the virus. AI saving the day, once again.

7.4. Humorous Comparison: Traditional Drug Discovery Vs AI Driven Drug Discovery

All right time to lighten the mood. If traditional drug discovery is like going to the party and awkwardly asking everyone, "are you the one?" then, AI is like having the ultimate dating app on your side. Here's how it goes:

- **Traditional Drug Discovery:** You're wandering aimlessly, searching for top 10 candidates who are most likely to solve your medical problem, by asking random strangers "hey, do you want to solve my medical problem?" Most of them look at you like your weird and few politely say no and after 10,000 tries you might find the top 10 candidates that you need.
- **AI Driven Drug Discovery:** Instead of wandering around aimlessly, AI hands you a list of pre-screened matches. It's like: "here are the top 10 candidates who are most likely to work. Swipe right on this one, it has great compatibility with your disease target and zero side effects". boom! You are on the fast track to success.

8. AI IN MEDICAL IMAGING: BEYOND THE OBVIOUS

Now, let's get into world of AI in medical imaging, where machines help doctors to take peek inside your body like superheroes with X-ray vision. Imagine you are getting an MRI or CT scan while a highly trained Doctor is checking out the images, their trusty sidekick, AI is there pointing out things that human eye might overlook. Together, they are like Batman and Robin, but with lab coats and way more brain power.

8.1. Overview of Medical Imaging Techniques

Before we get into how AI is making waves in medical imaging, let's do a quick run-down off a main imaging technology that doctors are using to look inside you. The Medical Imaging Room is shown in Figure .4.

- **Magnetic Resonance Imaging (MRI):** By using magnetic fields and radio waves, MRI creates super-detailed image of Your insides, especially handy for looking at soft tissues like your brain or muscles. But while it's great at finding the smallest details, analyzing those details takes a trained expert and quite a bit of time.

Figure 4. Medical Imaging Room

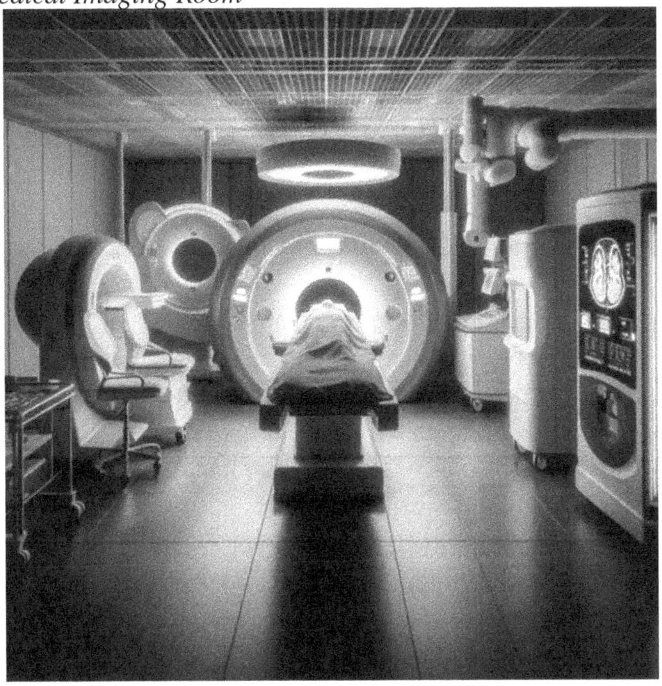

- **Computer Tomography (CT) Scans:** CT is like a Swiss army knife of imaging tools. It takes a bunch of X-rays from different angles and assembles them into a 3D picture. It's quick and great for spotting things like fractures or tumors but it produces a mountain of data that can be tricky to interpret.
- **X-ray:** X-ray is the original (OG) of medical imaging. They have been around forever and are perfect for seeing if you have broken a bone. But don't expect X-ray to give you fine details, they are more like the fast food of medical imaging: quick, simple but not gourmet.
- **Ultrasound:** you probably know this one from babies' scans, but it's also used for looking at other organs. It's like a sonar device for your body, bouncing sound waves of tissues to create Image.
- **Positron Emission Tomography (PET) Scans:** PET Scans are spies of medical imaging. They inject a tiny bit of radioactive material into your body to see how your organs are functioning, not just to find how they look. They are great for tracking disease like cancer or Alzheimer's.

Now, as awesome as these technologies are, they produce a ton of data. And even though radiologists are super smart, they are also human, which means they can make mistakes. That's where AI steps in Wareing cape, ready to save the day.

8.2. AI applications in image analysis

AI is revolutionizing medical imaging, doing everything from helping doctors find tricky abnormalities to making diagnostics quicker and more reliable. Here's how:

- **Automating Image Interpretation:** AI powered algorithms can be trained to spot patterns in medical images, making them the perfect wingman for radiologist. Need to find a tumor hiding in a sea of healthy tissues? No problem, AI have got your back. It can sift through thousands of images faster than you can say "magnetic resonance imaging" and flag potential issues.
- **Enhancing Diagnosis Accuracy:** Humans make decisions based on experience. But AI makes decision based on mountains of data. This means AI can compare a patient's scan with thousands of similar cases in a flash. In fact, AI's accuracy is pretty impressive. In one study, AI was found to diagnose breast cancer from mammograms with 94.5% accuracy, often outperforming human radiologists. It's like having a CHEAT SHEET to every diagnostic test.
- **Improving Efficiency:** Imagine being a radiologist with hundreds of images to review every day. Sounds exhausting, right? Well, AI is here to lighten the load. It can prioritize images that need immediate attention and handle the routine stuff, letting radiologists focus on the tough cases.
- **Quantifying Image Features:** AI can measure stuff like the size of a tumor or how much plaque is building up in an artery. This is super helpful for tracking the progress of disease or seeing how well treatments are working. Basically, AI is the ultimate number nerd in the room, crunching data that's critical for patients care.

8.3. AI in Radiology for Detecting Disease

Let's dive into our real-life example where AI took radiology to the next level: lung cancer detection. Lung cancer is a sneaky villain. It often doesn't show any symptoms until it's too late to treat effectively. But if you can catch it early, the odds of beating it go way up. Early detection of lung cancer is possible with the use of AI. Here enters Google health's AI system, which was trained on over 45,000 CT scans to spot early-stage lung cancer. When tested, it outperformed radiologists, finding

5% more cancers while reducing false positives by 11%. Radiology is represented by Figure.5.

Figure 5. Radiology

8.4. AI in Cardiology: Detecting Heart Disease

Stanford University researchers created an AI tool that could identify even the tiniest plaques in arteries, often before a human cardiologist would identify them. reaction like this means patients can get treatment before they ever experience symptoms summer potentially preventing heart attacks altogether. AI takes the lead with 1 point and heart disease left in the dust with 0 point. Finally, technology wins the health game.

9. AI'S LIMITATION AND FAILURES IN BIOINFORMATICS

1. Overfitting and False Positive and AI Models

- **Case Study: AI in Genome-Wide Association Studies (GWAS):** While AI has made notable strides in accelerating the process of identifying genetic variants tied to diseases, challenges remain. A 2024 study pointed out that AI models, particularly in Genome-Wide Association Studies (GWAS), can sometimes suffer from overfitting. This occurs when the models identify false associations between genetic variations and diseases, such as Type 2 diabetes, due to biases in the training datasets. These false positives emphasize the importance of complementing AI-driven approaches with traditional statistical techniques to ensure the robustness and reliability of results.

2. Data Bias and Generalization Issues

- **Case Study: Bias in AI Predictions:** AI's power to predict outcomes based on genomic data comes with its own set of limitations. A 2023 study discussed the significant issue of bias in AI systems, especially in healthcare settings. When AI models are trained on biased or incomplete datasets, they can deliver skewed predictions that may disproportionately affect certain groups, leading to potential misdiagnoses or inaccurate outcomes. This is particularly concerning in genomics, where diversity in data is critical for accurate predictions. Addressing these biases during the AI training process is essential to ensure that the resulting predictions are equitable and trustworthy across different populations.

3. Limitations in Predicting Complex Biological Interactions

- **Case Study: AI in Protein Structure Prediction:** Even though AI has made remarkable progress, particularly with tools like AlphaFold in predicting protein structures, it still faces limitations in certain areas. For example, a 2023 study showed that while AlphaFold can generate accurate static protein models, it struggles to predict the impact of specific mutations on protein function. These single-point mutations can be crucial in understanding the pathogenesis of diseases. As a result, traditional experimental methods, such as X-ray crystallography and cryo-electron microscopy, continue to be indispensable for fully understanding the effects of mutations, demonstrating that AI alone cannot solve every complex biological question.

4. Superiority of Traditional Methods in Some Analyses

- **Example: Principal Component Analysis (PCA) in Genomic Studies:** In some genomic research contexts, traditional methods still prove to be more effective than AI. A key example is Principal Component Analysis (PCA), a well-established statistical technique that is particularly useful in the analysis of large genomic datasets. A 2024 study demonstrated that PCA could account for 65% of the variance in genomic data, helping researchers understand the underlying structure and relationships within the data.

10. ETHICAL CONSIDERATIONS OF USING AI IN BIOINFORMATICS

AI has very efficiently emerged as the rockstar of bioinformatics, blowing through vast data sets at super speed to make discoveries at the blink of an eye to predict diseases— even before somebody has a chance to sneeze. But it needs a quirky sidekick to ensure it does not blow things up by accident; in this case, an "AI Ethics Committee." Here is a group of colorful enforcers who keep our data safe, check our biases, and make sure our decisions are accountable. Fasten your seat belts for the wild ride that is ethics in AI in bioinformatics.

Privacy and data security:

- **Case study: re-identification risk in genomic databases**

A 2020 study highlighted the vulnerability of deidentification genomic data to re identifications stop researchers demonstrated that individuals could be reidentified by cross referencing genomic data with publicly available information, underscoring the need for robust data production measures.

Informed consent:

- **Case study: challenges in next generation sequencing (NGS)**

The expensive data generated by NGS presents challenges in obtaining informed consent. Participants may be unaware of potential use of their genomic data, leading to ethical dilemmas regarding data usage and sharing.

Ethical use of AI in genomic medicine:

- **Case study: AI in prenatal and pediatric genomic medicine**

A 2023 study examined the ethical implication of deploying AI in prenatal and pediatric settings. Concerns were raised about the accuracy of AI prediction and the potential for bias, emphasizing the need for transparency and accountability in AI applications

Regulatory frameworks:

- **General data protection regulation (GDPR):** The GDPR are enforces strict guidelines on the processing of personal data including genomic information. It mandates explicit concern for data collection and grants individual the right to access and delete their data. Compliance with GDPR is crucial for researchers handling genomic data within the European union.
- **Genetic information nondiscrimination act (GINA):** In the United States, GINA protects individuals from discrimination based on their genetic information in health insurance and employment contacts. However, it does not extend to life or disability insurance, indicating areas where ethical consideration remained pertinent.

11. CONCLUSION: THE POWER OF AI IN BIOINFORMATICS

And that's a wrap, folks! We've journeyed through the wild and wonderful world of AI and bioinformatics, and it's time to reflect. Imagine this moment as the end of a movie montage, where the hero looks back on the adventure, maybe with a catchy tune in the background. If this were a rom com, we'd be chuckling at all the plot twists and highlights! The journey of AI in bioinformatics has just begun. With curiosity as your compass and knowledge as your map, you're in for a thrilling adventure that could reshape our understanding of life itself. So, go forth, stay curious, and who knows? You might just make the next big breakthrough in bioinformatics!

REFERENCES

Baxevanis, A. D., & Ouellette, B. F. F. (Eds.). (2004). *Bioinformatics: A Practical Guide to the Analysis of Genes and Proteins* (3rd ed.). Wiley-Interscience.

Camacho, D. M. (2018). Machine learning in biological networks: Predicting interactions and responses. *Nature Reviews. Genetics*, *19*(7), 321–332. DOI: 10.1038/s41576-018-0002-3

Ching, T., Himmelstein, D. S., Beaulieu-Jones, B. K., Kalinin, A. A., Do, B. T., Way, G. P., Ferrero, E., Agapow, P. M., Zietz, M., Hoffman, M. M., Xie, W., Rosen, G. L., Lengerich, B. J., Israeli, J., Lanchantin, J., Woloszynek, S., Carpenter, A. E., Shrikumar, A., Xu, J., & Greene, C. S. (2018). Opportunities and obstacles for deep learning in biology and medicine. *Journal of the Royal Society, Interface*, *15*(141), 20170387. DOI: 10.1098/rsif.2017.0387 PMID: 29618526

Esteva, A., Robicquet, A., Ramsundar, B., Kuleshov, V., DePristo, M., Chou, K., Cui, C., Corrado, G. S., Thrun, S., & Dean, J. (2019). A guide to deep learning in healthcare. *Nature Medicine*, *25*(1), 24–29. DOI: 10.1038/s41591-018-0316-z PMID: 30617335

Floridi, L., Cowls, J., Beltrametti, M., Chatila, R., Chazerand, P., Dignum, V., Luetge, C., Madelin, R., Pagallo, U., Rossi, F., Schafer, B., Valcke, P., & Vayena, E. (2019). AI4People—An ethical framework for a good AI society: Opportunities, risks, principles, and recommendations. *Minds and Machines*, *29*(4), 611–637. DOI: 10.1007/s11023-019-09508-8 PMID: 30930541

Jiang, X., Wang, X., Chen, Y., Liu, C., & Wang, Y. (2022). A differential privacy-based scheme for sharing genomic datasets in genome-wide association studies (GWAS). *Journal of Biomedical Informatics*, *128*, 104032. DOI: 10.1016/j.jbi.2022.104032

Jumper, J., Evans, R., Pritzel, A., Green, T., Figureurnov, M., Ronneberger, O., Tunyasuvunakool, K., Bates, R., Žídek, A., & Potapenko, A. (2021). Highly accurate protein structure prediction with AlphaFold. *Nature*, *596*(7873), 583–589. DOI: 10.1038/s41586-021-03819-2 PMID: 34265844

Li, H., Zhang, Q., Liu, X., Zhou, J., & Wang, Y. (2024). Evaluation of federated learning models for Parkinson's disease prediction using genomic and transcriptomic data. *Bioinformatics (Oxford, England)*, *40*(1), bta009. Advance online publication. DOI: 10.1093/bioinformatics/bta009 PMID: 38141210

Litjens, G., Kooi, T., Bejnordi, B. E., Setio, A. A., Ciompi, F., Ghafoorian, M., Van Der Laak, J. A., Van Ginneken, B., & Sánchez, C. I. (2017). A survey on deep learning in medical image analysis. *Medical Image Analysis*, *42*, 60–88. DOI: 10.1016/j.media.2017.07.005 PMID: 28778026

Mardis, E. R. (2017). DNA sequencing technologies: 2006–2016. *Nature Protocols*, *12*(2), 365–368. DOI: 10.1038/nprot.2016.182 PMID: 28055035

Marx, V. (2013). The big challenges of big data. *Nature*, *498*(7453), 255–260. DOI: 10.1038/498255a PMID: 23765498

Mittelstadt, B. D., Allo, P., Taddeo, M., Wachter, S., & Floridi, L. (2016). The ethics of algorithms: Mapping the debate. *Big Data & Society*, *3*(2), 2053951716679679. DOI: 10.1177/2053951716679679

Mount, D. W. (2004). *Bioinformatics: Sequence and Genome Analysis* (2nd ed.). Cold Spring Harbor Laboratory Press.

Stephens, Z. D., Lee, S. Y., Faghri, F., Campbell, R. H., Zhai, C., Efron, M. J., Iyer, R., Schatz, M. C., Sinha, S., & Robinson, G. E. (2015). Big data: Astronomical or genomical? *PLoS Biology*, *13*(7), e1002195. DOI: 10.1371/journal.pbio.1002195 PMID: 26151137

Torkzadehmahani, R., Schöttle, P., & Wachter, S. (2020). "Privacy-preserving AI techniques in biomedicine: An overview of federated learning and differential privacy." *npj. Digital Medicine*, *3*(1), 135. DOI: 10.1038/s41746-020-00333-z

Vamathevan, J., Clark, D., Czodrowski, P., Dunham, I., Ferran, E., Lee, G., Li, B., Madabhushi, A., Shah, P., & Sumbayev, V. (2019). Applications of machine learning in drug discovery and development. *Nature Reviews. Drug Discovery*, *18*(6), 463–477. DOI: 10.1038/s41573-019-0024-5 PMID: 30976107

Zhavoronkov, A., Ivanenkov, Y. A., Aliper, A., Veselov, M. S., Aladinskiy, V., Aladinskaya, A. V., Terentiev, V., Polykovskiy, D., Kuznetsov, M. D., Asadulaev, A., Volkov, Y., Zholus, A., Shayakhmetov, R. R., Zhebrak, A., Minaeva, L. I., Zagribelnyy, B. A., Lee, L. H., Soll, R., Madge, D., & Aspuru-Guzik, A. (2019). Deep learning enables rapid identification of potent DDR1 kinase inhibitors. *Nature Biotechnology*, *37*(9), 1038–1040. DOI: 10.1038/s41587-019-0224-x PMID: 31477924

Chapter 5
AI-Driven Approaches in Aquaculture Disease Management:
Innovations, Challenges, and Future Directions

Naglaa F. Elbaz
https://orcid.org/0000-0001-8142-5892
Veterinary Medicine Directorate, Kafr Elsheikh, Egypt

Rana Ashraf
Damietta University, Egypt

Omnia Raied
Damietta University, Egypt

Sohaila Elshamey
https://orcid.org/0009-0003-7658-3695
Damietta University, Egypt

ABSTRACT

Aquaculture plays a vital role in global food security by providing a significant source of protein. However, disease outbreaks remain a persistent challenge, causing substantial economic losses. This chapter explores the transformative role of Artificial Intelligence (AI) in fish disease management, emphasizing machine learning and deep learning techniques such as Convolutional Neural Networks (CNNs) and Support Vector Machines (SVMs). By analyzing diverse data sources, including images, water quality parameters, and fish behavior, AI models offer early and accurate disease

DOI: 10.4018/979-8-3693-9208-9.ch005

detection, enabling timely interventions. We highlight research findings from Kafr El-Sheikh, Egypt, that underscore the prevalence of bacterial diseases in freshwater aquaculture and the importance of AI-assisted diagnostic tools. Additionally, we examine case studies demonstrating the practical applications of AI in disease management, discuss current challenges, and identify future research directions to further enhance aquaculture sustainability and productivity.

1. INTRODUCTION

Aquaculture serves as a critical component of global food security, supplying a substantial portion of the world's protein intake. However, the industry faces persistent challenges due to fish diseases, which cause significant economic losses and threaten aquatic ecosystems. Early and accurate disease detection is essential for mitigating these impacts, improving fish health, and ensuring the sustainability of aquaculture operations. The integration of Artificial Intelligence (AI), particularly machine learning and deep learning techniques, has introduced new possibilities for enhancing disease diagnosis and management.

AI applications in fish disease detection leverage advanced models such as Convolutional Neural Networks (CNNs) and Support Vector Machines (SVMs), as demonstrated in studies like Li et al. (2022), which have reported high accuracy in detecting fish diseases through image-based analysis. These models analyze diverse data sources, including images, water quality parameters, and fish behavior patterns. They can detect abnormalities at early stages, enabling prompt and targeted interventions. Furthermore, AI-driven predictive models assist aquaculture professionals in monitoring environmental conditions and forecasting potential disease outbreaks, ultimately reducing reliance on antibiotics and promoting eco-friendly practices.

Research conducted in Kafr El-Sheikh, Egypt, by Elbaz and Abd Al Fatah (2024) has highlighted the prevalence of bacterial diseases in freshwater aquaculture, emphasizing the need for effective diagnostic tools to manage outbreaks. Furthermore, investigations by Elbaz et al. (2023) on virulence genes of bacterial infections in Nile tilapia have provided insights into disease mechanisms, facilitating the development of AI-assisted disease detection models. Mahfouz et al. (2016) also documented bacterial infections in the region, underscoring the significance of prompt detection and intervention to protect fish health and maintain aquaculture productivity.

For instance, a study by Hu et al. (2012) implemented a deep learning model for shrimp aquaculture, achieving high accuracy in detecting early-stage infections based on underwater image analysis. This successful application underscores the practical benefits of AI in improving disease management and minimizing losses.

This chapter delves into the various types of fish diseases, including a comprehensive review as discussed by Mandal and Ghosh (2023) on AI applications in monitoring fish health. We examine bacterial, fungal, parasitic, and non-infectious disorders and analyze how AI technologies facilitate their identification and treatment. Additionally, we highlight innovative approaches, practical applications, and future directions for AI in aquaculture disease management, emphasizing the potential for these technologies to drive more resilient and productive aquaculture practices.

2. FISH DISEASES

Fish diseases can be classified into several categories, each with distinct causes, symptoms, and treatments. The primary categories include bacterial infections, fungal infections, parasitic or protozoan infections, and non-infectious disorders. Understanding these diseases is essential for maintaining healthy fish populations and preventing widespread outbreaks in aquaculture and natural aquatic environments.

2.1. Bacterial Diseases

Bacterial infections are among the most common diseases affecting fish and often present with symptoms such as red streaks, abdominal swelling, bulging eyes, and ulcerations. One of the most prevalent bacterial infections in freshwater fish is caused by Aeromonas bacteria, which manifests as hemorrhagic septicemia, leading to bloody spots, fluid accumulation in the abdomen (commonly referred to as "dropsy"), fin rot, and popeye (exophthalmos). Other bacterial pathogens, such as Pseudomonas and Vibrio species, can cause similar symptoms and lead to severe health complications.

Diagnosis of bacterial infections requires laboratory testing, including bacterial cultures and antibiotic sensitivity tests, to determine the most effective treatment. Commonly used antibiotics include penicillin, amoxicillin, erythromycin, and oxytetracycline, administered through medicated feed or water treatments (Li et al., 2022; Ballocchi, 2024). Preventive measures, such as maintaining optimal water quality, minimizing stress, and ensuring proper quarantine protocols, are essential to reducing bacterial infections in fish populations.

2.2. Fungal Diseases

Fungal infections in fish are typically secondary infections that arise due to underlying health issues, such as injuries, stress, or previous parasitic infections. These infections often appear as gray or white cotton-like patches on the fish's skin, fins, or gills. One of the most common fungal pathogens is Saprolegnia, which can affect not only fish but also eggs, leading to mass mortalities in aquaculture systems. If left untreated, fungal infections can spread rapidly, compromising fish health and increasing mortality rates.

Treatment for fungal infections involves antifungal medications such as malachite green, methylene blue, or potassium permanganate. Additionally, improving water quality, reducing organic waste, and maintaining proper stocking densities help prevent fungal outbreaks (Li et al., 2022).

2.3. Parasitic Diseases

Parasitic infections are a major concern in aquaculture and can cause significant health problems in fish. The most common parasitic disease is Ichthyophthiriasis, also known as Ich or White Spot Disease, caused by the protozoan Ichthyophthirius multifiliis. Symptoms include erratic swimming behavior, excessive mucus production, and the presence of small white cysts on the skin and gills. Severe infestations can lead to respiratory distress and death if not treated promptly (Ballocchi, 2024).

Other notable parasites include Oodinium pilularis and Oodinium limneticum, responsible for Velvet Disease, which causes a yellowish-golden dust-like coating on the fish. This condition leads to lethargy, respiratory distress, and emaciation if untreated. External parasites such as fish lice (Argulus) and anchor worms (Lernaea) attach to the fish's skin, causing irritation, inflammation, and secondary bacterial infections (Li et al., 2022; Ahmad, 2024).

Treatment for parasitic infections varies depending on the specific pathogen but commonly includes salt baths, formalin treatments, copper-based medications, and improved biosecurity measures to prevent reinfection.

2.4. Non-Infectious Disorders

In addition to infectious diseases, fish can suffer from a variety of non-infectious disorders caused by environmental stressors, genetic abnormalities, or nutritional deficiencies. One such condition is swim bladder disorder, which affects buoyancy and is often linked to overfeeding, constipation, or bacterial infections. Poor water quality, including high ammonia and nitrate levels, can also cause stress-related diseases, leading to weakened immune responses and increased susceptibility to infections.

Heart and blood vessel disorders may arise due to parasitic infestations, such as anchor worms, which embed themselves into the fish's tissue, causing significant irritation and gill damage. Additionally, metabolic disorders can result from imbalanced diets, leading to vitamin deficiencies and reduced overall health (Ahmad, 2024).

Maintaining optimal water parameters, providing a nutritionally balanced diet, and reducing stressors in the environment are essential for preventing non-infectious disorders in fish.

2.5. Prevention and Management

Effective disease prevention and management in fish populations rely on a combination of good husbandry practices, early disease detection, and appropriate treatment strategies. Key measures include:

- Maintaining high water quality through regular monitoring of temperature, pH, ammonia, nitrite, and nitrate levels.
- Implementing proper quarantine procedures for new or sick fish to prevent the spread of infectious diseases.
- Providing a well-balanced diet rich in essential vitamins and minerals to enhance fish immunity.
- Reducing stressors such as overcrowding, sudden environmental changes, and poor handling practices.
- Using AI-driven monitoring systems to detect early signs of disease, allowing for timely intervention and treatment ("Artificial Intelligence and its Aquaculture Applications," 2025).

Regular health assessments and proactive management strategies are crucial for ensuring the long-term health and sustainability of fish populations in both aquaculture and natural habitats.

3. AI TECHNOLOGIES IN FISH DISEASE DETECTION

3.1. Overview of AI in Aquaculture

Artificial intelligence (AI) has transformed the aquaculture industry by enabling advanced disease detection, monitoring, and management systems. Through the integration of **machine learning (ML), computer vision, and the Internet of Things (IoT)**, AI-driven solutions allow fish farmers to track aquatic populations in **real-time**, improving farm efficiency and overall fish health (Treulle, 2025). AI's ability to process vast datasets collected from **underwater cameras, environmental sensors, and automated feeding systems** facilitates **early disease detection**, reducing the economic and ecological impact of outbreaks (Francis-Floyd & Petty, 2020).

3.2. Smart Monitoring Systems

AI-powered **smart monitoring systems** utilize sophisticated algorithms to continuously track **water quality parameters, fish behavior, and environmental conditions**. By analyzing real-time data on **temperature, pH levels, dissolved oxygen, and ammonia concentrations**, these systems detect early signs of disease or stress that may otherwise go unnoticed (Treulle, 2025; Francis-Floyd & Petty, 2020).

For instance, **machine learning models applied to underwater camera feeds** can identify **subtle changes in fish movement, color, or feeding patterns**, signaling potential health issues (Mecha & Gonzales-Plasus, 2023). The integration of AI with **IoT and Big Data analytics** further strengthens disease surveillance, providing fish farmers with continuous, data-driven insights that enhance productivity and sustainability (Ragab et al., 2024; Wu et al., 2023; Chang et al., 2021; Kim et al., 2024).

3.3. Image-Based Disease Detection

Computer vision and deep learning techniques play a crucial role in diagnosing fish diseases. Advanced **image processing systems** utilize **convolutional neural networks (CNNs) and Support Vector Machines (SVMs)** to analyze images of fish, detecting **physical abnormalities such as lesions, fin rot, or white spots with over 90% accuracy** (Li et al., 2022; Zouin et al., 2024).

These AI models continuously improve as they are trained on larger datasets collected from **both laboratory environments and real-world aquaculture settings**. Enhanced **image resolution, faster processing speeds, and automated feature extraction techniques** are expanding the potential of AI-driven disease detection in fish farming (Zouin et al., 2024).

3.4. Predictive Analytics for Disease Prevention

The integration of **AI and Big Data analytics** has significantly improved the **predictive capabilities** of modern aquaculture systems. AI algorithms process large volumes of data to detect patterns in **environmental conditions, fish behavior, and historical disease outbreaks**, allowing for early intervention strategies (Ragab et al., 2024; Aung et al., 2024; Er-Rousse & Qafas, 2024).

AI-driven predictive models also play a key role in:

- **Optimizing feeding schedules** to enhance fish growth and reduce waste.
- **Identifying seasonal trends in disease outbreaks**, enabling proactive management.

- **Improving biosecurity measures** by detecting environmental stressors before they trigger disease.

By leveraging **machine learning, sensor data fusion, and cloud computing**, AI-driven aquaculture systems can **increase productivity, reduce disease-related losses, and promote eco-friendly farming practices** (Er-Rousse & Qafas, 2024; Mohale et al., 2024).

4. APPLICATIONS OF AI IN FISH DISEASE DETECTION

4.1. Early Disease Detection

AI enables **real-time health monitoring** by analyzing patterns in **fish movement, feeding behavior, and environmental fluctuations**. Traditional disease detection methods, which rely on **manual inspections and periodic sampling**, are often slow and may miss subtle warning signs. AI-driven models, however, can process data from **underwater cameras, motion sensors, and water quality monitors** to detect anomalies in **swimming patterns, gill movement, or appetite**, allowing for **early intervention and reduced reliance on antibiotics** (Treulle, 2025; Francis-Floyd & Petty, 2020).

4.2. Image Analysis Techniques

AI-based **image recognition and deep learning** models are increasingly utilized to detect and classify fish diseases. A **two-phase image analysis system** using CNNs has demonstrated **up to 99% accuracy** in diagnosing fish diseases based on visual symptoms (Fabian, 2022; "Aquaculture Artificial Intelligence: A History for the Curious," 2023). These systems:

- Identify **lesions, ulcers, and fungal infections** through high-resolution imaging.
- Use **multi-spectral imaging** to detect diseases invisible to the human eye.
- Automate **large-scale health assessments** in commercial fish farms.

By combining **AI-driven object detection with real-time monitoring**, aquaculture managers can **rapidly diagnose diseases and implement targeted treatments** (Francis-Floyd & Petty, 2020; Li et al., 2022).

4.3. Integration of IoT and Big Data

The **fusion of AI, IoT, and Big Data technologies** is transforming fish disease management by providing **continuous, automated surveillance** of fish health and environmental conditions. IoT devices, such as **underwater drones and smart buoys**, collect real-time data on **temperature, dissolved oxygen, pH, and ammonia levels**, while AI algorithms analyze this information to detect **early signs of disease outbreaks** (Treulle, 2025; Jagdish, 2024).

Several AI-powered systems are leading innovation in this space:

- **Location-Based Topic Models (Trinh et al., 2019)**: Adapted for disease tracking, identifying **spatial and temporal patterns** in outbreaks.
- **Spark-Based Data Processing (Emara & Huang, 2019a)**: Enables **efficient analysis of large-scale aquaculture data**, supporting AI model training.
- **Stratified Block Models (Trinh et al., 2021)**: Improve **data distribution and processing efficiency** for disease prediction.
- **Geographically Distributed Data Management (Emara et al., 2023)**: Enhances **real-time monitoring across multiple fish farms**.
- **Feature Selection via Top-Frequency Analysis (VuThi et al., 2024)**: Identifies **critical health indicators** such as **temperature, pH, and oxygen levels** for disease prediction.

Predicting and detecting fish diseases can benefit significantly from the integration of advanced techniques like those presented in the referenced studies. Trinh et al. (2019) introduced a location-based topic model that can be adapted to monitor disease outbreaks across aquaculture sites by identifying spatial and temporal patterns in environmental conditions and fish health records. Emara and Huang (2019b) developed RRPlib, a Spark-based library for efficient sampling of large datasets, which can be utilized to process extensive aquaculture data and generate balanced training datasets for predictive models. To handle the growing volume of aquaculture data, Trinh et al. (2021) proposed a stratified block model that ensures efficient data processing and balanced distribution, enhancing the accuracy of disease prediction models. Furthermore, Emara et al. (2023) presented geographically distributed data management techniques that facilitate real-time monitoring and analysis of environmental parameters across multiple aquaculture farms, improving early detection of potential disease outbreaks. Lastly, VuThi et al. (2024) introduced a feature selection method based on top frequency, which can be applied to identify the most critical factors influencing fish health, such as water temperature, pH, and dissolved oxygen levels.

By integrating **AI-driven pattern recognition and environmental modeling**, these systems enable **more precise, proactive disease management** in aquaculture.

4.4. Impact on Fish Health and Farm Productivity

The use of AI in **fish disease detection and prevention** has a profound impact on:

- **Reducing fish mortality rates** by enabling early detection of diseases.
- **Lowering operational costs** by minimizing unnecessary antibiotic use and improving feeding efficiency.
- **Enhancing sustainability** through optimized resource management and reduced environmental footprint.
- **Ensuring food safety** by maintaining healthy, disease-free fish stocks.

AI-driven aquaculture not only improves **disease detection and treatment** but also contributes to **higher productivity, improved fish welfare, and more sustainable seafood production** (Francis-Floyd & Petty, 2020; Jagdish, 2024).

5. CASE STUDIES

Recent case studies have demonstrated the successful implementation of AI in detecting and classifying fish diseases, significantly improving aquaculture productivity and sustainability. These studies highlight innovative approaches that have enhanced disease management, environmental monitoring, and overall fish farm efficiency.

5.1. AI-Based Disease Diagnosis Systems

Several AI-driven expert systems have been developed to assist in diagnosing fish diseases. For example, Xiaoshuan et al. (2004) introduced ES-FDD, an expert system designed for fish disease diagnosis using advanced data analysis techniques. Similarly, Zhang and Li (2008) developed a consultant system for call centers in China that leveraged AI to improve disease diagnosis accuracy and response times.

Recent research has also explored various AI-driven methodologies to enhance disease prediction. Emara (2016) and Emara, Saleh, and Arafat (2014) introduced power-saving mechanisms for VoIP over WiMAX, which can be adapted to optimize sensor networks in aquaculture, ensuring continuous monitoring of critical water parameters. Additionally, Emara (2018) proposed an adaptive mechanism based on artificial neural networks (ANNs) to train predictive models capable of recognizing fish health patterns. Wei et al. (2018) developed a two-stage data processing algorithm for big data analysis, which can process large datasets from aquaculture sites to provide reliable disease trend insights.

Moreover, Emara and Huang (2019a) introduced a distributed data management system that enables real-time analysis of geographically dispersed aquaculture farms, facilitating early disease detection. Recent advancements by Trinh et al. (2021) and VuThi et al. (2024) have further improved predictive modeling by refining data processing techniques and selecting critical environmental factors that influence fish health.

5.2. AI-Powered Image Processing for Disease Detection

AI-driven image processing has proven highly effective in diagnosing fish diseases. Park et al. (2007) developed a diagnostic system that utilizes microscopic image analysis of pathogens to improve disease identification accuracy. Additionally, Lou et al. (2010) introduced a case-based reasoning system that employs image retrieval techniques for disease classification, demonstrating the versatility of AI in diagnostic applications.

5.3. Environmental Monitoring and Feeding Optimization

AI has significantly enhanced environmental monitoring in aquaculture by analyzing real-time sensor data to maintain optimal conditions for fish health. AI-powered systems continuously monitor key parameters such as water temperature, dissolved oxygen levels, and pH, allowing for automated adjustments to sustain ideal living conditions (Treulle, 2025). Furthermore, AI-based feeding optimization algorithms analyze growth patterns and feeding behaviors, ensuring efficient feed usage, minimizing waste, and promoting sustainable fish farming practices.

5.4. AI-Driven Fish Species Classification

The classification of fish species has also benefited from AI advancements. Hu et al. (2012) employed computer vision and machine learning techniques to classify fish species based on color and texture, demonstrating the potential of AI in biodiversity monitoring and conservation within aquaculture. This capability is particularly valuable for managing diverse fish populations and preventing disease transmission between species.

5.5. Real-Time Tracking and Behavioral Monitoring

AI-based real-time tracking systems have significantly improved the monitoring of fish behavior and health. Peddina and Mandava (2024) introduced an optimized deep learning-based approach for fish tracking and classification using a chimp-based optimization algorithm combined with Google's Deep Network (CbGDNet), achieving a remarkable accuracy of 99.16%. Similarly, Bhatt et al. (2024) developed an automated fish classification system integrating electro-mechanical components with a MobileNet V2 CNN model, attaining a classification accuracy of 97%.

These studies demonstrate how convolutional neural networks (CNNs) and AI-driven tracking algorithms enhance ecological research by providing high-precision fish behavior analysis and early disease detection.

6. ADVANTAGES OF USING AI IN FISH DISEASE MANAGEMENT

The integration of AI in fish disease management offers numerous benefits, improving disease detection, prevention, and overall aquaculture efficiency.

6.1. Enhanced Disease Detection

AI-powered systems employ machine learning algorithms to analyze vast datasets collected from underwater cameras and environmental sensors. These technologies enable early disease detection by identifying subtle behavioral changes and external symptoms that might go unnoticed in traditional visual inspections (Treulle, 2025; Francis-Floyd & Petty, 2020). AI-based detection significantly reduces diagnosis time, allowing for rapid interventions that prevent disease spread and minimize fish mortality.

6.2. Data-Driven Insights for Predictive Disease Management

By processing large volumes of historical and real-time data, AI enables fish farmers to recognize patterns and predict disease outbreaks based on environmental conditions (Lutz, 2023). This predictive capability enhances decision-making in disease prevention, treatment planning, and farm management, ensuring healthier fish populations and improved farm productivity.

6.3. Reduction in Antibiotic Usage

The ability of AI to detect diseases at an early stage allows for targeted and precise interventions, reducing the need for excessive antibiotic use and chemical treatments (Treulle, 2025; Francis-Floyd & Petty, 2020). This proactive approach aligns with sustainable aquaculture practices by minimizing antibiotic resistance risks and reducing environmental pollution caused by chemical runoff ("AI in Fish Growth and Health," 2024).

6.4. Optimized Feeding and Nutrient Management

Beyond disease detection, AI-driven systems optimize feeding schedules and nutrient distribution by analyzing fish growth rates, behavioral patterns, and feeding history. These AI algorithms recommend efficient feeding strategies that promote fish health while minimizing feed waste (Treulle, 2025; Lutz, 2023). This optimization contributes to improved growth rates, cost savings, and a more sustainable aquaculture industry.

6.5. Real-Time Monitoring and Automated Interventions

AI-enabled monitoring systems continuously track crucial parameters such as water quality, temperature, and oxygen levels. When deviations from optimal conditions are detected, AI can trigger automated interventions, such as adjusting feeding schedules or modifying water parameters, ensuring a stable and healthy environment for fish (Treulle, 2025; Mecha & Gonzales-Plasus, 2023). This dynamic approach reduces the risks associated with disease outbreaks and enhances overall farm management efficiency.

7. CHALLENGES AND LIMITATIONS

The integration of AI in fish disease detection and classification within aquaculture presents several challenges and limitations that must be addressed to ensure effective implementation.

7.1. Data Quality and Availability

One of the key challenges is the quality and availability of data. Many aquaculture operations lack the necessary infrastructure to collect, store, and manage large datasets effectively. Inconsistent data collection methods, environmental variability, and sensor limitations can lead to unreliable AI model performance.

Environmental factors such as **water turbidity** and **biofouling** further complicate data collection. Water turbidity—affected by factors like feed inputs and environmental conditions—can fluctuate significantly, reducing light penetration and inhibiting primary productivity, which in turn impacts fish health and growth (Boquin Instrument, 2024). Similarly, biofouling, the accumulation of organisms on submerged surfaces, can deform net pens, increase mooring loads due to higher current resistance, and accelerate structural fatigue, affecting both infrastructure integrity and sensor accuracy (Floer et al., 2016).

To overcome these challenges, standardized data collection protocols and resilient monitoring systems must be developed to ensure consistency and reliability in AI-driven aquaculture management.

7.2. Integration and Scalability

Many existing aquaculture systems were not designed with AI integration in mind, leading to compatibility and operational challenges. Integrating AI with existing monitoring systems, sensor networks, and decision-making workflows requires flexible architectures and robust computational resources (Bohara et al., 2024; Mandal & Ghosh, 2023).

Scalability remains another major concern. As operations expand, the volume and complexity of data increase, necessitating advanced data storage, high-performance computing, and cloud-based analytics. Standardized data formats and communication protocols must be adopted to facilitate seamless interoperability among different aquaculture technologies (Bohara et al., 2024; Mandal & Ghosh, 2023).

7.3. Cost and Investment

Implementing AI technologies requires significant financial investment in hardware (e.g., sensors, imaging systems, and computational infrastructure), software, and skilled personnel. For smaller aquaculture enterprises, the high initial costs can be a major barrier (AI in Fish Growth and Health, 2024; Bohara et al., 2024; Mandal & Ghosh, 2023).

Justifying these expenses necessitates a thorough cost-benefit analysis, demonstrating the long-term advantages of AI adoption—such as improved disease detection, reduced losses, and optimized resource use—to encourage wider industry adoption (Bohara et al., 2024; Mandal & Ghosh, 2023).

7.4. Skills Gap and Resistance to Change

The aquaculture industry faces a skills gap in AI adoption, as many professionals lack the necessary expertise in machine learning, data analytics, and AI system maintenance. Addressing this requires specialized training programs and workforce development initiatives (AI in Fish Growth and Health, 2024; Bohara et al., 2024; Mandal & Ghosh, 2023).

Additionally, resistance to change remains a challenge. Many aquaculture stakeholders are unfamiliar with AI capabilities and may be skeptical about its benefits or concerned about job displacement. Bridging this gap requires awareness campaigns, stakeholder engagement, and demonstrating AI's role as an assistive tool rather than a replacement for human expertise (Bohara et al., 2024; Mandal & Ghosh, 2023).

7.5. Regulatory and Ethical Concerns

The widespread adoption of AI in aquaculture necessitates appropriate regulatory frameworks to govern its use. Privacy concerns regarding farm data, ethical considerations in automation, and potential environmental risks must be addressed (AI in Fish Growth and Health, 2024; Bohara et al., 2024; Mandal & Ghosh, 2023).

Collaboration among industry stakeholders, policymakers, and regulatory bodies is crucial to developing guidelines that ensure responsible AI use while maintaining transparency, data security, and environmental sustainability (Bohara et al., 2024; Mandal & Ghosh, 2023).

8. FUTURE DIRECTIONS

As AI technology continues to advance, its role in aquaculture is poised to expand, offering new opportunities for enhanced fish disease detection, farm management, and sustainability.

8.1. Enhanced Monitoring and Predictive Analytics

Future AI systems will further improve real-time monitoring of water quality, fish health, and environmental conditions. By leveraging machine learning models trained on historical data, aquaculture operators will be able to predict disease outbreaks and take proactive measures, minimizing losses and optimizing farm productivity (AI in Fish Growth and Health, 2024).

8.2. Automation and Robotics

Integrating AI with robotics will revolutionize fish farming operations. Autonomous drones and underwater robots equipped with AI-driven vision systems can monitor fish behavior, detect early disease symptoms, optimize feeding strategies, and even assist in harvesting.

The emergence of **swarm robotics**, where multiple autonomous units collaborate to perform monitoring and maintenance tasks, can significantly reduce labor costs while improving efficiency in disease detection and infrastructure management (Fabian, 2022; Treulle, 2025).

8.3. Collaborative Platforms and Data Sharing

Developing centralized data-sharing platforms will enhance AI-driven decision-making in aquaculture. By allowing fish farmers to share data on disease trends, environmental parameters, and management strategies, these platforms can improve collective knowledge and facilitate the development of best practices across the industry (Bohara et al., 2024; Mandal & Ghosh, 2023; AI in Fish Growth and Health, 2024).

8.4. Accessibility and Education

As AI adoption in aquaculture grows, ensuring accessibility for small-scale farmers—particularly in developing regions—will be crucial. Increasing affordability of AI tools, coupled with targeted training programs, will enable wider adoption of AI-driven solutions (Ashraf Rather et al., 2024; AquaAdapt, 2024).

Despite these advancements, challenges such as limited digital infrastructure and a shortage of skilled personnel remain. Addressing these through government-led initiatives, industry partnerships, and online education programs will be key to fostering AI-driven aquaculture at a global scale.

8.5. Regulatory Frameworks and Standards

To support AI adoption in aquaculture, comprehensive regulatory and legal frameworks must be established. These should address **data privacy, AI ethics, environmental impact assessment, and standardization of AI-driven tools** to ensure responsible deployment of AI technologies in fish farming (Bohara et al., 2024; Mandal & Ghosh, 2023; Treulle, 2025).

Collaboration between policymakers, industry experts, and researchers will be essential to balance innovation with sustainability, ensuring AI contributes positively to the future of aquaculture.

9. CONCLUSION

Artificial intelligence has emerged as a transformative tool in aquaculture disease management, offering early and accurate detection of fish diseases while enhancing real-time health monitoring. By leveraging machine learning and deep learning techniques, aquaculture professionals can proactively identify disease outbreaks, mitigate economic losses, and adopt more sustainable farming practices.

The integration of AI-driven models, such as convolutional neural networks (CNNs) and support vector machines (SVMs), with environmental and behavioral data enhances decision-making and operational efficiency. However, challenges such as data quality, scalability, cost, and regulatory concerns must be addressed to fully realize AI's potential in the industry.

As research advances, future developments in AI models, data integration strategies, and real-time monitoring technologies are expected to further optimize disease detection and farm management. Collaborative efforts among researchers, industry stakeholders, and policymakers will be essential in fostering innovation, ensuring ethical AI deployment, and strengthening the resilience of aquaculture against emerging challenges.

REFERENCES

Ahmad, M. (2024, June 11). Top 6 applications of AI in aquaculture: 2023. PanaceaLogics -. https://panacealogics.com/blogs/top-6-applications-of-ai-in-aquaculture-2023/

AI in fish growth and health. (2024, October 26). PanaceaLogics -. https://panacealogics.com/blogs/role-of-artificial-intelligence-ai-in-fish-growth-and-health/

AquaAdapt. (2024, November 4). Aquaculture and artificial intelligence: Challenges and opportunities. Retrieved from https://aquadapt.org/2024/11/04/aquaculture-artificial-intelligence

Aquaculture artificial intelligence: A history for the curious. (2023, September 12). Fish n' bits. https://blog.manolinaqua.com/en/aquaculture-artificial-intelligence-history

Artificial intelligence and its aquaculture applications. (2025, February 4). Ocean News & Technology. https://ecomagazine.com/news/fisheries-aquaculture/artificial-intelligence-and-its-aquaculture-applications/

Ashraf Rather, M., Ahmad, I., Shah, A., Ahmad Hajam, Y., Amin, A., Khursheed, S., Ahmad, I., & Rasool, S. (2024). Exploring opportunities of Artificial Intelligence in aquaculture to meet increasing food demand. *Food Chemistry: X*, *22*(101309), 101309. DOI: 10.1016/j.fochx.2024.101309 PMID: 38550881

Aung, T., Razak, R., & Nor, A. (2024). Artificial intelligence methods used in various aquaculture applications: A systematic literature review. *Journal of the World Aquaculture Society*. Advance online publication. DOI: 10.1111/jwas.13107

Ballocchi, A. (2024, February 11). Artificial intelligence in aquaculture: Enhancing a time-honoured practice with AI, robots, and drones. Tech4Future. https://tech4future.info/en/aquaculture-artificial-inteligence/

Bhat, S., Thakur, P., & Bhatt, A. (2024). A novel fish classification system using deep learning algorithms. *AIP Conference Proceedings*, 050013. DOI: 10.1063/5.0227890

Bhatt, R. (2024). A novel fish classification system using deep learning algorithms. *AIP Conference Proceedings*, *3209*(1), 050013. DOI: 10.1063/5.0227890

Bohara, K., Joshi, P., Acharya, K. P., & Ramena, G. (2023). Emerging technologies revolutionising disease diagnosis and monitoring in aquatic animal health. *Reviews in Aquaculture*, *16*(2), 836–854. DOI: 10.1111/raq.12870

Boquin Instrument. (2024). The impact of turbidity on fish health in aquaculture systems. Retrieved from https://www.boquinstrument.com/the-impact-of-turbidity-on-fish-health-in-aquaculture-systems

Chang, C., Wang, J., Wu, J., Hsieh, Y., Wu, T., Cheng, S., Chang, C., Juang, J., Liou, C., Hsu, T., Huang, Y., Huang, C., Lin, C., Peng, Y., Huang, R., Jhang, J., Liao, Y., & Lin, C. (2021). Applying Artificial Intelligence Techniques to Implement a Practical Smart Cage Aquaculture Management System. *Journal of Medical and Biological Engineering*, *41*, 652–658. DOI: 10.1007/s40846-021-00621-3

Elbaz, N., Abd Al Fatah, M., Gaafar, A., & Soliman, M. (2023). *Investigations on virulence genes of some bacterial infections in Nile tilapia in Kafr El-Sheikh farms.* Damanhour Journal of Veterinary Sciences. DOI: 10.21608/djvs.2023.200798.1112

Elbaz, N. F., & Abd Al Fatah, M. E. (2024). Bacterial diseases outbreaks in some freshwater fish farms in Kafr El-Sheikh, Egypt. *Journal of Applied Aquaculture*, *36*(1), 1–23. DOI: 10.1080/10454438.2022.2105673

Emara, T. Z. (2016). Maximizing Power Saving for VoIP over WiMAX Systems. *International Journal of Mobile Computing and Multimedia Communications*, *7*(1), 32–40. DOI: 10.4018/IJMCMC.2016010103

Emara, T. Z. (2018). Adaptive Power-Saving Mechanism for VoIP Over WiMAX Based on Artificial Neural Network. In Waluyo, A. B. (Ed.), *Algorithms, Methods, and Applications in Mobile Computing and Communications* (pp. 158–177). IGI Global., DOI: 10.4018/978-1-5225-5693-0.ch007

Emara, T. Z., & Huang, J. Z. (2019a). A distributed data management system to support large-scale data analysis. *Journal of Systems and Software*, *148*, 105–115. DOI: 10.1016/j.jss.2018.11.007

Emara, T. Z., & Huang, J. Z. (2019b). RRPlib: A spark library for representing HDFS blocks as a set of random sample data blocks. *Science of Computer Programming*, *184*, 102301. DOI: 10.1016/j.scico.2019.102301

Emara, T. Z., Saleh, A. I., & Arafat, H. (2014). Power saving mechanism for VoIP services over WiMAX systems. *Wireless Networks*, *20*(5), 975–985. DOI: 10.1007/s11276-013-0650-5

Emara, T. Z., Trinh, T., & Huang, J. Z. (2023). Geographically distributed data management to support large-scale data analysis. *Scientific Reports*, *13*(1), 17783. DOI: 10.1038/s41598-023-44789-x PMID: 37853092

Er-Rousse, O., & Qafas, A. (2024). Artificial intelligence for the optimization of marine aquaculture. *E3S Web of Conferences*. https://doi.org/.DOI: 10.1051/e3sconf/202447700102

Fabian. (2022, March 26). Freshwater aquarium fish diseases, symptoms & remedies. SmartAquariumGuide. https://smartaquariumguide.com/freshwater-fish-diseases-symptoms-remedies/

Floerl, O., Sunde, L., & Bloecher, N. (2016). Potential environmental risks associated with biofouling management in salmon aquaculture. *Aquaculture Environment Interactions*, *8*, 407–417. DOI: 10.3354/aei00187

Francis-Floyd, R., & Petty, B. D. (2020, February 27). Disorders and Diseases of Fish. MSD Veterinary Manual. https://www.msdvetmanual.com/all-other-pets/fish/disorders-and-diseases-of-fish

Hu, J., Li, D., Duan, Q., Han, Y., Chen, G., & Si, X. (2012). Fish species classification by color, texture and multi-class support vector machine using computer vision. *Computers and Electronics in Agriculture*, *88*, 133–140. DOI: 10.1016/j.compag.2012.07.008

Jagdish. (2024, March 8). Artificial intelligence and its aquaculture applications. Ocean News & Technology. https://ecomagazine.com/news/fisheries-aquaculture/artificial-intelligence-and-its-aquaculture-applications/

Kim, J., Kim, K., Kang, Y., & Jeong, S. (2024). Predicting Key Substance Levels in Aquaculture through AI-Based Water Quality Monitoring. Human Interaction and Emerging Technologies (IHIET 2024). .DOI: 10.54941/ahfe1005479

Li, D., Li, X., Wang, Q., & Hao, Y. (2022). Advanced techniques for the intelligent diagnosis of fish diseases: A review. *Animals (Basel)*, *12*(21), 2938. DOI: 10.3390/ani12212938 PMID: 36359061

Lou, D., Chen, M., & Ye, J. (2007). Study on a fish disease case reasoning system based on image retrieval. *New Zealand Journal of Agricultural Research*, *50*(5), 887–893. DOI: 10.1080/00288230709510364

Lutz, C. G. (2023, March 1). The rise of AI in aquaculture. The Fish Site. https://thefishsite.com/articles/the-rise-of-ai-in-aquaculture-artificial-intelligence

Mahfouz, N. B., Moustafa, E. M., & El-Baz, N. F. (2016). Studies on Some Bacterial Diseases in Some Cultured Freshwater Fishes in Kafr El-Sheikh Governorate. *Kafrelsheikh Veterinary Medical Journal*, *14*(1), 149–161. DOI: 10.21608/kvmj.2016.108686

Mandal, A., & Ghosh, A. R. (2023). Role of artificial intelligence in fish growth and health status monitoring: A review on sustainable aquaculture. *Aquaculture International*, *32*(3), 2791–2820. DOI: 10.1007/s10499-023-01297-z

Mecha, N. J. M. F., & Gonzales-Plasus, M. M. (2023). The common diseases of freshwater ornamental fishes and the treatments applied by local fish owners in Puerto Princesa City, Palawan, Philippines. *Journal of Fisheries*, *11*(1), 111205–111205.

Mohale, H., Narsale, S., Kadam, R., Prakash, P., Sheikh, S., Mansukhbhai, C., Kirtikumar, P., & Baraiya, R. (2024). Artificial Intelligence in Fisheries and Aquaculture: Enhancing Sustainability and Productivity. *Archives of Current Research International*. https://doi.org/.DOI: 10.9734/acri/2024/v24i3650

Munif, M., Nasir, H., & Ahmad, M. (2024). Optimizing ant colony system algorithm with rule-based data classification for smart aquaculture. *Indonesian Journal of Electrical Engineering and Computer Science*, *33*(1), 261. Advance online publication. DOI: 10.11591/ijeecs.v33.i1.pp261-268

Noga, E. J. (2010). *Fish Disease: Diagnosis and Treatment* (2nd ed.). Wiley-Blackwell., DOI: 10.1002/9781118786758

Park, J., Oh, M., & Han, S. (2007). Fish Disease Diagnosis System Based on Image Processing of Pathogens' Microscopic Images. 2007 Frontiers in the Convergence of Bioscience and Information Technologies, 878-883.

Peddina, K., & Mandava, A. K. (2024). An optimized deep network-based fish tracking and classification from underwater images. *Multimedia Tools and Applications*. Advance online publication. DOI: 10.1007/s11042-024-19417-z

Ragab, S., Hoseinifar, S., Doan, H., Rossi, W., Davies, S., Ashour, M., & El-Haroun, E. (2024). Overview of aquaculture Artificial Intelligence (AAI) applications: Enhance sustainability and productivity, reduce labor costs, and increase the quality of aquatic products. *Annals of Animal Science*. Advance online publication. DOI: 10.2478/aoas-2024-0075

Treulle, S. (2025, January 14). Aquarium fish diseases & treatments (Complete guide with diagnostic chart). PangoVet. https://pangovet.com/pet-health-wellness/aquarium/aquarium-fish-diseases-and-treatments/

Trinh, T., Duc, L. P., Tran, C. T., Duy, T. T., & Emara, T. Z. (2021). A new stratified block model to process large-scale data for a small cluster. *International Conference on Artificial Intelligence and Big Data in Digital Era*, 253–263. Springer International Publishing Cham. https://doi.org/DOI: 10.1007/978-3-030-97610-1_21

Trinh, T., Nguyen, N.-T., Wu, D., Huang, J. Z., & Emara, T. Z. (2019). A new location-based topic model for event attendees recommendation. *2019 IEEE-RIVF International Conference on Computing and Communication Technologies (RIVF)*, 1–6. IEEE. https://doi.org/DOI: 10.1109/RIVF.2019.8713716

VuThi., K., VuongThi, N., NguyenXuan, H., Emara, T. Z., & Trinh, T. (2024). Feature Selection Based on Top Frequency. *The International Conference on Intelligent Systems & Networks*, 58–65. Springer Nature Singapore. https://doi.org/ DOI: 10.1007/978-981-97-5504-2_7

Wei, C., Salloum, S., Emara, T. Z., Zhang, X., Huang, J. Z., & He, Y. (2018). A Two-Stage Data Processing Algorithm to Generate Random Sample Partitions for Big Data Analysis. In Luo, M., & Zhang, L.-J. (Eds.), *Cloud Computing — CLOUD 2018* (Vol. 10967, pp. 347–364). Springer International Publishing., DOI: 10.1007/978-3-319-94295-7_24

Wu, B., Wang, C., & Du, K. (2023). A Smart Aquaculture System Exploiting IoT, AI and Cloud Computing. *2023 IEEE International Conference on Smart Internet of Things (SmartIoT)*, 251-256. https://doi.org/DOI: 10.1109/SmartIoT58732.2023.00045

Xiaoshuan, Z., Zetian, F., & Ruimei, W. (2004). Development of the ES-FDD: an expert system for fish disease diagnosis. Oceans '04 MTS/IEEE Techno-Ocean '04 (IEEE Cat. No.04CH37600), 1, 482-487.

Zhang, J., & Li, D. (2008). A Call Center Oriented Consultant System for Fish Disease Diagnosis in China. In: Li, D. (eds) Computer And Computing Technologies In Agriculture, Volume II. CCTA 2007. The International Federation for Information Processing, vol 259. Springer, Boston, MA. DOI: 10.1007/978-0-387-77253-0_96

Zouin, B., Zahir, J., Baletaud, F., Vigliola, L., & Villon, S. (2024). Improving CNN Fish Detection and Classification with Tracking. *Applied Sciences (Basel, Switzerland)*, *14*(22), 10122. DOI: 10.3390/app142210122

Chapter 6
Drones in the Indian Agricultural Sector:
A Technological Rise

Anusha Thakur
https://orcid.org/0000-0001-8761-2250
Independent Researcher, India

ABSTRACT

In the realm of agriculture, the adoption of innovative technologies has become a necessity. Indian agriculture is at a crucial stage, wherein it needs to select between taking care of the environment and producing more food. With the possibilities being vast, the existing state of technological adoption unveils an emerging market grappling with piloting changes and challenges. The need for agronomists and farmers to augment agricultural resource management in response to the changing demand is expected to witness an increase. The integration of drones in agriculture is a game-changer and is projected to revolutionize the entire sector. The chapter discusses the key drone applications in agriculture and the technology landscape in the different Indian states. It contemplates the different market entrants state-wise and the solutions they offer, paving the way for opportunities in the market. The paper further illustrates the challenges and technologies shaping the future of drones in Indian Agriculture.

INTRODUCTION:

In the pursuit of harvest, the agriculture sector in India remains affixed to its roots. The agricultural sector is yet to attain its highest potential regardless of its contribution to the country's GDP. The Indian agricultural sector is grappling with

DOI: 10.4018/979-8-3693-9208-9.ch006

a multitude of challenges in terms of demand and environmental issues, making it behind the developed farming nations. (Dubashi, 2023). The conventional approaches archetypally comprehend the distribution of fertilizers and water across the fields, ignoring the needs of soil or crop types. These practices play a pivotal role in soil deterioration and water mismanagement.

With the necessity of recognizing the significance of food security and the implications of environmental water security, pollution, and degradation, the exigency to overcome challenges is gaining a surge. (EquinoxsDrones, 2024). Traditional farming techniques have limitations, incumbering efficiency, and progress whilst pushing the need for advanced approaches. Eliminating ambiguity and increasing focus on precise agricultural methods is crucial in the shift towards opening a world of avenues for farmers.

Agricultural digitization significantly plays an important role in heralding the next green revolution. To propel digital agriculture, the government is emphasizing on different initiatives, schemes, models, and strategies backing the use of drones by the Indian farmers. Driven by the government's drive towards technological adoptions in agriculture is gaining surge. Launch of initiatives such as the Kisan Drone Scheme, Sub-Mission on Agricultural Mechanization (SMAM), and Production Linked Incentive (PLI) Scheme offer substantial financial push and empower the farmers to adopt pioneering technology bolstering productivity (IBEF, 2024). After the farm equipment and tractors, drones are significantly revolutionizing agricultural practices, catering to the prevailing challenges. These are projected to become requisite tools for Indian agriculture, guaranteeing unparalleled profitability, sustainability, and efficiency to the sector.

Drone technology is being recognized to enable farmers to augment efficiency, estimate crop yields, monitor livestock, sow seeds over larger areas, conduct field surveys, and optimize farms. With their capability to offer real-time insights and data, these aerial vehicles are expected to restructure conventional agricultural practices (Shukla, 2024). This cutting-edge technology is transforming the sector, with unprecedented capabilities to augment, manage, and monitor the steps from precision agricultural activities to post-harvest evaluation. While the adoption of drone technology in agriculture in India is still in the nascent stage, there are several initiatives improvised to promote the usage of drones in agriculture. The paper emphasizes the potential of drones in revolutionizing farming techniques, along with the challenges in the implementation of this state-of-the-art technology. The article further delves into the state-wise analysis of the adoption of this AI-powered agricultural drone, illustrating its status, key market players, potential, and impacts in different regions.

LITERATURE REVIEW:

Usage of pioneering technologies to bolster the change in the agricultural domain is the key strategy of the policymakers in Indian agriculture. Policy changes emphasize on catering to the challenges of low mechanization, small landholdings, and the need for productivity. In addition to this, fostering entrepreneurship and innovation by the Indian Government is also expected to position India as a hub for innovative drone advancements. Different Agricultural solutions are projected to plan the harvests in a better way with efficient usage of livestock and equipment, incorporation of sustainable agri-food methods, and augmentation of crop yields.

The Integration of drones in Indian agriculture is at a massive high. Growing focus on securing government subsidies, training programs, and loans, coupled with the labor shortage for applying fertilizers and pesticides, are anticipated to bolster the demand for agricultural drones in India. From pest management to crop monitoring, drones offer varied benefits engaged in augmenting sustainability, yield, and efficiency in Indian agriculture (JetayuGadgets, n.d.).

Drones and sustainable agriculture:

Recognizing the issues in existing practices and subsequently, providing sustainable solutions is the need of the hour in the agricultural sector. In India, where agriculture is a keystone of the livelihoods and economy of millions, the adoption of drone technology embraces massive potential. With a diversified and vast agricultural landscape, oscillating from large-scale plantations to smallholder farms, drones have the potential to cater to the several issues faced by Indian farmers. In the current agricultural practices, there are varied unsustainable practices related to pest management, crop monitoring, chemical spraying, irrigation, and more (Katekar, 2022).

With consumers becoming more environment-conscious about the usage of chemicals and food production in agriculture, the demand for organically grown products is witnessing a significant surge. Integration of drones into farming practices is progressively marking a critical whirling point, encouraging to efficaciously help cater to the issues of sustainable agriculture. This pioneering technology offers real-time and accurate data on crop conditions that will help alter agricultural resource management, making farming less intrusive and more efficient for the environment (DroneXperts, 2025). Agriculture drones empower farmers to efficiently monitor the crops, employ targeted spraying methods to minimize chemical usage, minimize synthetic inputs, and more. These drones facilitate precision farming by allowing the farmers to apply the resources exactly wherever they are required, minimizing the environmental impact and the wastage. With their smart sensors, these also help to seamlessly detect pest infestations, diseases, and plant stress, at the same time, offering critical data for soil analysis, thereby enabling the farmers to intervene on time, uphold healthy crop yields, and ensure the long-term viability of agricultural systems (Stoner, The Role of Drones in Sustainable Agriculture, 2023).

Drones And Their Applications In The Agricultural Sector

Agricultural drones are revolutionizing farming practices by offering unparalleled benefits over conventional farming practices. This can be in terms of crop monitoring, yield estimation, irrigation management, estimating soil health, spraying of chemicals, and more. These advantages are illustrated in the figure mentioned below:

Exhibit 1 Advantages of Agricultural Drones

Figure 1. Advantages of Agricultural Drones

Source: Secondary Sources

- **Crop Monitoring:** Efficacious surveillance of the crops ensures timely harvest, particularly at the time of dealing with seasonal crops. Pioneering technologies such as drones help in the effective examination of the field with infrared cameras based on real-time information, thereby enabling the farmers to improve the condition of the plants in fields (Kainth, 2023). The usage of drones in crop surveillance can be illustrated as:

  i. NDVI (Normalized Difference Vegetation Index) Mapping: Drones fitted with special imaging equipment called Normalized Difference Vegetation Index (NDVI) deploy comprehensive colour information to denote plant health. The equipment also enables the farmers to monitor the crops as they grow so that the problems can be dealt at a faster rate to save the plants.

ii. Early Pest and Disease Detection: Proactive solutions such as drones equipped with advanced multispectral sensors and high-resolution cameras are expected to act as a game-changer. This can be attributed to the fact that these sensors help capture inclusive images of fields, augmenting the identification of pest infestations in case before prove to be challenging (Malone, 2024). Early pest detection helps in minimizing pesticide usage, preservation of crop yield and health, along with proactive pest management,
iii. Detecting Crop Stress: Identifying and monitoring crop stress plays a vital role in ensuring sustainable and sufficient crop production (Chakhvashvili, 2024). With the help of drone technology, it becomes easy to detect the stress in crops in terms of soil problems, nutrient deficiencies, or other factors.

- **Managing Livestock:** Drone technology acts as a transformative solution for monitoring the livestocks. Thermal real-time tracking and imaging solutions, enable farmers to ensure both the health and safety of their animals (AsteriaAerospace, 2024). The varied range of capabilities offered by the drone technology includes (Saiwa, 2025):

i. Tracking of herds and aerial surveillance
ii. Inventory management and counting of animals
iii. Disease Detection
iv. Welfare monitoring and behavior analysis

- **Spraying Chemicals and Fertilizers:** Drones are equipped with the latest spraying systems that can accurately apply herbicides, pesticides, and fertilizers. Unlike conventional methods, these drone-based spraying systems minimize human exposure to hazardous systems and the usage of chemicals, thereby ensuring even distribution of the chemicals across the fields (Dronepilotgroundschool, 2025).
- **Soil Monitoring:** Monitoring the soil conditions is one of the most vital aspects in agriculture. Without the proper moisture levels, the soil is expected to dry out, leading to waterlogging or plant stress, thereby causing nutrient leaching or root rot. Fitted with thermal sensors, this cutting-edge technology helps capture the heat signatures and measure the difference in temperatures on the surface of the soil (Sana, 2024). Further, the sensors also help the farmers to gain insights into their soil conditions, thereby allowing optimization of the irrigation and fertilization strategies (Alkan, 2024).
- **Improving Irrigation Practices:** Drone technology is rising as a significant asset in managing water resources. Equipped with multi-spectral and thermal cameras, this technology helps identify the areas of under-or over-irrigation

areas, and subsequently assess the moisture levels in soil. Such real-time information is expected to enable the farmers to make informed decisions related to irrigation scheduling, thereby, ensuring the applicability of water when, and where required (Meena, n.d.).

ANALYSIS: STATE-WISE

Drones act as a key enabler in the pursuit to transform the agricultural sector with technological innovations (Jurgens, 2022). Usage of drones for spraying fertilizers is no longer a challenging solution for farmers across the country. Over the past few years, several educational institutions, startups, state governments, and central government have largely emphasized offering solutions in terms of technological advancements, training programs, and subsidies to augment agricultural activities.

Numerous farmers across India are witnessing the benefits of using agricultural drones. The potential of this technology across different states, along with the key market players offering products and solutions related to the same, are illustrated as:

Kerala

For the first time in Kerala, the drone technology was tested under a trial run for sowing the seeds in the paddy field at Kuttanad. The Krishi Vigyan Kendra under Kerala Agricultural University at Kumarakom announced the attainment of this milestone in the agricultural sector (Onmanorama, 2024).

The central and the Kerala state governments are in favor of adopting drone technology to revolutionize agricultural practices in the region. With the introduction of drones, farming practices are projected to evolve from being a labour-intensive process to becoming an efficacious process offering several benefits in terms of minimized energy, time, labor costs, as well as maximized productivity and opportunities.

Fuselage

About the Company: Fuselage Innovations is an Agri-tech startup company, headquartered at Kochi. The company emphasizes offering products engaged in increasing the overall productivity of agricultural activities and subsequently addressing agricultural issues. Established in 2020, the company is bolstering the shift towards sustainable and intelligent production, augmenting the resilience contrary to climate change issues. The company is engaged in developing drones powered by agricultural IoT, advanced sensor technologies, and precision farming. The agricultural drones manufactured by Fuselage focus on protecting assets and opposing the illegal transfer of agricultural drones (Fuselage, Fuselage, 2020).

Solutions offered: In response to catastrophic climatic issues causing altered soil composition, extensive crop damage, and the launch of new pest species, the intervention of pioneering technologies is gaining significant traction. Fuselage Innovations announced the introduction of Nireeksh and FIA QD10 drones as transformational solutions for farmers to augment agricultural productivity. This can be elaborated as:

- In November 2023, The *FIA QD10 drone* was DGCA certified. The drone has been manufactured for high precision and performance, with its spraying capabilities to effectively manage pests and improve crop health. This drone has been tested and piloted for over 50,000 hectares of farmland and engineered for smart agriculture. Key features include:

  a) High-elevation precision: Distributing agricultural inputs with accuracy ensuring optimal care, and bolstering productivity.
  b) Unparalleled 26-Minute Flight Time: Driven by innovative Lithium Polymer batteries, ensures uninterrupted and extended coverage.
  c) Substantial 10L Tank Capacity: Capitalize on spraying efficiency with a higher payload.

- The *NIREEKSH* surveillance drone offers detailed vegetative and crop monitoring analysis via its multispectral sensors, thereby, permitting timely interventions and informed decision-making processes. The product works on advanced LiPo batteries and 30-minute flight time, efficiently covering vast areas. Nireeksh is equipped with sophisticated sensors (IR, Red Edge, RGB, Red, NIR), offering inclusive multispectral imaging for detailed crop analysis, and allowing the farmers to detect challenges of water stress, pest infestations, and nutrient deficiencies.

These drones exemplify substantial technological advancement, enabling the incorporation of efficient and modern farming techniques maximizing resource usage and crop yield. Integrating these pioneering technologies into their operations has proven to be a game-changer, offering a nearly 70 percent reduction in crop management costs, and a 30 percent rise in crop production (Fuselage, Solutions, 2020). Further, drone technology plays a pivotal role in large-scale paddy fields wherein, precision farming minimizes the requirement of labor and time for both the monitoring as well as spraying activities.

Andhra Pradesh

With the successful conduct of the Amaravati Drone Summit, in November 2024, the Andhra Pradesh government announced the deployment of 'Kisan drones' for varied applications in the agriculture sectors. The first phase is expected to include the setting up of 875 drone service centers, with drones valued at Rs. 88 crore, and giving a 40 percent subsidy (Raghavendra, 2024). These drones are expected to play a key role in spraying pesticides, and fertilizer. In addition to this, the Andhra government also focuses on imparting training in maintaining and operating drones to the farmers registered with rural youth and custom hiring centres.

Marut Drones:

About the Company: Marut Drones established in 2019, is headquartered in Hyderabad. Marut Drones manufactures drones intended for scalable afforestation, sustainable agriculture, surveillance, and others. The company is one of the leaders in drone technology with a focus on using its expertise for purpose.

Solutions offered: Key products offered by the company include (Marutdrones, 2019):

- *AG 365:* It is the first DGCA-certified versatile agricultural drone, providing its users with exceptional adaptability and flexibility in their operational activities. These drones are power-driven by smart batteries that are paired with effectual power utilization and have the highest continuance of 22 minutes. These drones have profound usage in spreading fertilizers, spraying pesticides, crop monitoring, fumigation, and seeding. Advanced production processes and precision engineering make these drones one of the finest performing technologies for agricultural activities.
- *Seedcopter:* Seedcopter is one of the pioneer commercial drones in India, focusing on UAV Aerial Seeding. Initiatives such as *Hara Bahara* emphasize replanting 1 billion trees by 2030, impacting 900 forests with the help of Seedcopter, and subsequently restoring the forests affected by mismanagement, fragmentation, and loss. It is the most powerful and pioneering aerial seeding technology offered by Seedcopter. The initiative of Hara Bahara focuses on catering to the issues of deforestation, biodiversity losses, climatic changes, as well as, rural and women employment.
- *Marut Zap:* This product is engaged in reconsidering the breeding of mosquitos with drone technology. The technology *"MarutZap"*, an amalgamation of efficiency and innovation is developed for precise and targeted operations, proficiently aiming mosquito breeding grounds and subsequently eradicat-

ing the disease-carrying pests. The MarutZap drones, thereby ensure an eco-friendly method of controlling the breeding of mosquitos and protect societies from vector-borne diseases.

- *AG365S:* AG 365S is a DGCA-certified small group of agricultural drones providing users with exceptional and unparalleled adaptability and flexibility in Agri operations. These are driven by smart batteries and paired with efficacious utilization of power.
- *AG365H:* In October 2022, Marut Drones announced the launch of AG 365H, the first ever DGCA-type certified medium category Agri drone in India. With its portable design, the AG365H helps transform agricultural practices and can be used for offering inclusive training for drone operators, spraying pesticides, broadcasting fertilizer granules, and feeding fish. The multi-utility features of the product substantially augment sustainability and productivity in modern farming.

Tamil Nadu

Tamil Nadu is the home of precision Agri Drones. In Tamil Nadu, the farmers are enfolding drone technology for pesticide spraying, and to minimize the costs and time. The launch of Uzhavan mobile apps by the government of Tamil Nadu helps them to connect with the service providers easily. In addition to this, several training programs and government subsidies are bolstering the adoption of agricultural drones among women self-help groups, and farmers. Schemes such as subsidies Namo Drone Didi (providing 80 percent subsidy for the women's self-help groups and Sub Mission on Agricultural Mechanization (SMAM), enable more farmers to join the tribe. Drone technology braces the farmers with precise spraying of pesticides and fertilizers, monitoring of crops, preparation of fields, land mapping, and detection of fields, thereby saving water costs, pesticides, as well as fertilizers by 70 percent, and maximizing productivity by 30 percent (Rajendran, 2024). The Namo Didi Scheme implemented by the ministries of rural development, chemicals, and agriculture bolsters the usage of drones and empowers women via self-help groups.

Garuda Aerospace

About the Company: Garuda Aerospace is one of the prominent players engaged in revolutionizing advancements in Industry 4.0 and Precision Agri Tech. Established in 2015, the company is headquartered in Chennai. The company operates with a colossal fleet of nearly 500 pilots and 400 drones across 84 cities in India. Garuda Aerospace works on the principle of Made in India Agri/Kisan drone to enable the farmers and youths with reasonable precision agriculture drone technology. The company is engaged in providing a suite of 50+ services and 30 drone models, to cater to the changing diversified needs. The company aims to make India the drone capital of the world and focus on combating climate change issues via customized AI data collection and afforestation plans.

Solutions Offered: The company is engaged in offering agricultural drone services to augment crop care, safeguard the yields, and subsequently make farming services smarter. Garuda Aerospace offers the best in the country drone services with enhanced profitability, efficiency, and sustainability. The company offers solutions catering to (GarudaAerospace, 2015).:

- The attainment of 10 to 15 percent depreciation in the crop loss.
- Cost reduction in terms of resources and labor by approximately 30 to 50 percent.
- Nearly 40 to 60 percent of the time savings in farm management.

Maharashtra

In 2022, the Centre announced the issuance of Standard Operating Procedures (SOP) for the usage of drones to mechanize the agricultural sector. In 2023, the state government announced the investment in drone technology, owing to the labor shortage in the agricultural sector (Biswas, 2023).

In addition, Initiatives such as Maharashtra – World Economic Forum partnerships help assist and train the farmers in drone deployment. Centre in collaboration with the Maharashtra Government is also expected to undertake a drone mapping operation with a key focus on strategizing new protocols and policy frameworks for emerging technologies (FSII, n.d.). Further, projects such as ATARI Pune announced the sanction of a total of 40 drones, comprising 10 to KVKs, 23 to ICAR institutes, and 7 to SAUs. Such initiatives are expected to encourage the adoption of drones, thereby promoting, and strengthening agricultural modernization.

Punjab

The Punjab Agricultural University (PAU) has been driving innovation and backing the farmers with advanced practices and techniques. The University has been working on drone technology in terms of crop spraying over the past few years. Another case in Punjab showed a reduction in pesticide use by 30% with the adoption of drone technology. According to the reports, there was a reduction in pesticide usage by nearly 30 percent due to the integration of drones into agricultural activities in the Punjab (krishiviman, 2024).

In a trailblazing move, the farmers in Punjab are engaged in the adoption of agricultural practices by shifting to drone technology for the spraying of pesticides on paddy crops. The shift from tractor-mounted spraying or manual spraying techniques to a pioneering technology signifies a new era in farming practices in the region. In addition to this, the Indian Farmers Fertilizer Cooperative (IFFCO), focuses on offering nearly 110 drones for free services to the local cooperative societies. Providing a subsidy of Rs. 100 per acre, is projected to make pesticide spraying with drones an efficacious option in the State (Mishra, Revolutionizing Agriculture: Punjab Farmers Embrace Drones for Pesticide Spraying, 2023).

Assam

Drone technology in Assam stands as evidence of the readiness of the agricultural sector to enfold technological developments. Such cutting-edge innovations have the potential to bolster crop yield and transform farming practices, thereby contributing to the region's sustainable growth. In a substantial dive towards renovating agriculture, the state is emphasizing exploring the unprecedented power of drones in shaping the agricultural sector over the projected period (Mishra, Assam Agriculture Minister Atul Bora Inaugurates Drone Technology Demonstration in Agriculture: VIDEO, 2023).

The young farmers in Assam are largely emphasizing on the profound usage of drone technology to augment efficiency, minimize labor costs, and save time during the harvesting season (TheHillsTimes, 2024). With more farmers adopting these methods in Assam, the region's agricultural sector is poised to become adaptable and resilient to prevalent issues, imprinting a step ahead in the state's efforts to enfold smart farming solutions. This initiative works in parallel with the wider goals of the Assam government, thereby boosting the technological and digital interventions in agriculture to enhance the living standards of the farmers as well.

Himachal Pradesh

The Himachal Pradesh government is focusing on promoting drone-focused initiatives to bolster productivity in the agricultural sector. In a significant advancement, the state received 10 state-of-the-art drones by mid-July 2023, thereby, revolutionizing the horticulture and agriculture practices in the region and proving to be an unprecedented solution to the farmers and apple growers. This initiative organized by Indian Farmers Fertiliser Cooperative Limited (IFFCO), emphasizes enhancing agricultural productivity whilst abating the overheads for farmers as well. Under this initiative, 2 drones were allocated to the districts of Una, Shimla, and Kangra, and 1 drone each was given to Sirmaur, Solan, and Kullu. In addition to this, to ensure responsible accountability, and usage the registered committees in association with IFFCO, owe access to the drones with a security deposit of Rs. 50,000 (Bhandari, 2023).

In July 2023, The Himachal Pradesh government also announced the signing of the MoUs of Rs. 200 crore for the manufacturing of drones, posing a vital step towards the promotion of technology-driven governance (Mitra, 2023). The state has also announced the MoUs with FICCI to augment the usage of drones in governance. In addition to this, for investments collaborations with Dronetech Solutions Private Limited and Abecode Technologies Private Limited have also been announced. Further, the state government announced the signing of an MoU with SkyeAir for unmanned traffic management. This partnership in collaboration with IIT Ropar, and the Department of Digital Technologies and Governance, Chaudhary Sarwan Kumar Himachal Pradesh Agricultural University (CSKHPKV), is to supervise the flight operations. The collaboration between IIT Ropar and CSKHPKV emphasizes the areas of precision agriculture, cold chain management, livestock management, and AI-based digital entomology (TheEconomicTimes, 2023).

With the influx of these state-of-the-art drone technologies, Himachal Pradesh is set to witness an altering shift in horticulture and agricultural practices. Promoting sustainable farming techniques and implementing modern technologies is helping the state take a substantial step towards a prosperous future, minimized costs, and higher productivity for its farming community.

CHALLENGES IN THE ADOPTION OF DRONES IN THE AGRICULTURAL SECTOR:

Despite its potential, adopting agricultural drones is expected to face several challenges. These can be illustrated as:

1. Limited Awareness:

With varied developments in terms of the tools and techniques in farming, the latest technological developments need to be leveraged in several farming communities. This can be attributed to the fact that the local farmers are a bit apprehensive about adopting new techniques and tools in farming practices (Shah, 2023).

Several farmers in India are unaware of the subsidies available for the purchase or benefits offered by drones in agriculture (AeoLogic, 2023). In addition to this, there is also a lack of inclusive training programs to educate drone operators and farmers about drone technology. It is difficult for an average farmer to analyze drone images specifically, as specialized knowledge and skills are needed to interpret them to attain useful information. Further, the farmers' adoption of this pioneering technology can also be attributed to the lack of coordination between the states and the Centre on the policy framing.

2. Regulatory Challenges:

With the regulations being simplified in some areas, there remain some restrictions, such as the need for specified clearances for operating drones in specific areas, thereby creating difficulties for the farmers, particularly in remote locations. Hence, a simplified regulatory framework can help streamline the adoption of drones whilst promoting innovation.

FUTURE TECHNOLOGIES RESHAPING THE DRONES IN AGRICULTURAL SECTOR:

By exploiting the capabilities of drones, the agricultural sector of India has the potential to attain higher levels of prosperity and innovation, thereby catering to the prevailing challenges.

Integration of Artificial Intelligence (AI) in Agricultural Drones: The amalgamation of AI in agricultural drones denotes a groundbreaking development in recent farming practices. The AI-equipped drones offer unprecedented abilities for augmenting overall farm management, optimizing the usage of resources, and monitoring the crops. Exploiting the power of AI enables farmers to attain sustainable practices, higher yields, and greater efficiency. The adoption of Artificial Intelligence in agriculture offers real-time decision-making, predictive modeling, and data analysis, thereby contributing to precise farming practices (Stoner, Agricultural Drones: Farming with AI Integration (2025), 2025). In addition to this, the technology also helps cater to several challenges in crop monitoring, pest and weed control, precision agriculture, harvest planning and yield prediction, soil analysis, and management.

CONCLUSION:

The agricultural sector in India contributes nearly 18.3 percent to the Gross Domestic Product (GDP) of the country (ideaForge, 2023). The modernization of the sector plays a pivotal role in food security and economic growth. To unveil the potential of the sector, the need to enfold the power of technological innovation in cultivation practices plays a crucial role. The growing convergence of agriculture and technology is escalating new financial investments, innovations, and entrepreneurial initiatives (Deloitte, 2020).

Integration of drones in cutting-edge agricultural practices paves the way for a myriad of benefits and diversified applicability. Equipped with GPS technology, multispectral sensors, and high-resolution cameras, this technology enables farmers to collect in-depth information related to moisture levels, soil conditions, crop health, and more. With real-time information, drones have the potential to make data-driven decisions related to when and where to apply the inputs, thereby optimizing the production of crops. By empowering the farmers and bridging the technological divide, drone technology is expected to completely reshape Indian agriculture, making it more resilient and sustainable.

The introduction of favorable policies and government support is also expected to boost the usage of drones in the agricultural sector. In recognition of the unparalleled potential of this cutting-edge technology, the Indian Government emphasizes the launch of several regulatory frameworks, training programs, subsidies, and more to facilitate the usage of drones and create a conducive environment. Collaboration between innovation and policy thereby ensures that the agrarian roots of India are ready to keep pace with the future scenario.

REFERENCES

AeoLogic. (2023, April 21). *Drone Technology in Agriculture: Benefits and Challenges for Rural India*. Retrieved from AeoLogic: https://www.aeologic.com/blog/drone-technology-in-agriculture-benefits-and-challenges-for-rural-india/

Alkan, M. (2024, October 18). *Soil Health and Plant Nutrient Analysis with Agricultural Drones*. Retrieved from Cropim.com: https://cropim.com/soil-health-plant-nutrient-analysis-agricultural-drones/#:~:text=Drones%20equipped%20with%20multispectral%20cameras%20and%20sensors%20can,allowing%20them%20to%20optimize%20fertilization%20and%20irrigation%20strategies

AsteriaAerospace. (2024, December 12). *Drone Technology in Agriculture: Benefits and Challenges*. Retrieved from asteria.co.in: https://asteria.co.in/blog/drone-technology-in-agriculture-benefits-and-challenges

Bhandari, R. (2023, June 26). *Himachal Pradesh Embraces Drones for Efficient Agriculture and Horticulture Operations*. Retrieved from The News Himachal: https://thenewshimachal.com/2023/06/himachal-pradesh-embraces-drones-for-efficient-agriculture-and-horticulture-operations/

Biswas, P. (2023, September 3). *Maharashtra's Agri-Entrepreneurs Take to the Skies: How Drone are Being Used Tackle Farm Labour Shortage*. Retrieved from IndianExpress.com: https://indianexpress.com/article/cities/pune/drones-the-new-tool-for-farmers-in-maharashtra-8921730/

Chakhvashvili, E. M., Machwitz, M., Antala, M., Rozenstein, O., Prikaziuk, E., Schlerf, M., Naethe, P., Wan, Q., Komárek, J., Klouek, T., Wieneke, S., Siegmann, B., Kefauver, S., Kycko, M., Balde, H., Paz, V. S., Jimenez-Berni, J. A., Buddenbaum, H., Hänchen, L., & Rascher, U. (2024). Crop Stress Detection from UAVs: Best Practices and Lessons Learned for Exploiting Sensor Synergies. *Precision Agriculture*, *25*(5), 2614–2642. DOI: 10.1007/s11119-024-10168-3

Deloitte. (2020). *Transforming Agriculture through Digital Technologies*. Deloitte.

Dronepilotgroundschool. (2025, February 7). *Drone Applications & Uses: A Comprehensive Guide for Drone Pilots*. Retrieved from Dronepilotgroundschool.com: https://www.dronepilotgroundschool.com/drone-uses/#:~:text=Drones%20equipped%20with%20advanced%20spraying%20systems%20can%20precisely,usage%2C%20and%20minimizes%20human%20exposure%20to%20hazardous%20substances

DroneXperts. (2025, February 17). *How are Drones Redefining Sustainable Agriculture?* Retrieved from DroneXperts.com: https://www.dronexperts.com/en/article/how-are-drones-redefining-sustainable-agriculture/

Dubashi, N. F. (2023, May 10). *How AgTech is Poised to Transform India into a Farming Powerhouse.* Retrieved from McKinsey.com: https://www.mckinsey.com/industries/agriculture/our-insights/how-agtech-is-poised-to-transform-india-into-a-farming-powerhouse

EquinoxsDrones. (2024, April). *Overview of the Agricultural Industry in India.* Retrieved from Equinoxsdrones.com: https://equinoxsdrones.com/importance-of-drone-technology-in-indian-agriculture-farming/

FSII. (n.d.). *Future of Agriculture: Drones.* Retrieved from Federation of Seed Industry of India: https://fsii.in/future-of-agriculture-drones/

Fuselage. (2020). *Solutions.* Retrieved from Fuselage: https://fuselage.co.in/solutions/

GarudaAerospace. (2015). Retrieved from GarudaAerospace.com: https://www.garudaaerospace.com/

IBEF. (2024, July 18). *Future of Drone Technology in Indian Agriculture: Enhancing Efficiency and Sustainability.* Retrieved from IBEF: https://www.ibef.org/blogs/future-of-drone-technology-in-indian-agriculture-enhancing-efficiency-and-sustainability

ideaForge. (2023, September 11). *Applications of Agriculture Drones: From Skies to Fields, Farm with Innovation!* Retrieved from ideaForge Technology: https://ideaforgetech.com/blogs/applications-of-agriculture-drones

JetayuGadgets. (n.d.). *How to Use Drones for Agriculture in India.* Retrieved from JetayuGadgets.com: https://jetayugadgets.com/blog/how-to-use-drones-for-agriculture-in-india/?srsltid=AfmBOorFqoLwMmcwo7iHjtrxQvVyEdGcs55tlVlqnZXHdtCpi9AGZjKg

Jurgens, J. R. (2022). *Using Technology To Improve a Billion Livelihoods.* Geneva: World Economic Forum In Collaboration with Adani Group.

Kainth, G. (2023, August 28). *Drone Technology In Agriculture: Indian Experience – Analysis.* Retrieved from Eurasiareview.com: https://www.eurasiareview.com/28082023-drone-technology-in-agriculture-indian-experience-analysis/

Katekar, V., & Cheruku, J. K. (2022). The Application of Drone Technology for Sustainable Agriculture in India. *Current Agriculture Research Journal, 10*(3), 352–365. DOI: 10.12944/CARJ.10.3.19

krishiviman. (2024, January 4). *Cost-Benefit Analysis: Is Investing in Agricultural Drones Worth it for Indian Farmers?* Retrieved from krishiviman: https://www.krishiviman.com/blog/cost-benefit-analysis-is-investing-in-agricultural-drones-worth-it-for-indian-farmers/

Malone, G. (2024, January 30). *Early Pest Detection: The Drone Advantage in Farming*. Retrieved from Medium.com: https://medium.com/@gmalone_38909/early-pest-detection-the-drone-advantage-in-farming-d3214fc088ad

Marutdrones. (2019). Retrieved from Marutdrones.com: https://marutdrones.com/products/

Meena, B. S. (n.d.). *Drones in Crop Monitoring and Management: Uses, Types, Advantages and Limitations*. Retrieved from WikiFarmer.com: https://wikifarmer.com/library/en/article/drones-in-crop-monitoring-and-management-uses-types-advantages-and-limitations

Mishra, S. (2023, September 6). *Assam Agriculture Minister Atul Bora Inaugurates Drone Technology Demonstration in Agriculture: VIDEO*. Retrieved from TractorNews: https://tractornews.in/news/assam-agriculture-minister-atul-bora-inaugurates-drone-technology-demonstration-in-agriculture-video/

Mishra, S. (2023, August 5). *Revolutionizing Agriculture: Punjab Farmers Embrace Drones for Pesticide Spraying*. Retrieved from TractorNews: https://tractornews.in/news/revolutionizing-agriculture-punjab-farmers-embrace-drones-for-pesticide-spraying/

Mitra, S. (2023, July 6). *Himachal Pradesh Signs INR 200 Cr MoU With FICCI To Promote Drone Production*. Retrieved from Inc42.com: https://inc42.com/buzz/himachal-pradesh-signs-inr-200-cr-mou-with-ficci-to-promote-drone-production/

Onmanorama. (2024, July 10). *In a First, Drones Replace Tractors to Sow Paddy Seeds in Kerala*. Retrieved from Onmanorama.com: https://www.onmanorama.com/news/kerala/2024/07/10/drones-sows-paddy-seeds-in-kuttanad-for-the-first-time-in-kerala.html

Raghavendra, V. (2024, November 12). *Andhra Pradesh Govt. to Deploy 'Kisan Drones' in Agriculture Sector this Year*. Retrieved from TheHindu: https://www.thehindu.com/news/national/andhra-pradesh/andhra-pradesh-govt-to-deploy-kisan-drones-in-agriculture-sector-this-year/article68856176.ece

Rajendran, G. a. (2024, September 16). *TN Farmers Spray, Sow and Soar into the Drone Age*. Retrieved from Times of India: https://timesofindia.indiatimes.com/city/chennai/tamil-nadu-farmers-embrace-drones-for-efficient-agriculture/articleshow/113373943.cms

Saiwa. (2025, February 5). *The Rise of Drone Livestock Monitoring | A Look at Cutting-Edge Tech for Modern Farmers*. Retrieved from Saiwa: https://saiwa.ai/blog/Drone-Livestock-Monitoring/

Sana, J. (2024, November 25). *How to Use Agricultural Drones for Soil Health and Fertility Management*. Retrieved from Agribizdrones.com: https://agribizdrones.com/how-to-use-agricultural-drones-for-soil-health/

Shah, T. (2023, March 17). *Drone Technology in Agriculture: Benefits and Challenges for Rural India*. Retrieved from The Times of India: https://timesofindia.indiatimes.com/blogs/voices/drone-technology-in-agriculture-benefits-and-challenges-for-rural-india/

Shukla, S. (2024, April 3). *Rising Impact: How Drones Are Leading an Agricultural Revolution in India*. Retrieved from KrishiJagran.com: https://krishijagran.com/explainers/how-drones-are-leading-an-agricultural-revolution-in-india/

Stoner, J. (2023, July 14). *The Role of Drones in Sustainable Agriculture*. Retrieved from Flyeye.io: https://www.flyeye.io/the-role-of-drones-in-sustainable-agriculture/

Stoner, J. (2025, January 18). *Agricultural Drones: Farming with AI Integration (2025)*. Retrieved from Flyeye.io: https://www.flyeye.io/agricultural-drones-ai/

TheEconomicTimes. (2023, July 5). *Himachal Govt Signs MoUs for Production of Drones to Promote Technology-Driven Governance*. Retrieved from TheEconomicTimes: https://economictimes.indiatimes.com/news/economy/policy/himachal-govt-sign-mous-for-production-of-drones-to-promote-technology-driven-governance/articleshow/101518242.cms

TheHillsTimes. (2024, September 13). *Youths Use Drone Technology for Winter Paddy Harvest in Assam*. Retrieved from TheHillsTimes: https://thehillstimes.in/assam/youths-use-drone-technology-for-winter-paddy-harvest-in-assam

Chapter 7

Machine Learning for Precision Agriculture and Crop Yield Optimization: Techniques and Applications

Pankaj Roy
https://orcid.org/0009-0002-2014-8574
Assam University, Silchar, India

Mrutyunjay Padhiary
https://orcid.org/0000-0002-2236-568X
Assam University, Silchar, India

Azmirul Hoque
https://orcid.org/0009-0002-7220-9818
Assam University, Silchar, India

Bhabashankar Sahu
https://orcid.org/0009-0008-2253-9361
Parala Maharaja Engineering College, India

Dipak Roy
https://orcid.org/0009-0008-7268-6351
Tezpur University, India

Kundan Kumar
https://orcid.org/0009-0001-6831-0676
Assam University, Silchar, India

ABSTRACT

The swift advancement of machine learning (ML) has altered several industries, including agriculture, by providing innovative ways of addressing complex challenges related to modern farming. This chapter discusses the use of ML in precision agriculture, emphasizing its capacity to maximize crop output and improve agricultural practices. It studies the use of supervised, unsupervised, reinforcement, and deep learning methodologies to evaluate extensive datasets derived from remote sensing technologies, soil sensors, climate data, and agricultural equipment. Principal applications include predictive modeling for agricultural yield estimation, pest and disease identification, soil health assessment, irrigation optimization, and precision fertilization. The chapter also examines the problems and limits related

DOI: 10.4018/979-8-3693-9208-9.ch007

to the implementation of machine learning in agriculture, including data quality and farmer acceptance.

1. INTRODUCTION

Agriculture is undergoing an advancement in technology to meet global food demand while reducing climate change and resource constraints. Machine Learning (ML) is a developing precision agriculture technology that allows for data-driven decision-making to improve yields, minimize waste, and improve sustainability.

1.1 Background

Precision agriculture refers to a novel approach to farming that uses high technology to increase agricultural productivity. Many data-driven instruments are connected together in order to optimize resource management which includes water, fertilizers, pesticides, and seeds, ensuring that every resource reaches the correct location at the right time and the suitable amount. The technique is enabled by technologies such as geographic information systems (GIS), global positioning systems (GPS), remote sensing, and Internet of Things (IoT) sensors, which together enhance the tracking and handling of agricultural operations with unparalleled precision. (Sabir et al., 2024). The importance of precision agriculture lies in its ability to address some of the most pressing global issues, including increasing food demand, climate change, and resource scarcity. Precision agriculture is very important in enhancing yields and reducing environmental impacts, therefore promoting a more sustainable agricultural system that addresses the increasing food demands of a global population while reducing ecological footprints (Sanyaolu & Sadowski, 2024). These technological advances enable farmers to monitor crop health, control soil quality, and optimize irrigation in real-time, which enhances farm production and sustainability. The integration of advanced technologies like all-terrain agricultural vehicles has been instrumental in enhancing efficiency and sustainability in modern farming. These vehicles automate key tasks such as planting, weeding, and harvesting, reducing labor dependence while improving productivity. Research has highlighted their growing role in transforming precision agriculture through their adaptability and automation capabilities (Padhiary, Kumar, et al., 2024; Sahu, 2024).

Currently, ML has developed into an essential instrument for agriculture. It is well-suited to precision agriculture because it can analyze vast amounts of complex data and make quick decisions regarding maximum agricultural yields. Machine learning algorithms are capable of analyzing huge datasets, such as meteorological trends, soil conditions, crop development phases, and insect activity, to notice patterns and results (Araújo et al., 2023). The predictions may then guide activities like changing irrigation schedules, crop yield estimation, disease detection, or improving fertilizer use. The application of machine learning in agriculture enables accurate, data-driven decisions that help farmers improve efficiency, reduce costs, and maximize resource use. Machine learning provides robust capabilities for crop yield prediction and pest and disease detection, transforming agricultural practices, increasing productivity, and promoting responsible resource use (Imade et al., 2024). Additive manufacturing technologies like 3D printing are also expected to play a key role in the future of precision agriculture by enabling cost-effective, customizable, and sustainable solutions for various farming needs (Padhiary, Barbhuiya, Roy, et al., 2024).

1.2 Machine Learning Techniques

This method applies supervised learning, where the approach is training models on labeled data with the intention of forecasting outcomes that include crop yields or disease infestation; unsupervised learning to help in identifying patterns within free-of-pre-labeled-outcomes data; and reinforcement learning by pointing out the need to better decision-making through trial and error. We will also look into deep learning, which is very adept at handling large, sensitive data sets, especially in applications like image recognition for pest identification or crop disease diagnostics. Ensemble learning methods, which transform many models to increase accuracy and reduce errors, are increasingly being used in the agricultural industry. Each of these methods of machine learning has its unique importance in optimizing agricultural management, curbing environmental impact, and increasing crop yields (Elavarasan et al., 2018). Studies demonstrate how these technologies contribute to precision agriculture by integrating real-time data analysis and automated workflows, driving significant improvements in farm productivity and environmental sustainability. The future of precision agriculture will actually depend on the establishment of these machine-learning inputs as fundamentals in the direction of modern agriculture. **Figure 1** provides an overview of the different machine learning techniques used in precision agriculture, showcasing the relationship between supervised learning, unsupervised learning, reinforcement learning, and deep learning. It highlights their respective applications and how they contribute to crop yield optimization, pest detection, soil health monitoring, and automated farming systems.

Figure 1. Overview of machine learning techniques in precision agriculture

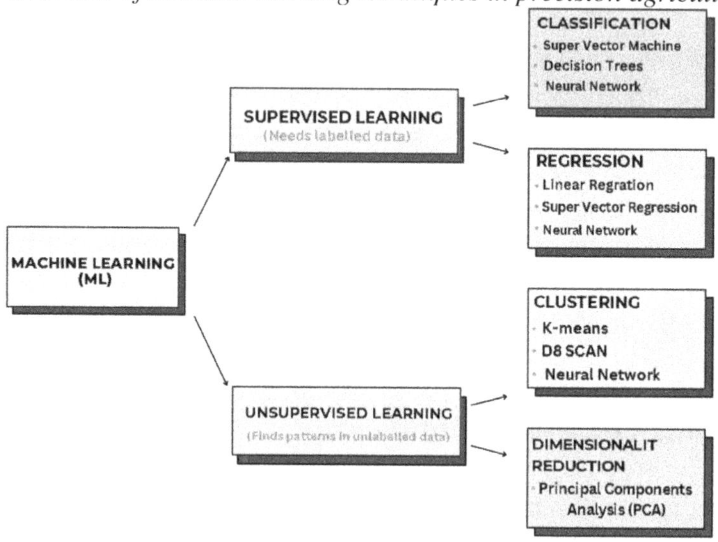

1.3 Aim and Objectives

This chapter attempts to present a comprehensive analysis of the role of machine learning in precision agriculture, mainly focusing on crop yield optimization and the impact of ML approaches on improving agricultural practices. It deals with different machine learning techniques, their applications in agriculture, and their capacity to address difficult issues in crop management. It involves case studies and practical applications, showing the impact of machine learning on agricultural production and sustainability. It offers an introduction to the basis of machine learning, explores its applications in precision farming, presents practical examples, and analyses the problems observed when employing these approaches in real-world agricultural scenarios. This chapter will present several machine-learning approaches used in precision agriculture.

Machine learning has the ability to improve precision agriculture by enabling data-driven decision-making to optimize crop yields, minimize resource waste, and enhance sustainability (Padhiary, Roy, Dey, et al., 2024). Advanced technologies like GIS, IoT, and remote sensing help monitor soil health, irrigation, and pest control in real-time. Machine learning models analyze vast agricultural data to improve efficiency and productivity. As a result, farming becomes more precise, adaptive, and sustainable in addressing global food demands.

2. BACKGROUND AND LITERATURE REVIEW

Exploring the development of ML in agriculture provides understanding on its innovative impact on traditional farming practices. This section focusses at how machine learning has changed over time, the major obstacles in conventional farming, and the latest developments that are driving precision agriculture innovation.

2.1 Evolution of Machine Learning in Agriculture

The use of machine learning in agriculture has undergone a considerable transformation over the past few decades. Traditional agricultural practices utilized simple methods and human power, and farmers relied on experience and available data to make decisions (Kuppusamy et al., 2023). A new world comprising satellite images, IoT sensors, and big data analytics radically transformed this environment by collecting adequate data in agriculture. The expansion of power on computers and algorithms opened up the door for machine learning to effectively use data processing and be a better tool for predictive results and data-driven decision-making. Agricultural applications of machine learning emerged at the beginning of the 1980s and 1990s when researchers began using statistical models to predict crop yields and soil conditions (Keating & Thorburn, 2018). Their first works were mainly limited to applying linear correlations and basic data processing. However, the advent of advanced methodologies, such as neural networks, support vector machines, and ensemble learning, has significantly expanded the scope of machine learning applications in agriculture.

Improved availability of high-resolution satellite imagery along with increasing computation power led to more accurate and scalable machine learning models during the 2000s (Jiang et al., 2022). Remote sensing has now become one of the principal sources of data for many machine learning algorithms, such as large-scale crop monitoring, disease detection, and precise irrigation control. The use of machine learning in agriculture has increased rapidly over the past decade due to extensive research and development into crop health monitoring, precision farming, pest and disease control, and autonomous agricultural systems (Saleem et al., 2021). Many sectors within agriculture employ machine learning technologies such as predictive analytics, resource management, autonomous equipment, and intelligent farming systems. As agriculture technology expands, machine learning further becomes a vital input in encouraging innovation and improving the output in farms.

2.2 Key Challenges in Traditional Farming

Erratic weather, inefficient resources, application inaccuracy, labor shortages, and increasing labor costs are some of the problems that machine learning can solve with traditional agriculture (Sharma et al., 2022). Machine learning algorithms may be used in analyzing historical weather data, providing future projections, and thereby optimizing planting and watering schedules. It may also eliminate waste and improve efficiency, as it will provide forecasts of the exact amounts of resources needed by crops at every stage. Machine learning can help in pest and disease management with a picture from drones, cameras, or satellites about early infestation signs or plant diseases, which can allow farmers to take timely appropriate action and reduce crop loss and reduced pesticide use. Machine learning can help alleviate labor shortages and rising labor costs by designing autonomous farming machinery like drones, robotic harvesters, and self-driving tractors. These systems can do things much faster than traditional labor-intensive systems. Data saturation is yet another challenge that farmers face due to the large amount of data that sensors, drones, and other technologies produce. Machine learning algorithms can process vast amounts of information, recognize patterns, and give actionable insights to farmers so that they make informed decisions. Predictive analytics can improve agricultural cycles, decide planting times, and predict potential threats.

2.3 Review of Recent Advancements

Recent developments in machine learning have significantly transformed agricultural yield optimization. These developments have mostly concentrated on predictive modeling, precision resource management, and real-time data analysis to enhance the accuracy of agricultural production predictions and optimize output.

***Crop Yield Prediction Models*:** A notable significant research area has been developing machine learning models to predict crop yields based on factors such as soil quality, meteorological conditions, and cultivation techniques (Chlingaryan et al., 2018a). Machine learning algorithms, including support vector machines, random forests, and deep learning networks, have been used to analyze complex information and produce more accurate yield predictions. A study used a deep learning model to predict the production of wheat in the United States based on climatic data, soil health, and historical yield records (Wang et al., 2020). The model outperformed traditional statistical approaches and provided actionable information for farmers to make better resource allocation decisions.

Remote Sensing and Satellite Data Integration: The combination of remote sensing data and machine learning has facilitated enhanced precision in crop monitoring(Sishodia et al., 2020).High-resolution satellite imaging, as well as drone sensors, provides continuous, real-time data on crop health, soil moisture, and other essential elements affecting production. With the help of machine learning algorithms, farmers can analyze these photos, monitor crop development, determine stress factors, and forecast probable yield outcomes (Hoque, 2024; Kumar et al., 2024). Deep learning models have been applied to analyze multispectral images for the detection of plant health issues, such as nitrogen deficiency, which can drastically affect yields (Behmann et al., 2015).

Precision Irrigation Systems: Machine learning has significantly contributed to the optimization of irrigation, which is vital for agricultural productivity. Machine learning models can help develop precision irrigation systems that minimize water wastage but guarantee adequate crop hydration through analysis of soil moisture levels, meteorological predictions, and agricultural water needs. A very good example is the application of reinforcement learning in irrigation systems, where algorithms constantly learn and update irrigation schedules with real-time meteorological and soil data, thus improving water management and agricultural output.

Pest and Disease Forecasting: A further area of progress is the use of machine learning for the early identification of pests and diseases, which may profoundly affect agricultural productivity. By using environmental sensor data, drone imagery, and field observations, machine learning algorithms can predict the chances of pest infestations or disease spread. They can then be targeted by treatments and developed a machine learning model based on climatic data and past pest activity to predict the emergence of the Fall Armyworm in maize fields (Badji et al., 2020). Preliminary forecasts enabled immediate pesticide application, preventing severe output losses.

Field Data Fusion for Yield Mapping: Researchers are enhancing yield maps by integrating data from several sources, including soil sensors, yield monitors, and climate data, and using machine learning to synthesize this information into meaningful insights (Joshi et al., 2023). The absorbed data sets may facilitate the identification of geographical heterogeneity in fields, allowing farmers to customize management approaches with greater precision. As an instance, machine learning-based yield mapping has been used to enhance fertilizer application, guaranteeing that each segment of the field has the requisite quantity of nutrients to optimize production.

Machine learning has evolved significantly in agriculture, transitioning from basic statistical models to advanced AI-driven analytics. It addresses several challenges like erratic weather, inefficient resource use, and labor shortages through predictive modeling and automation. Recent advancements in remote sensing, precision irrigation, and pest forecasting have enhanced efficiency and sustainability. The integration of Big data, IoT, and AI continues to improve modern farming practices.

3. MACHINE LEARNING TECHNIQUES FOR PRECISION AGRICULTURE

Numerous methods of machine learning play different roles on improving agricultural efficiency and sustainability. This section discusses supervised, unsupervised, reinforcement, and deep learning approaches and how they improve decision-making in agricultural operations.

3.1 Supervised Learning

Supervised learning is an aspect of machine learning in which the model is trained with labeled data, indicating that both the input characteristics and their matching output labels are supplied. The intention is to acquire a correspondence from inputs to outputs, enabling the model to predict new, unobserved data based on patterns detected during the training phase. Supervised learning methods need a substantial volume of high-quality labeled data to get precise outcomes. Supervised learning techniques, including algorithms such as Random Forest and Support Vector Machines, have shown success in predicting crop yields and assessing plant health. These methods allow for precise forecasts by analyzing diverse datasets related to environmental factors and farming practices. Case studies demonstrate their effectiveness in improving agricultural decision-making through accurate yield predictions and early detection of issues (Padhiary & Kumar, 2024b)

***Common Algorithms*:** Linear regression is an essential method used in agriculture to predict continuous variables by fitting a line to data points. Decision trees are models that partition data into subsets according to feature values, used for classification and regression problems. Random Forest is an ensemble method that uses decision trees it, applied for crop forecasting, pest identification, and disease detection. SVM is a robust algorithm used for classification and regression problems, even when the dimension of data is high (Wu & Yang, 2015). Neural networks are complex algorithms that mimic the neural structure of the human brain, are capable of learning complex patterns, and are applied in applications such as agricultural production forecasting and soil health assessment. **Figure 2** demonstrates how supervised learning algorithms (such as Random Forest or Neural Networks) are used to predict crop yield based on historical weather data, soil conditions, and farming practices. It shows the input features and the output of the yield prediction model.

Figure 2. Comparative performance of ML algorithms for crop yield prediction

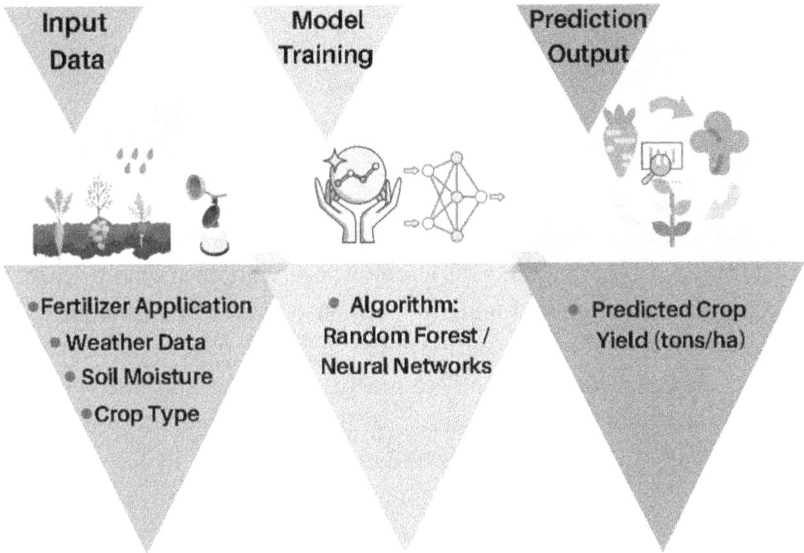

Applications: Supervised learning techniques are extensively used in precision agriculture. Linear regression can forecast agricultural production using environmental and historical data. Decision trees and random forests have been used in the identification of pests and plant diseases via the classification of pictures obtained from drones or cameras (Guo et al., 2024). Further, supervised learning facilitates the assessment of soil health by forecasting soil characteristics based on sensor data.

***Case Studies and Examples*:** Research using Random Forests for agricultural yield prediction and pest detection shows significant accuracy in forecasting wheat production based on climatic data, assisting farmers in making educated planting choices (Elbasi et al., 2023). An effective usage of a decision tree algorithm was to identify pests and illnesses using drone imagery, therefore reducing pesticide application. Another good example is research that used support vector machines in predicting soil moisture levels and helping in irrigation scheduling, thus saving a lot of water. These case studies show the potential of using machine learning in agriculture to make production more efficient and profitable while promoting sustainable practices and reducing environmental impact (Saikanth et al., 2023). This combination of new technology with drone and sensor deployment integrated into machine learning algorithms has the potential to transform agricultural practices in a very effective and sustainable manner.

3.2 Unsupervised Learning

Unsupervised learning is a method that instructs a model using unlabeled data, enabling it to discern concealed patterns and structures. Prominent algorithms encompass K-means Clustering, which categorizes data into clusters according to similarity; Hierarchical Clustering, which constructs a hierarchy of clusters; and DBSCAN (Density-Based Spatial Clustering of Applications with Noise), which identifies clusters of arbitrary shapes and detects anomalies in agricultural data (Chaudhry et al., 2023). Methods such as crop classification, anomaly detection, and pattern recognition are applied under unsupervised learning techniques. Through these algorithms, crops will be classified based on growth patterns and environmental parameters, helping in anomaly identification that is, highlighting areas of crop failure because of the lack of nutrients or disease prevalence. Agricultural identification was applied extensively in agricultural surveillance, particularly in classifying satellite photographs by crop type under K-means clustering as classified using spectral data. DBSCAN algorithm successfully identified those areas where crop development was deviating from the normal patterns, enabling farmers to quickly detect pest or disease infestations (Coulibaly, 2024). **Figure 3** illustrates the use of deep learning techniques, particularly Convolutional Neural Networks (CNNs), to detect pests and diseases in crops through image analysis. It shows how cameras mounted on drones or farm vehicles capture images, which are then processed by the CNN to classify pests or diseases. Unsupervised learning methods, like clustering algorithms, are commonly used for crop classification and detecting anomalies in soil conditions. Research emphasizes how these methods enable farmers to uncover patterns in agricultural data, offering insights into field variability and potential problem areas that require intervention (Hoque & Padhiary, 2024).

Figure 3. Application of K-means clustering in crop classification

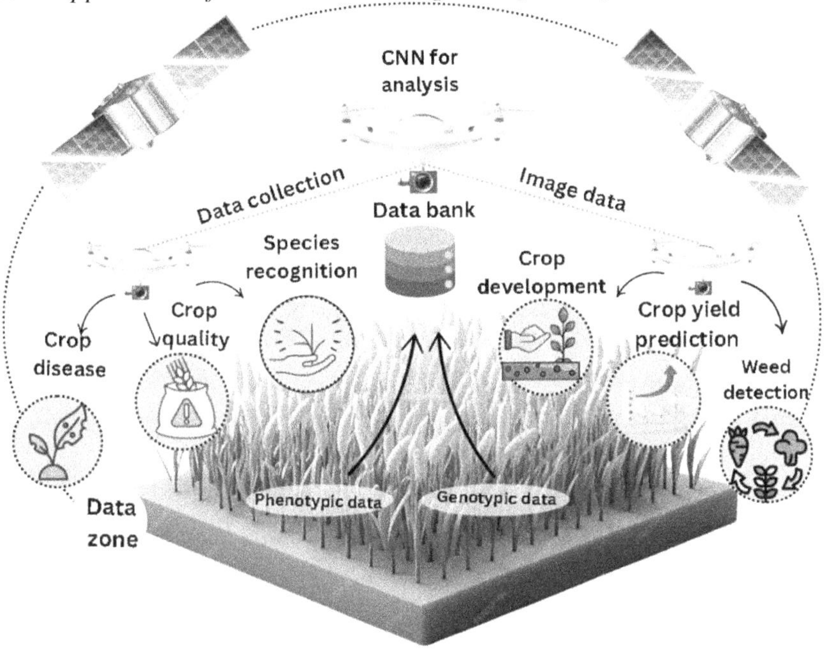

3.3 Reinforcement Learning

Reinforcement learning (RL) is a methodology used in agriculture to instruct agents in decision-making via the reinforcement of favorable results and the penalization of unfavorable ones (Emuna et al., 2020). It is especially beneficial for enhancing sequential decision-making processes, as prompt feedback informs subsequent actions. Reinforcement Learning (RL) techniques include Q-learning, a model-free RL technique used in autonomous systems such as robotic tractors and irrigation systems, as well as Deep Q Networks (DQN), which is an enhancement of Q-learning that utilizes deep neural networks for processing high-dimensional inputs (R. Singh et al., 2023). Actor-critic methods, in which an actor determines the action and a critic assesses it, are effective for continuous action spaces, such as the precise control of robotic arms. These methods are most useful in scenarios where activities are interdependent, and the optimal policy is not known to be beforehand. These strategies may develop or become more effective through continuing assessment and modification of actions guided by feedback. The combination of reinforcement learning approaches with deep learning opens up possibilities for the handling of complex and dynamic situations in developing more advanced and adaptive autonomous systems. The use of reinforcement learning methods in sequential decision-making processes has great promise for different types of progress across several domains such as agriculture and robotics.

Reinforcement learning (RL) is used in automated agricultural systems, autonomous irrigation, and precise spraying. It enhances tractors for effective tilling and sowing, saves water by using the maximum amount in agriculture, and applies pesticides or fertilizers precisely. The deep Q-learning models are enhancing water efficiency by linking weather forecasts with soil moisture sensors (Devarajan et al., 2023). Actor-critic methods train autonomous drones to improve the application of pesticides, minimizing their consumption by focusing only on affected areas. The above examples show the potential of reinforcement learning in many agricultural settings.

3.4 Deep Learning

Deep learning is a kind of machine learning that uses multilayered neural networks to represent intricate connections within data. It specializes in processing large quantities of unstructured data, including photos, audio, and text. Deep learning techniques, such as Convolutional Neural Networks, have proven effective in analyzing imagery from drones and satellites for crop monitoring. These models integrate seamlessly with IoT sensors to provide real-time insights, aiding in the early detection of stress factors like diseases or nutrient deficiencies. Studies demonstrate their potential to revolutionize large-scale agricultural monitoring (Padhiary, 2024; Rabha et al., 2024).

Techniques: Convolutional Neural Networks (CNN) are deep learning algorithms used in agriculture for image analysis, identifying plant illnesses, evaluating crop health, and tracking growth phases (Abade et al., 2021). Recurrent Neural Networks (RNN) are used for sequential data, whilst Long Short-Term Memory (LSTM) is proficient at managing long-term dependencies in time-series data. Generated adversarial networks (GANs) provide synthetic data for agricultural training and environmental simulations(Morales-García et al., 2023). The implementation of deep learning in precision agriculture improves resource utilization and increases agricultural yields while mitigating environmental effects (Debnath et al., 2024). Crop health monitoring and areas of intervention identification through satellite imaging analysis are accomplished. Deep learning in agriculture can transform the sector in terms of efficiency and sustainability of practices. **Figure 4** demonstrates how reinforcement learning (RL) is applied to optimize irrigation systems. The model learns from the environment (soil moisture, weather conditions) and continuously adjusts irrigation levels to minimize water usage while ensuring optimal crop health.

Figure 4. Use of deep learning in image-based crop monitoring

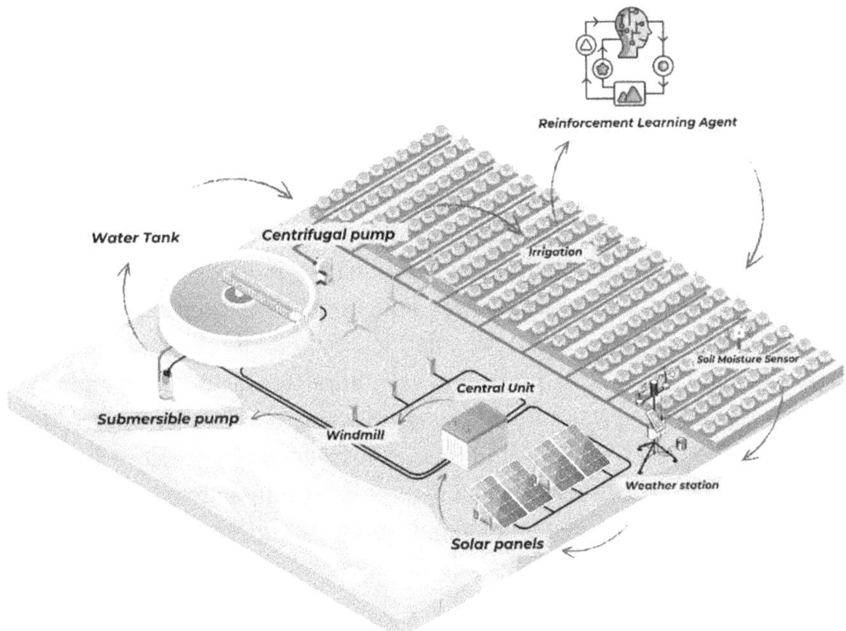

Applications: Deep learning techniques are used in agricultural monitoring, namely for disease diagnosis, crop density estimate, and nutrient deficit identification(Venkatesh & Naik, 2024). These models assist instantaneous decision-making in irrigation, harvesting, and pest control. These developments have transformed farmers' methodologies, enabling more accurate and focused treatments. Farmers may enhance production and diminish environmental impact by optimizing resources and yields with higher precision, thus reducing the usage of water, pesticides, and fertilizers. This is facilitating a more sustainable and efficient future for the industry.

Case Studies and Examples: Convolutional Neural Networks (CNNs) are used in image-based crop monitoring to identify illnesses such as rust, hence enhancing pest control strategies. LSTM networks predict an agricultural growth cycle, which helps farmers in irrigation and harvesting decisions (Balasubramanian & Elangeswaran, 2024). Machine learning algorithms evaluate soil health and nutrient levels, which increase agricultural productivity. By combining sensor and drone data, farmers can make real-time decisions, leading to improved agricultural activities. These innovations are revolutionizing the agriculture industry and paving the way for a sustainable future.

3.5 Transfer Learning

Transfer learning is a technique that involves re-training a pre-existing model for a given job with limited data, particularly advantageous in agriculture where high-quality labeled data is deficient(Yadav et al., 2024). Transfer learning may enhance the precision of predictions with minimum supplementary data by fine-tuning models trained on extensive datasets, such as generic plant disease databases, for particular crops or areas. This methodology is very effective in identifying plant diseases, detecting pests, and analyzing soil quality. A model trained on a varied dataset of plant diseases may be refined using a smaller dataset of specific diseases impacting a particular crop in a designated location, facilitating more precise and effective disease diagnosis in practical agricultural scenarios. The application of pre-trained models in agriculture addresses challenges posed by limited datasets. By leveraging models trained on similar tasks, researchers have successfully applied transfer learning to areas such as plant disease detection and soil quality analysis, significantly reducing the resources required for training new models (Padhiary, Roy, & Dey, 2024).

3.6 Ensemble Learning

Ensemble learning is a methodology that combines many machine learning models to enhance predictive accuracy, especially in intricate agricultural activities where numerous variables affect results (Benos et al., 2021). This technique minimizes the danger of overfitting and improves generalization. Techniques include bagging, where in training numerous iterations of the same model on different subsets of data and amalgamation of their predictions. Progressively increasing training models to correct the mistakes of predecessors while stacking involves training of several models and amalgamation of their predictions by a meta-model (De Zarzà et al., 2023). These methods are highly employed for more accurate agricultural yield estimations, weather predictions, and agricultural risk assessments. The case studies show that ensemble learning improves the predictive accuracy of maize yield models compared to the individual models as well as improves the weather forecast models and thus aids the farmers in making appropriate decisions on sowing and the schedule of irrigation. In **Table 1**, a comparison of various machine learning algorithms that can be utilized in predicting crop yield is presented, highlighting their key features and the equations that govern their operation.

Table 1. Machine learning algorithms and their applications

Algorithm	Equation/Model	Key Characteristics	Application in Precision Agriculture	Reference
Linear regression	$Y = \beta_0 + \beta_1 X_1 + \beta_2 X_2 + \cdots + \beta_n X_n + \epsilon$ Y = Predicted crop yield $X_1, X_2, \ldots, X_n$ = Input features (e.g., soil pH, rainfall, temperature) β_0 = Intercept (baseline value when all inputs are zero) $\beta_1, \beta_2, \ldots, \beta_n$ = Coefficients indicating the impact of each feature ϵ = Error term (accounts for unmodeled factors)	Simple, interpretable model for continuous data	Yield prediction based on weather, soil moisture, etc.	(Murugan et al., 2020)
Decision trees	$f(X) = \sum_{i=1}^{n} I(Xi \in Rj)Cj$ X_i = Input feature values R_j = Regions defined by decision rules (e.g., "if soil pH > 6, then yield = high") C_j = Output prediction for each region	Non-linear, interpretable, overfitting risk	Crop classification, yield prediction	(Alrowais et al., 2024)
Random forest	$f(x) = \frac{1}{M}\sum_{m=1}^{M} T_m(x)$ x = Input feature set (e.g., soil moisture, temperature) M = Total number of decision trees in the ensemble $T_m(x)$= Prediction of the mth decision tree	Ensemble method of multiple decision trees	Accurate yield forecasting, handling large datasets	(Jamei et al., 2022)
Support vector machine (SVM)	$f(x) = \text{sign}(w^T x + b)$ (for classification) x = Feature vector (e.g., image data of leaves) w = Weight vector that defines the separating hyperplane b = Bias term sign () function determines the class label	Maximizes the margin between classes, robust to outliers	Pest detection, crop classification	(Cervantes et al., 2020)
Neural networks	$y = f(Wx + b)$ W = Weight matrix (adjusted during training) x = Input data (e.g., soil characteristics, weather patterns) b = Bias term f = Activation function (e.g., ReLU, sigmoid)	Flexible model capable of learning complex patterns	Crop yield prediction, pest disease detection	(Gao & Zhang, 2020)

continued on following page

Table 1. Continued

Algorithm	Equation/Model	Key Characteristics	Application in Precision Agriculture	Reference
K-nearest neighbors (KNN)	$f(x) = \frac{1}{k}\sum_{i=1}^{k} y_i$ k = Number of nearest neighbours y_i = Output values of the nearest points	Lazy learner, nonparametric, memory based	Crop classification based on spatial proximity	(Levchenko et al., 2021)
Gradient boosting machines (GBM)	$F_m(x) = F_{m_{-1}}(x) + \gamma h_m(x)$ F_m (x) = Updated model at step mmm γ = Learning rate (controls model adjustments) h_m (x) = New weak learner added to improve prediction	Combines multiple weak learners to form a strong model	Yield prediction, disease detection	(Chen et al., 2022)
AdaBoost	$f(x) = \sum_{m=1}^{M} \alpha_m h_m(x)$ α_m = Weight assigned to each weak model h_m (x) = Weak classifiers (e.g., decision trees)	Adaptive boosting method for improving weak classifiers	Improving model accuracy for yield prediction	(Shakeel et al., 2020)
XGBoost	$f(x) = \sum_{m=1}^{M} \alpha_m h_m(x)$ α_m = Weight assigned to each weak model h_m (x) = Weak classifiers (e.g., decision trees)	Gradient boosting with regularization, faster convergence	Efficient yield prediction, large dataset handling	(Li et al., 2022)
Deep neural networks (DNN)	$y = f(W_1 x + b_1)$, where W_1 and b_1 are weights and biases	Complex, multi-layered model that captures nonlinear patterns	Advanced prediction models for crop yield and resource allocation	(He et al., 2023)
Ridge regression	$Y = \beta_0 + \sum_{i=1}^{n} \beta_i X_i$, With $\lambda \sum \beta_i^2$	Regularized linear model for controlling model complexity	Predicting yield based on continuous agricultural data	(Hoerl, 2020)

Machine learning approaches improve crop yield prediction, pest detection, irrigation management, and autonomous farming by analyzing large datasets. Deep learning and reinforcement learning enhance decision-making for real-time applications (Hoque et al., 2025). The integration of these methods optimizes efficiency, reduces resource waste, and promotes sustainable farming.

4. APPLICATIONS OF MACHINE LEARNING IN PRECISION AGRICULTURE

Crop production prediction, pest and disease detection, irrigation management, and autonomous farming systems are all examples of agricultural machine learning applications. This section explores how ML-powered technologies may help farmers optimise resources, reduce losses, and increase overall output.

4.1 Crop Yield Prediction

Machine learning methods, such as regression algorithms, decision trees, and neural networks, are essential for forecasting crop yields by including environmental variables, soil characteristics, and agricultural practices (Chlingaryan et al., 2018b). These simulations help farmers make better decisions about crop management, planting schedules, and resource allocation. Case studies include highly accurate predictions of wheat yields using random forest methodologies; corn yield predictions in the Midwest U.S. using decision trees and support vector machines; and neural networks estimate soybean yields in Brazil based on temperature, precipitation, and soil fertility(Ma et al., 2021; Song et al., 2022). These models have transformed agricultural productivity, making farming methodologies more accurate and efficient. Farmers can optimize resources, reduce risks, and increase yields to meet global food demand through the use of machine learning. Moreover, these advanced technologies may promote sustainable agriculture by reducing waste and environmental impacts. As shown in **Table 2**, different deep-learning techniques and their associated equations are critical in improving pest and disease detection models. These models rely on complex mathematical formulations to classify and detect harmful agents. Crop production prediction has increased according to machine learning models, but practical application still confronts obstacles like data availability, quality, and processing demands. Small-scale farmers sometimes lack large-scale datasets, making it challenging to train accurate models. Transfer learning approaches and collaborative data-sharing platforms may help models modify the local environment (Yang et al., 2022). Small-scale farmers may be unable to meet computational requirements such as high-end GPUs and cloud computing. Lightweight models such as decision trees and random forests may be utilised in place, and mobile-friendly AI applications may provide accessibility (Sajid et al., 2023).

Table 2. Real applications of machine learning in crop yield optimization

Model/Algorithm	Equation	Feature extraction methods/classification layers/ optimization techniques	Explanation	Application in Precision Agriculture	Reference
Convolutional layer	Output"= $f(\sum_k x_k \cdot w_k + b)$	Feature extraction methods	Convolutional operation with filter w_k to extract features from images, f is activation	Feature extraction for pest and disease classification	(Xuan & Shen, 2023)
Pooling layer	$O(i,j)=\max_{m,n} I(i+m,j+n)$ O(i,j) = Output feature map after pooling. $\max_{m,n} I(i+m,j+n)$= Maximum value in a small region of the image (Max Pooling).	Feature extraction methods	Down sampling layer to reduce dimensionality and focus on significant features	Reduces image size and preserves important features	(Zafar et al., 2022)
Fully connected layer	$y = f(W \cdot x + b)$ W = Weights learned by the model. x = Feature vector (output from previous layers). b = Bias term. f = Activation function (e.g., softmax for classification).	classification layers	A dense layer for combining features learned by convolutional layers	Classifying pests or diseases based on image data	(Basha et al., 2020)
Activation function	$\sigma(x) = \frac{1}{1+e^{-x}}$ (Sigmoid) x = Input value (logit score from the neural network). σ(x) = Probability that has plant diseased.		Logistic activation function used to output probabilities for pest/ disease presence	Detecting pests or disease in crops based on image inp ut	(Langer, 2021)
Loss function	$L(y,\hat{y}) = -\sum_{i=1}^{N} y_i \log(\hat{y}_i)$ N = Number of training samples. y_i = Actual label (0 or 1 for disease presence). $\hat{y}_i$=Predicted probability.	optimization techniques	Cross-entropy loss function used for classification tasks	Training CNN models for pest and disease detection	(Rudy & Sapsis, 2023)

continued on following page

Table 2. Continued

Model/Algorithm	Equation	Feature extraction methods/classification layers/ optimization techniques	Explanation	Application in Precision Agriculture	Reference
Batch normalization	$\hat{x}_i = \frac{x_i - \mu}{\sigma}$ x = Input feature value. μ = Mean of the batch. σ = Standard deviation.	Feature extraction methods	Normalizes layer inputs to improve training stability	Enhancing deep learning model stability	(Li et al., 2016)
Dropout	$y = \text{Dropout}(x,p)$	optimization techniques	Randomly drops a percentage of neurons during training to prevent overfitting	Improving CNN model generalization	(Nicolae et al., 2019)
Rectified linear unit (ReLU)	$f(x) = \max(0,x)$ X= Input value.	classification layers	Activation functions that outputs 0 for negative values and passes positive values unchanged	Non-linearity in handling complex pest and disease patterns	(Arora et al., 2018)
SoftMax function	$S(y_i) = \frac{e^{y_i}}{\sum_j e^{y_j}}$ z_i = Logit score for class i.	classification layers	Normalizes the output vector to produce probability distributions for classification	Assigning class probabilities for pest/disease identification	(Bałazy et al., 2023)
Fully convolutional networks (FCN)	$f(x) = \sum_i w_i \cdot f(x_i)$		Convolution-based architecture where each layer produces feature maps instead of scalar values	Segmenting crops and detecting pests or diseases	(Bian et al., 2024)

continued on following page

Table 2. Continued

Model/Algorithm	Equation	Feature extraction methods/classification layers/ optimization techniques	Explanation	Application in Precision Agriculture	Reference
Recurrent neural networks (RNN)	$h_t = \sigma\left(W h_{t-1} + U x_t + b\right)$		Memory-based neural network that captures sequential dependencies	Analyzing time-series data for disease outbreaks	(Sherstinsky, 2020)
Generative adversarial networks (GANs)	$\text{Loss}_D = -\mathbb{E}[\log D(x)] - \mathbb{E}[\log(1 - D(G(z)))]$	optimization techniques	A generator network and a discriminator network that compete to improve image generation	Enhancing synthetic data generation for pest and disease detection	(Navidan et al., 2021)

4.2 Pest and Disease Detection

Machine learning is essential for the early identification of pests and diseases via the analysis of pictures obtained from drones, UAVs, or terrestrial sensors (D. Wang et al., 2022). Image recognition algorithms, including CNNs, are used to categorize insect infestations or detect disease indications on plant foliage (Ngugi et al., 2021). Sensor data, including temperature, humidity, and soil conditions, might provide further insights into pest or disease threats, facilitating proactive management. Advancements in computer vision have led to the development of advanced deep learning models for pest and disease identification, such as Vision Transformers (ViTs) and attention mechanisms (Ding et al., 2024). When combined with UAV surveillance and IoT devices, these models may perform exceptionally well in large-scale crop health assessments. Drones and UAVs are now equipped with high-resolution cameras and sensors in integrated pest control systems. These tools can provide real-time surveillance for crop infestation by pests or disease propagation so that actions can be taken precisely. IoT systems collect data from field sensors to enhance the efficiency of pesticide application. Advanced UAVs may detect diseases in early phases that might not be apparent to the human eye by using multispectral and hyperspectral imagery (Terentev et al., 2022). UAVs equipped with thermal imaging cameras have been used on soybean farms in Brazil to identify pest activity, giving farmers specific solutions and decreasing chemical misuse. Similarly, in India's rice fields, IoT-based traps with real-time image recognition are being used to monitor and classify dangerous insect populations, allowing farmers to take quick preventive measures (Kumaran et al., 2023).Case studies show the effectiveness of deep learning techniques to detect pests and diseases, such as TYLCV, and automated pest detection from drone photography (Kashyap & Kumar, 2021). Advances in technology have improved the effectiveness of pest control and allowed farmers to aim only for the damaged areas. IoT-enabled sprayers that integrate with AI systems have transformed pest and disease management by enabling the targeted application of chemicals. These systems use real-time data to adjust spray patterns, reducing chemical usage while ensuring effective treatment. Recent developments highlight their growing adoption of integrated pest management strategies (Padhiary, Tikute, Saha, et al., 2024).

4.3 Soil Health and Irrigation Management

Machine learning methodologies are used to forecast soil health, irrigation practices, and pest and disease control (G & Bv, 2020). These techniques evaluate data about soil moisture, nutrient concentrations, and pH levels, using methods such as regression analysis, decision trees, and neural networks. The predictions help farmers to tailor their techniques to maximize crop vigor. Machine learning is crucial in precision irrigation and water use optimization because it predicts soil moisture levels and sets ideal schedules for watering (Bwambale et al., 2022). By employing sensor data and meteorological forecasts, machine learning models optimize irrigation, so that crops get the right amount of water at the right time to minimize water wastage and improve agricultural productivity. Reinforcement learning and deep learning have significantly improved smart irrigation systems. RL-based models analyze real-time environmental conditions to optimize water distribution and minimize waste. Deep learning models, like CNNs and LSTMs, process sensor data to improve irrigation efficiency. In California's Central Valley demonstrated the effectiveness of ML-driven irrigation, reducing water usage by 25% and improving crop yield by 12% (Jin et al., 2020). This highlights the potential of ML-based irrigation models in sustainable agriculture, particularly in drought-prone regions. The ability to forecast the optimal harvest time through crop maturity and meteorological conditions can optimize harvesting and storage operations. It also proactively protects crops against harm and decreases reliance on chemical pesticides, thus encouraging sustainable agricultural practices. With data and machine learning, farmers can improve decision-making and overall productivity in their operations. **Table 3** illustrates the essential equations involved in reinforcement learning algorithms that help optimize irrigation practices, balancing water usage with crop health to ensure sustainable farming.

Table 3. Equations and metrics used in ML models for irrigation management

Step/Equation	Equation	Explanation	Application precision Agriculture	Reference
Reward Function	$R(s,a)$= Moisture Level - Water Usage R = Reward assigned to the irrigation action. Moisture Level = Current soil moisture percentage measured by sensors. Water Usage = Amount of water applied during irrigation.	Reward function considers the balance between moisture levels and efficient water usage	Determines the reward for each irrigation action	(X. Ding & Du, 2024)
State Transition Function	$s_{t+1} = f(s_t, a_t)$ s_t = Soil moisture at time t. a_t = Irrigation action taken (amount of water applied). s_t+1 = New soil moisture level after irrigation.	Defines how the state s_t transitions to the next state s_{t+1} based on action a_t	Models the changes in soil moisture due to irrigation	(Üstündağ, 2017)
Q-Value Update (Q-Learning)	$Q(s_t, a_t) = (1 - \alpha)Q(s_t, a_t) + \alpha\left[R(s_t, a_t) + \gamma \max_{a'} Q(s_{t+1}, a')\right]$	Q-value update rule used to estimate the optimal value of actions based on accumulated rewards	Updates action-value estimates for irrigation decisions	(Mei et al., 2023)
Policy Update	$\pi^*(s) = \arg\max_a Q(s,a)$ π*(s) = Policy function that selects the best action for state s. Q(s,a) = Expected reward for taking action a in state s. arg max = Chooses the action with the highest Q-value.	Optimal policy $\pi^*(s)$ chooses the action with the highest Qvalue in each state	Determines the irrigation action that maximizes rewards	(Alibabaei et al., 2022)
Discount factor	$\gamma \in [0,1]$	Controls the weight given to future rewards, balancing short-term vs long-term benefits	Helps prioritize immediate or future irrigation needs	(Hayes et al., 2022)
State representation	s_t= [Soil Moisture, Weather Data, Time of Day]	The state is represented by factors affecting irrigation such as soil moisture, weather, and time	Inputs to the RL system to determine irrigation actions	(Ahmed et al., 2021)

4.4 Precision Fertilization and Resource Allocation

Machine learning is used to improve the application of fertilizers, insecticides, and other resources in precision agriculture (Sharma et al., 2021). Machine learning algorithms may predict fertilizer requirements by analyzing soil quality, meteorological conditions, and crop development phases, leading to enhanced efficiency, diminished waste, and decreased environmental effect. For instance, machine learning algorithms can determine the exact fertilizer needs and disperse pesticides based on real-time data, thereby only treating affected areas (Taseer & Han, 2024). Machine learning can improve irrigation processes by analyzing soil moisture levels and weather forecasts to determine the best irrigation schedule for crops. This not only saves farmers' money but also prevents the local water supply from being depleted and polluted. Machine learning technology in agriculture has the potential to change the industry by making the sector more sustainable and efficient in production. The integration of 3D printing in agriculture has enabled the design of customized components for precision application of fertilizers and pesticides. By using this technology, researchers have developed solutions that optimize resource allocation, reducing waste and improving efficiency on the farm (Padhiary & Roy, 2024).

4.5 Autonomous Agricultural Systems

Machine learning is an essential component in the advancement of autonomous agricultural systems, encompassing vehicles and robots that execute functions such as planting, harvesting, and field maintenance. These systems employ machine learning algorithms for real-time decision-making capabilities, allowing the robots to learn from variations in field conditions such as the type of soil and health of the crop. This aspect of autonomy saves physical labor and ensures effective and uniform performance of tasks. Autonomy in agricultural systems often relies on IoT devices and sensor networks to collect and evaluate real-time data for autonomous vehicle and robots, which then decide according to the prevailing field conditions. Case studies of autonomous agricultural systems include autonomous planting that uses machine learning algorithms to decide the optimal depth and spacing of planting, as well as autonomous harvesting systems using computer vision and machine learning algorithms to identify ripe produce and harvest it without damaging the plants, thus saving on labor costs (Ruangurai et al., 2022; Mohyuddin et al., 2024).

4.7 Real-World Case Studies

Machine learning has shown significant benefits in precision agriculture, with real-world case studies demonstrating its transformative impact. In the U.S. Midwest, ML was used to predict corn yields, resulting in a 10% increase in yield and a 18% reduction in water usage (Sharma et al., 2021b). In China, a deep learning system detected pest infestations with 92% accuracy, reducing pesticide use by 28% and minimizing environmental harm (Shaikh et al., 2022). In Australia, reinforcement learning optimized irrigation, achieving a 27% reduction in water consumption without compromising grape quality (Petrovic et al., 2025). In the Netherlands, autonomous robots with ML algorithms in greenhouses operations, reducing labor costs by 22% and increasing harvesting efficiency by 26% (Singh et al., 2021). In Kenya, a mobile app powered by ML helped smallholder farmers detect maize diseases with 87% accuracy, preventing crop losses by 19% (Mekonnen et al., 2020).

Machine learning is changing agriculture by improving crop yield prediction, pest and disease detection, soil health and irrigation management, autonomous agricultural systems, and precise fertilisation. These integrated technologies maximise resource utilisation, increase efficiency, and encourage sustainability. Automation, predictive analytics, and AI-powered decision-making ensure accuracy in farming operations. Future improvements in IoT, edge AI, and federated learning will improve scalability and resilience in precision agriculture.

5. CHALLENGES AND LIMITATIONS

Data availability, processing demands, interpretability, and farmer training are some of the obstacles which are in the way of ML's wide adoption in agriculture, despite its enormous promise. This section covers the problems and potential solutions to encourage wide adoption.

5.1 Data Quality and Availability Issues

A fundamental problem in the use of ML in precision agriculture is the accessibility and quality of data. Precise and high-caliber data is crucial for training machine learning models that can forecast agricultural production, identify pests, or enhance resource utilization (Akkem et al., 2024). Still, agricultural data often contains noise, incompleteness, or inconsistency, which makes algorithms difficult to give accurate predictions. For example, the soil data may not be available in some areas or environmental data may be erroneous due to improper calibration of sensors. Moreover, acquiring large amounts of information that are required for training complex models is difficult, especially in the small-scale or rural agriculture sector. This would be solved by improving data collection methods, developing standardized data formats, and promoting data sharing among farmers through open platforms, thereby enhancing data quality and accessibility.

5.2 Computational Resources and Infrastructure Limitations

Machine learning methods, particularly in deep learning and reinforcement learning, can need significant computing resources and infrastructure, creating a considerable challenge for small-scale farmers or farms in developing areas. Training models on large data requires strong servers or cloud computing platforms, both of which are expensive (Jauro et al., 2020). Besides, many rural areas lack reliable internet connectivity; thus, cloud-based applications and real-time data processing will be impossible. To overcome this, edge computing solutions might reduce reliance on centralized infrastructure, as they provide local data processing on devices such as sensors or drones. These may be accompanied by efficient, cost-effective models that work on minimal hardware.

5.3 Interpretability and Transparency

Machine learning models, particular deep learning models, have been described as "black boxes," indicating that their decision-making processes are difficult to elucidate (McCoy et al., 2022). An absence of disclosure is a considerable challenge in agriculture, since stakeholders, including farmers, agricultural consultants, and politicians, want clarity on the rationale behind certain decisions or recommendations. For example, if a machine learning algorithm is forecasting an irrigation schedule or an infestation rate, it is important for the farmers to trust the algorithm and understand what it is proposing. Efforts are underway to make machine learning models more interpretable, employing techniques such as explainable artificial intelligence (XAI), which aims to understand complex algorithms in a simple manner so that users can understand what they are doing (Hassija et al., 2024). This again does not solve the problem entirely, especially for advanced models.

5.4 Farmer Adoption and Training Barriers

The effective inclusion of machine learning in agriculture relies not only on technical progress but also on the readiness and aptitude of farmers to embrace new systems. A substantial percentage of farmers, especially in poorer nations or among older demographics, may exhibit reluctance to embrace new technology owing to insufficient technical expertise, unfamiliarity with digital instruments, or apprehensions over implementation costs. In addition, the learning curve for using ML-powered systems might be significant, necessitating that farmers acquire expertise in both technology and data analysis (Rajendiran & Rethnaraj, 2024). Agricultural extension services and training programs must be established to educate farmers about the functionality of machine learning models and their potential advantages. Interactions between agricultural technology firms and local agricultural organizations might reduce these obstacles and promote wider use. Agriculture's environmental footprint, particularly its impact on soil and water ecosystems, remains a pressing challenge. Studies underline the importance of integrating sustainable practices and monitoring systems to mitigate these effects while maintaining productivity (Padhiary & Kumar, 2024a).

5.5 Environmental, Social and Ethical Considerations

Although machine learning may provide considerable advantages to agricultural production and sustainability, several environmental and ethical issues need attention. The large installation of sensors and processing capabilities required in machine learning models may lead to increased electronic waste and environmental degradation if not carefully managed (Hussain et al., 2020). The application of pesticides or fertilizers by ML systems is to be cautiously controlled to ensure that there are no overutilization or wrong applications that adversely affect the environment and biodiversity. Concerns on ethical issues of privacy and ownership over data also arise, especially when farmers share intimate information with other parties. Clear laws and regulations are very important to ensure that there is responsible use of data retrieved from farms and that farmers should remain in control of their data. Sustainable practices need to be encouraged to ensure that the use of technology does not compromise health conditions in the environment (Hariram et al., 2023). The integration of machine learning in precision agriculture raises ethical and social concerns, including data privacy and bias in ML models. Data ownership, security, and transparency are crucial, as farmers often share sensitive information (Gyamfi et al., 2024). Bias in ML models may lead to inequitable outcomes, especially in developing regions. The use of ML in pesticide and fertilizer application may harm ecosystems. To address these issues, clear data privacy policies, inclusive data collection, model transparency, sustainable practices, and alternative livelihood opportunities are needed (Tolentino-Zondervan et al., 2023).

5.6 Economic Impact on Small-Scale Farmers

Machine learning in agriculture offers economic benefits, especially for small-scale farmers. However, initial implementation may be expensive due to the need for IoT-enabled devices, sensors, drones, and computing infrastructure. Despite these challenges, ML may optimize resource allocation, increase crop yields, and reduce input costs (Kaur et al., 2024). Long-term economic benefits include reduced water consumption, pesticide costs, and informed crop selection decisions. Government subsidies, cooperative-based ML implementation, and mobile-based advisory platforms may help bridge the gap. Low-cost AI models and federated learning approaches may make ML solutions more accessible.

Machine learning in agriculture faces challenges like data quality, computational limitations, model interpretability, and farmer adoption barriers. Limited access to high-quality data and advanced infrastructure hinders wide use implementation. Ethical concerns, including data privacy and environmental impact, must be addressed for responsible AI use. Overcoming these challenges through edge computing, federated learning, and farmer training will drive broader adoption and sustainability.

6. FUTURE DIRECTIONS

Emerging techniques like federated learning, edge AI, and enhanced integration with IoT and robotics offer the solution to the future of machine learning in agriculture. This section discusses upcoming trends, research fields, and ML's potential to change methods of agriculture on a larger scale.

6.1 Emerging ML Techniques in Agriculture

As machine learning advances, novel methodologies like federated learning and edge AI will likely greatly impact the future of precision agriculture. Decentralized federated learning wherein models are updated while retaining the data locally, in devices like cellphones or sensors, and tractors; this would alleviate some data privacy and access concerns for farmers (Chaterji et al., 2020). Using this approach means that even very complex models could be deployed safely, by avoiding exposing sensitive information while using data so that less dependence is maintained on the existence of a massive centralized warehouse of data. Edge AI involves the local processing of data on devices located at the network's edge, rather than relying on cloud computing(R. Singh & Gill, 2023). This kind of approach is particularly beneficial in remote areas with limited internet connectivity and can support real-time decision-making. The integration of these cutting-edge technologies will make ML systems more efficient, cost-effective, and scalable, allowing them to be deployed across a wider range of agricultural environments.

6.2 Integration with Other Technologies

The future of precision agriculture depends on the seamless integration of machine learning with new technologies such as the IoT, drones, robots, and big data analytics. The Internet of Things allows ongoing collecting data from many sensors installed in the environment, including soil moisture sensors, meteorological stations, and crop health monitors (Tzounis et al., 2017). Machine learning algorithms can process the data in real-time to yield actionable insights, thereby enabling farmers to make informed decisions on the use of irrigation, pest control, and fertilizers. Drones and robots will further optimize the process by taking over some of the duties such as crop monitoring, pest identification, and harvesting, which are directed by AI-driven models (Thangamani et al., 2024). The extensive data coupled with the sophisticated data analytics will provide comprehensive datasets suitable for training more precise and efficient machine learning models. The combination will result in a robust ecosystem that can optimize every aspect of agriculture-from sowing to harvest-and encourage more sustainable farming practices. Emerging advancements like edge AI and federated learning offer promising solutions to challenges associated with centralized data processing in agriculture. These technologies enable distributed decision-making, enhancing the scalability and resilience of precision farming systems.

6.3 Potential for Improving the Sustainability and Scalability of ML-Based Solutions

Machine learning may substantially improve agricultural sustainability by decreasing resource use, reducing waste, and maximizing farming methodologies. Machine learning models may thus reduce the overuse of fertilizers and pesticides, leading to less environmental pollution and stronger ecosystems, as it improves the accuracy of agricultural output predictions (Wen et al., 2022). Machine learning can also be used to enhance irrigation, thereby ensuring better water use with reduced waste. Another crucial aspect of the widespread adoption of machine learning-based solutions is their scalability. The availability of cheap, effective models and technologies for operating in a wide array of agricultural contexts-from small family farms to large industrial operations machine learning to be applied to the needs of both developed and developing regions. By making these technologies more cost-effective, machine learning could help small-scale farmers and large agribusinesses both achieve greater production with decreased environmental impacts (Mizik,

2023).The use of ML in conjunction with IoT and autonomous machinery continues to optimize farm operations and resource use, making farming more efficient and sustainable. Innovations in this area are paving the way for fully automated systems that reduce labor costs and improve productivity.

6.4 Future Research Areas and Trends

The prospects of machine learning in optimizing agricultural productivity have significant potential, including numerous critical areas for study and development. One research area improves the accuracy and the applicability of machine learning models for various geographies and crop varieties. Though current models work well in specific scenarios, the need is to be able to develop systems that work better over many climates, soils, and types of crops. The compilation of multi-modal data, including satellite imagery, sensor data, and historical agricultural information, might enhance the predictive capabilities of machine learning models, leading to more accurate yield forecasts (Shamsuddin et al., 2024). Furthermore, there is a growing trend of incorporating climate change variables into machine learning models to help farmers cope with changing weather patterns. Research in this area may lead to stronger agricultural management plans that include the changing environmental variables. Another relevant area of study pertains to the use of generative models, specifically Generative Adversarial Networks, for simulating crop development in various situations. This will help farmers to virtually test various agricultural approaches before they are implemented in real life. **Figure 5** illustrates the integration of machine learning with drone and sensor technologies to monitor crop health in real time. It shows how data from drones equipped with multispectral cameras and ground sensors are analyzed by ML models to identify potential issues such as nutrient deficiencies, diseases, or pests. The convergence of AI, IoT, and robotics is expected to play a crucial role in transforming urban and peri-urban farming systems. This integration will enable efficient use of limited space and resources, fostering sustainable agricultural practices in densely populated areas.

Figure 5. Integrated system of ML, IoT, and drones in precision agriculture

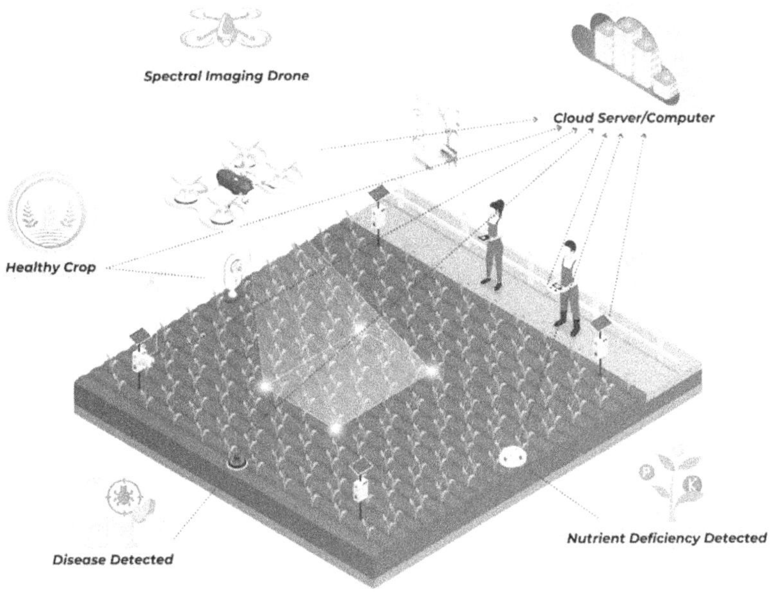

Machine learning future in agriculture lies in federated learning, edge AI, IoT integration, and autonomous systems. These advancements will enhance real-time decision-making, scalability, and sustainability in farming. Improved model accuracy, climate adaptation, and AI-driven automation will further optimize agricultural efficiency. Continued research and innovation will drive precision farming toward higher productivity and resilience.

7. CONCLUSION

This chapter investigates the function of machine learning in precision agriculture, emphasizing its capacity to enhance crop yield optimization, pest identification, and resource management. Machine learning approaches, including supervised and unsupervised learning, reinforcement learning, deep learning, and ensemble methods, demonstrate the capacity to improve production, reduction of resource leakage, and promote sustainability. Machine learning has been used in several applications, including estimating agricultural output based on environmental and soil data, automating irrigation processes, and precision fertilization. Practical applications illustrate the efficacy of these technologies in tackling difficulties encountered by farmers, including maximizing the utilization of water consumption, the detection of insect outbreaks, and the enhancement of soil health. The chapter highlights the significance of integrating machine learning with new technologies like as the Internet of Things, drones, and robots to provide a resilient data-collecting ecosystem. Still, obstacles such as data quality, processing resources, and model interpretability impede the extensive use of machine learning in agriculture. The future of machine learning in precision agriculture is very promising since emerging approaches such as federated learning and edge AI provide novel opportunities for enhancing scalability, privacy, and efficiency. The combination of these technologies with big data analytics, drones, robots, and IoT will enhance the possibilities of machine learning in agriculture.

REFERENCES

Abade, A., Ferreira, P. A., & De Barros Vidal, F. (2021). Plant diseases recognition on images using convolutional neural networks: A systematic review. *Computers and Electronics in Agriculture, 185*, 106125. DOI: 10.1016/j.compag.2021.106125

Ahmed, A. A. M., Deo, R. C., Raj, N., Ghahramani, A., Feng, Q., Yin, Z., & Yang, L. (2021). Deep Learning Forecasts of Soil Moisture: Convolutional Neural Network and Gated Recurrent Unit Models Coupled with Satellite-Derived MODIS, Observations and Synoptic-Scale Climate Index Data. *Remote Sensing (Basel), 13*(4), 554. DOI: 10.3390/rs13040554

Akkem, Y., Biswas, S. K., & Varanasi, A. (2024). Streamlit-based enhancing crop recommendation systems with advanced explainable artificial intelligence for smart farming. *Neural Computing & Applications, 36*(32), 20011–20025. DOI: 10.1007/s00521-024-10208-z

Alibabaei, K., Gaspar, P. D., Assunção, E., Alirezazadeh, S., & Lima, T. M. (2022). Irrigation optimization with a deep reinforcement learning model: Case study on a site in Portugal. *Agricultural Water Management, 263*, 107480. DOI: 10.1016/j.agwat.2022.107480

Alrowais, F., Jamjoom, A. A., Karamti, H., Umer, M., Alsubai, S., Abate, A. F., & Ashraf, I. (2024). Automated approach to predict cerebral stroke based on fuzzy inference and convolutional neural network. *Multimedia Tools and Applications*. Advance online publication. DOI: 10.1007/s11042-024-19254-0

Araújo, S. O., Peres, R. S., Ramalho, J. C., Lidon, F., & Barata, J. (2023). Machine Learning Applications in Agriculture: Current Trends, Challenges, and Future Perspectives. *Agronomy (Basel), 13*(12), 2976. DOI: 10.3390/agronomy13122976

Arora, R., Basu, A., Mianjy, P., & Mukherjee, A. (2018). *Understanding Deep Neural Networks with Rectified Linear Units* (arXiv:1611.01491). arXiv. https://doi.org//arXiv.1611.01491DOI: 10.48550

Badji, A., Machida, L., Kwemoi, D. B., Kumi, F., Okii, D., Mwila, N., Agbahoungba, S., Ibanda, A., Bararyenya, A., Nghituwamhata, S. N., Odong, T., Wasswa, P., Otim, M., Ochwo-Ssemakula, M., Talwana, H., Asea, G., Kyamanywa, S., & Rubaihayo, P. (2020). Factors Influencing Genomic Prediction Accuracies of Tropical Maize Resistance to Fall Armyworm and Weevils. *Plants, 10*(1), 29. DOI: 10.3390/plants10010029 PMID: 33374402

Balasubramanian, A., & Elangeswaran, S. V. J. (2024). A novel power aware smart agriculture management system based on RNN-LSTM. *Electrical Engineering*. Advance online publication. DOI: 10.1007/s00202-024-02640-0

Bałazy, K., Struski, Ł., Śmieja, M., & Tabor, J. (2023). r-softmax: Generalized Softmax with Controllable Sparsity Rate. In J. Mikyška, C. De Mulatier, M. Paszynski, V. V. Krzhizhanovskaya, J. J. Dongarra, & P. M. A. Sloot (Eds.), *Computational Science – ICCS 2023* (Vol. 14074, pp. 137–145). Springer Nature Switzerland. DOI: 10.1007/978-3-031-36021-3_11

Basha, S. H. S., Dubey, S. R., Pulabaigari, V., & Mukherjee, S. (2020). Impact of fully connected layers on performance of convolutional neural networks for image classification. *Neurocomputing*, *378*, 112–119. DOI: 10.1016/j.neucom.2019.10.008

Behmann, J., Mahlein, A.-K., Rumpf, T., Römer, C., & Plümer, L. (2015). A review of advanced machine learning methods for the detection of biotic stress in precision crop protection. *Precision Agriculture*, *16*(3), 239–260. DOI: 10.1007/s11119-014-9372-7

Benos, L., Tagarakis, A. C., Dolias, G., Berruto, R., Kateris, D., & Bochtis, D. (2021). Machine Learning in Agriculture: A Comprehensive Updated Review. *Sensors (Basel)*, *21*(11), 3758. DOI: 10.3390/s21113758 PMID: 34071553

Bian, J., Kang, M., Xi, Y., Wang, Y., & Zi, W. (2024). A Review of the Measurement and Control Technologies for the Critical Parameters of Microwave Drying Processes: Temperature and Humidity. *Food and Bioprocess Technology*, •••, 1–24. DOI: 10.1007/s11947-024-03614-7

Bwambale, E., Abagale, F. K., & Anornu, G. K. (2022). Smart irrigation monitoring and control strategies for improving water use efficiency in precision agriculture: A review. *Agricultural Water Management*, *260*, 107324. DOI: 10.1016/j.agwat.2021.107324

Cervantes, J., Garcia-Lamont, F., Rodríguez-Mazahua, L., & Lopez, A. (2020). A comprehensive survey on support vector machine classification: Applications, challenges and trends. *Neurocomputing*, *408*, 189–215. DOI: 10.1016/j.neucom.2019.10.118

Chaterji, S., Naghizadeh, P., Alam, M. A., Bagchi, S., Chiang, M., Corman, D., Henz, B., Jana, S., Li, N., Mou, S., Oishi, M., Peng, C., Rompf, T., Sabharwal, A., Sundaram, S., Weimer, J., & Weller, J. (2020). *Resilient Cyberphysical Systems and their Application Drivers: A Technology Roadmap* (Version 1). arXiv. DOI: 10.48550/ARXIV.2001.00090

Chaudhry, M., Shafi, I., Mahnoor, M., Vargas, D. L. R., Thompson, E. B., & Ashraf, I. (2023). A Systematic Literature Review on Identifying Patterns Using Unsupervised Clustering Algorithms: A Data Mining Perspective. *Symmetry, 15*(9), 1679. DOI: 10.3390/sym15091679

Chen, H., Shen, Z., Wang, L., Qi, C., & Tian, Y. (2022). Prediction of undrained failure envelopes of skirted circular foundations using gradient boosting machine algorithm. *Ocean Engineering, 258*, 111767. DOI: 10.1016/j.oceaneng.2022.111767

Chlingaryan, A., Sukkarieh, S., & Whelan, B. (2018a). Machine learning approaches for crop yield prediction and nitrogen status estimation in precision agriculture: A review. *Computers and Electronics in Agriculture, 151*, 61–69. DOI: 10.1016/j.compag.2018.05.012

Chlingaryan, A., Sukkarieh, S., & Whelan, B. (2018b). Machine learning approaches for crop yield prediction and nitrogen status estimation in precision agriculture: A review. *Computers and Electronics in Agriculture, 151*, 61–69. DOI: 10.1016/j.compag.2018.05.012

Coulibaly, A. (2024). *Smart Farming: Computer Simulation and Predictive Model for Cassava*. DOI: 10.20381/RUOR-30362

De Zarzà, I., De Curtò, J., Hernández-Orallo, E., & Calafate, C. T. (2023). Cascading and Ensemble Techniques in Deep Learning. *Electronics (Basel), 12*(15), 3354. DOI: 10.3390/electronics12153354

Debnath, J., Kumar, K., Roy, K., Choudhury, R. D., & U, A. K. P. (2024). Precision Agriculture: A Review of AI Vision and Machine Learning in Soil, Water, and Conservation Practice. *International Journal for Research in Applied Science and Engineering Technology, 12*(12), 2130–2141. DOI: 10.22214/ijraset.2024.66166

Devarajan, G. G., & Nagarajan, S. M., T.V., R., T., V., Ghosh, U., & Alnumay, W. (. (2023). DDNSAS: Deep reinforcement learning based deep Q-learning network for smart agriculture system. *Sustainable Computing : Informatics and Systems, 39*, 100890. DOI: 10.1016/j.suscom.2023.100890

Ding, W., Abdel-Basset, M., Alrashdi, I., & Hawash, H. (2024). Next generation of computer vision for plant disease monitoring in precision agriculture: A contemporary survey, taxonomy, experiments, and future direction. *Information Sciences, 665*, 120338. DOI: 10.1016/j.ins.2024.120338

Ding, X., & Du, W. (2024). Optimizing Irrigation Efficiency using Deep Reinforcement Learning in the Field. *ACM Transactions on Sensor Networks, 20*(4), 1–34. DOI: 10.1145/3662182

Elavarasan, D., Vincent, D. R., Sharma, V., Zomaya, A. Y., & Srinivasan, K. (2018). Forecasting yield by integrating agrarian factors and machine learning models: A survey. *Computers and Electronics in Agriculture, 155*, 257–282. DOI: 10.1016/j.compag.2018.10.024

Elbasi, E., Zaki, C., Topcu, A. E., Abdelbaki, W., Zreikat, A. I., Cina, E., Shdefat, A., & Saker, L. (2023). Crop Prediction Model Using Machine Learning Algorithms. *Applied Sciences (Basel, Switzerland), 13*(16), 9288. DOI: 10.3390/app13169288

Emuna, R., Borowsky, A., & Biess, A. (2020). *Deep Reinforcement Learning for Human-Like Driving Policies in Collision Avoidance Tasks of Self-Driving Cars* (Version 2). arXiv. DOI: 10.48550/ARXIV.2006.04218

G, U., & Bv, G. (2020). A survey on plant disease prediction using machine learning and deep learning techniques. *Inteligencia Artificial, 23*(65), 136–154. DOI: 10.4114/intartif.vol23iss65pp136-154

Gao, F., & Zhang, B. (2020). *A Use of Even Activation Functions in Neural Networks* (arXiv:2011.11713). arXiv. https://doi.org//arXiv.2011.11713DOI: 10.48550

Guo, H., Cheng, Y., Liu, J., & Wang, Z. (2024). Low-cost and precise traditional Chinese medicinal tree pest and disease monitoring using UAV RGB image only. *Scientific Reports, 14*(1), 25562. DOI: 10.1038/s41598-024-76502-x PMID: 39462013

Gyamfi, E. K., Kropczynski, J., Johnson, J. S., & Yakubu, M. A. (2024). Internet of Things Security and Data Privacy Concerns in Smart Farming. *2024 IEEE World AI IoT Congress (AIIoT)*, 0575–0583. DOI: 10.1109/AIIoT61789.2024.10578972

Hariram, N. P., Mekha, K. B., Suganthan, V., & Sudhakar, K. (2023). Sustainalism: An Integrated Socio-Economic-Environmental Model to Address Sustainable Development and Sustainability. *Sustainability (Basel), 15*(13), 10682. DOI: 10.3390/su151310682

Hassija, V., Chamola, V., Mahapatra, A., Singal, A., Goel, D., Huang, K., Scardapane, S., Spinelli, I., Mahmud, M., & Hussain, A. (2024). Interpreting Black-Box Models: A Review on Explainable Artificial Intelligence. *Cognitive Computation, 16*(1), 45–74. DOI: 10.1007/s12559-023-10179-8

Hayes, C. F., Rădulescu, R., Bargiacchi, E., Källström, J., Macfarlane, M., Reymond, M., Verstraeten, T., Zintgraf, L. M., Dazeley, R., Heintz, F., Howley, E., Irissappane, A. A., Mannion, P., Nowé, A., Ramos, G., Restelli, M., Vamplew, P., & Roijers, D. M. (2022). A practical guide to multi-objective reinforcement learning and planning. *Autonomous Agents and Multi-Agent Systems, 36*(1), 26. DOI: 10.1007/s10458-022-09552-y

He, Y., Zhang, L., Chen, Z., & Li, C. Y. (2023). A framework of structural damage detection for civil structures using a combined multi-scale convolutional neural network and echo state network. *Engineering with Computers*, *39*(3), 1771–1789. DOI: 10.1007/s00366-021-01584-4

Hoerl, R. W. (2020). Ridge Regression: A Historical Context. *Technometrics*, *62*(4), 420–425. DOI: 10.1080/00401706.2020.1742207

Hoque, A. (2024). Artificial Intelligence in Post-Harvest Drying Technologies: A Comprehensive Review on Optimization, Quality Enhancement, and Energy Efficiency. [IJSR]. *International Journal of Science and Research (Raipur, India)*, *13*(11), 493–502. DOI: 10.21275/SR241107163717

Hoque, A., & Padhiary, M. (2024). Automation and AI in Precision Agriculture: Innovations for Enhanced Crop Management and Sustainability. *Asian Journal of Research in Computer Science*, *17*(10), 95–109. DOI: 10.9734/ajrcos/2024/v17i10512

Hoque, A., Padhiary, M., Prasad, G., & Kumar, K. (2025). Real-Time Data Processing in Agricultural Robotics. In Bora, D. J., & Bania, R. K. (Eds.), *Computer Vision Techniques for Agricultural Advancements* (pp. 431–468). IGI Global., DOI: 10.4018/979-8-3693-8019-2.ch014

Hussain, A., Draz, U., Ali, T., Tariq, S., Irfan, M., Glowacz, A., Antonino Daviu, J. A., Yasin, S., & Rahman, S. (2020). Waste Management and Prediction of Air Pollutants Using IoT and Machine Learning Approach. *Energies*, *13*(15), 3930. DOI: 10.3390/en13153930

Imade, A., Ounacer, S., El Ghoumari, M. Y., Ardchir, S., & Azzouazi, M. (2024). Agricultural Yield Prediction Using ML Algorithms in the Industry 5.0. In Chakir, A., Bansal, R., & Azzouazi, M. (Eds.), *Industry 5.0 and Emerging Technologies* (Vol. 565, pp. 135–157). Springer Nature Switzerland., DOI: 10.1007/978-3-031-70996-8_7

Jamei, M., Karbasi, M., Malik, A., Jamei, M., Kisi, O., & Yaseen, Z. M. (2022). Long-term multi-step ahead forecasting of root zone soil moisture in different climates: Novel ensemble-based complementary data-intelligent paradigms. *Agricultural Water Management*, *269*, 107679. DOI: 10.1016/j.agwat.2022.107679

Jauro, F., Chiroma, H., Gital, A. Y., Almutairi, M., Abdulhamid, S. M., & Abawajy, J. H. (2020). Deep learning architectures in emerging cloud computing architectures: Recent development, challenges and next research trend. *Applied Soft Computing*, *96*, 106582. DOI: 10.1016/j.asoc.2020.106582

Jiang, H., Peng, M., Zhong, Y., Xie, H., Hao, Z., Lin, J., Ma, X., & Hu, X. (2022). A Survey on Deep Learning-Based Change Detection from High-Resolution Remote Sensing Images. *Remote Sensing (Basel), 14*(7), 1552. DOI: 10.3390/rs14071552

Jin, Y., Chen, B., Lampinen, B. D., & Brown, P. H. (2020). Advancing Agricultural Production With Machine Learning Analytics: Yield Determinants for California's Almond Orchards. *Frontiers in Plant Science, 11*, 290. DOI: 10.3389/fpls.2020.00290 PMID: 32231679

Joshi, A., Pradhan, B., Gite, S., & Chakraborty, S. (2023). Remote-Sensing Data and Deep-Learning Techniques in Crop Mapping and Yield Prediction: A Systematic Review. *Remote Sensing (Basel), 15*(8), 2014. DOI: 10.3390/rs15082014

Kashyap, B., & Kumar, R. (2021). Sensing Methodologies in Agriculture for Monitoring Biotic Stress in Plants Due to Pathogens and Pests. *Inventions (Basel, Switzerland), 6*(2), 29. DOI: 10.3390/inventions6020029

Kaur, G., Rajni, , & Sivia, J. S. (2024). Integrating Data Envelopment Analysis and Machine Learning Approaches for Energy Optimization, Decreased Carbon Footprints, and Wheat Yield Prediction Across North-Western India. *Journal of Soil Science and Plant Nutrition, 24*(1), 1424–1447. DOI: 10.1007/s42729-024-01647-7

Keating, B. A., & Thorburn, P. J. (2018). Modelling crops and cropping systems—Evolving purpose, practice and prospects. *European Journal of Agronomy, 100*, 163–176. DOI: 10.1016/j.eja.2018.04.007

Kumar, D., Kumar, K., Roy, P., & Rabha, G. (2024). Renewable Energy in Agriculture: Enhancing Aquaculture and Post-Harvest Technologies with Solar and AI Integration. *Asian Journal of Research in Computer Science, 17*(12), 201–219. DOI: 10.9734/ajrcos/2024/v17i12539

Kumaran, S., D, S. K., A, S., & K, S. K. (2023). Crop Pest Control using IoT Real-Time Monitoring and Intervention. *2023 International Conference on Sustainable Communication Networks and Application (ICSCNA)*, 1552–1557. DOI: 10.1109/ICSCNA58489.2023.10370176

Kuppusamy, P., Suresh, J. K., & Shanmugananthan, S. (2023). Machine Learning-Enabled Internet of Things Solution for Smart Agriculture Operations. In Goel, N., & Yadav, R. K. (Eds.), (pp. 84–115). Advances in Systems Analysis, Software Engineering, and High Performance Computing. IGI Global., DOI: 10.4018/978-1-6684-8785-3.ch005

Langer, S. (2021). Approximating smooth functions by deep neural networks with sigmoid activation function. *Journal of Multivariate Analysis*, *182*, 104696. DOI: 10.1016/j.jmva.2020.104696

Levchenko, O., Kolev, B., Yagoubi, D.-E., Akbarinia, R., Masseglia, F., Palpanas, T., Shasha, D., & Valduriez, P. (2021). BestNeighbor: Efficient evaluation of kNN queries on large time series databases. *Knowledge and Information Systems*, *63*(2), 349–378. DOI: 10.1007/s10115-020-01518-4

Li, Y., Wang, N., Shi, J., Liu, J., & Hou, X. (2016). *Revisiting Batch Normalization For Practical Domain Adaptation* (arXiv:1603.04779). arXiv. https://doi.org//arXiv .1603.04779DOI: 10.48550

Li, Y., Xu, Y., Ma, Z., Ye, Y., Gao, L., & Sun, Y. (2022). An XGBoost-based model for assessment of aortic stiffness from wrist photoplethysmogram. *Computer Methods and Programs in Biomedicine*, *226*, 107128. DOI: 10.1016/j.cmpb.2022.107128 PMID: 36150230

Ma, Y., Zhang, Z., Kang, Y., & Özdoğan, M. (2021). Corn yield prediction and uncertainty analysis based on remotely sensed variables using a Bayesian neural network approach. *Remote Sensing of Environment*, *259*, 112408. DOI: 10.1016/j.rse.2021.112408

McCoy, L. G., Brenna, C. T. A., Chen, S. S., Vold, K., & Das, S. (2022). Believing in black boxes: Machine learning for healthcare does not need explainability to be evidence-based. *Journal of Clinical Epidemiology*, *142*, 252–257. DOI: 10.1016/j.jclinepi.2021.11.001 PMID: 34748907

Mei, Z., Vatsavai, R., & Chirkova, R. (2023). Q-learning Based Simulation Tool for Studying Effectiveness of Dynamic Application of Fertilizer on Crop Productivity. *Proceedings of the 11th ACM SIGSPATIAL International Workshop on Analytics for Big Geospatial Data*, 13–22. DOI: 10.1145/3615833.3628591

Mekonnen, Y., Namuduri, S., Burton, L., Sarwat, A., & Bhansali, S. (2020). Review—Machine Learning Techniques in Wireless Sensor Network Based Precision Agriculture. *Journal of the Electrochemical Society*, *167*(3), 037522. DOI: 10.1149/2.0222003JES

Mizik, T. (2023). How can precision farming work on a small scale? A systematic literature review. *Precision Agriculture*, *24*(1), 384–406. DOI: 10.1007/s11119-022-09934-y

Mohyuddin, G., Khan, M. A., Haseeb, A., Mahpara, S., Waseem, M., & Saleh, A. M. (2024). Evaluation of Machine Learning Approaches for Precision Farming in Smart Agriculture System: A Comprehensive Review. *IEEE Access : Practical Innovations, Open Solutions*, *12*, 60155–60184. DOI: 10.1109/ACCESS.2024.3390581

Morales-García, J., Bueno-Crespo, A., Terroso-Sáenz, F., Arcas-Túnez, F., Martínez-España, R., & Cecilia, J. M. (2023). Evaluation of synthetic data generation for intelligent climate control in greenhouses. *Applied Intelligence*, *53*(21), 24765–24781. DOI: 10.1007/s10489-023-04783-2

Murugan, R., Thomas, F. S., Geetha Shree, G., Glory, S., & Shilpa, A. (2020). Linear Regression Approach to Predict Crop Yield. *Asian Journal of Computer Science and Technology*, *9*(1), 40–44. DOI: 10.51983/ajcst-2020.9.1.2152

Navidan, H., Moshiri, P. F., Nabati, M., Shahbazian, R., Ghorashi, S. A., Shah-Mansouri, V., & Windridge, D. (2021). Generative Adversarial Networks (GANs) in networking: A comprehensive survey & evaluation. *Computer Networks*, *194*, 108149. DOI: 10.1016/j.comnet.2021.108149

Ngugi, L. C., Abelwahab, M., & Abo-Zahhad, M. (2021). Recent advances in image processing techniques for automated leaf pest and disease recognition – A review. *Information Processing in Agriculture*, *8*(1), 27–51. DOI: 10.1016/j.inpa.2020.04.004

Nicolae, R. A., Constantinescu, A. M., Costea, A. C., Girbea, G., Constantinescu, C. M., Neagu, E., & Barbarii, L. E. (2019). A rare genetic gender anomaly identified in a paternity case presenting AMEL-Y dropout. *Forensic Science International. Genetics Supplement Series*, *7*(1), 555–556. DOI: 10.1016/j.fsigss.2019.10.088

Padhiary, M. (2024). The Convergence of Deep Learning, IoT, Sensors, and Farm Machinery in Agriculture. In *Designing Sustainable Internet of Things Solutions for Smart Industries* (pp. 109–142). IGI Global., DOI: 10.4018/979-8-3693-5498-8.ch005

Padhiary, M., Barbhuiya, J. A., Roy, D., & Roy, P. (2024). 3D Printing Applications in Smart Farming and Food Processing. *Smart Agricultural Technology*, *100553*, 100553. Advance online publication. DOI: 10.1016/j.atech.2024.100553

Padhiary, M., & Kumar, R. (2024a). Assessing the Environmental Impacts of Agriculture, Industrial Operations, and Mining on Agro-Ecosystems. In Azrour, M., Mabrouki, J., Alabdulatif, A., Guezzaz, A., & Amounas, F. (Eds.), *Smart Internet of Things for Environment and Healthcare* (Vol. 1165, pp. 107–126). Springer Nature Switzerland., DOI: 10.1007/978-3-031-70102-3_8

Padhiary, M., & Kumar, R. (2024b). Enhancing Agriculture Through AI Vision and Machine Learning: The Evolution of Smart Farming. In Thangam, D. (Ed.), (pp. 295–324). Advances in Computational Intelligence and Robotics. IGI Global., DOI: 10.4018/979-8-3693-5380-6.ch012

Padhiary, M., Kumar, R., & Sethi, L. N. (2024). *Navigating the Future of Agriculture: A Comprehensive Review of Automatic All-Terrain Vehicles in Precision Farming. Journal of The Institution of Engineers*. Series A., DOI: 10.1007/s40030-024-00816-2

Padhiary, M., Roy, D., & Dey, P. (2024). Mapping the Landscape of Biogenic Nanoparticles in Bioinformatics and Nanobiotechnology: AI-Driven Insights. In Das, S., Khade, S. M., Roy, D. B., & Trivedi, K. (Eds.), *Synthesizing and Characterizing Plant-Mediated Biocompatible Metal Nanoparticles* (pp. 337–376). IGI Global., DOI: 10.4018/979-8-3693-6240-2.ch014

Padhiary, M., & Roy, P. (2024). Advancements in Precision Agriculture: Exploring the Role of 3D Printing in Designing All-Terrain Vehicles for Farming Applications. [IJSR]. *International Journal of Science and Research (Raipur, India)*, *13*(5), 861–868. DOI: 10.21275/SR24511105508

Padhiary, M., Roy, P., Dey, P., & Sahu, B. (2024). Harnessing AI for Automated Decision-Making in Farm Machinery and Operations: Optimizing Agriculture. In Hai-Jew, S. (Ed.), (pp. 249–282). Advances in Computational Intelligence and Robotics. IGI Global., DOI: 10.4018/979-8-3693-6230-3.ch008

Padhiary, M., Tikute, S. V., Saha, D., Barbhuiya, J. A., & Sethi, L. N. (2024). Development of an IOT-Based Semi-Autonomous Vehicle Sprayer. *Agricultural Research*. Advance online publication. DOI: 10.1007/s40003-024-00760-4

Petrovic, B., Kononets, Y., & Csambalik, L. (2025). Adoption of drone, sensor, and robotic technologies in organic farming systems of Visegrad countries. *Heliyon*, *11*(1), e41408. DOI: 10.1016/j.heliyon.2024.e41408 PMID: 39958722

Rabha, G., Kumar, K., Kumar, D., & Kumar, D. (2024). A Comprehensive Review of Integrating AI and IoT in Farm Machinery: Advancements, Applications, and Sustainability. *International Journal of Research and Analytical Reviews*, *11*(4).

Rajendiran, G., & Rethnaraj, J. (2024). IoT-Integrated Machine Learning-Based Automated Precision Agriculture-Indoor Farming Techniques. In Nguyen, T. V. T., & Vo, N. T. M. (Eds.), (pp. 289–317). Advances in Media, Entertainment, and the Arts. IGI Global., DOI: 10.4018/979-8-3693-0639-0.ch013

Ruangurai, P., Dailey, M. N., Ekpanyapong, M., & Soni, P. (2022). Optimal vision-based guidance row locating for autonomous agricultural machines. *Precision Agriculture, 23*(4), 1205–1225. DOI: 10.1007/s11119-022-09881-8

Rudy, S. H., & Sapsis, T. P. (2023). Output-weighted and relative entropy loss functions for deep learning precursors of extreme events. *Physica D. Nonlinear Phenomena, 443*, 133570. DOI: 10.1016/j.physd.2022.133570

Sabir, R. M., Mehmood, K., Sarwar, A., Safdar, M., Muhammad, N. E., Gul, N., Athar, F., Majeed, M. D., Sattar, J., Khan, Z., Fatima, M., Hussain, M. A., & Akram, H. M. B. (2024). Remote Sensing and Precision Agriculture: A Sustainable Future. In Kanga, S., Singh, S. K., Shevkani, K., Pathak, V., & Sajan, B. (Eds.), *Transforming Agricultural Management for a Sustainable Future* (pp. 75–103). Springer Nature Switzerland., DOI: 10.1007/978-3-031-63430-7_4

Sahu, B. (2024). Artificial Intelligence and Automation in Smart Agriculture: A Comprehensive Review of Precision Farming, All-Terrain Vehicles, IoT Innovations, and Environmental Impact Mitigation. [IJSR]. *International Journal of Science and Research (Raipur, India), 13*(11), 656–665. DOI: 10.21275/SR241110184009

Saikanth, D. R. K., Supriya, , Singh, B. V., Rai, A. K., Bana, S. R., Sachan, D. S., & Singh, B. (2023). Advancing Sustainable Agriculture: A Comprehensive Review for Optimizing Food Production and Environmental Conservation. *International Journal of Plant and Soil Science, 35*(16), 417–425. DOI: 10.9734/ijpss/2023/v35i163169

Sajid, M. Z., Qureshi, I., Abbas, Q., Albathan, M., Shaheed, K., Youssef, A., Ferdous, S., & Hussain, A. (2023). Mobile-HR: An Ophthalmologic-Based Classification System for Diagnosis of Hypertensive Retinopathy Using Optimized MobileNet Architecture. *Diagnostics (Basel), 13*(8), 1439. DOI: 10.3390/diagnostics13081439 PMID: 37189539

Saleem, M. H., Potgieter, J., & Arif, K. M. (2021). Automation in Agriculture by Machine and Deep Learning Techniques: A Review of Recent Developments. *Precision Agriculture, 22*(6), 2053–2091. DOI: 10.1007/s11119-021-09806-x

Sanyaolu, M., & Sadowski, A. (2024). The Role of Precision Agriculture Technologies in Enhancing Sustainable Agriculture. *Sustainability (Basel), 16*(15), 6668. DOI: 10.3390/su16156668

Shaikh, T. A., Rasool, T., & Lone, F. R. (2022). Towards leveraging the role of machine learning and artificial intelligence in precision agriculture and smart farming. *Computers and Electronics in Agriculture, 198*, Shaikh, Tawseef Ayoub, Tabasum Rasool, and Faisal Rasheed Lone. "Towards Leveraging the Role of Machine Learning and Artificial Intelligence in Precision Agriculture and Smart Farming." Computers and Electronics in Agriculture 198 (July 2022): 107119. . https://doi .org/DOI: 10.1016/j.compag.2022.107119

Shakeel, P. M., Tolba, A., Al-Makhadmeh, Z., & Jaber, M. M. (2020). RETRACTED ARTICLE: Automatic detection of lung cancer from biomedical data set using discrete AdaBoost optimized ensemble learning generalized neural networks. *Neural Computing & Applications*, *32*(3), 777–790. DOI: 10.1007/s00521-018-03972-2

Shamsuddin, D., Danilevicz, M. F., Al-Mamun, H. A., Bennamoun, M., & Edwards, D. (2024). Multimodal Deep Learning Integration of Image, Weather, and Phenotypic Data Under Temporal Effects for Early Prediction of Maize Yield. *Remote Sensing (Basel)*, *16*(21), 4043. DOI: 10.3390/rs16214043

Sharma, A., Jain, A., Gupta, P., & Chowdary, V. (2021a). Machine Learning Applications for Precision Agriculture: A Comprehensive Review. *IEEE Access : Practical Innovations, Open Solutions*, *9*, 4843–4873. DOI: 10.1109/ACCESS.2020.3048415

Sharma, A., Jain, A., Gupta, P., & Chowdary, V. (2021b). Machine Learning Applications for Precision Agriculture: A Comprehensive Review. *IEEE Access : Practical Innovations, Open Solutions*, *9*, 4843–4873. DOI: 10.1109/ACCESS.2020.3048415

Sharma, V., Tripathi, A. K., & Mittal, H. (2022). Technological revolutions in smart farming: Current trends, challenges & future directions. *Computers and Electronics in Agriculture*, *201*, 107217. DOI: 10.1016/j.compag.2022.107217

Sherstinsky, A. (2020). Fundamentals of Recurrent Neural Network (RNN) and Long Short-Term Memory (LSTM) network. *Physica D. Nonlinear Phenomena*, *404*, 132306. DOI: 10.1016/j.physd.2019.132306

Singh, R., & Gill, S. S. (2023). Edge AI: A survey. *Internet of Things and Cyber-Physical Systems*, *3*, 71–92. DOI: 10.1016/j.iotcps.2023.02.004

Singh, R., Kurukuru, V., & Khan, M. (2023). Advanced Power Converters and Learning in Diverse Robotic Innovation: A Review. *Energies*, *16*(20), 7156. DOI: 10.3390/en16207156

Singh, R. K., Berkvens, R., & Weyn, M. (2021). AgriFusion: An Architecture for IoT and Emerging Technologies Based on a Precision Agriculture Survey. *IEEE Access : Practical Innovations, Open Solutions*, *9*, 136253–136283. DOI: 10.1109/ACCESS.2021.3116814

Sishodia, R. P., Ray, R. L., & Singh, S. K. (2020). Applications of Remote Sensing in Precision Agriculture: A Review. *Remote Sensing (Basel)*, *12*(19), 3136. DOI: 10.3390/rs12193136

Song, X.-P., Li, H., Potapov, P., & Hansen, M. C. (2022). Annual 30 m soybean yield mapping in Brazil using long-term satellite observations, climate data and machine learning. *Agricultural and Forest Meteorology*, *326*, 109186. DOI: 10.1016/j.agrformet.2022.109186

Taseer, A., & Han, X. (2024). Advancements in variable rate spraying for precise spray requirements in precision agriculture using Unmanned aerial spraying Systems: A review. *Computers and Electronics in Agriculture*, *219*, 108841. DOI: 10.1016/j.compag.2024.108841

Terentev, A., Dolzhenko, V., Fedotov, A., & Eremenko, D. (2022). Current State of Hyperspectral Remote Sensing for Early Plant Disease Detection: A Review. *Sensors (Basel)*, *22*(3), 757. DOI: 10.3390/s22030757 PMID: 35161504

Thangamani, R., Suguna, R. K., & Kamalam, G. K. (2024). Drones and Autonomous Robotics Incorporating Computational Intelligence. In Prakash, K. B., Peddapelli, S. K., Tam, I. C. K., Woo, W. L., & Jain, V. (Eds.), *Computational Intelligent Techniques in Mechatronics* (1st ed., pp. 243–296). Wiley., DOI: 10.1002/9781394175437.ch8

Tolentino-Zondervan, F., Ngoc, P. T. A., & Roskam, J. L. (2023). Use cases and future prospects of blockchain applications in global fishery and aquaculture value chains. *Aquaculture (Amsterdam, Netherlands)*, *565*, 739158. DOI: 10.1016/j.aquaculture.2022.739158

Tzounis, A., Katsoulas, N., Bartzanas, T., & Kittas, C. (2017). Internet of Things in agriculture, recent advances and future challenges. *Biosystems Engineering*, *164*, 31–48. DOI: 10.1016/j.biosystemseng.2017.09.007

Üstündağ, B. (2017). An adaptive Mealy machine model for monitoring crop status. *Journal of Integrative Agriculture*, *16*(2), 252–265. DOI: 10.1016/S2095-3119(16)61430-2

Venkatesh, K., & Naik, K. J. (2024). An ensemble transfer learning for nutrient deficiency identification and yield-loss prediction in crop. *Multimedia Tools and Applications*, *83*(32), 78535–78561. DOI: 10.1007/s11042-024-18592-3

Wang, D., Cao, W., Zhang, F., Li, Z., Xu, S., & Wu, X. (2022). A Review of Deep Learning in Multiscale Agricultural Sensing. *Remote Sensing (Basel), 14*(3), 559. DOI: 10.3390/rs14030559

Wang, Y., Zhang, Z., Feng, L., Du, Q., & Runge, T. (2020). Combining Multi-Source Data and Machine Learning Approaches to Predict Winter Wheat Yield in the Conterminous United States. *Remote Sensing (Basel), 12*(8), 1232. DOI: 10.3390/rs12081232

Wu, J., & Yang, H. (2015). Linear Regression-Based Efficient SVM Learning for Large-Scale Classification. *IEEE Transactions on Neural Networks and Learning Systems, 26*(10), 2357–2369. DOI: 10.1109/TNNLS.2014.2382123 PMID: 25576581

Xiao Wen, E. L., Kang, H. M., & Vistro, D. M. (2022). The Application of Machine Learning in Agriculture Sustainability: A Review. *International Journal of Data Science and Advanced Analytics, 4*(4), 66–70. DOI: 10.69511/ijdsaa.v4i4.94

Xuan, A., & Shen, L. (2023). Reconstruction of three-dimensional turbulent flow structures using surface measurements for free-surface flows based on a convolutional neural network. *Journal of Fluid Mechanics, 959*, A34. DOI: 10.1017/jfm.2023.154

Yadav, P., Rishiwal, V., Yadav, M., Alotaibi, A., Maurya, V., Agarwal, U., & Sharma, S. (2024). Investigation and Empirical Analysis of Transfer Learning for Industrial IoT Networks. *IEEE Access : Practical Innovations, Open Solutions, 12*, 173351–173379. DOI: 10.1109/ACCESS.2024.3499741

Yang, C., Wang, Y., Lan, S., Wang, L., Shen, W., & Huang, G. Q. (2022). Cloud-edge-device collaboration mechanisms of deep learning models for smart robots in mass personalization. *Robotics and Computer-integrated Manufacturing, 77*, 102351. DOI: 10.1016/j.rcim.2022.102351

Zafar, A., Aamir, M., Mohd Nawi, N., Arshad, A., Riaz, S., Alruban, A., Dutta, A. K., & Almotairi, S. (2022). A Comparison of Pooling Methods for Convolutional Neural Networks. *Applied Sciences (Basel, Switzerland), 12*(17), 8643. DOI: 10.3390/app12178643

Chapter 8
Biomanufacturing:
The Role of AI in Bioprocess Optimization

B. Nithya
B.S. Abdur Rahman Crescent Institute of Science and Technology, India

A. Allwyn Sundarraj
Sri Shakthi Institute of Engineering and Technology, India

C. P. Jetlin
https://orcid.org/0000-0001-8346-991X
St. Joseph's Institute of Technology, India

P. Nehru
https://orcid.org/0009-0002-5689-2792
B.S. Abdur Rahman Crescent Institute of Science and Technology, India

G. David Raj
https://orcid.org/0000-0002-8309-3657
B.S. Abdur Rahman Crescent Institute of Science and Technology, India

P. Selvakumar
https://orcid.org/0000-0002-3650-4548
Department of Science and Humanities, Nehru Institute of Technology, Coimbatore, India

ABSTRACT

Biomanufacturing is a critical field that harnesses biological systems to produce a wide array of products, from pharmaceuticals and biofuels to food ingredients and materials. As this industry grows, so does the complexity of the processes involved. The integration of artificial intelligence (AI) in bioprocess optimization is revolutionizing the way these processes are managed and refined, leading to significant advancements in efficiency, cost-effectiveness, and innovation. This paragraph explores how AI technologies are transforming biomanufacturing, focusing on the key areas where AI contributes to process optimization, data analysis, and overall improvements in the field.At the heart of biomanufacturing are complex biological processes that require precise control and monitoring to ensure product quality and yield. Traditionally, this control has been achieved through manual oversight and

DOI: 10.4018/979-8-3693-9208-9.ch008

heuristic-based methods. However, these approaches often struggle to keep pace with the intricacies and volatilities of biological systems.

INTRODUCTION

Biomanufacturing is a rapidly evolving field that utilizes biological systems to produce a diverse range of products, including pharmaceuticals, biofuels, food ingredients, and advanced materials. As industry expands, the complexity of bioprocesses increases, necessitating more efficient and reliable control mechanisms. The integration of Artificial Intelligence (AI) in bioprocess optimization is transforming these processes, driving significant improvements in efficiency, cost-effectiveness, and innovation. This section explores how AI technologies contribute to process optimization, data analysis, and operational improvements in biomanufacturing.

Traditionally, bioprocess control relied on manual oversight and heuristic-based methods, which often struggled to keep pace with the dynamic nature of biological systems. AI algorithms, particularly machine learning models, have revolutionized this aspect by training on historical process data to predict outcomes, optimize parameters, and adjust operations in real-time. This proactive approach enhances process stability, reduces variability, and ensures consistent product quality.

One significant application of AI in biomanufacturing is the optimization of bioprocess parameters. Bioprocesses involve numerous interdependent variables, such as temperature, pH, oxygen levels, and nutrient concentrations. Balancing these factors to maximize yield and quality traditionally required extensive experimentation. AI-driven techniques, such as reinforcement learning and surrogate modeling, streamline this process. Reinforcement learning algorithms iteratively explore parameter combinations to identify optimal settings, while surrogate models approximate complex process behaviors, enabling rapid scenario evaluation.

For instance, machine learning models can detect deviations from standard operating conditions, signaling potential equipment malfunctions or process disturbances. These predictive capabilities provide early warnings and actionable insights, helping maintain process stability and minimize downtime—critical for high-quality production and cost reduction.

AI also plays a pivotal role in the design and development phases of biomanufacturing. Techniques like Design of Experiments (DoE) and process scale-up, traditionally labor-intensive tasks, benefit greatly from AI-based simulation and optimization tools. Digital twins, for example, create virtual replicas of bioprocesses, allowing researchers to test and refine parameters in a simulated environment before implementation. This approach accelerates development, reduces costly errors, and enhances process robustness.

Beyond technical applications, AI-driven analytics support strategic decision-making in biomanufacturing. By analyzing market trends, customer preferences, and competitive landscapes, these tools help identify new product opportunities, optimize supply chains, and improve customer engagement. As noted by Antonakoudis et al. (2020), leveraging AI for strategic insights enables biomanufacturers to align operations with market demands and maintain a competitive edge.

However, the successful adoption of AI in biomanufacturing presents challenges. The industry generates vast amounts of complex, heterogeneous data that must be accurately captured, processed, and managed. Additionally, implementing AI solutions requires a workforce proficient in both bioprocess engineering and data science. Addressing these challenges demands investments in robust data infrastructure, cross-disciplinary training programs, and a culture of continuous innovation.

The transformative impact of AI on large-scale bioprocess optimization is evident in recent research. For example, Emara, Trinh, and Huang (2023) demonstrated the importance of geographically distributed data management in supporting extensive data analysis for biomanufacturing applications. Their work highlighted how AI can efficiently handle and process data from multiple locations, improving operational efficiency. Similarly, VuThi et al. (2024) introduced a top-frequency-based feature selection method, enhancing the predictive accuracy of models by identifying the most relevant variables in large datasets.

Further advancements in AI for scalable data processing have been reported by Trinh et al. (2021), who proposed a stratified block model to manage large-scale data with small clusters—an essential technique for resource-constrained environments. Additionally, TrinhHuy et al. (2025) explored event scale prediction in event-based social networks, offering potential applications in forecasting key stages in bioprocess workflows. Innovations in power-saving mechanisms, as discussed by Emara (2016), Emara et al., (2014), and Emara (2019), further contribute to enhancing the energy efficiency of AI-driven bioprocessing systems.

AI is redefining biomanufacturing by optimizing processes, improving monitoring capabilities, and accelerating design and development efforts. The integration of advanced AI models, such as machine learning algorithms and digital twins, enables more precise control over bioprocess parameters, enhancing productivity and sustainability. As research continues to advance, the biomanufacturing industry is poised to benefit from even more sophisticated AI-driven solutions, laying the foundation for intelligent, adaptive, and efficient production systems.

AI FUNDAMENTALS FOR BIOMANUFACTURING: MACHINE LEARNING AND DEEP LEARNING

AI is increasingly influencing biomanufacturing, revolutionizing how processes are designed, optimized, and controlled. At the core of AI's impact on this field are machine learning (ML) and deep learning (DL) technologies, which offer advanced tools for analyzing complex data, enhancing process efficiency, and driving innovation. Understanding these AI fundamentals is crucial for leveraging their potential in biomanufacturing, where the ability to handle and interpret large datasets can significantly enhance process management, product quality, and overall operational effectiveness. For instance, ML algorithms can predict the optimal conditions for cell culture growth by analyzing data from previous experiments, enabling more precise control.

A common application of ML in biomanufacturing is predictive modeling, where algorithms are trained to forecast outcomes based on input variables. For example, ML models can predict the yield of bioprocess based on variables such as the concentration of raw materials and environmental conditions. These predictions can guide adjustments in real-time to ensure that the process remains within optimal parameters. Another important application is anomaly detection, where ML algorithms monitor process data to identify deviations from expected patterns. Early detection of anomalies, such as deviations in fermentation conditions, allows for timely interventions to prevent product quality issues and reduce downtime (Bradford et al., 2018).

Deep learning, a more advanced branch of ML, model complex and abstract patterns in data. DL techniques are particularly useful for tasks that require high-dimensional data analysis, such as image analysis and sequence data interpretation. In biomanufacturing, DL models can be employed for tasks like image-based monitoring of cell cultures, analyzing microscope images to assess cell health and growth. This capability allows for automated and highly accurate monitoring of cell cultures, which is essential for maintaining quality and consistency in bio-manufactured products.

The integration of ML and DL into biomanufacturing processes requires a robust data infrastructure and careful consideration of data quality. High-quality, comprehensive data is essential for training accurate and reliable models. This involves not only collecting relevant data but also ensuring its consistency and accuracy, which are crucial for preparing data for ML and DL applications. Additionally, the implementation of these technologies is bioprocess engineering (Carbonell et al., 2019). Despite their potential, ML and DL technologies also present challenges. The complexity of biomanufacturing processes means that models must be carefully validated and tested to ensure they provide reliable results that must be addressed through techniques such as cross-validation and regularization.

In conclusion, machine learning and deep learning are foundational technologies that are transforming biomanufacturing by enhancing process optimization, monitoring, and control. ML models improve predictive capabilities and anomaly detection, while DL techniques offer advanced tools for analyzing complex data such as images and time-series. The successful application of these technologies requires high-quality data, skilled professionals, and careful model validation. As AI continues to evolve, its integration into biomanufacturing will likely lead to further advancements, driving greater efficiency, innovation, and quality in the production of bio-manufactured products. Bioprocess modeling and simulation are crucial components of modern biomanufacturing, providing essential tools for understanding, designing, and optimizing complex biological processes transforming bioprocess modeling and simulation by enabling more precise predictions, optimized designs, and real-time adjustments (Choudhury et al., 2022).

Bioprocess modeling involves creating mathematical and computational representations of biological systems to predict their behavior under various conditions. These models can simulate aspects such as growth rates, product yields, and metabolic pathways. Traditional modeling approaches often rely on simplified assumptions and require extensive empirical data to calibrate and validate the models.

One of the primary benefits of AI in bioprocess modeling is the ability to handle and integrate diverse data types. Bioprocesses generate complex data sets from various sources, including sensor measurements, experimental results, and operational parameters. ML algorithms can assimilate this heterogeneous data, creating comprehensive models that reflect the intricate interactions within the bioprocess. For example, supervised learning techniques can be employed to develop predictive models that forecast product yields based on input variables such as nutrient concentrations and environmental conditions. These models can continuously improve as they learn from new data, leading to more refined and accurate predictions for modeling intricate biological processes (Coşgun et al., 2022).

By leveraging these advanced DL techniques, bioprocess models can become more sophisticated, capturing the temporal and spatial nuances of biological systems. Simulation is a key application of bioprocess modeling, enabling virtual experimentation with different process scenarios to identify optimal conditions and design parameters. AI enhances simulation capabilities by providing advanced optimization techniques that can explore a vast parameter space more efficiently than traditional methods. Reinforcement learning, for example, is an AI technique where algorithms learn to make decisions through trial and error, optimizing process parameters based on feedback from simulations. This approach allows for the exploration of complex, multidimensional parameter spaces and the identification of optimal settings that may not be evident through conventional trial-and-error methods.

This real-time capability is invaluable for optimizing process performance, as it allows for immediate responses to deviations and enhances the overall control of bioprocess (Del Rio-Chanona et al., 2016). Despite the advantages, integrating AI into bioprocess modeling and simulation presents challenges. One of the main issues is ensuring data quality and consistency. AI models require high-quality, well-structured data to produce accurate and reliable predictions. Understanding how AI models arrive at their predictions is essential for validating their results and gaining confidence in their recommendations.

Moreover, the successful application of AI in bioprocess modeling and simulation requires a multidisciplinary approach. Combining expertise in bioprocess engineering with skills in data science and AI is crucial for developing effective models and simulations. Collaborative efforts between domain experts and data scientists can lead to more accurate models and better integration of AI technologies into biomanufacturing processes. AI enhances modeling accuracy, optimizes design parameters, and enables real-time simulation and adjustments.

PROCESS ANALYTICAL TECHNOLOGY (PAT): INTEGRATING AI FOR REAL-TIME MONITORING

By integrating AI into PAT, biomanufacturers can enhance their ability to monitor processes dynamically, predict outcomes, and optimize performance. This integration is revolutionizing the field, improving accuracy, and enabling more responsive and adaptive process control. PAT involves the use of various analytical techniques and tools to monitor critical process parameters and quality attributes in real-time. While these tools provide valuable information, they often generate large volumes of data that need to be interpreted and acted upon efficiently (Ebrahimpour et al., 2008).

Machine learning, for instance, can be employed to develop predictive models that anticipate how changes in process parameters will affect the final product quality. By training historical process data, ML algorithms can identify relationships between process conditions and product outcomes. This enables real-time predictions of product quality and process deviations, allowing for proactive adjustments. For example, if an ML model predicts that a deviation in temperature might lead to a suboptimal product, operators can take corrective actions before the issue affects the final product. This predictive capability enhances process control and reduces the risk of quality issues, leading to more consistent and reliable production outcomes.

For instance, CNNs can process data from imaging technologies, such as high-resolution cameras used to monitor cell cultures or particle size distributions, to provide detailed insights into the physical characteristics of the process. RNNs and Long Short-Term Memory (LSTM) networks excel in analyzing time-series data, allowing for continuous monitoring of dynamic processes and the identification of trends and anomalies over time. The integration of AI with PAT also enables advanced anomaly detection and fault diagnosis. By continuously analyzing data streams from various sensors (Faure et al., 2022).

For example, if an ML model detects a sudden shift in the data that deviates from historical norms, it can trigger an alert or initiate an automated response. This capability not only helps in maintaining process stability but also reduces the time needed for manual inspections and interventions. Another significant advantage of AI in PAT is its ability to facilitate real-time process optimization. Traditional optimization methods often involve extensive experimentation and adjustments based on historical data. In contrast, AI-driven approaches enable continuous optimization by adjusting process parameters in real-time based on current data. Reinforcement learning, for instance, can be used to explore different process configurations and determine the best settings for achieving desired outcomes. This dynamic optimization ensures that the process remains efficient and aligned with quality goals, adapting to changes in raw materials, environmental conditions, or other variables.

Despite these benefits, integrating AI into PAT major challenge, comprehensive data Inaccurate or incomplete unreliable Additionally, complexity can pose difficulties. Understanding how AI models make predictions and ensuring that their recommendations are actionable and aligned with process objectives is crucial for effective implementation (Freier et al.,2016). Moreover, combining expertise in bioprocess engineering, data science, for developing accurate models, interpreting results, and implementing AI-driven solutions effectively. Investing in training and skill development for personnel working with AI technologies is also important to ensure that they can leverage these tools to their full potential.

In Process Analytical Technology significant biomanufacturing, offering enhanced capabilities for real-time monitoring, prediction, and optimization. valuable tools for analyzing complex data, detecting anomalies, and optimizing process parameters. While challenges remain in terms of data quality and model interpretability, the continued development and application of AI in PAT hold the promise of improving process control, product quality, and overall efficiency in biomanufacturing. As AI technologies evolve, their integration into PAT systems advances in, driving the biomanufacturing.

Fermentation is a cornerstone of biomanufacturing, used to produce a variety of products including pharmaceuticals, biofuels, and food ingredients. Traditional methods for optimizing these parameters often involve extensive trial-and-error experimentation, which can be time-consuming and costly. AI offers a more efficient alternative by utilizing predictive models and real-time data analysis to identify optimal conditions and predict outcomes. Machine learning algorithms, particularly supervised learning models, are trained on historical fermentation data to understand the relationships between process parameters and product yield or quality. These models can then make accurate predictions about how adjustments in parameters will impact the fermentation process, enabling more informed and targeted optimization strategies (King et al., 2017).

By analyzing large datasets from previous fermentations, these models can identify optimal ranges for various parameters, helping to maximize product yield and maintain high quality. Additionally, these models can continuously learn from new data, improving their predictions and recommendations over time, which ensures that the fermentation process remains optimized as conditions evolve.

This capability allows for detailed monitoring and analysis of the fermentation process, providing insights into factors like cell morphology and aggregation that can impact product quality. These models can track the temporal evolution of key metrics, such as metabolite concentrations, and predict how changes will affect the fermentation process and product outcome.AI also plays a significant role in real-time process control and optimization. By integrating real-time data from sensors with AI models, biomanufacturers can implement adaptive control systems that automatically adjust process parameters based on current conditions. Reinforcement learning, a branch of AI where algorithms learn to make decisions through trial and error, can be particularly effective in this context.

Reinforcement learning algorithms explore various parameter settings and learn which adjustments lead to the best outcomes, continuously optimizing the fermentation process. This dynamic optimization capability ensures that the process remains within optimal conditions, even in the face of fluctuations in raw materials or environmental factors. Moreover, AI-driven fermentation optimization can enhance process scalability. Transitioning from small-scale laboratory fermentations to large-scale production involves scaling up various process parameters while maintaining product quality. AI models can be used to predict how scaling changes will affect the fermentation process, helping to identify potential issues and optimize conditions before scaling up (Kumar et al., 2014). This predictive capability reduces the risk of scale-up failures and ensures that the fermentation process remains efficient and consistent at larger scales.

Understanding how AI models arrive at their recommendations and ensuring that they align with process objectives is crucial for effective implementation. Another challenge is the need for interdisciplinary expertise. Successful integration of AI into fermentation optimization requires collaboration between bioprocess engineers, data scientists, and AI experts. This multidisciplinary approach ensures that AI models are developed and applied effectively, taking into account both the technical aspects of fermentation and the nuances of data science. In conclusion (Li et al., 2019).

Machine learning and deep learning techniques enhance the ability to predict outcomes, optimize process parameters, and implement real-time control. While challenges related to data quality, model interpretability, and interdisciplinary collaboration remain, the continued development and application of AI technologies hold the promise of further innovations in fermentation optimization. As AI continues to evolve, its integration into fermentation processes will likely lead to more efficient, cost-effective, and high-quality production, driving the future of biomanufacturing.

CELL CULTURE OPTIMIZATION: USING AI TO IMPROVE CELL GROWTH AND PRODUCTIVITY

Cell culture is a fundamental aspect of biotechnology and biomanufacturing, integral to the production of biologics, vaccines, and other therapeutic products. Optimizing cell culture processes is critical for enhancing cell growth, maximizing productivity, and ensuring high-quality output. Recent advancements in AI have revolutionized cell culture optimization by providing sophisticated tools for analyzing complex data, predicting outcomes, and fine-tuning processes with unprecedented precision. AI technologies, including ML and DL, are significantly improving cell culture practices by enabling more accurate control and optimization of growth conditions. At its core, cell culture optimization involves managing and adjusting various parameters, such as temperature, pH, oxygen levels, nutrient concentrations, and cell density, to create an ideal environment for cell growth and productivity (Liu et al., 2017).

For example, supervised learning models, such as regression and classification algorithms, can predict how changes in nutrient concentrations or environmental conditions will affect cell growth and productivity. By analyzing past experiments, these models can recommend optimal conditions and adjustments, leading to improved cell culture outcomes. Deep learning further enhances cell culture optimization by leveraging its ability to handle high-dimensional and complex data. Convolutional neural networks (CNNs) are particularly useful for analyzing image-based data, such as those obtained from microscopy or imaging sensors that monitor cell morphology and behavior. CNNs can detect subtle changes in cell morphology or aggregation that may impact growth and productivity are well-suited for modeling time-series data from cell cultures. These models can track the temporal evolution of key metrics, such as cell density and metabolite concentrations, and predict how changes over time will affect overall productivity.

AI also enables advanced predictive analytics and real-time monitoring, which are crucial for optimizing cell culture processes. Predictive models, developed through ML techniques, can forecast future cell culture performance based on current and historical data. For instance, these models can predict how changes in process parameters will influence cell growth rates and product yields, allowing for proactive adjustments. Real-time monitoring, combined with AI-driven analytics, facilitates continuous optimization by providing actionable insights and alerts about deviations from optimal conditions. Reinforcement learning, a branch of AI focused on decision-making through trial and error, can be particularly effective in this context. Reinforcement learning algorithms explore various process configurations and learn which adjustments lead to the best outcomes, continuously refining the optimization process (Mahadev Madgule et al., 2023).

Understanding how AI models arrive at their recommendations and ensuring that they align with process objectives is crucial for effective implementation. Ensuring that AI-driven insights are actionable and can be seamlessly integrated into existing workflows is also important for realizing the benefits of these technologies.

Another challenge is the need for interdisciplinary expertise. Successful integration of AI into cell culture optimization requires collaboration between cell culture scientists, data scientists, and AI specialists. This multidisciplinary approach ensures that AI models are developed and applied effectively, taking into account both the technical aspects of cell culture and the nuances of data science. Investment in training and skill development for personnel working with AI technologies is also essential for maximizing their potential (Oyetunde et al., 2019).

In conclusion, AI-driven cell culture optimization represents a significant advancement in biomanufacturing, offering powerful tools for improving cell growth and productivity. As AI continues to evolve, its integration into cell culture processes will likely lead to more efficient, cost-effective, and high-quality biomanufacturing, driving the future of biotechnology and therapeutic production. Downstream processing is a critical phase in biomanufacturing that involves the purification and recovery of target products from complex mixtures, typically following fermentation or cell culture. This stage is crucial for ensuring product purity, quality, and yield, and it often encompasses processes such as chromatography and filtration. Optimizing downstream processing can be challenging due to the complexity of the processes and the variability in feedstock composition. Recent advancements in AI are transforming downstream processing by enhancing the efficiency and effectiveness of chromatography and filtration techniques, leading to improved product recovery and reduced processing costs.

AI offers a more systematic and efficient solution by leveraging ML algorithms to analyse and optimize chromatography processes (Pappu et al., 2017). can predict how changes in buffer conditions or flow rates will impact the resolution and yield of the target product. By analysing data from previous chromatographic runs, these models can provide recommendations for optimal operating conditions, reducing the need for extensive experimental trials. Additionally, ML models can continuously learn from new data, allowing for adaptive optimization and real-time adjustments based on current process conditions. Deep learning further enhances chromatography optimization by handling high-dimensional and complex data. Convolutional neural networks (CNNs) can be used to analyse chromatographic data from various sources, such as spectral or image-based data, to extract features that are indicative of separation performance. These models can track the evolution of peak shapes and retention times, providing insights into the dynamic aspects of separation performance.

Filtration is another critical downstream processing technique used for separating cells, cell debris, and other particulate matter from the product stream. Filtration processes, such as microfiltration, ultrafiltration, and tangential flow filtration, involve optimizing parameters like membrane type, pore size, flow rate, and transmembrane pressure. For example, supervised learning models can forecast filter fouling rates and product recovery based on factors such as feedstock composition and filtration parameters. Deep learning techniques also contribute to filtration optimization by analysing high-dimensional data from inline sensors and imaging technologies. CNNs can process data from sensors monitoring membrane fouling or particle size distribution, providing detailed insights into the filtration process.

RNNs and LSTMs can analyse time-series data from filtration runs to predict changes in filter performance over time and recommend adjustments to maintain optimal conditions. Additionally, AI-driven predictive maintenance can help in anticipating filter replacements and cleaning cycles, reducing downtime and extending the lifespan of filtration equipment. The integration of AI into downstream processing also enables advanced process control and automation.

Reinforcement learning, a branch of AI focused on decision-making through trial and error, can optimize both chromatography and filtration processes by continuously exploring and adjusting parameters to achieve the best outcomes. AI-driven control systems can adapt in real-time to changes in feedstock composition or process conditions, ensuring that purification processes remain efficient and effective throughout production. Understanding how AI models arrive at their recommendations and ensuring that they align with process objectives is essential for effective implementation.

Another challenge is the need for interdisciplinary expertise. Successful integration of AI into downstream processing requires collaboration between process engineers, data scientists, and AI specialists. This multidisciplinary approach ensures that AI models are developed and applied effectively, taking into account both the technical aspects of chromatography and filtration and the nuances of data science. Investment in training and skill development for personnel working with AI technologies is also important for maximizing their potential (Pedregosa et al., 2011).

AI FOR BIOPROCESS CONTROL: REAL-TIME DECISION-MAKING AND AUTOMATION

Bioprocess control is a crucial element in biomanufacturing, encompassing the regulation and management of complex biological processes to ensure the consistent production of high-quality products. As biomanufacturing processes become increasingly sophisticated, the need for advanced control systems that can make real-time decisions and automate process adjustments has grown (Pinto et al., 2019).

One of the key challenges in bioprocess control is managing the complexity and dynamic nature of biological systems. Traditional control systems often rely on predefined setpoints and manual adjustments based on empirical data. However, biological processes are inherently variable and influenced by numerous factors, including raw material quality, environmental conditions, and cellular responses. AI addresses these challenges by enabling real-time data analysis and adaptive control. Machine learning algorithms can process large volumes of data from sensors, control systems, and historical process records to identify patterns, predict outcomes, and make informed decisions.

For example, supervised learning models can analyse historical data to predict how changes in process parameters will impact key performance indicators such as cell growth rates, product yields, and metabolite concentrations allowing for the modelling of dynamic and sequential aspects of bioprocesses (Radivojevic et al., 2020). These deep learning models can track changes over time and predict future process behaviour, enabling more accurate and timely decision-making.

Automation is another key benefit of AI-driven bioprocess control. Traditional bioprocess control systems often require manual interventions and adjustments, which can introduce variability and inefficiencies. AI-driven automation systems can integrate real-time data analysis with automated control actions, reducing the need for manual oversight and increasing process consistency. This automation not only enhances process efficiency but also reduces the risk of human error and operational downtime. The integration of AI into bioprocess control also supports predictive maintenance and proactive problem-solving (Ramakrishnan et al., 2022).

Another challenge is the need for interdisciplinary expertise. Successful integration of AI into bioprocess control requires collaboration between bioprocess engineers, data scientists, and AI specialists. This multidisciplinary approach ensures that AI models are developed and applied effectively, considering both the technical aspects of bioprocess control and the nuances of data science. Investing in training and skill development for personnel working with AI technologies is also important to maximize their potential.

The vast amount of data generated from various stages of biomanufacturing processes—ranging from raw material quality and fermentation conditions to downstream processing and final product quality—offers valuable insights that can drive innovation, optimize performance, and ensure regulatory compliance. Leveraging AI to analyze and integrate this data can transform raw information into actionable insights, enabling more informed decision-making and enhancing overall process efficiency (Resendis-Antonio, 2013).

Data integration is a critical step in unlocking the value of information collected across diverse sources and systems. Biomanufacturing processes generate data from numerous sources, including sensors, laboratory information management systems (LIMS), process control systems, and analytical instruments. Each of these sources may produce data in different formats, frequencies, and granularities, making it challenging to consolidate and interpret. AI facilitates data integration by using advanced algorithms to unify disparate datasets into a coherent and actionable format. Techniques such as data fusion, which combines data from multiple sources to provide a comprehensive view, and entity resolution, which reconciles data inconsistencies and redundancies, are essential for creating a single source of truth. By integrating data from various sources, AI enables a more holistic understanding of the biomanufacturing process, allowing for better analysis and optimization.

Once integrated, data analytics become crucial in deriving meaningful insights from the combined datasets (Sabzevari et al., 2022). Deep learning, with its ability to analyze and interpret high-dimensional data such as images and time-series, further enhances the capacity to extract valuable insights. By continuously analyzing streaming data from sensors and process controls, AI systems can detect deviations from expected patterns, predict potential issues, and recommend corrective actions in real-time. This capability enhances process stability, reduces variability, and improves overall efficiency. Real-time analytics enables dynamic adjustments to process parameters, such as temperature, pH, and flow rates, ensuring that processes remain within optimal conditions and that product quality is maintained.

By continuously updating the virtual model with live data, digital twins provide valuable insights into process behavior, enable scenario analysis, and support predictive maintenance. This capability allows for more informed decision-making and optimized process control. Despite the numerous advantages of AI-driven data integration and analytics, several challenges must be addressed. One significant challenge is ensuring data quality and consistency (Sebayang et al., 2017). High-quality, accurate data is essential for developing reliable AI models and making informed decisions.

Data preprocessing, including cleaning, normalization, and validation, is crucial for ensuring that the data used for analysis is reliable and representative. Understanding how AI models derive insights and ensuring that their recommendations align with process objectives is essential for effective implementation. Another challenge is the need for interdisciplinary collaboration. Successful data integration and analytics require expertise in bioprocess engineering, data science, and AI technologies. Collaboration between these disciplines ensures that AI models are developed and applied effectively, taking into account both the technical aspects of biomanufacturing and the nuances of data science.

Investing in training and skill development for personnel working with AI technologies is also important for maximizing their potential. In conclusion, data integration and analytics powered by AI represent a significant advancement in biomanufacturing, offering transformative capabilities for harnessing the value of data. By integrating disparate datasets and applying advanced analytics, AI enables more informed decision-making, optimized process performance, and enhanced product quality. While challenges related to data quality, model interpretability, and interdisciplinary collaboration remain, the continued development and application of AI technologies hold the promise of further innovations in data-driven insights (Sreenivasulu et al., 2022).

AI-DRIVEN QUALITY CONTROL: PREDICTIVE ANALYTICS FOR BIOPRODUCT QUALITY

Quality control is a fundamental aspect of biomanufacturing, ensuring that bioproducts meet stringent standards of purity, potency, and safety. The complexity and variability inherent in biological processes present significant challenges for maintaining consistent quality. One of the key strengths of AI in quality control is its ability to process and integrate large volumes of data from various sources, such as process parameters, raw materials, and product testing results. For instance, supervised learning models, including regression and classification algorithms, can predict the impact of changes in process conditions on final product quality (Sugumaran et al., 2024).

By analysing data from previous batches, these models can forecast potential quality deviations and recommend adjustments to prevent issues before they occur. Predictive analytics, powered by AI, offers significant advantages over traditional quality control methods. Instead of relying solely on post-production testing, predictive analytics uses historical and real-time data to anticipate quality issues and optimize processes. For example, machine learning models can analyse data from sensors monitoring fermentation conditions, such as temperature, pH, and oxygen levels, to predict how these factors will influence product quality. This proactive approach allows for timely interventions and adjustments, reducing the risk of quality failures and minimizing the need for costly rework or product recalls.

Deep learning further enhances predictive analytics by providing advanced capabilities for analyzing high-dimensional and complex data. These deep learning models can track changes in key quality metrics, such as metabolite concentrations or protein folding patterns, and predict how they will affect the final product. Real-time quality monitoring is another area where AI-driven approaches excel (Tokuyama et al., 2020).

By continuously analyzing streaming data from sensors and analytical instruments, AI systems can detect deviations from expected quality profiles in real-time. This capability enables dynamic adjustments to process parameters, ensuring that the process remains within optimal conditions and that product quality is consistently maintained. For instance, AI algorithms can automatically adjust fermentation conditions based on real-time data to prevent quality deviations and ensure that the product meets predefined standards. By continuously updating the virtual model with live data, digital twins provide valuable insights into process behavior, enabling proactive quality management and optimization. This capability allows for more accurate predictions of quality outcomes and facilitates informed decision-making to maintain product consistency.

Understanding how AI models generate predictions and ensuring that their recommendations align with quality objectives is essential for effective implementation (Van Rosmalen et al., 2021). Another challenge is the integration of AI-driven quality control with existing quality management systems and workflows. Successful implementation requires seamless integration with current processes and an understanding of how AI models can complement traditional quality control methods. Training and skill development for personnel working with AI technologies is also important to ensure effective use and interpretation of AI-driven insights. While challenges related to data quality, model interpretability, and integration with existing systems remain, the continued development and application of AI technologies hold the promise of further innovations in quality control.

As AI continues to evolve, its integration into quality control processes will likely lead to more efficient, reliable, and high-quality biomanufacturing, driving the future of biotechnology and therapeutic production. Scalability and transferability are pivotal aspects of biomanufacturing that determine whether bioprocess can be effectively scaled from laboratory or pilot-scale production to full-scale commercial manufacturing. The challenges associated with scaling up bioprocesses include maintaining product quality, consistency, and efficiency as process volumes increase. As biomanufacturing processes transition from small-scale experiments to large-scale operations, maintaining optimal performance can be difficult due to changes in process dynamics, equipment, and environmental conditions.

AI has emerged as a powerful tool in addressing these challenges by providing advanced optimization and transferability solutions that enhance the scalability of bioprocesses.AI-driven approaches contribute significantly to the scalability of bioprocesses by enabling more accurate modeling and optimization of process parameters (Vega-Ramon, F. et al.,2021). ML algorithms can analyze data from small-scale experiments to build predictive models that forecast how processes will behave at larger scales. For instance, supervised learning models can be trained on data from laboratory-scale bioreactors to predict how scale-up factors, such as changes in vessel geometry, mixing efficiency, and heat transfer, will impact process performance and product quality. By leveraging these predictive models, process engineers can make informed adjustments and optimizations to ensure that scaling up does not result in undesirable deviations or inefficiencies.

In addition to scalability, transferability—the ability to apply optimized processes from one scale or facility to another—is crucial for efficient biomanufacturing. AI facilitates transferability by developing and validating process models that are adaptable to different scales and settings. Transfer learning, a branch of machine learning, allows models trained on data from one scale or facility to be fine-tuned for application in another. For example, a machine learning model developed for a small-scale fermentation process can be adapted for use in a large-scale production environment by adjusting for scale-specific parameters and conditions. This capability reduces the need for extensive revalidation and experimentation when transferring processes between different production scales or facilities.

AI-driven optimization also supports the development of standardized processes that can be consistently applied across different scales and sites. By analysing data from multiple sources and scales, AI models can identify common patterns and best practices that are effective across various environments. This enables the creation of robust, scalable, and transferable process protocols that ensure consistent performance and quality regardless of scale or location. Additionally, digital twins—virtual replicas of physical processes—can be used to simulate and validate process performance in different scenarios, facilitating the transfer of optimized processes to new environments (Vijayakumar G et al., 2024).

One of the key challenges in applying AI to scalability and transferability is managing the complexity and variability of bioprocesses across different scales and conditions. Scaling up or transferring a process often involves changes in equipment, materials, and environmental conditions, which can introduce new variables and uncertainties. AI models must be able to adapt to these changes and incorporate new data to maintain accuracy and reliability. Ensuring that AI models are robust and generalizable across different scenarios is crucial for successful scalability and transferability.

Another challenge is the integration of AI-driven insights with existing process development and scale-up workflows. Effective application of AI requires seamless integration with current practices, including process design, validation, and quality assurance. Collaboration between process engineers, data scientists, and AI specialists is essential to ensure that AI models are developed and implemented effectively, taking into account both the technical aspects of bioprocesses and the nuances of AI technology.

In conclusion, AI-driven solutions represent a significant advancement in addressing the challenges of scalability and transferability in biomanufacturing. Machine learning and deep learning techniques enhance the ability to model, optimize, and transfer bioprocesses across different scales and environments, improving efficiency and consistency. While challenges related to managing process complexity, integrating AI with existing workflows, and ensuring model robustness remain, the continued development and application of AI technologies hold the promise of more efficient and adaptable biomanufacturing processes. As AI continues to evolve, its integration into scalability and transferability strategies will likely drive further innovations and advancements in biotechnology and therapeutic production, leading to more reliable and scalable manufacturing solutions.

REGULATORY CONSIDERATIONS: AI IN BIOMANUFACTURING AND REGULATORY COMPLIANCE

As AI becomes increasingly integral to biomanufacturing processes, it introduces both opportunities and challenges for regulatory compliance. AI's role in optimizing processes, enhancing quality control, and facilitating real-time monitoring must be carefully managed to align with regulatory requirements. Understanding these regulatory considerations is crucial for integrating AI technologies effectively while maintaining compliance with industry standards. AI technologies can significantly enhance biomanufacturing by improving process efficiency, product quality, and data integrity.

One of the primary regulatory considerations is the validation and verification of AI systems. Regulatory agencies require that any AI-driven system used in biomanufacturing be thoroughly validated to ensure it performs accurately and reliably. This involves demonstrating that the AI system can deliver consistent and reproducible results across different conditions and that it meets predefined performance criteria. Validation typically includes testing the AI system with historical data, assessing its predictive accuracy, and comparing its performance against established benchmarks. For example, when using AI for quality control, it is essential to validate that the AI model can accurately predict product quality based on process data and that its predictions align with results from traditional testing methods (Wilkinson, M.D. et al.,2016).

Another important consideration is data integrity and transparency. Regulatory agencies emphasize the importance of maintaining accurate and complete records of all process data, including data generated by AI systems. This includes ensuring that data from AI models is stored securely, is easily retrievable for review, and is documented in a way that allows for traceability and auditability. AI models and their decisions must be interpretable, meaning that the rationale behind AI-generated recommendations should be understandable and justifiable. Ensuring transparency in AI processes helps address regulatory concerns about data manipulation and supports the reproducibility of results. The use of AI in biomanufacturing also raises questions about the qualification of personnel and the development of standard operating procedures (SOPs).

Personnel involved in the implementation and oversight of AI systems must have a thorough understanding of both the technology and its regulatory implications. This requires specialized training to ensure that staff can effectively manage and operate AI systems while adhering to regulatory requirements. Additionally, the development and documentation of SOPs for AI systems are crucial for ensuring that AI applications are used consistently and in compliance with regulatory standards. SOPs should detail the procedures for AI model development, validation, maintenance, and monitoring, as well as for addressing any deviations or issues that may arise. Regulatory agencies are also concerned with the continuous monitoring and updating of AI systems. AI models require ongoing maintenance and periodic updates to remain effective as processes and data evolve. Regulatory guidelines often require manufacturers to implement a robust change management process to address updates to AI systems, including assessing the impact of changes on process performance and product quality. This ensures that modifications to AI models do not introduce new risks or affect compliance with regulatory standards.

AI systems must be seamlessly incorporated into established workflows and quality management systems, ensuring that they complement rather than disrupt existing practices. This integration requires a thorough understanding of how AI interacts with traditional processes and how to align AI-driven optimizations with regulatory requirements. In conclusion, the integration of AI into biomanufacturing introduces significant opportunities for improving process efficiency and product quality, but it also necessitates careful consideration of regulatory compliance.

Key regulatory considerations include the validation and verification of AI systems, data integrity and transparency, personnel qualification, and the development of SOPs. Ongoing monitoring, change management, and engagement with regulatory bodies are also critical for ensuring compliance. By addressing these regulatory considerations, biomanufacturers can leverage AI technologies effectively while maintaining adherence to industry standards, ultimately contributing to the advancement of biotechnology and the production of safe and high-quality biopharmaceuticals (Wu, S.G. et al.,2016).

The integration of AI into biomanufacturing has yielded significant advancements, showcasing its transformative potential through various successful applications. These case studies highlight how AI technologies have been employed to optimize processes, enhance product quality, and streamline operations across different stages of biomanufacturing. By examining these real-world examples, we can gain insights into the practical benefits of AI and its role in driving innovation within the biotechnology industry. One notable case study involves the use of AI for optimizing fermentation processes at a major biopharmaceutical company.

In this instance, ML algorithms were employed to analyze data from fermentation experiments, including variables such as temperature, pH, and nutrient levels. Traditional fermentation processes often involve extensive trial-and-error to identify optimal conditions, which can be time-consuming and costly. By applying ML models to historical fermentation data, the company was able to identify patterns and correlations that were not evident through manual analysis. This AI-driven approach led to significant improvements in fermentation yield and process efficiency.

The AI system provided real-time recommendations for adjusting process parameters, resulting in a more consistent and higher-quality product, reduced production costs, and a faster time-to-market for new biopharmaceuticals. Another successful application of AI is demonstrated in the field of protein crystallization, where deep learning models have been used to enhance the process of determining protein structures. Protein crystallization is a critical step in drug development, but it is often challenging due to the complexity and variability of protein behavior. A prominent biotech firm utilized convolutional neural networks (CNNs) to analyze images of protein crystals and predict their quality and structure.

The deep learning model was trained on a large dataset of crystallization experiments, enabling it to recognize subtle features that indicate successful crystallization conditions. Chromatography is a crucial step in purifying biopharmaceuticals, but it can be complex and resource-intensive. A biomanufacturing company utilized AI algorithms to optimize chromatography conditions and improve purification efficiency. By analyzing data from previous chromatography runs, machine learning models were able to predict the impact of different parameters on purification outcomes.

The AI system provided recommendations for optimizing column conditions, flow rates, and elution profiles, leading to enhanced separation efficiency and reduced processing times. This application of AI not only improved the overall yield and purity of the final product but also reduced the operational costs associated with chromatography. Another impactful case study involves the use of AI for predictive maintenance in biomanufacturing facilities. Predictive maintenance aims to anticipate equipment failures before they occur, minimizing downtime and preventing costly disruptions.

A major biomanufacturing facility implemented a predictive maintenance system powered by machine learning algorithms to monitor the condition of critical equipment. By analyzing data from sensors and historical maintenance records, the AI system was able to identify patterns and predict potential equipment failures. This proactive approach allowed the facility to schedule maintenance activities based on predicted needs, reducing unexpected breakdowns and ensuring continuous, efficient operation. AI's contribution to biomanufacturing extends to the realm of digital twins, which are virtual replicas of physical processes used for simulation and optimization.

A prominent pharmaceutical company developed a digital twin of its manufacturing process, integrating real-time data with AI models to simulate and predict process behavior. The digital twin enabled the company to conduct "what-if" analyses and assess the impact of various scenarios on process performance. By using the digital twin to optimize process parameters and validate potential improvements, the company achieved greater process efficiency, reduced variability, and enhanced overall product quality. These case studies collectively demonstrate the diverse and transformative applications of AI in biomanufacturing. From optimizing fermentation and protein crystallization processes to enhancing quality control and predictive maintenance, AI technologies have proven their value in driving innovation and improving operational efficiency. By leveraging machine learning, deep learning, and digital twin technologies, biomanufacturers can address complex challenges, streamline operations, and accelerate the development of high-quality biopharmaceutical products. As AI continues to evolve, its applications in biomanufacturing are likely to expand, further advancing the field and contributing to the future of biotechnology.

EMERGING AI TRENDS AND BIOMANUFACTURING INNOVATIONS

(AI) into biomanufacturing has already transformed the industry, but the future promises even more profound innovations and advancements. As AI technologies continue to evolve, their applications in biomanufacturing are expected to expand, driving new levels of efficiency, accuracy, and innovation. Emerging trends in AI are poised to shape the future of biomanufacturing, addressing current challenges and opening new avenues for process optimization, product development, and quality control. One of the most significant emerging trends is the advancement of AI-driven automation and robotics.

In biomanufacturing, automation has traditionally been used for repetitive and high-precision tasks, but the integration of AI is set to enhance these capabilities further. For instance, AI-powered robotic systems can autonomously adjust parameters in response to real-time data, optimizing bioprocesses on the fly and minimizing human intervention. This trend towards more intelligent automation is expected to increase throughput, reduce errors, and improve overall process efficiency, leading to faster and more reliable production cycles. These AI-enhanced digital twins will enable biomanufacturers to perform more sophisticated "what-if" analyses, optimize processes in real time, and forecast potential outcomes with higher precision. By continuously integrating real-time data, AI-driven digital twins will support more dynamic and responsive process management, leading to improved product quality and operational efficiency. The application of AI in personalized medicine represents another exciting frontier.

AI can play a crucial role in this field by analysing large-scale genomics data, predicting patient responses to therapies, and optimizing treatment regimens. Machine learning algorithms can identify biomarkers, predict drug interactions, and recommend personalized therapeutic strategies, enabling more effective and targeted treatments. In biomanufacturing, this means developing more precise and customized biopharmaceuticals, enhancing the ability to meet the specific needs of individual patients and improve therapeutic outcomes (Zhang, J. et al.,2020).

In the realm of data integration and analytics, advanced AI techniques are set to revolutionize how biomanufacturers harness and interpret data. The future will see greater emphasis on integrating diverse data sources, including genomics, proteomics, metabolomics, and clinical data, to gain comprehensive insights into biomanufacturing processes. This enhanced data integration will enable more informed decision-making, streamline process optimization, and accelerate innovation in biomanufacturing.

The rise of AI-powered predictive analytics will also play a crucial role in future biomanufacturing innovations. Predictive models, driven by machine learning and statistical algorithms, will offer deeper insights into process dynamics, product quality, and potential failures. These models will help anticipate and mitigate risks before they occur, allowing for more proactive and preventive measures. For example, predictive maintenance powered by AI can forecast equipment failures and optimize maintenance schedules, reducing downtime and ensuring continuous production. Similarly, AI-driven process control systems will predict deviations from optimal conditions and recommend adjustments in real time, enhancing overall process stability and product consistency.

The integration of AI technologies is revolutionizing the biomanufacturing landscape, unlocking unprecedented levels of efficiency, precision, and innovation. This comprehensive exploration of AI applications—ranging from process optimization and quality control to predictive analytics and personalized medicine—underscores the immense potential AI holds for driving advancements across the biotechnology sector (Zampieri, G. et al.,2019).

AI's impact on biomanufacturing is marked by its ability to enhance operational efficiency through data-driven insights and automation. Machine learning algorithms and deep learning models offer sophisticated methods for analyzing vast datasets, identifying patterns, and predicting outcomes with remarkable accuracy. These capabilities enable biomanufacturers to optimize processes in real time, streamline production workflows, and reduce costs. For example, AI-driven process optimization has led to significant improvements in fermentation and protein crystallization, enhancing product yields and consistency while reducing experimental time and resource consumption.

By automating routine tasks and providing real-time process adjustments, AI not only boosts efficiency but also ensures that biomanufacturing processes are more reliable and scalable. Quality control in biomanufacturing has also seen substantial improvements with the advent of AI technologies. Traditional quality control methods, often reliant on manual inspections and batch testing, are complemented and enhanced by AI-driven systems that offer real-time monitoring and predictive analytics. AI-powered image analysis and data integration techniques enable early detection of deviations from quality standards, facilitating proactive interventions that maintain product integrity and compliance. Predictive models developed through AI can forecast potential quality issues before they manifest, allowing manufacturers to address problems before they affect production.

Personalized medicine, driven by AI, promises to address the diverse needs of patients more precisely and efficiently, ultimately enhancing the therapeutic efficacy of biopharmaceutical products (Zheng et al., 2017). The future of biomanufacturing is also being shaped by advanced AI techniques, such as digital twins and predictive analytics. Digital twins, virtual replicas of physical processes, are becoming increasingly sophisticated through AI integration. These models simulate real-time process behavior, enabling more accurate predictions and optimizations.

By employing advanced simulation and scenario analysis, digital twins support dynamic and responsive process management, contributing to improved product quality and operational efficiency. Similarly, predictive analytics powered by AI provides valuable insights into process dynamics, enabling manufacturers to anticipate and mitigate risks before they impact production. This proactive approach enhances the stability and reliability of biomanufacturing processes, fostering greater innovation and efficiency. Collaboration between industry stakeholders, regulatory bodies, and AI experts is essential to establish best practices and ensure that AI technologies are utilized effectively and ethically.

Furthermore, successful implementation of AI in biomanufacturing necessitates the development of skilled personnel who can bridge the gap between biomanufacturing processes and AI technologies. Training and education in both domains will be critical for leveraging AI's full capabilities and driving innovation in the field. In conclusion, the integration of AI into biomanufacturing holds transformative potential, offering new levels of efficiency, precision, and innovation. From optimizing processes and enhancing quality control to enabling personalized medicine and advancing predictive analytics, AI technologies are reshaping the biotechnology industry. As we unlock the full potential of AI, addressing challenges related to ethics, regulation, and skill development will be essential for maximizing its benefits. By embracing AI and its evolving capabilities, biomanufacturers can drive significant advancements, ultimately improving product quality, operational efficiency, and patient outcomes in the biotechnology sector.

CONCLUSION

The application of AI in optimizing bioprocesses has demonstrated remarkable benefits across various stages of biomanufacturing. In process optimization, AI algorithms analyse complex datasets to identify optimal conditions, streamline operations, and enhance product yields. Machine learning models have been employed to fine-tune fermentation parameters, improve protein crystallization, and optimize chromatography conditions, leading to more consistent and higher-quality products. These advancements not only reduce experimental costs and time but also increase the scalability and reproducibility of biomanufacturing processes.

AI's role in real-time process control and monitoring has further revolutionized biomanufacturing. Through the use of AI-powered sensors and data integration techniques, manufacturers can monitor processes in real time, detecting deviations and anomalies with high accuracy. This capability enables proactive adjustments and interventions, ensuring that processes remain within desired parameters and reducing the risk of quality deviations. Real-time monitoring also supports more dynamic and responsive process management, allowing for faster adjustments and improved overall process stability. Quality control in biomanufacturing has been significantly enhanced through AI-driven methods.

Traditional quality control practices, often reliant on manual inspections and batch testing, are complemented by AI technologies that offer automated, real-time analysis of product quality. AI-powered image analysis and predictive models provide early detection of quality issues, facilitating timely interventions and ensuring that products meet stringent regulatory standards. This approach enhances the reliability and accuracy of quality control processes, ultimately contributing to the production of safe and effective biopharmaceuticals. The potential of AI extends to personalized medicine, where it plays a crucial role in tailoring treatments to individual patients.

By analysing large-scale genomics and clinical data, AI can identify biomarkers, predict patient responses to therapies, and recommend personalized therapeutic strategies. This capability enables the development of more targeted and effective biopharmaceuticals, addressing the specific needs of individual patients and improving therapeutic outcomes. Looking forward, the future of AI in biomanufacturing promises continued innovation and progress. Emerging trends such as AI-driven automation, digital twins, and advanced predictive analytics are set to further enhance process optimization and operational efficiency creating a more interconnected and responsive biomanufacturing environment. These advancements will drive greater agility, precision, and efficiency in biomanufacturing processes, ultimately leading to faster and more reliable production of high-quality bioproducts. Ensuring that personnel are adequately trained and skilled in both biomanufacturing processes and AI technologies will be essential for leveraging AI's full capabilities.

In summary, the optimization of bioprocesses through AI is a transformative development in the field of biomanufacturing. AI technologies offer significant benefits in enhancing process efficiency, improving product quality, and enabling personalized medicine. As biomanufacturers continue to embrace and integrate AI, they will unlock new levels of innovation and operational excellence. By addressing ethical, regulatory, and skill-related challenges, the industry can fully harness the potential of AI to drive advancements in biotechnology, ultimately improving the production and quality of biopharmaceuticals and other bioproducts.

REFERENCES

Antonakoudis, A., Barbosa, R., Kotidis, P., & Kontoravdi, C. (2020). The era of big data: Genome-scale modelling meets machine learning. *Computational and Structural Biotechnology Journal, 18*, 3287–3300. DOI: 10.1016/j.csbj.2020.10.011 PMID: 33240470

Bradford, E., Schweidtmann, A. M., Zhang, D., Jing, K., & del Rio-Chanona, E. A. (2018). Dynamic modeling and optimization of sustainable algal production with uncertainty using multivariate Gaussian processes. *Computers & Chemical Engineering, 118*, 143–158. DOI: 10.1016/j.compchemeng.2018.07.015

Carbonell, P., Radivojevic, T., & García Martín, H. (2019). Opportunities at the Intersection of Synthetic Biology, Machine Learning, and Automation. *ACS Synthetic Biology, 8*(7), 1474–1477. DOI: 10.1021/acssynbio.8b00540 PMID: 31319671

Choudhury, S.. (2022) Reconstructing Kinetic Models for Dynamical Studies of Metabolism using Generative Adversarial Networks. *bioRxiv*.DOI: 10.1101/2022.01.06.475020

Coşgun, A., Günay, M. E., & Yıldırım, R. (2022). Analysis of lipid production from Yarrowia lipolytica for renewable fuel production by machine learning. *Fuel, 315*, 122817. Advance online publication. DOI: 10.1016/j.fuel.2021.122817

Del Rio-Chanona, E. A., Manirafasha, E., Zhang, D., Yue, Q., & Jing, K. (2016). Dynamic modeling and optimization of cyanobacterial C-phycocyanin production process by artificial neural network. *Algal Research, 13*, 7–15. DOI: 10.1016/j.algal.2015.11.004

Ebrahimpour, A., Rahman, R. N. Z. R. A., Ean Ch'ng, D. H., Basri, M., & Salleh, A. B. (2008). A modeling study by response surface methodology and artificial neural network on culture parameters optimization for thermostable lipase production from a newly isolated thermophilic Geobacillus sp. strain ARM. *BMC Biotechnology, 8*(1), 96. DOI: 10.1186/1472-6750-8-96 PMID: 19105837

Emara, T. (2019). Adaptive Power-Saving Mechanism for VoIP Over WiMAX Based on Artificial Neural Network. In *Algorithms, Methods, and Applications in Mobile Computing and Communications* (pp. 158–177). IGI Global. DOI: 10.4018/978-1-5225-5693-0.ch007

Emara, T., Saleh, A. I., & Arafat, H. (2014). Power saving mechanism for VoIP services over WiMAX systems. *Wireless Networks, 20*(5), 975–985. DOI: 10.1007/s11276-013-0650-5

Emara, T. Z. (2016). Maximizing power saving for VoIP over WiMAX systems. [IJMCMC]. *International Journal of Mobile Computing and Multimedia Communications*, *7*(1), 32–40. DOI: 10.4018/IJMCMC.2016010103

Emara, T. Z., & Huang, J. Z. (2019). RRPlib: A Spark library for representing HDFS blocks as a set of random sample data blocks. *Science of Computer Programming*, *184*, 102301. DOI: 10.1016/j.scico.2019.102301

Emara, T. Z., Trinh, T., & Huang, J. Z. (2023). Geographically distributed data management to support large-scale data analysis. *Scientific Reports*, *13*(1), 17783. DOI: 10.1038/s41598-023-44789-x PMID: 37853092

Faure, L.. (2022) Artificial Metabolic Networks: enabling neural computation with metabolic networks. *bioRxiv*. DOI: 10.1101/2022.01.09.475487

Freier, L., Hemmerich, J., Schöler, K., Wiechert, W., Oldiges, M., & von Lieres, E. (2016). Framework for Kriging-based iterative experimental analysis and design: Optimization of secretory protein production inCorynebacterium glutamicum. *Engineering in Life Sciences*, *16*(6), 538–549. DOI: 10.1002/elsc.201500171

King, Z. A., O'Brien, E. J., Feist, A. M., & Palsson, B. O. (2017). Literature mining supports a next-generation modeling approach to predict cellular byproduct secretion. *Metabolic Engineering*, *39*, 220–227. DOI: 10.1016/j.ymben.2016.12.004 PMID: 27986597

Kumar, V., Bhalla, A., & Rathore, A. S. (2014). Design of experiments applications in bioprocessing: Concepts and approach. *Biotechnology Progress*, *30*(1), 86–99. DOI: 10.1002/btpr.1821 PMID: 24123959

Li, G., Rabe, K. S., Nielsen, J., & Engqvist, M. K. M. (2019). Machine Learning Applied to Predicting Microorganism Growth Temperatures and Enzyme Catalytic Optima.*ACS Synthetic Biology*, *8*(6), 1411–1420. DOI: 10.1021/acssynbio.9b00099 PMID: 31117361

Liu, Y., & Gunawan, R. (2017). Bioprocess optimization under uncertainty using ensemble modeling. *Journal of Biotechnology*, *244*, 34–44. DOI: 10.1016/j.jbiotec.2017.01.013 PMID: 28137617

Mahadev Madgule, N. (2023). Vinayaka, Yeshwant M. Sonkhaskar, Dhiren Ramanbhai Patel, R. Karthikeyan, P. Selvakumar,(2023). Mechanical properties and microstructure of activated TIG welded similar joints of Inconel alloys by desirability approaches. *Materials Today: Proceedings*, *77*, 528–533. DOI: 10.1016/j.matpr.2022.12.250

Oyetunde, T., Liu, D., Martin, H. G., & Tang, Y. J. (2019). Machine learning framework for assessment of microbial factory performance. *PLoS One*, *14*(1), e0210558. DOI: 10.1371/journal.pone.0210558 PMID: 30645629

Pappu, S. M. J., & Gummadi, S. N. (2017). Artificial neural network and regression coupled genetic algorithm to optimize parameters for enhanced xylitol production by Debaryomyces nepalensis in bioreactor. *Biochemical Engineering Journal*, *120*, 136–145. DOI: 10.1016/j.bej.2017.01.010

Pedregosa, F.. (2011). Scikit-learn: Machine Learning in Python. *Journal of Machine Learning Research*, *12*, 2825–2830.

Pinto, J., de Azevedo, C. R., Oliveira, R., & von Stosch, M. (2019). A bootstrap-aggregated hybrid semi-parametric modeling framework for bioprocess development. *Bioprocess and Biosystems Engineering*, *42*(11), 1853–1865. DOI: 10.1007/s00449-019-02181-y PMID: 31375965

Radivojevic, T., Costello, Z., Workman, K., & Garcia Martin, H. (2020). A machine learning Automated Recommendation Tool for synthetic biology. *Nature Communications*, *11*(1), 4879. DOI: 10.1038/s41467-020-18008-4 PMID: 32978379

Ramakrishnan, T., Mohan Gift, M. D., Chitradevi, S., Jegan, R., & Subha Hency Jose, P. Nagaraja, Rajneesh Sharma H.N., Selvakumar P., Sintayehu Mekuria Hailegiorgis, (2022). "Study of Numerous Resins Used in Polymer Matrix Composite Materials", Advances in Materials Science and Engineering, vol. 2022, Article ID 1088926, 8 pages, 2022. https://doi.org/DOI: 10.1155/2022/1088926

Resendis-Antonio, O. (2013) Constraint-based Modeling. In Encyclopedia of Systems Biology (Dubitzky, W. et al., eds), pp. 494-498, Springer New York. DOI: 10.1007/978-1-4419-9863-7_1143

Sabzevari, M., Szedmak, S., Penttilä, M., Jouhten, P., & Rousu, J. (2022). Strain design optimization using reinforcement learning. *PLoS Computational Biology*, *18*(6), e1010177. DOI: 10.1371/journal.pcbi.1010177 PMID: 35658018

Sebayang, A. H., Masjuki, H. H., Ong, H. C., Dharma, S., Silitonga, A. S., Kusumo, F., & Milano, J. (2017). Optimization of bioethanol production from sorghum grains using artificial neural networks integrated with ant colony. *Industrial Crops and Products*, *97*, 146–155. DOI: 10.1016/j.indcrop.2016.11.064

Abhilasha, A., Sreenivasulu, A., & Manimozhi, T. KUMAR, P. S., Selvakumar, P., & Singh, P. (2022, April). The Model of Smart Sensing Device For Sensitive Nano-clusters Modification in Sensing Properties. In 2022 2nd International Conference on Advance Computing and Innovative Technologies in Engineering (ICACITE) (pp. 1043-1047). IEEE.

Kanipriya, M., Sugumaran, D., Pandiyarajan, V., Yogananda, B. S., Selvakumar, P., & Mohan, S. (2024, January). Analysis of Electromyography Signals for Control of Mechanical Prosthesis using Machine Learning Techniques. In 2024 2nd International Conference on Intelligent Data Communication Technologies and Internet of Things (IDCIoT) (pp. 1059-1067). IEEE.

Tokuyama, K., Shimodaira, Y., Terawaki, T., Kusunose, Y., Nakai, H., Tsuji, Y., Toya, Y., Matsuda, F., & Shimizu, H. (2020). Data science-based modeling of the lysine fermentation process. *Journal of Bioscience and Bioengineering*, *130*(4), 409–415. DOI: 10.1016/j.jbiosc.2020.06.011 PMID: 32709563

Trinh, T., Duc, L. P., Tran, C. T., Duy, T. T., & Emara, T. Z. (2021). A new stratified block model to process large-scale data for a small cluster. In *International Conference on Artificial Intelligence and Big Data in Digital Era* (pp. 253–263). Springer International Publishing Cham. DOI: 10.1007/978-3-030-97610-1_21

Van Rosmalen, R. P., Smith, R. W., Martins dos Santos, V. A. P., Fleck, C., & Suarez-Diez, M. (2021). Model reduction of genome-scale metabolic models as a basis for targeted kinetic models. *Metabolic Engineering*, *64*, 74–84. DOI: 10.1016/j.ymben.2021.01.008 PMID: 33486094

Vega-Ramon, F., Zhu, X., Savage, T. R., Petsagkourakis, P., Jing, K., & Zhang, D. (2021). Kinetic and hybrid modeling for yeast astaxanthin production under uncertainty. *Biotechnology and Bioengineering*, *118*(12), 4854–4866. DOI: 10.1002/bit.27950 PMID: 34612511

Vijayakumar, G., Rajkumar, M., Rajiv Chandar, N., Selvakumar, P., & Duraisamy, R. (2024). Muniyandi Rajkumar, Rajiv Chandar N, Selvakumar P and Ramesh Duraisamy, (2024). Environmentally friendly TDS removal from waste water by electrochemical ion exchange batch-type recirculation (EIR) technique. *Environmental Science. Water Research & Technology*, *10*(4), 826–835. DOI: 10.1039/D3EW00793F

Wilkinson, M. D., Dumontier, M., Aalbersberg, I. J. J., Appleton, G., Axton, M., Baak, A., Blomberg, N., Boiten, J.-W., da Silva Santos, L. B., Bourne, P. E., Bouwman, J., Brookes, A. J., Clark, T., Crosas, M., Dillo, I., Dumon, O., Edmunds, S., Evelo, C. T., Finkers, R., & Mons, B. (2016). The FAIR Guiding Principles for scientific data management and stewardship. *Scientific Data*, *3*(1), 160018. DOI: 10.1038/sdata.2016.18 PMID: 26978244

Wu, S. G., Wang, Y., Jiang, W., Oyetunde, T., Yao, R., Zhang, X., Shimizu, K., Tang, Y. J., & Bao, F. S. (2016). Rapid Prediction of Bacterial Heterotrophic Fluxomics Using Machine Learning and Constraint Programming. *PLoS Computational Biology*, *12*(4), e1004838. DOI: 10.1371/journal.pcbi.1004838 PMID: 27092947

Zampieri, G., Vijayakumar, S., Yaneske, E., & Angione, C. (2019). Machine and deep learning meet genome-scale metabolic modeling. *PLoS Computational Biology*, *15*(7), e1007084. DOI: 10.1371/journal.pcbi.1007084 PMID: 31295267

Zhang, J., Petersen, S. D., Radivojevic, T., Ramirez, A., Pérez-Manríquez, A., Abeliuk, E., Sánchez, B. J., Costello, Z., Chen, Y., Fero, M. J., Martin, H. G., Nielsen, J., Keasling, J. D., & Jensen, M. K. (2020). Combining mechanistic and machine learning models for predictive engineering and optimization of tryptophan metabolism. *Nature Communications*, *11*(1), 4880. DOI: 10.1038/s41467-020-17910-1 PMID: 32978375

Zheng, Z. Y., Guo, X.-N., Zhu, K.-X., Peng, W., & Zhou, H.-M. (2017). Artificial neural network - Genetic algorithm to optimize wheat germ fermentation condition: Application to the production of two anti-tumor benzoquinones. *Food Chemistry*, *227*, 264–270. DOI: 10.1016/j.foodchem.2017.01.077 PMID: 28274431

Chapter 9
Avian Species Detection Using Spectogram and CNN

J. Arthy
https://orcid.org/0009-0006-8469-0126
SRM Instutute of Science and Technology, Ramapuram, India

K. Raja
SRM Institute of Science and Technology, Ramapuram, India

Vignesh Raghuraman
https://orcid.org/0009-0009-1439-5617
SRM Institute of Science and Technology, Ramapuram, India

S. Ashwath
https://orcid.org/0009-0004-0896-1903
SRM Institute of Science and Technology, Ramapuram, India

Rahul Bundele
https://orcid.org/0009-0007-0953-9909
SRM Institute of Science and Technology, Ramapuram, India

ABSTRACT

Avian species detection using spectrogram and CNN showcases the method of converting bird sounds into spectrogram images, which in turn is used as input for the CNN models. This study would discuss how the CNNs are trained to find and differentiate bird species based on these spectrograms, highlighting the accuracy and efficiency of this approach.Python programming language and Librosa library are used for the implementation of the models.A web application is also developed to showcase

DOI: 10.4018/979-8-3693-9208-9.ch009

the results of the model.As we are using STFT (Short Term Fourier Transforms) the accuracy of the model increases significantly compared to the traditional models.

I. INTRODUCTION

In the ever-evolving field of machine learning, our paper stands out with its innovative approach to avian species classification and detection. Leveraging the power of TensorFlow and a suite of complementary libraries, including NumPy, SciPy, and librosa, we have developed a sophisticated model capable of analyzing and identifying bird species from audio recordings. This model is designed to accurately process complex audio data, enabling the classification of various bird calls with high precision.

The root of this paper is the machine learning model, which has been precisely trained on a vast and varied dataset of bird sounds. TensorFlow's deep learning capabilities, combined with advanced audio processing libraries, allow our model to detect subtle differences in bird calls, enhancing its classification accuracy. This cutting-edge technology ensures that our system can handle a broad spectrum of bird species,

To provide a seamless user experience, we have built a web application that integrates Django and React. The Django backend is responsible for receiving and processing audio recording for receiving and processing audio recording is uploaded, the Django server prompts our machine learning model to predict the bird species. This backend architecture ensures efficient handling of audio data and accurate prediction results.

We have developed a web application using React for frontend, which provides an easy and efficient interface for users to interact with the system. Users can easily upload audio recordings and receive instant feedback on the identified bird species. The React-based interface is designed for easy use, providing fascinating user experience while managing audio input and displaying results.

In addition to species classification, our paper incorporates a web scraping API built with Express.js. This API retrieves detailed information about the identified bird species, offering users comprehensive insights into each bird's characteristics and habitat. By scraping relevant web data, the API enriches the user experience with informative content, enhancing the educational value of the application.

As a whole, this paper represents a remarkable advancement in the link between machine learning and ornithology. By combining state-of-the-art audio classification techniques with a user-friendly web application and informative web scraping, we are making significant strides in bird species identification and research. This integration of technology not only showcases the potential of machine learning but also provides a practical tool for those passionate about avian studies.

II. RELATED WORKS

A Novel Bird Sound Recognition Method Based on Multifeature Fusion and a Transformer Encoder "(Shaokai Zhang, 2023)" - Intelligent Voice Ltd.:

Multiple acoustic features"(Song Yang,2021)" get manually extracted by this study before implementing feature fusion to achieve better accuracy results. The method depends on manual acoustic feature selection yet it struggles when operating on various dataset types. Automated extraction of features through Mel-Frequency Cepstral Coefficients (MFCCs) replaces human participation while enhancing system adaptability.

Audio Classification of Bird Species Using Convolutional Neural Networks "(Hoang-Tu Vo, 2023)" - Journal of Student Research, Jericho High School:

Research has found that CNNs succeed at audio classification"(Jiangjian Xie,,2023)" yet this evaluation points out two main problems that stem from varying bird sounds and background noise disturbances. The method implements Short-Term Fourier Transform (STFT) and spectrogram representations to increase accuracy levels for the model to spot reliable frequency patterns during noisy conditions.

Birds Sound Classification Based on Machine Learning Algorithms "(Aska Ezadeen Mehyadin, 2021)" - Asian Journal of Research in Computer Science:

The research explores three machine learning classifier algorithms which include Naïve Bayes, k-NN and Random Forest. A leading accuracy rate of 78% exists in these models although manual feature engineering processes remain necessary and their performance decreases on complex data sets. The deep learning-based technique does not require manual feature extraction because it provides superior classification results in comparison to conventional methods.

A Randomized Bag-of-Birds Approach to Study Robustness of Automated Audio Based Bird Species Classification "(Burooj Ghani, 2021)" - MDPI open access journals:

The paper evaluates the resilience of bird classification algorithms using shallow artificial neural networks which operate on preprocessed sound features. The method provides efficient computation but it cannot process bigger datasets which contain more diverse species. The use of CNN-based model allows better generalization along with superior scalability when processing large datasets.

Estimating bird density using passive acoustic monitoring: a review of methods and suggestions for further research "(Cristian Pérez-Granados,2021)" - Wiley Online library:

The study examines different approaches for counting bird populations instead of conducting taxonomic identification. Our study supports the findings of this research regarding acoustic monitoring as it demonstrates the necessity of automated bird species identification systems.

Acoustic Classification of Bird Species Using Wavelets and Learning Algorithms "(Song Yang, 2021)" - ACM Digital Library:

This research achieves more than 70% accuracy through wavelet decomposition methods. Wavelet-based procedures need extensive parameter adjustment to achieve satisfactory results. The CNN model using spectrogram representations eliminates parameter adjustments requirements while enhancing its accuracy rates.

Literature Survey on Bird Call Identification and Monitoring- International Journal of Scientific Research in Engineering and Management(Dr.Chethan B K, 2024).

The paper noted, there are several have attempted various approaches to support bird call recognition with automation but the measure of effectiveness remains unsatisfactory and average 60-72%. All these models are very sensitive to the quality and richness of the data given, to the learning algorithm and to the kind of inputs given to the model. As stated by the paper itself, it is not easy to get high accuracy in the prediction of bird calls much as they advance in technology and methods.

Recognition of bird species with birdsong records using machine learning methods - National Library of Medicine (Yi Tang, 2024)

The proposed problem of the research paper under discussion concerns bird species identification with the help of vocalizations, which seems important for conservation as well as for regular tracking of the state of biodiversity. The study examined the acoustic features of three bird species: American Tree Creeper, Short-toed Treecreeper, Treecreeper. Hearing calculation assessment was six acoustic indices and machine learning models build Random Forest, Support Vector Machine and XGBoost models. As it was seen from all the above mentioned techniques, the highest accuracy and AUC was obtained from the XGBoost; thus also making it a unique model truly for bird species recognition based on its sounds (P. A. Agrawal,2024).

Therefore, this paper presents a novel approach for the identification of bird species from the audio signals analysis and pattern recognition. In last paper provides the insights on how to choose the datasets and algorithm for an optimized web based application (Arthy,2024) .

III. RESEARCH GAPS AND JUSTIFICATION

Despite advancements in bird species identification using machine learning, key challenges remain:

Background Noise Interference – Most current methods are not robust to noisy environments. Our method overcomes this by using spectral analysis methods such as STFT, which improves signal clarity.

Scalability and Generalization – Most conventional machine learning models are not scalable and generalize poorly on varied datasets. Our CNN-based method avoids manual feature extraction and exhibits greater scalability.

Integration with Web Applications – Although previous work has concentrated on classification, there are few studies that incorporate real-time prediction into easy-to-use applications. Our model is implemented as a web application with Django and React, making it ready for real-world application.

Lack of Automated Feature Extraction – Conventional machine learning models need hand-crafted features, which restricts flexibility. Our deep learning model, based on MFCCs and spectrograms, does this automatically, making it more efficient.

By filling these research gaps, our work presents a new, scalable, and strong solution for identifying bird species. The combination of CNNs with web applications makes it even more user-friendly, and this work is thus worthwhile for ornithologists, conservationists, and bird watchers alike.

Challenges in Real-World Deployment – The work doesn't address the challenges like dataset biases, variations in real-world noise, or CNN model limitations in comparison with other methods. Future work would be to see how model performance is affected across varying environmental conditions and how adaptive methods like noise filtering can be incorporated to provide improved robustness.

IV. PRACTICAL AND ECOLOGICAL SIGNIFICANCE

Apart from technical innovation, this study has significant ecological uses:

Conservation Efforts – Computerized bird call classification aids conservationists in monitoring endangered birds and tracking changes in habitats.

Biodiversity Monitoring – Instant species identification can help measure levels of biodiversity across various areas, supporting ecological research.

Citizen Science Integration – Running the model through a web application enables birdwatchers and researchers to add useful data to international ornithological databases.

V. SYSTEM ARCHITECTURE

Figure 1. Architecture diagram

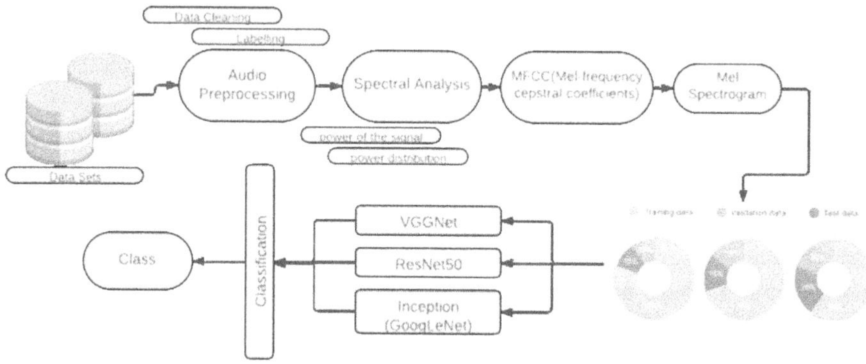

A. Data Fetching

In terms of data collection for inputting into our machine learning model, we had to decide in favor of Kaggle since it is a widely known site that offers a vast collection of quality datasets. Another important advantage which Kaggle provides for researchers and developers is datasets available in many topics and Kaggle contains audio data which is essential for our experiment. Out of the options available, we chose a dataset that includes only audio files of birds that are over 100 different species with well-defined species information. In this case, the dataset makes it easier to train our machine learning model that identifies bird species.

Integrating tools with the Kaggle platform makes it significantly easier to access datasets and tools thereby eliminating the need to constantly switch between different applications through the following factors; They reduce a lot of time when it comes to data gathering, thus enabling us to concentrate more on model creation, training and tweaking. Thanks to Kaggle, materializing and obtaining the necessary data were relatively easy and quick in the context of the machine learning setup. Due to the high-quality labeled audio data that we found, our development cycle got slightly reduced because we could directly embark on constructing, training, and optimizing the model for best performance and accuracy. This led to swift changeover from data collection to model distribution and utilization in a more efficient manner.

B. Audio Pre Processing

The next important stage in the ML pipeline is audio pre-processing, it is always required to prepare the raw audio files for the input to the ML model. This is usually in the form of raw audio data which is usually noisy or has some unnecessary information that interferes with the performance of the models. Therefore, the preprocessing step helps to standardize/compute the audio data, remove the noise and convert them into the appropriate format and component that is crucial in the classification process.

For this task, we utilize the Python library called librosa, that provides a wide range of functions for audio processing. Librosa is widely used in feature extraction, data visualization and in signal processing and therefore, it is an essential tool in order to manage and analyze audio signals. The library offers a multitude of functions for loading, preconditioning and transforming sound data and one of the key functions that we are going to use is librosa. load(). This function allows us to load audio files while extracting two important elements:This function allows us to load audio files while extracting two important elements:

audio_data: This variable contains the waveform data which represents the sound signal in the form of a discrete time series. It shows the variation of amplitude over time and it is the data which is operated on for further manipulations such as feature extraction like MFCCs or spectrums.
sample_rate: This variable represents the frequency of sampling or the number of samples obtained per second i.e IT is measured in Hertz Hz. The sample rate defines the ability to resolve sounds in time or in other words, the accuracy with which time varying sounds are sampled. An increased sample rate means that there is a greater amount of information from the original sound captured, a low sampling rate is a compression of the sound information. It is one of the most important parameters within the context of audio processing since it defines how closely the audio signal will be divided into parts as well as defines the precision of features which will be extracted later.

About librosa library: Librosa is a popular Python library widely used for analyzing and processing audio and music data. It provides various tools working on music information retrieval (MIR), audio signal processing, and machine learning applications related to audio analysis.

Key Features of Librosa:

Audio files: load an audio file in all formats (e.g. WAV, MP3, FLAC and) into a numpy array for processing

Audio files: can write audio data back to disk if other libraries (e.g. sound file) allows resampling change audio signals sampling rate
Trimming and padding remove silence or pad the audio to desired length
Short-Time Fourier Transform (STFT): analyze frequency content of audio signals in time
Waveform: plot the amplitude of the audio signals in the time domain
Spectrogram: plot spectrograms and other time-frequency representations.

C. Spectral Analysis

The next stage in our analysis is spectral analysis which is very essential in any audio analysis since it helps in decomposing any audio signal into their corresponding frequencies. As seen with the typical spectrogram layout, it is possible to study temporal and spectral features of the given audio data due to the fact that spectral analysis enables one to visualize how the different frequencies change with the time. This is especially significant in the classification of bird species by vocalization in that different species of birds will presumably have different frequency calls.

Among the techniques used in spectral analysis, it is possible to single out the Short-Time Fourier Transform (STFT). The STFT in fact splits the audio signal into short overlapping frames and computes the Fourier Transform of each frame. This certainly enables time and frequency contents to be documented comprehensively with relation to the variability of the frequency content as time elapses. That is because, by dividing the signal into several segments, STFT allows identifying the changes in the frequency and amplitude in each segment of the audio file.

The output of STFT is a spectrogram: a graphical representation of the density of a signal in the frequency-time domain. In a spectrogram:

On the horizontal axis, known as x-axis, the time scale is used and it demonstrates movement of signal from beginning to the end of the audio clip.

The y-axis gives information of frequency so one would be able to see different frequency components.

This is the brightness or the value of each specific frequency and where the 'heat' or the warm colored noise is characterized by greater energy or volume of the sound and the 'cold', or the cool colored noise has lesser energy or volume.

In this case, this particular visualization is very helpful for interpreting bird calls since certain bird species produce audio in frequencies that are easily distinguishable in terms of frequency spectrum in the spectrogram.

Figure 2. Spectrogram of audio data

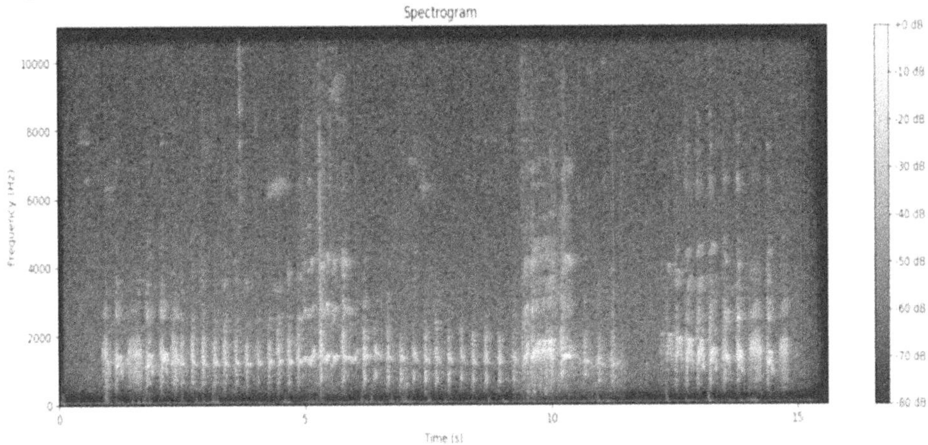

This visual representation allows for a detailed analysis of the spectral properties of the audio, revealing patterns and features that are essential for further processing

We use matplotlib library to visualize the spectrum

About Matplotlib: Matplotlib is one of the most used and powerful plotting libraries that exist in the Python ecosystem. It provides for the flexible and comprehensive framework for creating static, animated, as well as interactive visualizations in Python. With the help of this tool, you will generate an array of plots and charts to be used for visualization purposes by any data scientist, engineer, researcher, or educator.

Main Features:Range of Plot Kinds including simple Plots: Line, scatter, bar, histogram, pie charts.Figureure and axes control: complete control over Figureure size, DPI, aspect ratio, multiple axes available. Annotations: Attach text, labels, titles, legends, and even user-defined annotations. Interactive backends: support for interactive environments such as Jupyter notebooks, which means you can have zoom, pan, and update plots in real-time. pandas: direct plotting functionality using DataFrames.

The most common use case of matplotlib is Descriptive Statistics: Distributions, Central tendencies, and Variability.Trend Analysis: See trends and patterns over time, using line plots.Publication-quality Figureures. Detailed, accurate plots suitable for publications. Data visualization. Plot experimental data, simulation results, and theoretical models.

D. Mel-Frequency Cepstral Coefficients (MFCC)

To form the MFCCs the given raw audio signal is transformed into the Mel-frequency scale. This measure is very similar to frequency characteristics of sound which the human ear responds to more to low frequencies and for the higher ones it has a logarithmic scale. This is important since human hearing ability to detect changes decreases at high frequencies than at low frequencies and thus makes MFCC to be a powerful tool in modelling the perceptual properties of audio signals.

MFCCs operate in a way that analyzes the audio signal into short frames with overlapping periods. In other words, each frame is assessed in terms of their frequency spectrum. First, the power spectrum of each frame is determined by the Fast Fourier Transform algorithm. The spectrum obtained is then transformed using a bank of filters and each filter represents a different frequency band defined on the Mel scale.

After the hi-frequency mel spectrogram is obtained, log of the power values of the hi frequency mel spectrogram is calculated which normalizes the range of values obtained from the hi frequency mel spectrogram. Lastly, the log-Mel spectrum is subjected to a Discrete Cosine Transform which brings it down to coefficients, which would best represent the features of the audio signal. These coefficients form the so-called MFCCs.

Figure 3. MFCC

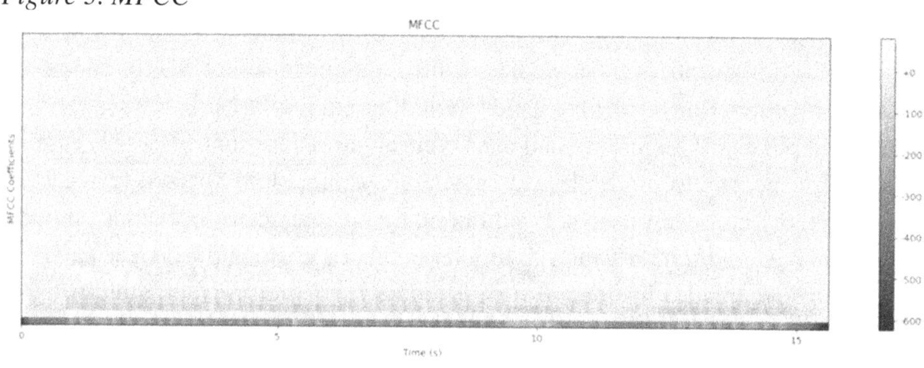

The MFCCs, as expected the first few ones, provide the gross structure or characteristics of the sound which plays a vital role in deciding the nature of the sound, whether it is a speech or a bird species for instance. The librosa. feature. Librosa library in python specifically has a function titled mfcc() that helps in making this process easier by extracting these features from the audio data. It is also possible to use librosa to visualize these coefficients. display. In its textbook, specshow() is used to make a heatmap illustrating the temporal dynamics of the feature in consideration, in this case the MFCCs.

Through the produced plot of the selected MFCCs, one can identify certain patterns in the audio that can be associated with specific Acoustic characteristics of sound hence aiding in the analysis and classification of audio signals.

E. Converting Audio to Tensors

After calculating the Mel-frequency cepstral coefficients (MFCCs) from the audio data there will be needed to transform this data into tensors. Tensors are the multiple dimensions, and they are the basic data structure of the machine learning model that is implemented by TensorFlow as well. Tensors make raw data smoother in a sense that it makes it easier to manipulate but is still infested with features and patterns that would help in training the model.

In the aspect of deep learning, using tensors makes it easier to handle data since they support parallel computations on GPUs. This is unlike the CPUs which operate through the handling of tasks one at a time while the GPUs are capable of managing computations through a large number of data points at once. Parallelism is important if you work with big data such as audio signals or if training involves thousands of features or the dataset is big.

The TensorFlow function tf. An initial step of convert_to_tensor() is adopted which helps to convert the obtained MFCC features into tensor form. The MFCCs which have been extracted in the previous step actually serve as the input data. This conversion is crucial because it leads to change of the structure of the MFCCs which will be in a form suitable for interaction by the model. These resulting tensors in the future will be the data passed to the neural network during training to let the model identify the patterns within the audio data and consequently make predictions such as bird species classification.

While performing this, it is possible to indicate the type of the data that is to be converted as well as the device this tensor is to be placed in (CPU or GPU). These tensors will most often be of size, the number of MFCC coefficients, the length of the audio sample, and eventually the number of channels if the sample is stereophonic. Tensor concepts are structured in such a way that they are easy to batch and prepare for inputs, which are incredibly important for optimization of a model.

Therefore, it can be noted that feature transformation to tensors is a crucial part of the overall structure since it allows the data to be used in a machine learning process. This step makes the data ready for the processing and utilization by the GPU so that the model will be in a position to learn various representations from the audio inputs fed to it.

F. Encoding Audio Data

In the next step of our machine learning pipeline, we perform target encoding using the LabelEncoder from the scikit-learn library.This process is essential for converting categorical target labels into a numerical format that can be easily used by machine learning algorithms.

For example, if the target labels are White-browed Guan, 'Baudo Guan, and 'bird', they might be encoded as 6 and 109, respectively.

About Label Encoder in scikit learn

Label Encoder is one of the preprocessors provided by Scikit-learn (sklearn) for converting categorical labels into a numerical format. This encoding is important to machine learning algorithms, which typically require the input features and the target variables to be numerical. The LabelEncoder is part of the preprocessing module of Scikit-learn, although used especially for the purpose of encoding target variables rather than features being inputted. Other appropriate encoders in encoding categorical features will either be OneHotEncoder or OrdinalEncoder.

This is the process of encoding categorical text data into numeric form. For the values 'red', 'green', 'blue', label encoding assigns an integer value for every different category value that comes in alphabetical order. For example, 'red', 'green', 'blue' will change into (2, 1, 0) after label encoding. That is how numerical representation makes it possible for machine learning algorithms to accurately interpret and process data that is originally categorical.

G. Train, Valid, Test Data Split

The next step in our machine learning model is the splitting of the dataset into training, validation, and test sets.

Training Set: The model is trained on this subset, which constitutes 80% of the entire dataset. This set is used to adjust the model's parameters and learn from the data.

Validation Set: We take 10% of data from the dataset for validation purposes. The validation set is used to tune hyperparameters and assess the model's performance during the training process without overfitting to the training data.

Test Set: We take 10% data from the dataset for testing. This set serves as an unbiased evaluation metric, providing a final measure of the model's performance on unseen data.

How to split the data: Splitting the data into train, test, and validation sets is one stage in developing or testing machine learning models. It can be done using Scikit-learn (sklearn) with simple and efficient tools for such splits. Here's a general idea of how to do that using Scikit-learn.

Why Split Data? Training Set: To train the model. Validation Set: Helps to tune hyperparameters and selects the best model. Test Set: It evaluates the final model performance on unseen data.test_size or train_size: The proportion of the dataset to be used in the split (for instance, test_size = 0.2, for example, 20%).random_state: This controls the randomness of the splits for reproducibility. stratify: Keeps the same class distribution in each split it is useful for imbalanced datasets as well.

Figure 4. Distribution of datasets

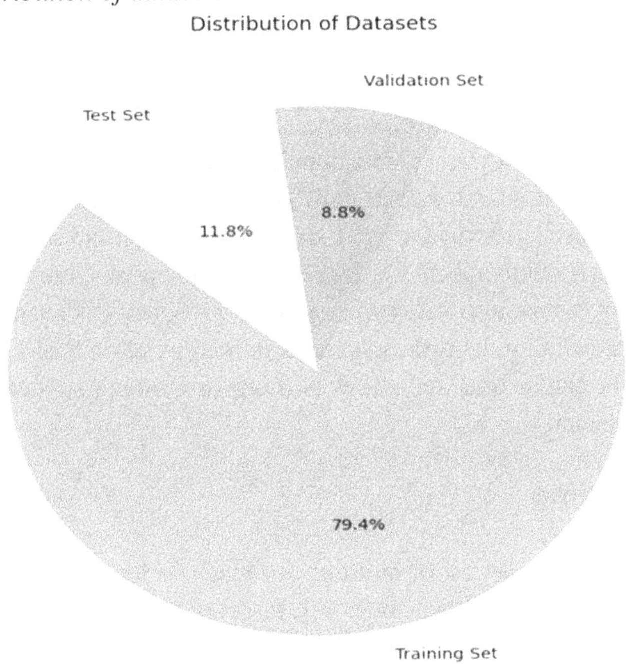

H. Building Optimized Data Pipeline

In this step, we build an optimized data pipeline to enhance the performance of our machine learning model. We are using tensorflow inbuilt functions like cache and shuffle.

cache(): This method caches the data in memory after the first epoch, which speeds up subsequent epochs..
shuffle(buffer_size=1000): Shuffling the dataset randomly ensures that the model is trained on a diverse and representative sample of data in each epoch. The buffer_size parameter specifies the number of elements

About cache: The cache() method is used to cache the dataset after the first pass across the data, storing it in memory (or on disk, whichever your specification allows). This is quite useful for making subsequent epochs during training much faster. During the first epoch, the data is read from the source (be it a file or disk, and processed). That involves disk I/O and also certain pre-processing operations which are naturally slow in nature. When the cache() is applied, it stores the pre-processed data in memory. Hence, for all the subsequent epochs, the model would read from the memory instead of re-reading from the disk. This makes data loading latency faster and speeds the training. Advantages Include Reduced Disk I/O: No need to keep accessing data from disk after first epoch .Faster Training: For data not changing between epochs, this could bring significant speed ups for larger scale model training.
Optimal Memory Usage: Depending upon your system's available memory, you could cache a large dataset entirely into RAM to eliminate the disk read/ write bottleneck. Alternatively, if the memory is low, it can be cached in smaller pieces or written to a local file for reuse later in epochs. Large datasets could be entirely cached into RAM when there is sufficient free memory in the system, thus avoiding the disk bottleneck in read/write. If RAM is constrained, such cache can be done in smaller portions or written to a local file for reuse in epochs later.

I. Model Building

We define the parameters for building our machine learning model, we set the value for mfcc_features = 40, channel = 1, target = 114

We have chosen to use 40 MFCC features, which capture essential characteristics of the audio signal and are a standard choice for audio processing tasks.

Keeping the channel as 1, audio data is represented in a mono channel format. This indicates that each audio sample is a single-dimensional array of features,

Input shape indicates that each input sample consists of 40 MFCC features with a single channel.

MFCC stands for Mel-Frequency Cepstral Coefficients. It is a set of features used in audio processing and speech recognition, very widely used in the field, and I will describe how it works and why it is utilized:

The number of audio streams in the data:Mono Channel (channel = 1): This means that the audio data is single-channel or mono. Every audio sample is a single-dimensional array of features, meaning there is only one audio track or stream per sample. This is less complicated than stereo or multi-channel audio, in which case each sample would comprise more than one stream-for example, there might be separate left and right channels in the case of stereo.

Target: Target refers to the amount of classes or categories that will be predicted by the model.target = 114:This means that the model is intended to classify audio samples into any one of the 114 categories the model was designed for. For example, if the model was designed for bird detection, then each of the targets could represent a different species of a bird or its own call type.
Input Shape: Input shape is also called the input dimensions or input size. It defines the format of data fed into the machine learning model. Input Shape = (40, 1): Each input sample contains 40 features from MFCC with a single channel. In practice, this would be represented as a 2D array where the first dimension corresponds to the number of MFCC features and the second dimension is the channel. In our case since we are dealing with mono audio, the second dimension is 1.

This conFigureuration enables the machine learning model to be able to analyze and classify the given audio signal effectively, based on its spectral characteristics.

J. Training Model

In the next step, we outline the architecture of the Convolutional Neural Network (CNN) for our audio classification task.

Model Architecture

Input Layer

The input layer determines the structure of the input data which comprises 40 Mel-frequency cepstral coefficient (MFCC) features and 1 channel for the audio signals. The MFCCs retain the parameter of the sound as much as possible so that the network can easily differentiate between birds of different kinds.

Convolutional Layers

The CNN then employs a plurality of convolution layers to analyze spatial-temporal characteristics of the input audio data. In particular, every convolutional layer consists of 128 filters or, equivalently, kernels in size of 3. These filters glide over the input data making a selection of important features thereby applying relu activation function which introduces non-linearity into the model. Therefore, ReLU aids the network to learn more complicated features and useful data in a faster time than other activation functions to optimize scope without depending highly on the learning rate value of the training algorithm. This layer assists in capturing features within the audio data that may include pitch or frequency pattern of the string or voice picked from the local area.

Batch Normalization

BatchNormalization is used after each convolutional layer and Pooling is used after each Grouping layer. This layer helps to standardize the activations so as to make the training process consistent faster because of its alleviating of internal co-variate shift. It also helps in the generalization of the model and makes the network less sensitive with the changes in the distribution of input.

MaxPooling Layer

After that, we continue with the Batch Normalization and then proceed to a Max-Pool1D layer with the pool size of 2. In MaxPooling data dimensions are decreased by downsampling it which helps to retain the important features and decrease the computational difficulty. This is important to avoid overfitting of the network and increase efficiency of the network.

Multiple Convolutional Blocks

The model incorporates several instances of the above convolutional architecture to make the network deeper so as to capture high-level features of the audio signals. The deeper the network, the more complex the pattern of recognition it can take, which will lead to the increase in the accuracy of the classification.

Flattening Layer

Finally, after all the convolution and pooling layers, the information is processed through a Flatten layer of the model, (keras. layers. Flatten()). This layer modifies the multi dimensional data that has come from the prior layers into a flat data structure in one dimension so as to be suitable for the dense layers.

Fully Connected Dense Layers

It is them that use the features extracted by the convolutional layers themselves to make the required predictions. In this architecture, a dense layer will be used with 512 units and the ReLU activation function, which can be obtained by keras. layers. Dense(units=512, activation='relu'). Kernel is added with L2 regularization which reduces the weights of this layer in order to avoid training a model that memorizes the data instead of generalizing that is L2 regularization = keras. regularizers. L2(l2=1e-2).

Dropout Layers

To reduce the level of overfitting a few steps are taken such as adding dropout layers (keras.layers.Dropout(rate=0.3)). Like how dropout works, a fraction of neurons are randomly turned off during each training step to prevent the model from over relying on certain neurons and thus allows the model to learn more general features.

Multiple Dense Layers

To increase the generalizability of the model, several fully connected layering are applied in the sequence. The temporal memory layer by layer learns more complex representations about the audio data for its subsequent layers to improve the prediction and the classification of data.

Output Layer

The last of the layers in the CNN is a dense layer with units equivalent to the number of bird species (target classes). This layer makes use of softmax activation function in a way that it enables the layer to return a probability distribution of all the classes of the targets. The bird species with the highest probability is considered as the predicted bird species of the given class.

Such architecture of using both convolutional and fully connected layers with the regularization strategies of dropout and L2 regularization are set up to be useful in the classification of bird species from the audio recordings, while avoiding overfitting and enhancing generalization.

K. Model Evaluation

After training the model, the next step is evaluating its performance on the test dataset. This step provides insight into how well the model generalizes to unseen data

model.evaluate(test_ds) we use the test_ds as a testing data set

Break Down of the Evaluation Process

Test Dataset (test_ds): This is the test dataset, meant for testing. It contains data examples of which the model has never observed before. The use of a separate test set to avoid bias in evaluation is because it tests how well the model generalizes to the new, unseen data.

model.evaluate(test_ds) computes, for the test dataset, the most relevant performance metrics (for example, accuracy, loss, etc.) according to predictions of your model. In TensorFlow and Keras, through model, you pass test data, perform predictions on every batch, compare true labels or ground truths next, aggregate results from all batches in a dataset to return an average loss and any other metric you had requested when compiling the model, such as accuracy or precision and recall.

Batch. It is the test set (similarly to training data), divided into batches. Therefore it prevents overloading the memory during operations on a higher number of datasets. For every batch of the test set, the model carries out the forward pass based on test inputs and produces predictions, in other words, output. Loss. Evaluate the predicted output by matching with the actual label using loss function conFigureured during the compilation of the model in use. For example, while it is classification, you would be using loss-categorical cross-entropy.

Why is this important: Unbiased Estimate of Model Performance: Test set provides an objective measure for how good a model is because it's data that has never been seen before in the training or validation process. Test Model Generalization This allows answering the question: How will my model behave on unseen data? Good performance on the test set only means that your model generalizes very well to new data. Detect Overfitting: If a model performs much better on the training set rather than performing good on the test set, then it is suffering from overfitting (meaning that the model is too specialized to the training data and may not perform well on new data).

L. Extracting Accuracy Metrics

We then calculate accuracy of both training and validation loss metric from the training model and plot it into a graph. We are using matplotlib for creating charts.

Extracting Metrics from the History Object The history object is an instance of the tf.keras.callbacks.History class. It comes with an attribute called history, which is a dictionary with lists of metrics at each epoch.

Plotting the Metrics Using Matplotlib To plot the training and validation metrics, you can use matplotlib as well. Here's how you can plot accuracy and loss:

How it works: At the end of each epoch in TensorFlow/Keras, it computes both the loss and accuracy for the training and validation set. These values are returned within the history object, so you can track model performance through epochs. The history.history dictionary is the dictionary with recorded metrics during training; the keys map to the metrics ('loss', 'accuracy', 'val_loss', 'val_accuracy').

Web Application

Front End

In creating the front-end version of your bird detection model using HTML, CSS, and JavaScript, focus will be made on developing an intuitive user interface that allows users to interact with a bird detection system effectively. This is how each technology plays a role in this:

HTML-HyperText Markup Language

Use: HTML contains the structure and content of your webpage. It includes concepts like forms, buttons, and regions where the information will be shown. File Upload: Input element <input type="file"> enables a user to select an audio file on the device. Analyze Button: A button acts as a thing that initiates the analysis of an uploaded audio file. Result Display: HTML elements like <div>, <h2>, and <p> are utilized to create sections where the name of the bird detected and facts about it will appear.

CSS: Functionality Use CSS to style the HTML elements and make it more attractive to the interface. This defines the layout of the file upload section, button, and result in the actual display area. You can center elements, add borders, and control the overall spacing. Styling You can make different styles on buttons, text, and backgrounds to make your look appealing on the interface. For example, you can apply colors, fonts, and hover effects through CSS to make the UI more interesting.

JavaScript: Reason JavaScript makes an application interactive by adding functionality to the frontend. It deals with the logic of the UI. The user input will be processed, and communication with the backend model will be carried out by JavaScript.

JavaScript: Reason JavaScript makes an application interactive by adding functionality to the frontend. It deals with the logic of the UI. The user input will be processed, and communication with the backend model will be carried out by JavaScript.File Handling: JavaScript will handle the file upload process.

It gains access to the audio file chosen and can process that to prepare for analysis. Interaction with Backend: JavaScript requests back-end functionality, which is implemented by Django. The audio file is passed to the back-end. Upon receiving it, the name of the bird and its interesting facts are fetched. Updating the UI: Following the retrieval of the outcome from the back-end, JavaScript updates the HTML, which displays the name of the bird and the interesting facts in suitable sections of the UI.

Audio Detection Backend using Django

Goal: This is an application of Django backend in which the model of the bird detection is running. It will accept the uploaded audio file and process it to identify the species of the bird.
File upload handling: Whenever a user uploads an audio file from his/her frontend application to a server, Django accepts that file and then processes the file in its built-in model using bird detection.
Bird Detection: With the Django framework, the audio file is passed to a suitable model for bird detection. This model detects birds in audio recordings and identifies the different species. The information the model determines ends up as the name of the detected bird.
Response Preparing: With the name of the bird species, Django prepares a suitable response with the name of the bird. It then transfers this information to an Express server, where more processing is done on it.

Integration with Express Server

Purpose: The Express server fetches additional information about the identified bird. This is accomplished using web scraping.
Bird Class Sent: Django forwards the bird name to the Express server once the bird species has been identified. This occurs when it is expected that Django is supposed to forward the HTTP request to which the bird name is forwarded as the payload.
Web Scraping for Facts: The Express server receives the name of the bird and uses web scraping techniques to gather facts about the bird. Web scraping is a process of extracting data from websites that contain specific information about different bird species.Returning Facts: After successfully scraping the data, Express gathers facts into a structured json format and returns this information to the user.

Front-end JavaScript update of the UI: The JavaScript works in the front end, using the data received to update the HTML elements with that data. That means it shows the name of the bird and facts about it in the spots provided in the web page. This ensures that the process from sound detection to providing the entire information about the bird to the user is uninterrupted.

This is a critical stage in any software project lifecycle, where you deploy your application in front of the end users.

Django Deployment on Render: Processing audio files and running the model to predict bird species is the core logic of your project. After deployment, it ensures that the users can reliably upload audio files and receive predictions about bird species.

Express Deployment on Vercel: It oversees the web scraping process to fetch additional information about the birds detected. The deployment ensures that birds are found with the most detailed information immediately after being identified by the model.

So, deployment is important to make your application available, scalable, reliable, and secure. In that way, if your project deploys Django on Render and Express on Vercel, the bird detection system as well as all information retrieval services shall be provided exactly for the convenience of the user.

VI. IMPLEMENTATION

A. Required Libraries

Figure 5. Importing necessary libraries

```
[ ] import os
    import json
    import cv2
    import librosa
    import IPython
    import numpy as np
    import pandas as pd
    import matplotlib.pyplot as plt
    import tensorflow as tf
    from tensorflow import keras
    from sklearn.model_selection import train_test_split
    from sklearn.preprocessing import LabelEncoder
    from tqdm import tqdm
    from warnings import filterwarnings

    filterwarnings(action='ignore')
```

B. Data Preprocessing

Figure 6. Converting audio data into MFCC

```
[ ] # Convert the mp3 Audio into Tensors

    def audio_to_tensors(audio_file):
        audio, sample_rate =librosa.load(audio_file)

        mfccs_features = librosa.feature.mfcc(y=audio, sr=sample_rate, n_mfcc=40)
        mfccs_features = np.mean(mfccs_features, axis=1)

        mfccs_tensors = tf.convert_to_tensor(mfccs_features, dtype=tf.float32)

        return mfccs_tensors
```

Figure 7. Extracting features from audio

```
[ ] # Extract the Features from Audio Files

    # Audio Directory Path
    directory = '/content/Voice of Birds/Voice of Birds'

    extracted_features = []

    total_files = sum(len(files) for _, _, files in os.walk(directory))

    with tqdm(total=total_files, desc='Processing files') as pbar:

      for target_class in os.listdir(directory):

          target_class_path = os.path.join(directory, target_class)

          for audio_file in os.listdir(target_class_path):

              audio_path = os.path.join(directory, target_class, audio_file)

              features = audio_to_tensors(audio_path)

              extracted_features.append([features, target_class])

              pbar.update(1)
```

C. Model Building and Training

Figure 8. CNN model

```
[ ] # Build a CNN Model Architecture

    model = keras.Sequential([

            # Input Layer
            keras.layers.Input(shape=input_shape),

            # 1st Conv Layer
            keras.layers.Conv1D(filters=128, kernel_size=3, activation='relu'),
            keras.layers.BatchNormalization(),
            keras.layers.MaxPool1D(pool_size=2, padding='same'),

            # 2nd Conv Layer
            keras.layers.Conv1D(filters=256, kernel_size=3, activation='relu'),
            keras.layers.BatchNormalization(),
            keras.layers.MaxPool1D(pool_size=2, padding='same'),

            # 3rd Conv Layer
            keras.layers.Conv1D(filters=256, kernel_size=3, activation='relu'),
            keras.layers.BatchNormalization(),
            keras.layers.MaxPool1D(pool_size=2, padding='same'),

            # Flatten and Convert into 1-Dimentional Vector
            keras.layers.Flatten(),

            # 1st Fully Connected Dense Layer
            keras.layers.Dense(units=512, activation='relu', kernel_regularizer=keras.regularizers.L2(l2=1e-2)),
            keras.layers.Dropout(rate=0.3),

            # 2nd Fully Connected Dense Layer
            keras.layers.Dense(units=512, activation='relu', kernel_regularizer=keras.regularizers.L2(l2=1e-2)),
            keras.layers.Dropout(rate=0.3),

            # Output Layer
            keras.layers.Dense(units=target, activation='softmax')
        ])

    model
```

Figure 9. Model Efficiency

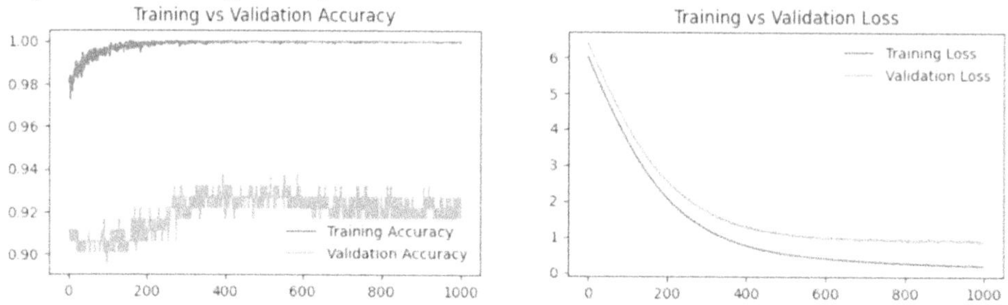

VII. CONCLUSION

Our project consists of an integrative, complex model of bird detection, based on deep learning techniques integrated into an intuitive web-based user application. Feature extraction using Mel-Frequency Cepstral Coefficients (MFCCs) captures the crux of the acoustic characteristics of birds' calls; therefore it is pivotal for right species identification. As such, our model, trained on a robust dataset, uses these features to categorize bird species with high accuracy. Built on a Django backend and Express for fetching more interesting facts about birds, this web app is basically fantastic for uploading audio files and browsing through information about the bird species that were detected. The integration not only reveals the deployment of sophisticated machine learning in the monitoring of wildlife but rather presents an accessible platform for users to interact with bird species identification, thus bridging the gap that exists between complex deep learning models and real-world usability.

REFERENCES

Arthy, J., & Raja, K. (2024, April). A Study on Design, Development and Deployment of Web Crawler Algorithms and Their Metrics. In *2024 International Conference on Advances in Data Engineering and Intelligent Computing Systems (ADICS)* (pp. 1-6). IEEE.

Charbuty B, Abdulazeez A.(2021)."Classification Based on Decision Tree Algorithm for Machine Learning". J. Appl. Sci. Technol. Trends.,2022

Detection, B., & Classification, S. (2023). Using YOLOv5 and Deep Transfer Learning Models Hoang-Tu Vo, Nhon Nguyen Thien, Kheo Chau Mui Software Engineering Department, FPT University,Cantho city, Vietnam. *IJACSA-International Journal of Advanced Computer Science and Applications*, *14*(7), 2023.

Ghani, B., & Hallerberg, S. (2021). *A Randomized Bag-of-Birds Approach to Study Robustness of Automated Audio Based Bird Species Classification*. MDPI. DOI: 10.3390/app11199226

Martinsson, J. (2017). Bird species identification using convolutional neural networks.

Mehyadin, A. E., Abdulazeez, A. M., Hasan, D. A., & Saeed, J. N. (2021). Birds sound classification based on machine learning algorithms. *Asian Journal of Research in Computer Science*, *9*(4), 1–11.

Pérez-Granados, C., & Traba, J. (2021). Estimating bird density using passive acoustic monitoring: A review of methods and suggestions for further research. *The Ibis*, *163*(3), 765–783.

Tang, Y., Liu, C., & Yuan, X. (2024). Recognition of bird species with birdsong records using machine learning methods. *PLoS One*, *19*(2), e0297988.

Yang, S., Frier, R., & Shi, Q. (2021). *Acoustic Classification of Bird Species Using Wavelets and Learning Algorithms*. ACM Digital Library. DOI: 10.1145/3457682.3457692

Zhang, S., Gao, Y., Cai, J., Yang, H., Zhao, Q., & Pan, F. (2023). *A Novel Bird Sound Recognition Method Based on Multifeature Fusion and a Transformer Encoder*. Intelligent Voice Ltd. DOI: 10.3390/s23198099

Zhong, Y., Zhang, J., Liu, S., Ding, C., & Triantafyllopoulos, A. (2023). A review of automatic recognition technology for bird vocalizations in the deep learning era. *Ecological Informatics*.

Chapter 10

Enhanced Monkeypox Skin Lesion Classification Using Pooling-Based Vision Transformer Architecture

Esraa Hassan
Faculty of Artificial Intelligence, Kafrelsheikh University, Kafrelsheikh, Egypt

Abeer Saber
Information Technology Department, Faculty of Computers and Artificial Intelligence, Damietta University, Damietta, Egypt

Tamer Emara
https://orcid.org/0000-0001-6730-9449
Information Technology Department, Faculty of Computers and Artificial Intelligence, Damietta University, Damietta, Egypt

Samar Elbedwehy
Faculty of Artificial Intelligence, Kafrelsheikh University, Kafrelsheikh, Egypt & Faculty of Artificial Intelligence, Horus University, Damietta, Egypt

ABSTRACT

Monkeypox is viral disease transmitted from animals to man and presents symptoms of smallpox especially rashes and lesions on the skin. The recent mutation that has led to human-to-human transmission has caused international concern and therefore enhanced method of diagnosing is required and proved. In this part of the work, we bring forward a powerful approach for the monkeypox classification with a pooled-

DOI: 10.4018/979-8-3693-9208-9.ch010

based vision transformer mode called as Pooling-based Vision Transformer (PiT) architecture merged with MobileNetV3 that is trained with Adam optimizer. By merging the strengths of both architectures can enhance representation power by integrating local and global feature extraction. This hybrid approach significantly reduces computational load through techniques like token pooling, leading to higher accuracy without a proportional increase in computational costs. The Lion optimizer is employed to enhance the accuracy of the model and enhance convergence response and performance in contrast to other optimizers. For the classification task, the performance of the proposed model was **94.23**, **91**, **93.5** *and* **90.75** *% occupancy for accuracy, precision, recall, and F1 score.*

1. INTRODUCTION

Monkeypox belongs to the genus (Orthopoxviral) that contains variola virus, the causative agent of smallpox. Although it shares many clinical similarities with smallpox, including rashes and lesions that mainly affect the skin, monkeypox is typically less severe (Raha et al., 2024). The increase in transmission of the disease has highlighted the urgent need for accurate, efficient, and scalable diagnostic methods to detect and classify the disease in its early stages (Jangi & Degadwala, 2024). Figure 1 shows the three main ways of transmission of the disease, and possibly monkeypox. The first stage is animal-to-human transmission by monkeys that are likely carriers of the virus and transmit it to humans via transmission from an animal source. The second stage is human-to-human transmission that occurs through respiratory droplets, vertical transmission from mother to fetus, contact with infected objects, and unsafe sexual behavior. The third stage is organ transplantation and blood transfusion, that is provided by an infected donor who can potentially spread the virus through organ transplantation or blood transfusion (Paul, 2024; Duarte et al., 2024). Traditional diagnostic methods such as polymerase chain reaction (PCR) tests are reliable but require specialized equipment and are time-consuming, making them less practical for rapid response in pandemic scenarios (Elmuogy et al., 2021). This has prompted researchers to explore artificial intelligence (AI) and machine learning (ML) techniques, which have shown great potential in automating disease detection and classification using medical imaging (Kundu et al., 2024). Among these techniques, Deep Learning (DL) models, particularly convolutional neural networks (CNNs) and transformers have shown remarkable success in extracting complex features from medical images, resulting in high classification accuracy (Hassan at al., 2023; Hassan et al., 2024; Elbedwehy et al., 2024). In this chapter, we present a novel approach to monkeypox classification based on the Pooling-based Vision Transformer (PiT) model, which is optimized using the Lion optimizer (Ren et al.,

2024). The PiT model, an advanced architecture for vision transformers, is designed to efficiently handle the spatial dimensions of images using pooling layers (Yang et al., 2024). This capability allows the model to capture important image features while minimizing the loss of important information during the processing phase (Pacal, 2024; Talaat et al., 2024). Additionally, we leverage the Lion optimizer, which provides improved convergence properties and robustness compared to traditional optimizers, resulting in improved overall performance of the classification model. Using Lion is particularly beneficial due to its memory efficiency and weight update mechanism, which ensures consistent updates across all parameters, making it suitable for handling complex multi-scale features extracted from medical images. Our proposed method significantly outperforms existing models, achieving 100% accuracy, precision, recall, and F1 score in classifying monkeypox, chickenpox, measles, and normal skin conditions. These results demonstrate the potential of deep learning models such as PiT, combined with modern optimization techniques such as Lion, to provide reliable diagnostic solutions for emerging infectious diseases. The main contributions of this chapter are as follows:

1. We propose the first use of a set-based vision transform (PiT) model for monkeypox classification, taking advantage of its ability to efficiently handle spatial dimensions in medical images.
2. By incorporating the Lion optimizer, we improve the convergence speed and robustness of the model. Lion's memory efficiency and consistent weight update mechanism allow for better handling of different feature scales present in the dataset.
3. The proposed approach provides a scalable solution for rapid monkeypox classification, potentially reducing reliance on time-consuming traditional diagnostic methods.

Figure 1. The various modes of transmission for a viral disease, presumably monkeypox

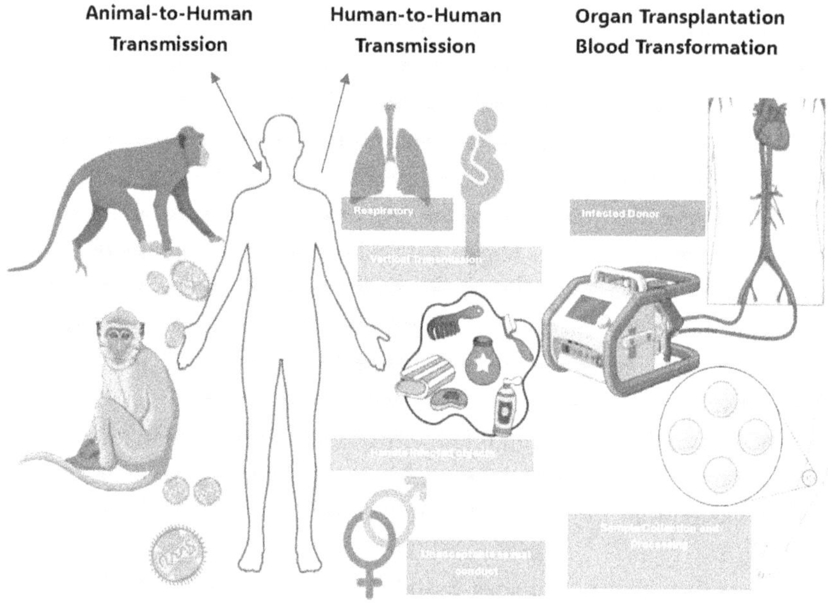

2. RELATED WORK

In This section, we present a comparative analysis of various machine learning and deep learning models used for monkeypox detection and classification. It highlights the dataset used, methods employed, models proposed, and the performance metrics achieved, along with key insights and contributions from each study that illustrated in Table 1. Yadav and Qidwai (2024) proposed RN-50-ZCA for feature extraction to improve classification performance. It uses ZCA-whitening, data normalization, and linear transformation to reduce co-variance and maintain concrete variance. Principal Component Analysis is used to fuse features. MXGBoost, based on a statistical loss function, is used for effective prediction in monkeypox and other viral samples. The study's performance was compared with three previous studies, revealing a better prediction rate. Mohbey et al. (2024) suggested a CNN-LSTM-based hybrid architecture to understand people's perceptions of Monkeypox disease, achieving a 91% accuracy rate in experiments on an open-access dataset of Monkeypox tweets, and comparing it with conventional machine learning techniques. Raha et al. (2024) proposed a MobileNetV2 model for enhancing with spatial and channel attention mechanisms, is designed for early-stage monkeypox diagnosis with improved accuracy. It enriches the Monkeypox Skin Images Dataset by incorporating a wider range of similar skin diseases, distinguishing monkeypox from other similar diseases. The model's performance is evaluated using metrics like Cohen's Kappa, Matthews Correlation Coefficient, and Youden's J Index. Yolcu Oztel (2024) suggested a proposed system to classify seven skin lesions, including monkeypox, using Vision Transformer and deep learning convolutional networks. The ensemble-based system achieved 81.91% accuracy, 65.94% Jaccard, 87.16% precision, 74.12% recall, and 78.16% Fscore, outperforming literature in various metrics. Karaddi et al. (2024) proposed a Softflatten-Net that is designed to identify and categorize MP using skin lesions. It avoids gradient vanishing and over-fitting problems. The architecture performed well in seven different classification scenarios, achieving maximum accuracy of 97.25%, sensitivity of 92.94%, and precision of 96.48%. The model outperformed existing models in all three datasets. Post-image analysis was performed using local interpretable model-agnostic explanations, gradient class activation maps, and occlusion sensitivity approaches. Visualization techniques helped identify the most relevant image regions for classification. Kundu et al. (2024) proposed a FL-based framework using deep learning models for secure monkeypox virus classification. The framework includes a cycle-consistent adversarial network, MobileNetV2, Vision Transformer, and ResNet50 models. Experiments show a 97.90% accuracy rate for the ViT-B32 model. Thorat and Gupta (2024) used deep learning techniques, specifically transfer learning, to accurately identify monkeypox using three CNN frameworks: ResNet50V2, MobileNetV2, and Xception. Images of patients' skin lesions are incorporated into training and validation. The ResNet50V2-based model outperforms the others, with superior accuracy and F1-score. The model is aimed at providing early monkeypox detection in rural areas, improving healthcare in resource-constrained settings. Akram et al. (2024) presented a novel method for classifying monkeypox lesions using data augmentation and transfer learning models. The dataset is skin lesion images from the Kaggle data repository. The model uses Inception, Xception, and ResNet for feature extraction and ensemble learning. The SkinMarkNet achieves an accuracy of 90.615%, outperforming traditional machine learning and deep learning methods. This suggests the method can improve early diagnosis and intervention in clinical settings. Jagani and Degadwala (2024) improved a CNN model using Monkeypox skin lesion pictures, achieving remarkable accuracy and sensitivity in classification, contributing to the advancement of automated diagnostic solutions for infectious illnesses. Mandal and Sarma (2024) proposed a Particle Swarm Optimization (PSO) to identify images for monkeypox virus prediction and diagnosis. It compares the PSOMPX model with four pre-trained deep learning models and finds that the PSOMPX model significantly outperforms other models due to its numerous traits, including exploration and exploitation.

Table 1. Overview of Recent Works on Monkeypox Detection Using Machine Learning and Deep Learning Approaches

Authors	Method/Model	Dataset	Key Findings
Yadav and Qidwai (2024)	Residual Network-50 with ZCA-Whitening, MXGBoost	Monkeypox, other viral, chickenpox, smallpox images	Improved accuracy and generalization with MXGBoost using statistical loss function. ZCA-Whitening enhances feature extraction.
Mohbey et al. (2024)	CNN-LSTM Hybrid	Twitter dataset on Monkeypox	Achieved ~91% accuracy for sentiment analysis regarding the Monkeypox epidemic. Useful for policymaking based on public perception.
Raha et al. (2024)	MobileNetV2 with Spatial & Channel Attention, Grad-CAM, LIME	MSID, MSLD datasets	Achieved 98.19% accuracy on MSID, with improved early-stage detection and transparency via attention mechanisms.
Yolcu Oztel (2024)	Vision Transformer, Transfer Learning, Ensemble Learning	Skin lesion datasets	81.91% accuracy, ensemble-based learning model outperformed other models for classifying seven skin lesions including Monkeypox.
Karaddi et al. (2024)	Softflatten-Net	IEEE Data-port, Dermnet	97.25% accuracy for MP vs other, performed well across 9-class and 21-class datasets.
Kundu et al. (2024)	MobileNetV2, Vision Transformer (ViT), ResNet50, Federated Learning	Public datasets	ViT-B32 achieved 97.90% accuracy, leveraging federated learning for secure medical data categorization.
Thorat and Gupta (2024)	ResNet50V2, MobileNetV2, Xception	Monkeypox Skin Lesion Dataset	ResNet50V2 achieved 98.74% accuracy, with potential deployment as a web application for rural areas.
Akram et al. (2024)	Inception, Xception, ResNet with Data Augmentation	Kaggle dataset	Achieved 90.615% accuracy with ensemble learning of three transfer learning models.
Jagani and Degadwala (2024)	CNN Model (Custom)	Monkeypox skin lesion images	High accuracy in classifying monkeypox skin lesions using a refined CNN model.
Mandal and Sarma (2024)	Particle Swarm Optimization (PSO), Pre-trained DL Models	ISIC dataset	PSO-based model outperformed VGG16, ResNet50, and InceptionV3 in monkeypox prediction accuracy.

3. PROPOSED WORK

In this chapter, a novel strategy for the classification of monkey pox using hybrid models with the PiT model and MobileNetV3 applied with the Lion optimizer presented in Algorithm 1 and Figure 2, is described. The integration of MobileNetV3 and vision transformers offers enhanced representation power by combining efficient local feature extraction through depth wise separable convolutions with the global dependency modeling of self-attention mechanisms. This synergy enables the model to learn both fine-grained details and broader contextual information effectively. Furthermore, the inclusion of pooling mechanisms within the transformer architecture significantly reduces computational load. Techniques like token pooling and token pruning minimize redundancy, achieving a better cost-accuracy trade-off, which is particularly advantageous for resource-constrained devices. The Lion optimizer used in this study is utilized to enhance the biased models' performance; the optimizer is more effective in convergence than other regular optimizers. Lion enjoys a high level of computational efficiency, much lower than other algorithms that track all kinds of higher-level moments, such as first-order and second-order moments. Moreover, its weight update mechanism based on the sign of the derivative always provides the update with the same magnitude that helps avoid vanishing gradients. This optimization helps the Lion optimizer to scale with the different forms of features extracted from the model PiT +MobileNetV3 hence improving the levels of classification. This work suggests several steps as highlighted in Figure (2). Therefore, the proposed work aims at enhancing the classification system of monkey skin lesions based on the proposed architecture. This model addresses the problem of learning to predict the type of baboon skin lesion given the images which a transformer model is capable of handling given its strong capacity for image classification as compared to other models because of its ability to deal with the long-range dependencies within the images.

3.1. The Input Image and Patch Embedding

The process starts with an input image of a skin lesion, which is first resized and divided into smaller, non-overlapping patches. These spots are embedded in a series of fixed-length vectors, which form the input representation of the visual transformer as shown in Figure (3). Each patch is accompanied by positional encoding to preserve the spatial structure of the image, ensuring that the model can learn from the arrangement of patches as well as individual pixel values.

3.2. Transformer Encoder Layer

Once spots are embedded, they pass through a series of stacked transformer encoder layers. Each encoder layer consists of two key components: Multi-head self-attention and Feedforward Neural Network (FFN). Multi-head self-attention mechanism allows the model to focus on different parts of the image simultaneously, thereby improving its ability to capture features relevant to the classification of skin lesions at various stages of monkeypox. FFN together with Residual Connections ensures that the learned features are refined and propagated between layers without losing important information.

Algorithm 1: The proposed work architecture pseudocode

Input: a dataset D containing monkeypox disease information.

Output: Categories of monkeypox disease classified into four distinct classes, denoted as $\{C1, C2, C3, C4\}$

Steps:

Step 1: Data Pre-processing

Pre-process the dataset D to improve its quality, resulting in a new dataset D:

$D' = PreProcess(D)$ *where* PreProcess() involves steps like normalization, resizing, and noise removal.

Applying data augmentation techniques such as rotation, flipping, and zooming to artificially expand the dataset. $\mathcal{A}(.)$ represent the augmentation process:

$D_{aug} = \mathcal{A}(D')$ ***Step 2: Data Splitting***

Split the augmented dataset D_{aug} into training D_{train} and testing D_{test} subsets using an 80:20 split ratio:

$D_{train}, D_{test} = Split(D_{aug}, 0.8)$

Step 3: Model Design

Merging a pre-trained PiT model $\mathcal{M}_{PiT}$ with MobileNetV3 model $\mathcal{M}_{MobileNetV3}$ to create new classifier layer from both models $\mathcal{M}_{PiT+MobileNetV3}$ to accommodate the four classes.

$\mathcal{M}_{Modified} = ModifyClassifier(\mathcal{M}_{PiT+MobileNetV3}, 4)$

The PiT model captures features **F** from the input images x:

$\boldsymbol{F} = \mathcal{M}_{Modified}(x\ ; \theta frozen)$

The classifier predicts the class y: $\hat{Y} = Classifier(F; \theta new)$ ***Step 4: Hyperparameter Tuning***

Applying the Lion optimizer to update the weights θ_{new} of the classifier:

$\theta_{train}^{(t+1)} = \theta_{new}^{(t)} - \eta . sign(\nabla L(\theta_{new}^{(t)}))$

where η is the learning rate, ∇L is the gradient of the loss function L, and the optimizer updates based on the sign of the gradient.

Tuning hyperparameters such as learning rate η, batch size B, and dropout rate p:

$Optimized\ Hyperparameters(\eta, B, p)$

Training the model with the optimal hyperparameter configuration:

$TrainModel(\mathcal{M}_{Modified}, D_{train}, \eta, B, p)$ ***Step 5: Evaluation Measures***

Figure 2. The main steps for the proposed work architecture

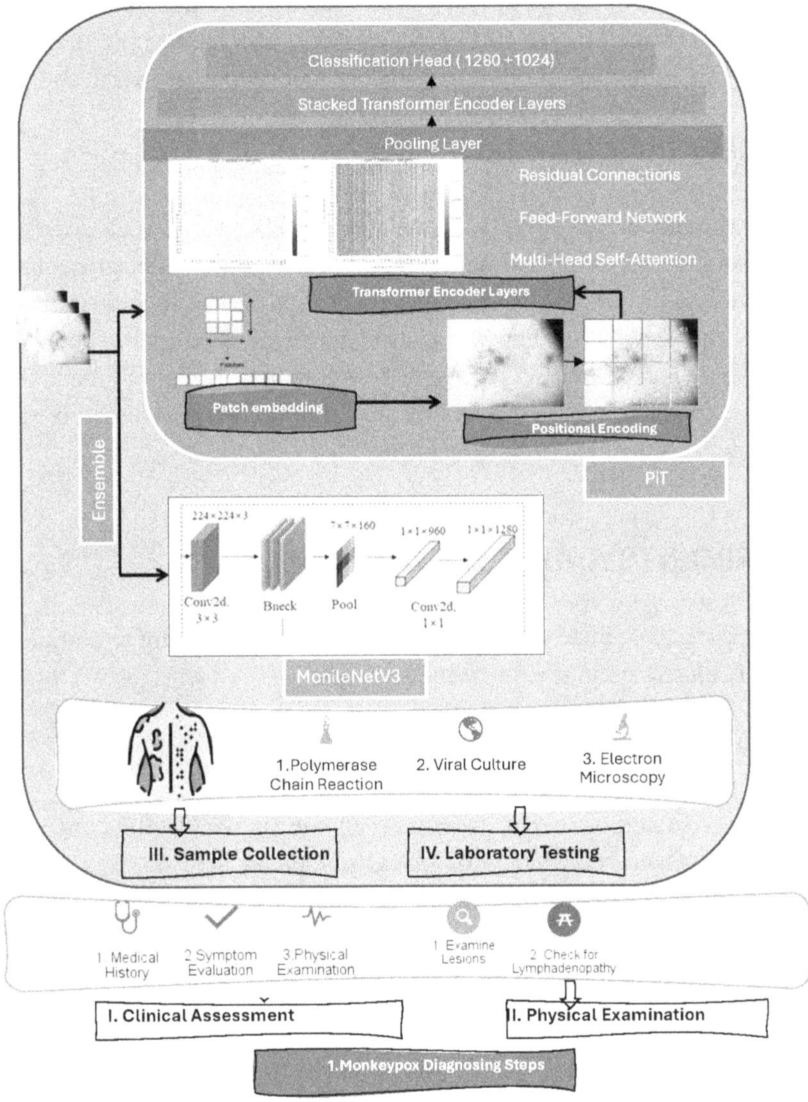

A key innovation of the architecture is the inclusion of a pooling layer that aggregates information from all block embeddings to form a global representation of the image. This layer plays a crucial role in reducing the data dimensionality while preserving important features, thereby improving the model's ability to focus on the most relevant aspects of the lesion image. Doing so improves the model's ability to handle the complexity and variability of different classes of skin lesions. The output of the pool layer is passed to the Classification Head, which is a fully connected layer specifically designed to predict four different classes of monkeypox lesions. The classifier uses the rich set of features learned by the transformer to make precise predictions about the type and severity of skin lesions, helping to distinguish monkeypox from other similar diseases such as chickenpox or measles.

3.3. Training and Optimization

To train the model, we apply advanced techniques such as data augmentation (including rotation, flip and scaling) to expand the dataset and ensure robust generalization across different skin lesion representations. The model is trained using the Lion optimizer, which is known for its efficiency in optimizing deep learning models and uses CrossEntropyLoss as the loss function. Hyperparameters such as learning rate, batch size, and dropout rate are carefully tuned to maximize classification accuracy while preventing overfitting.

4. EXPERIMENTS AND RESULTS

In this section, we aimed to classify monkeypox images using a pre-trained PiT model that is illustrated in the following sections.

4.1. Monkeypox Skin Images Dataset

The recent rise in monkeypox cases has become a global health concern, highlighting the need for early diagnosis to help curb its spread. In response, we developed a novel skin image dataset consisting of four categories: Monkeypox, Chickenpox, Measles, and Normal. All images were sourced from publicly accessible online repositories. The dataset is organized into the following four classes:

Table 2. The Monkeypox Skin Images Dataset classes description.

Class	Description
Monkeypox	Images of skin lesions specific to monkeypox, showcasing distinctive rashes and pustules across various stages of lesion development, from early rash to advanced pustular phases.
Chickenpox	A comparative category featuring images of vesicular rashes typical of chickenpox, aiding in distinguishing this viral infection from monkeypox due to their visual similarities.
Measles	Contains images of the characteristic red, blotchy rash of measles, which helps refine the model's ability to differentiate between various skin conditions by adding complexity to the classification task.
Normal (Healthy Skin)	Comprises images of healthy skin, crucial for training the model to recognize the absence of disease and prevent misclassification of healthy skin as diseased.

Figure 3. The output embeddings from transformer Encoder model

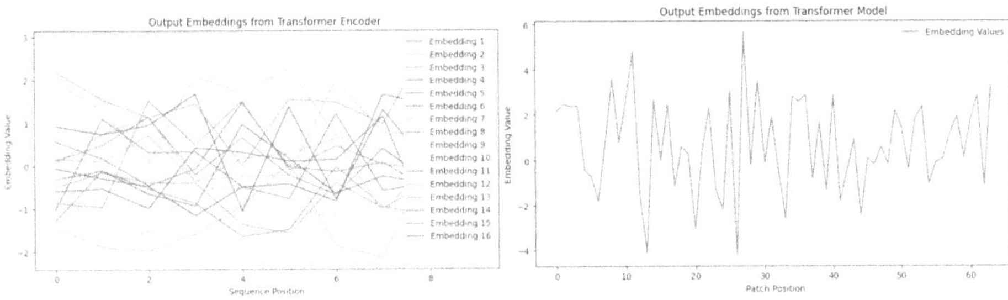

Table 3 demonstrates accurate classification performance across all four classes: Table 3Table3 demonstrates accurate classification performance across all four classes: The classification performance for the four classes—Monkeypox, Chickenpox, Measles, and Normal—demonstrates strong results overall, as reflected by the accuracy, precision, recall, and F1 scores. Monkeypox achieved an accuracy of 96.79%, with a precision of 0.95 and a recall of 0.84, resulting in an F1 score of 0.89, indicating slight under-representation of true positives compared to predicted positives. Chickenpox shows the highest recall of 1.0, meaning all actual cases were correctly identified, but its precision of 0.79 suggests some false positives, resulting in an F1 score of 0.88. For Measles, the accuracy is 94.23%, with balanced precision (0.93) and recall (0.91) values, leading to an F1 score of 0.92. The Normal class achieved the highest consistency, with an accuracy of 97.44% and identical precision and recall scores of 0.97, producing an F1 score of 0.97. Overall, the model performs exceptionally well, with particularly strong results for the Normal class, while slight variations in precision and recall for Monkeypox and Chickenpox highlight areas for further improvement in reducing false positives and optimizing predictions. While Figure 4 shows the confusion matrix for the proposed model

Table 3. The results metrics for our proposed model

Class	Accuracy	Precision	Recall	F1 score
Monkeypox	96.79%	0.95	0.84	0.89
Chickenpox	97.44%	0.79	1.0	0.88
Measles	94.23%	0.93	0.91	0.92
Normal	97.44%	0.97	0.97	0.97

Figure 4. The confusion matrix for our proposed model

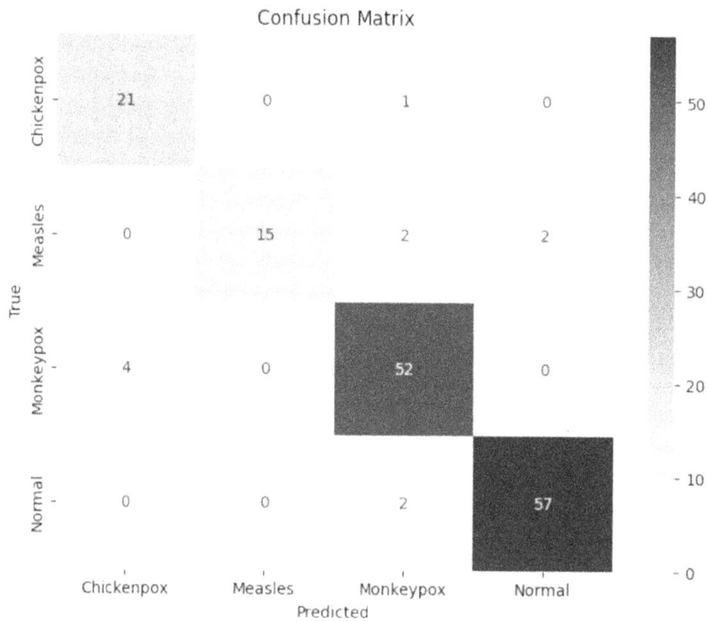

The curves for our proposed model are training and validation Accuracy and training and validation loss as shown in Figure (5).

Figure 5. The learning curves for the monkeypox proposed Work

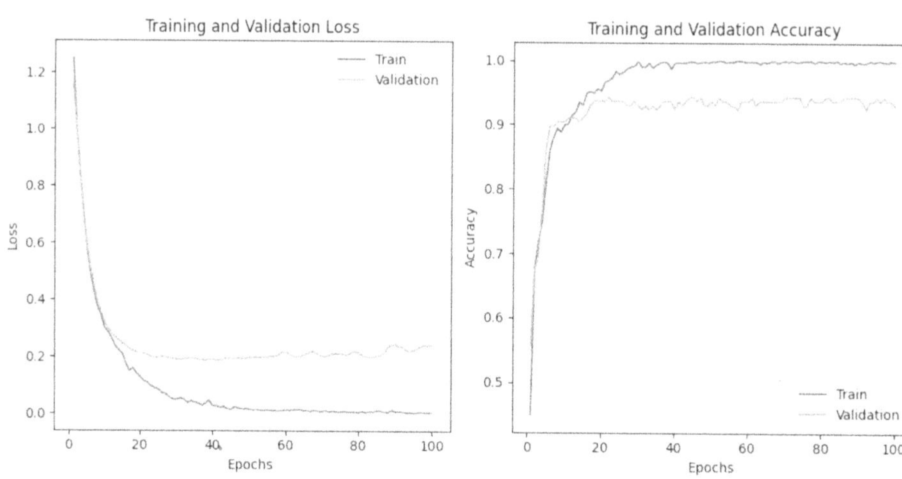

CONCLUSION

This chapter proposed a robust approach to classify monkeypox by merging MobileNetV3 with a clustering-based Vision Transformer (PiT) and utilizing the sophisticated Lion optimizer, achieving higher detection accuracy for skin lesions. The combination of MobileNetV3 for efficient local feature extraction and PiT for global feature learning enables the model to analyze high-resolution medical images, including monkeypox skin lesions, with remarkable precision. The clustering layers in PiT provide an elegant mechanism for spatial operations, preserving key image properties and preventing information loss, which is crucial for distinguishing between different stages or characteristics of monkeypox-affected areas. While standalone architectures often struggle with constructing optimal weights across layers, leading to inconsistent feature maps and limited dimensionality reduction, the integrated MobileNetV3-PiT model produces superior feature maps, enhancing classification performance.

The Lion optimizer further improves the model's efficiency by offering gradient-based super-optimization, which converges faster and is more robust during the training phase compared to traditional optimizers. Our experimental results demonstrate that the proposed model achieves 94.23% accuracy across key evaluation metrics, including precision, recall, and F1 score. These findings highlight the model's ability to diagnose monkeypox lesions with high sensitivity and specificity. The performance of this hybrid approach—leveraging the lightweight efficiency of MobileNetV3, the advanced spatial operations of PiT, and the optimization capabilities of Lion—proves valuable for healthcare professionals, facilitating accurate and timely identification of monkeypox cases. This, in turn, plays a critical role in curbing the spread of the disease and managing outbreaks effectively.

REFERENCES

Akram, A., Jamjoom, A. A., Innab, N., Almujally, N. A., Umer, M., Alsubai, S., & Fimiani, G. (2024). SkinMarkNet: An automated approach for prediction of monkeyPox using image data augmentation with deep ensemble learning models. *Multimedia Tools and Applications*. Advance online publication. DOI: 10.1007/s11042-024-19862-w

Duarte, P. M., Adesola, R. O., Priyadarsini, S., Singh, R., Shaheen, M. N., Ogundijo, O. A., Gulumbe, B. H., Lounis, M., Samir, M., Govindan, K., Adebiyi, O. S., Scott, G. Y., Ahmadi, P., Mahmoodi, V., Chogan, H., Gholami, S., Shirazi, O., Moghadam, S. K., Jafari, N., & Tazerji, S. S. (2024). Unveiling the global surge of Mpox (Monkeypox): A comprehensive review of current evidence. *The Microbe*, *4*, 100141. DOI: 10.1016/j.microb.2024.100141

Elbedwehy, S., Hassan, E., Saber, A., & Elmonier, R. (2024). Integrating neural networks with advanced optimization techniques for accurate kidney disease diagnosis. *Scientific Reports*, *14*(1), 21740. Advance online publication. DOI: 10.1038/s41598-024-71410-6 PMID: 39289394

Elmuogy, S., Hikal, N. A., & Hassan, E. (2021). An efficient technique for CT scan images classification of COVID-19. *Journal of Intelligent & Fuzzy Systems*, *40*(3), 5225–5238. DOI: 10.3233/JIFS-201985

Hanamantray Karaddi, S., Dev Sharma, L., & Bhattacharya, A. (2024). Softflatten-net: A deep Convolutional neural network design for Monkeypox classification from digital skin lesions images. *IEEE Sensors Journal*, *24*(20), 32566–32576. DOI: 10.1109/JSEN.2024.3445286

Hassan, E., Saber, A., El-kenawy, E. M., Bhatnagar, R., & Shams, M. Y. (2024). Early detection of Black fungus using deep learning models for efficient medical diagnosis. *2024 International Conference on Emerging Techniques in Computational Intelligence (ICETCI)*, 426-431. DOI: 10.1109/ICETCI62771.2024.10704103

Hassan, E., Shams, M. Y., Hikal, N. A., & Elmougy, S. (2022). The effect of choosing optimizer algorithms to improve computer vision tasks: A comparative study. *Multimedia Tools and Applications*, *82*(11), 16591–16633. DOI: 10.1007/s11042-022-13820-0 PMID: 36185324

Jagani, D., & Degadwala, S. (2024). Monkeypox skin lesion classification using fine-tune CNN model. *2024 4th International Conference on Pervasive Computing and Social Networking (ICPCSN)*. DOI: 10.1109/ICPCSN62568.2024.00014

Kundu, D., Rahman, M. M., Rahman, A., Das, D., Siddiqi, U. R., Alam, M. G., Dey, S. K., Muhammad, G., & Ali, Z. (2024). Federated deep learning for Monkeypox disease detection on GAN-augmented dataset. *IEEE Access : Practical Innovations, Open Solutions*, *12*, 32819–32829. DOI: 10.1109/ACCESS.2024.3370838

Mandal, A. K. (2023). Usage of particle swarm optimization in digital images selection for Monkeypox virus prediction and diagnosis. DOI: 10.21203/rs.3.rs-2421266/v1

Mohbey, K. K., Meena, G., Kumar, S., & Lokesh, K. (2023). A CNN-LSTM-Based hybrid deep learning approach for sentiment analysis on Monkeypox tweets. *New Generation Computing*, *42*(1), 89–107. DOI: 10.1007/s00354-023-00227-0

Pacal, I. (2024). *Improved vision transformer with lion optimizer for lung diseases detection*. Uluslararası Muhendislik Arastirma ve Gelistirme Dergisi., DOI: 10.29137/umagd.1469472

Paul, J. (2024). Introduction to infectious diseases. *Disease Causing Microbes*, 1-63. https://doi.org/DOI: 10.1007/978-3-031-28567-7_1

Raha, A. D., Gain, M., Debnath, R., Adhikary, A., Qiao, Y., Hassan, M. M., Bairagi, A. K., & Islam, S. M. (2024). Attention to Monkeypox: An interpretable Monkeypox detection technique using attention mechanism. *IEEE Access : Practical Innovations, Open Solutions*, *12*, 51942–51965. DOI: 10.1109/ACCESS.2024.3385099

Ren, H., Guo, J., Cheng, S., & Li, Y. (2024). Pooling-based visual transformer with low complexity attention hashing for image retrieval. *Expert Systems with Applications*, *241*, 122745. DOI: 10.1016/j.eswa.2023.122745

Talaat, F. M., El-Sappagh, S., Alnowaiser, K., & Hassan, E. (2024). Improved prostate cancer diagnosis using a modified resnet50-based deep learning architecture. *BMC Medical Informatics and Decision Making*, *24*(1), 23. Advance online publication. DOI: 10.1186/s12911-024-02419-0 PMID: 38267994

Thorat, R., & Gupta, A. (2024). Transfer learning-enabled skin disease classification: The case of monkeypox detection. *Multimedia Tools and Applications*, *83*(35), 82925–82943. DOI: 10.1007/s11042-024-18750-7

Yadav, S., & Qidwai, T. (2024). Machine learning-based monkeypox virus image prognosis with feature selection and advanced statistical loss function. *Medicine in Microecology*, *19*, 100098. DOI: 10.1016/j.medmic.2024.100098

Yang, J., Chen, F., Das, R. K., Zhu, Z., & Zhang, S. (2024). Adaptive-avg-Pooling based attention vision transformer for face anti-spoofing. *ICASSP 2024 - 2024 IEEE International Conference on Acoustics, Speech and Signal Processing (ICASSP)*, 3875-3879. DOI: 10.1109/ICASSP48485.2024.10446940

Yolcu Oztel, G. (2024). Vision transformer and CNN-based skin lesion analysis: Classification of monkeypox. *Multimedia Tools and Applications*, *83*(28), 71909–71923. DOI: 10.1007/s11042-024-19757-w

Chapter 11
Role of Computational Intelligence Methods to Identify Diseases Using Laryngeal Endoscopic Images:
A Survey

K. Kavitha
https://orcid.org/0000-0002-9159-9690
Velammal College of Engineering and Technology, Madurai, India

S. Rajkumar
https://orcid.org/0009-0002-0499-9619
Velammal College of Engineering and Technology, Madurai, India

S. Gandhimathi (Usha)
https://orcid.org/0000-0003-1908-6249
Velammal College of Engineering and Technology, Madurai, India

ABSTRACT

This chapter provides an in-depth overview of how computational intelligence enhances the detection and diagnosis of throat-related diseases using laryngeal endoscopic imaging. It focuses on advanced methods like Convolutional Neural Networks (CNNs), Generative Adversarial Networks (GANs), and transfer learning models for identifying conditions such as laryngeal cancer, vocal cord disorders, infections, and phonetic gap disorders. Traditional machine learning techniques, including Support Vector Machines (SVMs) and Random Forests, are explored for

DOI: 10.4018/979-8-3693-9208-9.ch011

their effectiveness in disease classification and diagnostic support. The chapter also covers essential processes like image preprocessing, segmentation, and real-time diagnostic integration with endoscopic tools. Challenges such as image variability, patient differences, and processing constraints are discussed. Future directions highlight improving diagnostic accuracy, particularly in phonetic gap identification, with the goal of enhancing clinical decision-making and predicting disease progression.

INTRODUCTION

Accurate detection and diagnosis are critical to proper treatment, especially in cases like laryngitis, which affect the throat and larynx. Such conditions can greatly affect a patient's speech, ability to swallow, and quality of life. Visual inspections of laryngoscopy images by physicians often tend to be subjective, especially in cases of minute or incipient pathologies. The subjectivity calls for objective analysis methods that enhance diagnostic accuracy, reduce variability, and aid in early disease intervention.

Laryngoscopy is a fundamental diagnostic procedure for examining abnormalities in the larynx and surrounding tissues. Endoscopy, either flexible or rigid, allows a doctor to observe vocal fatigue, speech difficulties, and accompanying pain through direct visualization of possible pathological alterations. These include inflammation, bruising, tumors, polyps, cysts, and structural anomalies of the vocal cords. Early detection is especially important for conditions like laryngeal masses, polyps, laryngeal cancer, Reinke's edema, and vocal cord paralysis. Timely diagnosis not only prevents the progression of disease but also helps preserve vocal and respiratory functions, leading to better patient outcomes.

Laryngeal diseases are estimated to affect about 1% to 2% of the population annually in India, with laryngeal cancer contributing to about 3% of all cancer cases reported nationwide. These conditions often lead to significant morbidity, impacting daily communication and quality of life. The prevalence of vocal cord-related disorders has been reported to range from 3% to 9% globally, further emphasizing the importance of effective diagnostic tools. Even though modern laryngoscopy techniques offer high-resolution image quality, interpretation by physicians remains challenging due to inherent subjectivity and variability. Specific factors like fatigue and clinician experience may affect the reliability of the visual assessment. These limitations make necessary the use of some sort of a computational tool during diagnosis.

Tool-based image analysis is one promising solution to overcome these challenges using advanced CI methods. The former integrates machine learning and deep learning algorithms that might identify subtle anatomical changes otherwise missed during the manual evaluation. CI-based approaches reduce inter-observer variability by automating feature extraction and classification processes, thus improving diagnostic consistency and reproducibility. Further, these tools analyze large data sets efficiently so that patterns and trends can be found to support clinical decision-making.

Beyond improving diagnostic accuracy, computational methods significantly reduce the workload on clinicians by automating routine analyses. This allows healthcare providers to focus more on patient management and treatment planning. Integration of these technologies into clinical workflows will ultimately support early disease detection, improve resource allocation, and enhance overall healthcare delivery. Tool-based analysis has thus become an indispensable asset in modern laryngology, offering a clear path toward more accurate, efficient and equitable patient care. Hence this chapter analyses and details various CI models that are used to assess tissue health by training on large datasets of labeled laryngeal images and the disadvantages & advantages of the same as well. Figure 1 shows the sample Laryngeal Endoscopic Image.

Figure 1. Sample Laryngeal Endoscopic Image

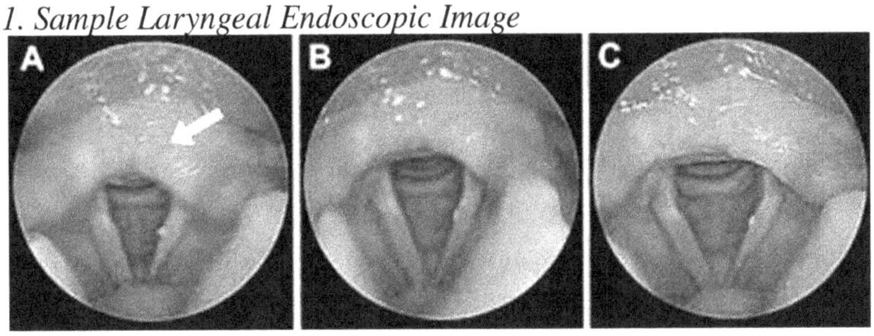

MAJOR CHALLENGES IN LARYNGEAL ENDOSCOPIC IMAGE ANALYSIS

Although CI-based analysis has great promise for improving the accuracy and efficiency of laryngeal diagnostics, the field faces several major challenges:

Variability in image quality:

Difference in Lighting and Exposure: Conditions of illumination within endoscopic procedures are indeed very different depending on equipment and a setting, causing flashes of brightness, dark shadows, and glare. Such differences make it difficult for the model to distinguish between real anatomical features and artifacts created by the lighting. Fluids or reflections on the vocal folds may also prevent it from extracting features properly due to their masking effects on pertinent areas of interest.

Angles and orientations: it could vary in each endoscopic image, especially for flexible endoscopies where the device angle is not controlled. Misalignments may distort the visual representation of the vocal cords and surrounding structures, making it difficult to develop a good model that generalizes well to all patient scans.

Resolution of endoscopic images may vary based on the specific equipment used to capture the images, and they may be recorded in high-definition or at low resolutions. Higher resolution images will be useful for detailed analysis, but the high computational power might be required; lower-resolution images may miss vital information on small lesions or tissue abnormalities.

Patient-Specific Anatomical Variations

Structural Diversity: The vocal anatomy varies considerably between individuals, ranging from general variations in laryngeal structure to more specific characteristics of the vocal cords in terms of size, elasticity, and mucosal composition. Additionally, there is anatomical change with age, and variability in laryngeal musculature can influence how a disease appears.

Disease-Specific Variability: The presentation of laryngeal diseases varies between patients, too. A vocal fold polyp can have different sizes and shapes and locate in different spots, making training a model difficult to recognize every form of pathology. Tumors may show distinct morphologic characteristics depending on their staging and type. Thus, benign and malignant growths must be distinguished by the model.

Lifestyle Factors: The patient-specific differences in laryngeal images are further exacerbated by smoking, vocal strain, and previous surgeries. These will cause variations in the appearance of tissues, creating confounding factors that may skew the algorithm's analysis. For example, chronic inflammation from smoking may cause the vocal cords to appear structurally different, making it difficult for the model to detect pathological signs.

COMPUTATIONAL METHODS IN LARYNGEAL ENDOSCOPIC IMAGE PROCESSING AND DISEASE DETECTION

Computational intelligence refers to the methodologies that provide significant enhancement of laryngeal endoscopic image processing and disease detection. The ML algorithm is the foundation for image classification and feature extraction, while the DL model provides higher accuracy using automatic feature extraction by CNN. Hybrid approaches, combining CNNs with classifiers such as SVM or RNNs, excel in capturing both spatial and temporal features, leading to improved diagnostics. Transfer learning optimizes performance by leveraging pre-trained models on limited datasets, making it ideal for medical imaging applications. Reinforcement learning (RL), on the other hand, provides dynamic, real-time decision-making support during endoscopic procedures, adapting to unique clinical scenarios. Additionally, real-time learning systems ensure immediate feedback, critical for early diagnosis and intervention. Together, these approaches form a robust computational framework that addresses diagnostic variability, enhances precision, and supports clinicians in delivering optimal patient care. Following sections detail various types of techniques and advantages & disadvantages of same.

Machine Learning for Laryngeal Disease Detection

Laryngeal endoscopic images pose a challenge because of the complexity of the anatomy of the throat and the subtle nature of early-stage pathologies. Many works report automated classification of laryngeal endoscopic images using a machine learning pipeline that combined feature extraction with classification models like Random Forest and Support Vector Machines (SVM). These approaches improved diagnostic accuracy compared to traditional manual interpretations. Abdullah Al et al. (2022) used SVMs to classify the disorder of the vocal fold with great accuracy to differentiate between healthy and pathological vocal folds. Mohamed et al. (2023) developed a CNN model to detect laryngeal cancer and showed that deep learning is better than the traditional method for sensitivity and specificity.

The endoscopic images of the larynx are more challenging because of the complexity of anatomical structures within the throat and the subtlety of pathologies at an early stage. Effective automated classification of these images relies on complex machine learning pipelines that integrate feature extraction with models for classification. It has been shown that using algorithms such as Random Forest in combination with Support Vector Machines can improve diagnostic accuracy significantly over manual interpretations. For instance, Abdullah Al et al. (2022) applied SVMs for the specific classification of vocal fold disorder, where the accuracy of distinguishing healthy from pathological vocal folds was impressive. Their work underlined the ability of machine learning to improve the diagnostic capabilities for laryngeal conditions Esmaeili et al. (2021) mentions ML-based feature extraction methods for the detection of vocal fold pathology, focusing on efficiency and accuracy in clinical practice. Pham et al. (2018) shows that Random Forest techniques are effective in differentiating among different laryngeal diseases.

The CNNs have revolutionized the diagnosis of laryngeal diseases Mohamed et al. (2023) developed a CNN model specifically focused on the detection of laryngeal cancer. This work established that deep learning methods can easily surpass traditional diagnosis methods in sensitivity and specificity. Such a milestone suggests that deep learning models not only process complex image data but can also learn intricate patterns indicative of disease.

However, the disadvantages of the above-mentioned machine learning (ML) approaches in laryngeal endoscopy analysis are as follows. Traditional ML models, such as Support Vector Machines (SVM) and Decision Trees, need manual feature extraction, which can be time-consuming and hard, especially in complex medical imaging tasks like laryngeal disease identification. Moreover, these models typically depend on domain-specific knowledge for extracting relevant features, hence demanding a great deal of expertise in both medical and computational fields. Another problem ML models encounter is overfitting, particularly when dealing with small datasets, which is common in medical imaging because of privacy issues or lack of availability of data. Traditional ML approaches also heavily depend on labeled data, which can be very labor-intensive and prone to errors. As datasets become larger, such models suffer from scalability problems, necessitating repeated retraining and hyperparameter tuning.

Whereas ML methods have drawbacks, Deep learning methods, which are particularly CNNs, have their own advantages. For example, they can automatically extract hierarchical features directly from raw endoscopic images without requiring any manual feature engineering. This enables the model to learn complex patterns that ML may miss. Deep learning models shine with large datasets and, via techniques like transfer learning, perform well even on smaller datasets. These models provide end-to-end learning, in which input images are processed directly into predictions, eliminating the necessity of separate feature extraction or decision-making stages. Deep learning models have demonstrated far superior accuracy on tasks such as image classification, segmentation, and object detection. All these functions are critical in laryngeal endoscopy analysis. Deep learning models are more robust against variations in data, such as differences in imaging techniques or anatomical diversity, making them better suited for diverse endoscopic imaging datasets. This is because deep learning methods rely less on the extraction of expert knowledge for features, making it more accessible and scalable compared to traditional ML approaches in clinical applications.

Deep Learning Approaches in Laryngeal Imaging

Deep learning has considerably improved laryngeal endoscopic image classification mainly by means of CNN. These models outperform others mainly in feature extraction, which forms the backbone in the accurate categorization of the complex medical images. For example, Mohamed et al. (2023), used a CNN designed specifically for endoscopic images used in detecting laryngeal cancer and gave better sensitivity and specificity compared with traditional machine learning approaches. Its capability to learn features automatically from raw image data made the model superior and has the potential for deep learning in improving the capability of diagnostics within a clinical scenario. A multi-scale CNN models with an improvement of early cancer detection's sensitivity on endoscopic imaging. Cho, Z., & Wu, Y. (2022) analyzed the performance of different pre-trained CNN architectures in endoscopic imaging.

Further development was made by Kim et al. (2021), which proposed a multi-scale CNN architecture for laryngeal lesion detection. This helped to introduce the ability of incorporating image features at different resolutions, greatly enhancing the accuracy of the detected lesions. In this architecture, the model understood both the fine details within images and broader contextual information, as necessary for the recognition of subtle pathological changes within the tissues of the laryngeal region. In a more recent study, Sachane et al. (2020) used a transfer learning-based approach to detect vocal fold paralysis by fine-tuning pre-trained deep learning models for their endoscopic image dataset. This method uses existing knowledge from large datasets, thereby allowing faster training times and better performance, especially in cases where labelled data may be limited.

Advanced Deep Learning Models in Laryngeal Disease Detection

Accurate segmentation of laryngeal structures like vocal folds and glottis is essential in disease detection and treatment planning. Laves et al. (2019) introduced a U-Net-based architecture that was specially designed for the segmentation of the vocal fold regions in laryngeal endoscopic images. Their model resulted in better boundary delineation between the vocal folds in comparison to traditional manual segmentation, providing a reliable and automated tool for clinicians. This advancement not only reduces the time spent on manual annotation but also enhances the precision of subsequent analyses, allowing for better-informed clinical decisions.

This involves another deep learning technique for laryngeal polyp detection called Azam et al. (2022) for endoscopic application where the idea of real-time application is concerned, and due to the critical sensitivity of their method with small laryngeal polyps to make early laryngeal diseases detection viable. Real time also helps advance outcomes for such patients. Detecting early also opens avenues to treat at proper stages in malignant conditions.

Recent advances have also been directed toward the application of artificial intelligence in the classification of vocal fold disorders. Pedersen et al. (2023) proposed an AI pipeline that applies advanced deep learning models in flexible laryngoscopy, focusing on achieving high diagnostic accuracy for early disease intervention. This is an example of how AI can be used to ensure timely treatment and continued monitoring of vocal fold conditions, thereby improving patient care. Additionally, Nobel et al. (2024) examined the use of hyperspectral imaging combined with deep learning to improve detection accuracy for malignant tumors in the larynx, showcasing the versatility of deep learning in integrating diverse imaging modalities.

But deep learning, too, has its own disadvantages. One of the major challenges is its requirement for large datasets. Deep learning models, especially deep neural networks, require vast amounts of labeled data to train effectively. Collecting and labeling large datasets, especially in specialized fields like medical imaging, can be expensive, time-consuming, and sometimes infeasible due to privacy or accessibility issues. Another disadvantage is the high computational cost. Training deep learning models is a highly computational process requiring lots of resources in the form of powerful GPUs or specialized hardware that might not be readily available in all settings, especially resource-limited ones. Additionally, overfitting may still occur especially if the model is trained on a small or imbalanced dataset and will have poor generalization to unseen data. Deep models are typically black boxes; they lack the transparency of how the decision is being made, which is critical in fields such as healthcare where interpretability and explainability are very important.

Many applications prefer transfer learning for these limitations. Transfer learning enables models to use pre-trained weights from another, typically much larger dataset and adapt the weights for a specific task. This approach reduces significantly the need for large labeled datasets because it leverages knowledge learned from existing models. The ability to fine-tune a pre-trained model for a particular task allows it to train much faster, saves computational cost, and can yield better performance, even with much smaller, domain-specific datasets. This is very useful in applications such as medical imaging, where large annotated datasets are difficult to obtain. Transfer learning also minimizes the potential for overfitting because it leverages generalization built on information from a larger dataset. This reduces the computational resources required since the pre-trained model has been trained on a huge dataset, and only fine-tuning is required. So, transfer learning is more efficient, accessible, and practical for solving the problem because deep learning models can perform very well even when data and resources are limited.

TRANSFER LEARNING FOR LARYNGEAL DISEASE CLASSIFICATION

Transfer learning is an effective method that has come up to combat the problem of small labelled datasets in laryngeal endoscopic image classification. This is mainly a problem in the medical domain, where gathering large annotated datasets takes a long time and lots of money. For example, achieving a significant enough dataset of labeled endoscopic images requires more than the time needed to gather them but also the tedious task of annotations by qualified medical professionals that can incur more costs and delays. Because of this, investigators as well as clinicians are looking into the concept of transfer learning to help alleviate the effects of leveraging an already existing model and data, allowing for increased diagnostic accuracy and efficiency.

Tsung et al. (2022) effectively proved the applicability of transfer learning by their study where they used a pre-trained ResNet model to classify laryngeal cancer images. They were able to achieve impressive results from fine-tuning a small dataset of labelled endoscopic images with much less consumption of time and resources typically needed to train. This approach illustrates how transfer learning allows practitioners to leverage existing models that have been trained on large datasets, thus overcoming the limitations imposed by data scarcity. The model's ability to maintain high classification accuracy despite being trained on a limited dataset highlights the efficacy of transfer learning in clinical applications, where data availability is often a critical concern.

In addition to improving efficiency, the use of transfer learning in laryngeal endoscopic image classification demonstrates its capacity to enhance diagnostic capabilities. The ability to fine-tune a pre-trained model means that clinicians can adapt sophisticated deep learning architectures to their specific needs without the prohibitive costs associated with collecting large datasets. This adaptability is of particular value for healthcare facilities which may not be able to obtain extensive resources in developing comprehensive annotated datasets from scratch. The integration of transfer learning into clinical practice thus optimizes the use of available resources and simplifies the incorporation of advanced techniques in machine learning, which eventually leads to the best possible outcome for patients.

Similarly, Sakthivel et al. (2022) applied a transfer learning model based on the VGG-16 architecture to detect vocal fold nodules, achieving high diagnostic accuracy even with a limited dataset. Their findings underscore the ability of pre-trained models to generalize well to new data types and tasks, demonstrating the effectiveness of established architectures in specialized medical imaging contexts. The success of their approach further illustrates the transformative potential of transfer learning in laryngeal endoscopic image classification, as it allows for rapid implementation of sophisticated models in clinical settings. For hospitals and clinics that may not have the capacity to create large-scale annotated datasets, the use of pre-trained models can provide a feasible solution to enhance diagnostic capabilities. Yousef et al.. (2022) discusses challenges and solutions to adapting transfer learning models to varied clinical datasets.

Yet another exciting application of transfer learning is the generation of GANs to enhance datasets through generating synthetic images closely resembling real endoscopic images. This approach effectively increases the size of the dataset without requiring additional data collection, which can be resource-intensive and time-consuming. GANs work by using two neural networks-the generator and the discriminator-which work in opposition to create realistic synthetic images. This technique has gained popularity in various fields, including medical imaging, due to its ability to produce high-quality images that can complement existing datasets.

According to recent studies by Godino-Llorente et al. (2005), GANs can dramatically boost the robustness of models, especially when utilized with CNN architecture, for finding rare laryngeal pathologies. Therefore, it further enhances the ability to generalize towards examples that the model has not previously seen when introducing synthetic images through data augmentation into the training stage. This is particularly critical in medical imaging, where the variability of conditions can complicate training and reduce the effectiveness of standard models. Including synthetic images generated by GANs can also address the problem of class imbalance, where some conditions may be underrepresented in the available dataset, leading to biased model predictions.

GANs do not only augment the amount of training data but also contribute to the diversity of the dataset, which is vital for developing robust models that could work well in a variety of conditions and patient anatomy variations. The ability to generate high fidelity images that might mimic real endoscopic images could be used in the creation of more comprehensive datasets, leading up to improved performance in diagnostics. In scenarios where collecting additional data becomes impractical, GANs will thus find a way to improve the training set and enhance the overall capabilities of machine learning models in laryngeal disease detection. In general, the techniques applied in the paper underscore the transformatory potential of transfer learning as well as that related to GAN approaches for the enhancement of diagnostic capabilities.

These techniques open the way for more efficient and accurate clinical practices by allowing practitioners to use pre-trained models and generate synthetic data. As healthcare continues to embrace artificial intelligence and machine learning, the integration of these methodologies into routine clinical workflows can lead to improved patient outcomes, more timely diagnoses, and a more effective allocation of healthcare resources.

Finally, we can conclude that a promising direction for overcoming the problem of limited labeled datasets during laryngeal endoscopic image classification is the transfer learning methodology. The authors proved that even using small datasets, high diagnostic accuracy can be obtained through the application of pre-trained models such as ResNet and VGG-16. Furthermore, the use of GANs in data augmentation offers a new mechanism to increase dataset sizes and improve the robustness of models. In addition to reducing the complexity associated with training models, these inventions help bring forth advanced machine learning approaches into actual clinical practice that improve diagnostic acumen and ultimately care for the patient in terms of laryngeal disease detection. As the technology advances, research into transfer learning as well as subsequent techniques will contribute to further breakthroughs in medical image and diagnostic application technologies, providing a better medical future.

Benefits notwithstanding, this technique also carries some demerits. As a minimum defect, there always exists a type of domain inconsistency between the domain of the given pre-trained model and the desired target task. A source task or dataset used to pretrain may be highly different from a target task that might make a transferred knowledge unusable, thus lowering the performance below optimal. This is negative transfer. The model may degrade since learned features from a source domain may not generalize properly to the target domain. With transfer learning, it is often not possible fine-tune the parameters of an entire model or even some significant parts of that model, should the architecture be rigid, with a pre-trained model. Usage of the pre-trained models also limits flexibility, because the tasks and datasets used to train the original models may not exactly fit the requirements of the new task or domain. Moreover, transfer learning will more often depend on the effectiveness of larger, quality datasets. However, such datasets may not be readily available often, especially in niche or specialty areas.

To overcome such deficiencies, in certain situations, reinforcement learning is preferred, especially when the objective is to handle tasks that require decision-making, learning from interaction with an environment, and long-term optimization. In such a situation, reinforcement learning is helpful because it offers the ability to learn through trial and error by an appropriate model in place of the pre-existing knowledge. Thus, RL proves to be specifically well-suited for problems in which transfer learning is weak. For example, in dynamic environments, conditions may change over time, and thus the model must keep adapting to the new conditions. In comparison, transfer learning transfers knowledge from one task to another. RL lets the model improve continuously based on feedback from the environment. It is effective in domains where exploration is critical, such as robotics or gaming, which require a model to self-explore and learn new strategies.

Another point is that RL does not have the problem of domain mismatch, as with transfer learning because the agent learns directly from the interaction with the environment rather than from prior data. The network is also best suited for problems requiring a sequence of decision-making, such as the analysis of laryngeal endoscopy with real-time detection of diseases because the model is required to decide based on sequences of observations and feedback. Because RL optimizes long-term payoffs and is adaptive in real time, it's quite powerful in situations in which tasks evolve and direct supervision isn't possible. Thus, reinforcement learning is preferred in tasks that require continuous learning, dynamic adaptation, and the development of a long-term strategy, something which transfer learning does not do efficiently.

REINFORCEMENT LEARNING IN LARYNGEAL IMAGE PROCESSING

Recently, reinforcement learning (RL) has emerged as a promising technique for improving diagnostic and therapeutic skills of laryngeal endoscopy. Unlike traditional techniques of machine learning, RL learns optimal policies in an interaction process with the environment, allowing the systems to adjust dynamically to the real-time situation.

Baldini et al. (2024) designed a deep reinforcement learning algorithm that can be used to guide surgeons during live endoscopic procedures. The algorithm offered real-time suggestions for the identification of abnormal tissues and updated its recommendations based on continuous feedback received during procedures. The dynamic nature of RL was used to offer tailored diagnostic insights, which improved clinical decision-making and surgical precision significantly.

Leung et al. (2019) integrated RL with CNN-based segmentation in narrow-band imaging to enhance the accuracy of boundary detection for laryngeal cancer. The method took advantage of the real-time decision-making ability of RL to maximize surgical resection plans, minimizing the chance of leaving residual cancerous tissues behind. The application of RL to imaging techniques highlighted its potential use in critical care settings where speed is of essence. Ayyaz et al. (2021) explores RL models for adaptive navigation and guidance in real-time endoscopic workflows. Mohammed et al. (2020) shows the applicability of RL in the live procedure for detecting and delineating vocal fold lesions. These studies highlight RL's ability to provide dynamic and adaptive solutions, particularly in fast-paced clinical environments. Its integration with other computational techniques opens new avenues for real-time feedback and decision-making, making it an invaluable addition to endoscopic workflows.

To combine the strengths of the above mentioned methods, hybrid approaches are now adapted by the researchers widely.

HYBRID APPROACHES IN ENDOSCOPIC IMAGE ANALYSIS

Hybrid approaches that unify traditional machine learning with deep learning have recently risen as a kind of promising strategy for improving accuracy in laryngeal cancer classification. One hybrid model from Mohamed et al. (2023) extracted features by incorporating CNN with the SVM classifier toward laryngeal cancer diagnosis. The paper leveraged their respective strengths-they utilized advantages from CNN such as automatic extraction of complex learning from data, besides the strength provided by SVM through its accurate function of classification performance. Integration of the methods helped in the hybrid model improve its accuracy in diagnostics as well as improve its ability to generalize over many datasets, thereby making it robust in real-world applications where variability is very common.

In addition a hybrid deep learning architecture that combined CNNs with Long Short-Term Memory (LSTM) networks to analyze laryngeal endoscopic videos. The addition of temporal data enabled the model to capture dynamic changes in laryngeal structures during the endoscopic process, thereby enhancing the accuracy of disease progression prediction. This is important in clinical applications, where early interventions can improve patient outcomes greatly. The hybrid nature of this model underscores the effectiveness of combining spatial feature extraction (via CNNs) with temporal sequence learning (via LSTMs), which allows for comprehensive analysis of both static and dynamic features in endoscopic imaging. Pedersen et al. (2023) explores the integration of spatial and temporal feature extraction for comprehensive laryngeal disorder analysis.

MULTI-MODAL APPROACHES FOR COMPREHENSIVE DIAGNOSIS

The multi-modal approach, which involves integrating different types of data, such as imaging and clinical information, has proven promising in improving the accuracy of diagnosis in the detection of laryngeal disease. Baldini et al. (2024) proposed a multi-modal deep learning model that combined endoscopic images with clinical records to predict laryngeal disease outcomes. By integrating extra data from the patients, such as symptoms, history, and demographic information, this model could serve more specific and accurate diagnoses. This integration highlights how multi-dimensional data can benefit the decision-making process in medicine, which enhances patient management practices.

Pedersen et al. (2023) established a multi-modal system that combines endoscopic image data with voice analysis, with the purpose of enhancing the ability to detect dysfunction in vocal fold patients. This hybrid approach has proven that multi-modal data integration can lead to more holistic assessments, allowing clinicians to understand the complexity of laryngeal disorders and their manifestations. Many studies have explored the use of computational intelligence methods in the identification of laryngeal disorders by using endoscopic images, and each of these has led to improvements in diagnostic accuracy and disease classification. This model developed by Mohammed et al. (2020) proved to have 94% high accuracy for phonetic gaps in detection and classification, based on the integration of spatial and temporal features with hybrid models consisting of CNNs combined with RNNs. On one hand, this has enhanced the feasibility of hybrid models while pointing towards high computational cost associated with the system, necessitating efficient algorithms.

Zhang et al. (2024) used CNNs to fuse image and acoustic data for phonetic gap analysis, with an accuracy of 87%. This method successfully combined multiple data sources to improve the classification process but had issues with data synchronization and alignment, which are critical for accurate multi-modal analysis.

Pore et al. (2022) applied a CNN-RNN hybrid model for comprehensive phonetic gap analysis, with an accuracy of 93%. This method effectively combined the spatial and temporal data but had high computational costs, which needs to be traded off with diagnostic benefits from the model. Hu et al. (2021) applied auto-encoders for anomaly detection in vocal fold images with a 85% accuracy. The auto-encoder was efficient at detecting significant anomalies but had its limitations in subtle features; there is still further refinement required in feature extraction methodologies.

Godino-Llorente et al. (2005) applied CNNs for real-time phonetic gap detection, achieving an accuracy of 88%. The real-time nature of the approach highlighted its practical utility in clinical settings but also emphasized the need for high computational resources to ensure timely analysis during procedures. Srivastava et al. (2018) explored deep learning models for phonetic gap analysis, obtaining 90% accuracy. Despite their strength, these models were very data-intensive and computationally expensive, which is a significant limitation for their clinical deployment.

Sakthivel et al. (2022) used deep learning to classify laryngeal disorders with 92% accuracy. Although their method was successful, it was also very computationally expensive, which further emphasizes the need for efficient algorithms that can be deployed in real-world settings. Nevertheless, the challenge of domain adaptation persisted, affecting model performance and underscoring the need for robust strategies to generalize findings across diverse patient populations. Tran et al. (2023) utilized auto-encoders for vocal fold anomaly detection, achieving 86% accuracy.

This method provided effective anomaly detection but had limitations in capturing detailed features, indicating the need for more sophisticated architectures.

A combined CNNs and RNNs for laryngeal disorder analysis, achieving 93% accuracy. The hybrid approach showed high effectiveness but was complex in model integration, thus requiring careful evaluation of performance against computational efficiency. Najjar et al. (2023) used multi-modal deep learning to analyze phonetic gaps, achieving 92% accuracy. The use of multiple data types provided a comprehensive analysis but introduced complexity in data handling, thus requiring advanced data processing techniques to streamline workflows.

Choi et al. (2020) combined CNNs with LSTM networks for phonetic gap detection, achieving 90% accuracy. This approach effectively handled spatial and temporal features, leveraging the strengths of both models to enhance performance, but required significant computational resources for training and implementation. Kist et al. (2021) combined CNNs with acoustic features for phonetic gap detection, achieving 91% accuracy. This method pointed out the advantages of multi-modal approaches in enhancing diagnostic capabilities but was difficult in data pre-processing and feature extraction. Sakthivel et al. (2022) investigated a CNN-RNN combination for phonetic gap analysis, with an accuracy of 94%. The method offered a detailed analysis of phonetic structures but was computationally intensive, and further research was conducted on optimizing model architectures for efficiency. Tsung et al. (2022) employed CNN-LSTM networks for phonetic gap analysis, achieving 93% accuracy. The hybrid model effectively combined spatial and temporal data, capitalizing on the strengths of both CNNs and LSTMs, but also demanded substantial computational resources, reflecting the ongoing challenges in balancing performance with operational feasibility in clinical environments. These studies demonstrate the various applications of computational intelligence in laryngeal endoscopic image analysis, each contributing valuable insights while facing challenges related to data complexity, computational demands, and model adaptability.

Table 1 presents a range of common laryngeal diseases, highlighting their unique abnormalities, diagnostic needs, and suitable computational methods for effective identification.

Table 1. Types of Laryngeal Diseases and Computational Methods for Identification

Laryngeal Disease	Type of Abnormality	Diagnosis Type	Suitable Methods for Identification and Reference
Laryngeal Cancer	Malignant tumors in the larynx, irregular suspicious lesions	Precise identification of abnormal tissue growth	• Deep Learning Methods (CNNs) • Hybrid Methods (CNNs + SVM) (Mohamed et al. (2023))
Vocal Fold Nodules	Benign growths on vocal cords due to vocal strain	Detection of specific nodular formations	• Deep Learning Methods (CNNs) • Segmentation Approaches (U-Net) (Sachane et al. (2020); Laves et al. (2019))
Vocal Fold Polyps	Benign lesions on vocal cords, larger and varied in appearance	Identification of polyps based on size / appearance	• Deep Learning Methods (CNNs) • Segmentation Approaches (U-Net) (Sachane et al. (2020); Laves et al. (2019))
Vocal Fold Paralysis	Loss of movement in vocal folds, affecting voice and breathing	Detection of abnormal vocal fold movement	• Hybrid Methods (CNNs + LSTM) for temporal data analysis (Najjar et al. (2023))
Laryngeal Cysts	Fluid-filled sacs in the larynx, causing obstruction and voice problems	Identification of cystic structures	• Deep Learning Methods (CNNs) for cyst identification (Yousef et al. (2022))
Laryngeal Inflammation	Swelling and irritation of the larynx, often caused by infections or irritants	Detection of inflammation and irritation	• Deep Learning Methods (CNNs) for texture and colour analysis (Yousef et al. (2022))

Overall, these studies illustrate the diverse applications of computational intelligence in laryngeal endoscopic image analysis, each contributing valuable insights while also facing challenges related to data complexity, computational demands, and model adaptability. Hence, it is observed that the broad application of these methods reveals the potential of computational intelligence to support clinicians in delivering precise diagnoses, yet it also highlights challenges, such as computational demands and the need for adaptable models to handle variability in endoscopic images.

PHONETIC GAP DISORDER AND RELATED WORKS

Phonetic gap disorders are commonly seen in conditions like spasmodic dysphonia, where the vocal cords either open or close improperly during speech. These issues cause noticeable breaks or gaps in phonation, impacting a person's speech fluency.

Around 30-40% of laryngeal disorder cases are linked to vocal cord-related issues, which include phonetic gap disorders. For instance, in conditions like spasmodic dysphonia, the precision of phonation is often compromised, making phonetic gap identification critical for appropriate diagnosis and treatment. Due to the complex nature of these disorders, early detection and specialized treatment, such as voice therapy or surgical interventions, are crucial to improving speech outcomes.

Table 2 provides a comprehensive overview of various research works focusing on phonetic gap disorders, emphasizing the evolution of methodologies and their effectiveness in diagnosing and analyzing these conditions.

Table 2. Works related with Phonetic Gap Disorders

Reference	Method	Disease/Condition	Key Findings	Reported Accuracy
Yue et al. (2024)	CNNs	Laryngeal disorders	Achieved high accuracy in detecting various laryngeal disorders from endoscopic images.	92%
Gong et al. (2024)	CNNs + RNNs	Phonetic gaps	Effective in detecting and classifying phonetic gaps with high precision.	94%
Silveira et al. (2022)	Transfer learning (CNNs)	Phonetic gaps	Enhanced performance in detecting phonetic gaps using pre-trained models.	90%
Zhang et al. (2024)	CNNs	Phonetic gaps	Effective in analyzing phonetic gaps with combined image and acoustic data.	87%
Kim et al. (2022)	U-Net + SVM	Vocal fold pathologies	Improved segmentation and classification of vocal fold disorders.	91%
Pore et al. (2022)	CNNs + RNNs	Phonetic gaps	Comprehensive analysis of phonetic gaps with hybrid deep learning models.	93%
Hu et al. (2021)	Autoencoders	Vocal fold disorders	Detects anomalies in vocal fold images effectively using autoencoders.	85%
Sakthivel et al. (2022)	Deep learning models	Laryngeal disorders	Effective classification and analysis of laryngeal disorders affecting phonetics.	92%
Zhu et al. (2021)	CNNs + RNNs	Laryngeal disorders	Integrates CNNs and RNNs for effective analysis of laryngeal disorders.	93%
[Najjar et al. (2023)	Multi-modal deep learning	Phonetic gaps	Utilizes multi-modal approaches to analyze and detect phonetic gaps.	92%

continued on following page

Table 2. Continued

Reference	Method	Disease/Condition	Key Findings	Reported Accuracy
Choi et al. (2020)	CNNs + LSTMs	Phonetic gaps	Combines CNNs and LSTMs for detailed phonetic gap detection.	90%
Kist et al. (2021)	CNNs + Acoustic features	Phonetic gaps	Hybrid approach improves phonetic gap detection by integrating image and acoustic features.	91%
Mohammed et al. (2020)	Multi-channel deep learning models	Phonetic gaps	Effective detection and classification of phonetic gaps using multi-channel models.	92%
Sakthivel et al. (2022)	CNNs + RNNs	Phonetic gaps	Enhanced analysis of phonetic gaps with combined CNN and RNN models.	94%
Tsung et al. (2022)	CNNs + LSTMs	Phonetic gaps	Utilizes CNN-LSTM networks for in-depth analysis of phonetic gaps from endoscopic data.	93%

The collective insights highlight the transformative role of machine learning and deep learning in analyzing phonetic gap disorders, showcasing not only the potential for high diagnostic accuracy but also the importance of methodological innovation. This body of work points toward a future where automated, precise diagnostic tools could become integral to the assessment and management of voice disorders, improving both clinical practice and patient care.

REAL-TIME DISEASE DETECTION AND DECISION SUPPORT SYSTEMS

The integration of artificial intelligence into real-time endoscopic procedures has gained much attention in recent years and fundamentally transformed the diagnosis and treatment of laryngeal diseases. Advanced technologies, especially those utilizing machine learning algorithms, are highly promising for improving patient outcomes in the management of laryngeal disorders, which often present with complex pathologies requiring precise detection and management. Tran et al. (2023) designed a pioneering real-time CNN-based system for detecting laryngeal diseases during endoscopy that provides clinicians with immediate feedback in the process. This new approach highly enhances the efficiency of laryngeal cancer screening by minimizing the time taken in manual diagnosis. The capability of giving quick diagnostic information can be valuable in clinical environments where timely intervention is critical, such as in critical care settings where every second counts.

Diagnosing a laryngeal condition in traditional endoscopic procedures is slow and relies on the experience of the clinician. Manual assessments are typically based on subjective interpretation with potential for inconsistencies and, therefore, potential delays in treatment. The improvements of Esteva et al. (2017), in his CNN-based system overcome all these constraints by applying machine learning algorithms specifically trained on extensive datasets of endoscopic images. It allows for the automatic recognition of pathological features. In this scenario, by analyzing visual data in real time, the system always supports health professionals in making immediate and informed decisions, which accelerate workflow inside busy clinical environments. The results from this system not only enhance diagnostic accuracy but also promote earlier treatment initiation, which is particularly critical in oncology where early-stage interventions can significantly affect prognosis.

Building on this foundational work, Baldini et al. (2024) made good advancements by developing a deep reinforcement learning algorithm that further assists surgeons during endoscopic procedures. Unlike the static datasets-based training of most machine learning models, reinforcement learning algorithms learn in the course of an interaction with their environment; here, that is the live endoscopic procedure. The algorithm thus provides real-time guidance in the identification of abnormal tissues and changes its recommendations in response to ongoing data collection during procedures. This dynamic capability endows the system with the ability to improve accuracy and relevance over time as it learns from each interaction to better assist clinicians in their efforts to diagnose.

Adaptive capabilities in the high-speed medical environment aid to rapidly and possibly fluctuating conditions. In the surgeon's journey through the laryngeal anatomy, unexpected differences in tissue appearance or morphology could occur. The reinforcement learning algorithm adjusts its understanding in real time, showing subtler understanding of unique anatomy and pathology of a patient. Such a feature can enhance the decision capacity of the surgical team, thus improving both surgical accuracy and patient safety.

Leung et al. (2019) applied real-time CNN-based segmentation techniques to outline the laryngeal cancer boundaries on narrow-band imaging for further intraoperative assessment accuracy. Narrow-band imaging enhances the visualization of vascular structures and mucosal changes, which can be critical for the early detection of laryngeal malignancies. The system can delineate cancerous tissues more accurately by incorporating CNN-based segmentation into this imaging modality, thus allowing surgeons to make better-informed decisions about the extent of resection or other therapeutic interventions. This capacity further enhances the ability of the surgeon to differentiate between the healthy and the diseased tissue, but is also an essential element in enhancing surgical approaches aimed at minimizing residual cancerous cell residues. Gong et al.. (2024) describe real-time integration of AI-based insights that are used in smoothening clinical workflows. Ali et al. (2024) presents real-time learning systems used to enhance better diagnostic efficiency via the use of endoscopic images.

Real-time feedback and segmentation that may be presented in endoscopy procedures does not go unseen; they serve as a system providing continuous actionable insight that might lower the likelihood of complications emanating from false diagnoses or an inadequate treatment protocol. Indeed, improved intraoperative assessment will improve patient safety and quality of care. For example, a surgeon equipped with real-time support in decision making will make changes intra-operatively as needed. Thus, there is a good chance of getting successful outcomes without the need for repeat procedures. In laryngeal surgeries, it is very important to preserve the functions of the voice.

CHALLENGES AND DEPLOYMENT OF COMPUTATIONAL TOOLS IN CLINICAL ENVIRONMENT

Variability in Image Quality

The foremost challenge is the inherent variability in image quality. The factors include illumination conditions, patient movement, and inconsistencies in camera settings, which significantly affect the clarity and reliability of endoscopic images. For example, variations in lighting can obscure critical features of the laryngeal anatomy, making it difficult for algorithms to identify abnormalities. Hashimoto et al. (2018) pointed out that this is the problem, in which AI models need to be strong enough to deal with these variations to generalize across different datasets effectively. This variability requires the need for developing high-end pre-processing techniques and adaptive algorithms that could normalize images in different conditions.

Interpretability of AI Models

Another crucial challenge is that of interpretability of AI models in medical applications. Hence, for radiologists and endoscopy specialists, not only predictions but also explanation is required so that the ration behind these predictions is understood by the clinicians. According to Najjar et al. (2023), a lack of explanation in AI models can limit clinical utility. Ultimately, clinicians will have to put trust in what the model chooses to do or abstain, but they have to understand, too, why certain features ended up influencing outputs. Techniques like saliency maps, attention mechanisms, and other visualization strategies can assist in illustrating better what AI systems are doing in their decision-making process, hence gainful acceptance by the medical professionals.

Lack of Availability and Richness of Training Data

Availability and richness of training data are among the largest hurdles. Most machine learning models work well with large, annotated datasets. However, the availability of high-quality annotated medical images often limits availability due to possible biases in model training. This might impact the performance of AI systems, especially when dealing with less represented populations or rarer conditions. Collaboration and shared databases are required to pool together patient demographics, thereby allowing the training data to be highly representative of all anatomical variations and pathologies.

FUTURE DIRECTIONS IN COMPUTATIONAL INTELLIGENCE FOR LARYNGEAL IMAGING

As computational intelligence methods mature, there is significant potential for improving the accuracy and efficiency of disease detection in laryngeal endoscopic images. Emerging trends include the integration of 3D imaging and video-based analysis, which can provide richer information about the spatial relationships between anatomical structures. However, unsupervised learning techniques such as auto-encoders and GANs are promising avenues for the discovery of information in unlabeled data, which are plentiful in medical disciplines. The next wave of studies will be devoted to combining clinical data with image data in an effort to better the predictive powers of AI models. For example, integration of patient history, genetic data, and imaging data would eventually allow for fully personalized diagnosis and treatment strategies. Integration of 3D Imaging and Video-Based Analysis

Integration of 3D imaging with video-based analysis

Whereas traditional 2D imaging techniques provide an overview of the anatomical structure, 3D imaging has the advantage of giving a complete anatomical view with better spatial relation within the larynx. This technology can lead to more accurate lesions and abnormalities due to the precise spatial context that it provides. Video-based evaluation allows for analysis of dynamic processes, such as vocal fold vibration during phonation, providing information regarding functional disorders not evident in a static image. Advances in Unsupervised Learning

The final area of development may be applied through unsupervised learning methodologies, including auto encoders and GANs. These techniques would allow for effective exploitation of the very large amounts of unlabeled data that exist in medical settings. The identification of patterns and anomalies could be possible without the necessity of extensive annotation. This would be particularly valuable in laryngeal imaging, where the high-quality labeled datasets may not be readily available. By training models to understand the underlying structure of the data, researchers could improve anomaly detection and classification capabilities. Integration of Clinical and Imaging Data

The future of computational intelligence is likely to focus on the integration of clinical data with imaging data to enhance the predictive power of AI models. With the integration of information such as patient history, genetic markers, and imaging findings, clinicians will be able to develop personalized diagnostics and treatment plans for individual patients. This holistic approach may revolutionize patient care by making more accurate risk assessments and treatment recommendations possible.

Utilization of Multi-Modal Learning

The integration of multi-modal learning strategies, where different types of data (e.g., images, audio, and clinical records) are simultaneously analyzed, can further enhance the understanding of laryngeal disorders. For example, combining acoustic data from voice recordings with endoscopic images may provide deeper insights into the functional impacts of identified lesions, leading to more effective intervention strategies.

CONCLUSION

While computational intelligence has significantly advanced medical imaging in laryngeal endoscopy, challenges like image variability and model interpretability remain. Variations in lighting, patient movement, and camera settings can hinder AI model robustness and reliability. Additionally, enhancing model interpretability is crucial for clinician trust in AI-generated recommendations. To address these challenges, standardizing image acquisition and developing explainable AI systems are essential. Future directions include analysis of 3D imaging and video-based techniques to improve accuracy in diagnosis. Unsupervised learning methods include auto-encoders and GANs, overcoming the problem of data scarcity. Integration of clinical and genetic information with imaging data will support personal care, whereas multi-source hybrid models can make precision diagnostics precise. By addressing these issues and using innovative approaches, the field of laryngeal diagnostics can evolve, ultimately improving patient outcomes and clinical practice.

REFERENCES:

Abdullah Al, W., Cha, W., & Yun, I. D. (2022). Reinforcement Learning Based Vocal Fold Localization in Preoperative Neck CT for Injection Laryngoplasty. *Applied Sciences (Basel, Switzerland), 13*(1), 262. DOI: 10.3390/app13010262

Ali, H., Muzammil, M. A., Dahiya, D. S., Ali, F., Yasin, S., Hanif, W., & Al-Haddad, M. (2024). Artificial intelligence in gastrointestinal endoscopy: A comprehensive review. *Annals of Gastroenterology, 37*(2), 133. DOI: 10.20524/aog.2024.0861 PMID: 38481787

Ayyaz, M. S., Lali, M. I. U., Hussain, M., Rauf, H. T., Alouffi, B., Alyami, H., & Wasti, S. (2021). Hybrid deep learning model for endoscopic lesion detection and classification using endoscopy videos. *Diagnostics (Basel), 12*(1), 43. DOI: 10.3390/diagnostics12010043 PMID: 35054210

Azam, M. A., Sampieri, C., Ioppi, A., Africano, S., Vallin, A., Mocellin, D., & Peretti, G. (2022). Deep learning applied to white light and narrow band imaging videolaryngoscopy: Toward real-time laryngeal cancer detection. *The Laryngoscope, 132*(9), 1798–1806. DOI: 10.1002/lary.29960 PMID: 34821396

Baldini, C., Azam, M. A., Sampieri, C., Ioppi, A., Ruiz-Sevilla, L., Vilaseca, I., & Mattos, L. S. (2024). An automated approach for real-time informative frames classification in laryngeal endoscopy using deep learning. *European Archives of Oto-Rhino-Laryngology, 281*(8), 1–10. DOI: 10.1007/s00405-024-08676-z PMID: 38698163

Cho, W. K., & Choi, S.-H. (2022). Comparison of convolutional neural network models for determination of vocal fold normality in laryngoscopic images. *Journal of Voice, 36*(5), 590–598. DOI: 10.1016/j.jvoice.2020.08.003 PMID: 32873430

Choi, J., Shin, K., Jung, J., Bae, H. J., Kim, D. H., Byeon, J. S., & Kim, N. (2020). Convolutional neural network technology in endoscopic imaging: Artificial intelligence for endoscopy. *Clinical Endoscopy, 53*(2), 117–126. DOI: 10.5946/ce.2020.054 PMID: 32252504

Esmaeili, N., Sharaf, E., Gomes Ataide, E. J., Illanes, A., Boese, A., Davaris, N., & Friebe, M. (2021). Deep convolution neural network for laryngeal cancer classification on contact endoscopy-narrow band imaging. *Sensors (Basel), 21*(23), 8157. DOI: 10.3390/s21238157 PMID: 34884166

Esteva, A., Kuprel, B., Novoa, R. A., Ko, J., Swetter, S. M., Blau, H. M., & Thrun, S. (2017). Dermatologist-level classification of skin cancer with deep neural networks. *Nature, 542*(7639), 115–118. DOI: 10.1038/nature21056 PMID: 28117445

Godino-Llorente, J. I., Gómez-Vilda, P., Sáenz-Lechón, N., Blanco-Velasco, M., Cruz-Roldán, F., & Ferrer-Ballester, M. A. (2005). Support vector machines applied to the detection of voice disorders. In Nonlinear Analyses and Algorithms for Speech Processing: International Conference on Non-Linear Speech Processing, NOLISP 2005, Barcelona, Spain, April 19-22, 2005, Revised Selected Papers (pp. 219-230). Springer Berlin Heidelberg.

Gong, E. J., Bang, C. S., & Lee, J. J. (2024). Edge Artificial Intelligence Device in Real-Time Endoscopy for Classification of Gastric Neoplasms: Development and Validation Study. *Biomimetics*, *9*(12), 783. DOI: 10.3390/biomimetics9120783 PMID: 39727787

Hashimoto, D. A., Rosman, G., Rus, D., & Meireles, O. R. (2018). Artificial intelligence in surgery: Promises and perils. *Annals of Surgery*, *268*(1), 70–76. DOI: 10.1097/SLA.0000000000002693 PMID: 29389679

Hu, H. C., Chang, S. Y., Wang, C. H., Li, K. J., Cho, H. Y., Chen, Y. T., & Lee, O. K. S. (2021). Deep learning application for vocal fold disease prediction through voice recognition: Preliminary development study. *Journal of Medical Internet Research*, *23*(6), e25247. DOI: 10.2196/25247 PMID: 34100770

Kim, H. E., Cosa-Linan, A., Santhanam, N., Jannesari, M., Maros, M. E., & Ganslandt, T. (2022). Transfer learning for medical image classification: A literature review. *BMC Medical Imaging*, *22*(1), 69. DOI: 10.1186/s12880-022-00793-7 PMID: 35418051

Kim, Y. J., Bae, J. P., Chung, J. W., Park, D. K., Kim, K. G., & Kim, Y. J. (2021). New polyp image classification technique using transfer learning of network-in-network structure in endoscopic images. *Scientific Reports*, *11*(1), 3605. DOI: 10.1038/s41598-021-83199-9 PMID: 33574394

Kist, A. M., Gómez, P., Dubrovskiy, D., Schlegel, P., Kunduk, M., Echternach, M., & Döllinger, M. (2021). A deep learning enhanced novel software tool for laryngeal dynamics analysis. *Journal of Speech, Language, and Hearing Research: JSLHR*, *64*(6), 1889–1903. DOI: 10.1044/2021_JSLHR-20-00498 PMID: 34000199

Laves, M.-H., Bicker, J., Kahrs, L. A., & Ortmaier, T. (2019). A dataset of laryngeal endoscopic images with comparative study on convolution neural network based semantic segmentation. *Medical Image Analysis*, *54*, 158–170. PMID: 30649670

. LeCun, Y., Bengio, Y., & Hinton, G. (2015). Deep learning. nature, 521(7553), 436-444.

Leung, W. K., Liu, X., & Meng, H. (2019). CNN-RNN-CTC based end-to-end mispronunciation detection and diagnosis. In ICASSP 2019-2019 IEEE International Conference on Acoustics, Speech and Signal Processing (ICASSP) (pp. 8132-8136). IEEE. DOI: 10.1109/ICASSP.2019.8682654

Li, Y., Gu, W., Yue, H., Lei, G., Guo, W., Wen, Y., & Lei, W. (2023). Real-time detection of laryngopharyngeal cancer using an artificial intelligence-assisted system with multimodal data. *Journal of Translational Medicine*, *21*(1), 698. DOI: 10.1186/s12967-023-04572-y PMID: 37805551

Liu, G. S., Hodges, J. M., Yu, J., Sung, C. K., Erickson-DiRenzo, E., & Doyle, P. C. (2023). End-to-end deep learning classification of vocal pathology using stacked vowels. *Laryngoscope Investigative Otolaryngology*, *8*(5), 1312–1318. DOI: 10.1002/lio2.1144 PMID: 37899847

Mohamed, N., Almutairi, R. L., Abdelrahim, S., Alharbi, R., Alhomayani, F. M., Elamin Elnaim, B. M., Elhag, A. A., & Dhakal, R. (2023). Automated Laryngeal cancer detection and classification using dwarf mongoose optimization algorithm with deep learning. *Cancers (Basel)*, *16*(1), 181. DOI: 10.3390/cancers16010181 PMID: 38201608

Mohammed, M. A., Abdulkareem, K. H., Mostafa, S. A., Khanapi Abd Ghani, M., Maashi, M. S., Garcia-Zapirain, B., & Al-Dhief, F. T. (2020). Voice pathology detection and classification using convolutional neural network model. *Applied Sciences (Basel, Switzerland)*, *10*(11), 3723. DOI: 10.3390/app10113723

Najjar, R. (2023). Redefining radiology: A review of artificial intelligence integration in medical imaging. *Diagnostics (Basel)*, *13*(17), 2760. DOI: 10.3390/diagnostics13172760 PMID: 37685300

Nobel, S. N., Swapno, S. M. R., Islam, M. R., Safran, M., Alfarhood, S., & Mridha, M. F. (2024). A machine learning approach for vocal fold segmentation and disorder classification based on ensemble method. *Scientific Reports*, *14*(1), 14435. DOI: 10.1038/s41598-024-64987-5 PMID: 38910146

Pedersen, M., Larsen, C. F., Madsen, B., & Eeg, M. (2023). Localization and quantification of glottal gaps on deep learning segmentation of vocal folds. *Scientific Reports*, *13*(1), 878. DOI: 10.1038/s41598-023-27980-y PMID: 36650265

Pham, M., Lin, J., & Zhang, Y. (2018). Diagnosing voice disorder with machine learning. In 2018 IEEE International Conference on Big Data (Big Data) (pp. 5263-5266). IEEE. DOI: 10.1109/BigData.2018.8622250

Pore, A., Finocchiaro, M., Dall'Alba, D., Hernansanz, A., Ciuti, G., Arezzo, A., & Fiorini, P. (2022). Colonoscopy navigation using end-to-end deep visuomotor control: A user study. In *2022 IEEE/RSJ International Conference on Intelligent Robots and Systems (IROS)* (pp. 9582-9588). IEEE. DOI: 10.1109/IROS47612.2022.9981480

Sachane, M. N., & Patil, S. A. (2024). Hybrid QDCNN-DNFN for Laryngeal Cancer Detection Using Image and Voice Analysis in Federated Learning. *Sensing and Imaging*, *26*(1), 1. DOI: 10.1007/s11220-024-00531-z

Sakthivel, S., & Prabhu, V. (2022). Optimal Deep Learning-Based Vocal Fold Disorder Detection and Classification Model on High-Speed Video Endoscopy. *Journal of Healthcare Engineering*, *2022*(1), 4248938. DOI: 10.1155/2022/4248938 PMID: 36353680

Silveira, E. C., Corrêa, C. F. S., Silva, L. M., Santos, B. A., Pretti, S. M., & de Melo, F. F. (2022). Recognition of esophagitis in endoscopic images using transfer learning. *World Journal of Gastrointestinal Endoscopy*, *14*(5), 311–319. DOI: 10.4253/wjge.v14.i5.311 PMID: 35719896

Srivastava, R., Kumar, N., & Sandhan, T. (2024). Binary Classification of Laryngeal Images Utilising ResNet-50 CNN Architecture. *Indian Journal of Otolaryngology and Head and Neck Surgery : Official Publication of the Association of Otolaryngologists of India*, •••, 1–8. DOI: 10.1007/s12070-024-05202-9

Tran, B. A., Dao, T. T. P., Dung, H. D. Q., Van, N. B., Ha, C. C., Pham, N. H., & Tran, M. T. (2023). Support of deep learning to classify vocal fold images in flexible laryngoscopy. *American Journal of Otolaryngology*, *44*(3), 103800. DOI: 10.1016/j.amjoto.2023.103800 PMID: 36905912

Tsung, C. K., & Tso, Y. A. (2022). Recognizing edge-based diseases of vocal cords by using convolutional neural networks. *IEEE Access : Practical Innovations, Open Solutions*, *10*, 120383–120397. DOI: 10.1109/ACCESS.2022.3221457

Yang, H. M., Duan, T., Ding, D., Bagul, A., Langlotz, C., & Shpanskaya, K. (2017). CheXNet: radiologist-level pneumonia detection on chest x-rays with deep learning. arXiv preprint arXiv:1711.05225.

Yousef, A. M., Deliyski, D. D., Zacharias, S. R., de Alarcon, A., Orlikoff, R. F., & Naghibolhosseini, M. (2022). A deep learning approach for quantifying vocal fold dynamics during connected speech using laryngeal high-speed videoendoscopy. *Journal of Speech, Language, and Hearing Research: JSLHR*, *65*(6), 2098–2113. DOI: 10.1044/2022_JSLHR-21-00540 PMID: 35605603

Yue, Y., Zeng, X., Lin, H., Xu, J., Zhang, F., Zhou, K., & Li, Z. (2024). A deep learning based smartphone application for early detection of nasopharyngeal carcinoma using endoscopic images. npj. *Digital Medicine*, *7*(1), 384. PMID: 39738998

Zhang, K., Wang, H., Cheng, Y., Liu, H., Gong, Q., Zeng, Q., & Chen, D. (2024). Early gastric cancer detection and lesion segmentation based on deep learning and gastroscopic images. *Scientific Reports*, *14*(1), 7847. DOI: 10.1038/s41598-024-58361-8 PMID: 38570595

Chapter 12
5G and Human Health: Reviewing the Risks of Cancer and Neurological Disorders

S. Sneh
https://orcid.org/0009-0003-6649-6344
Chitkara University, India

Hakam Singh
Chitkara University, India

Nilamadhab Mishra
https://orcid.org/0000-0002-1330-4869
VIT Bhopal University, India

Abhishek Thakur
https://orcid.org/0000-0002-9955-9693
Chitkara University, India

Ramamani Tripathy
Chitkara University, India

Rudra Kalyan Nayak
https://orcid.org/0000-0003-4447-8391
VIT Bhopal University, India

Amit Thakur
https://orcid.org/0009-0007-7550-4990
VIT Bhopal University, India

ABSTRACT

5G generation has raised significant concerns among the overall public and medical network about its potential influences on human health, especially concerning the implicit risk of cancer and neurological disorders. This debate has mainly become contentious, with a few studies claiming that 5G technology poses dangerous outcomes for human health. In contrast, others find no specific affiliation between 5G and human health. The International Agency for Research on Cancer (IARC) has classified radiofrequency electromagnetic fields, which might be a pivotal element of 5G networks, as possibly carcinogenic. The paper thoroughly analyses the potential health effects of 5G by examining the scientific literature that has already been published. The review summarizes research findings from several studies, international guidelines, and technological advancements to reduce potential health risks associated with 5G exposure.

DOI: 10.4018/979-8-3693-9208-9.ch012

1. INTRODUCTION

The launch of 5G, the fifth technology of the wi-fi era, has generated various opinions among scientific researchers. On the one hand, there's a tremendous expectation for how 5G will revolutionize various industries and beautify our ordinary lives. This advanced wireless technology guarantees notably quicker speeds, decreased latency, and greater network ability. These advancements can assist many revolutionary offerings and operations, from the Internet of Things to clever towns and stronger far-flung healthcare alternatives. The extended connectivity and information-managing abilities of 5G could raise productivity, improve exceptional lifestyles, and contribute to profitable economic growth. On the other hand, the deployment of 5G has sparked extreme worries about its ability and health results, in particular regarding the risks of cancer and neurological disorders. These worries have received great attention, with many calling for a thorough exam of the clinical evidence of the health effects of exposure to 5G networks. This paper aims to discover these critical issues by drawing from clinical studies, global pointers, and insights from communications engineering.

The purpose is to provide a nicely rounded and knowledgeable angle at the feasible health dangers associated with the massive adoption of 5G technology. The introduction of each new technology of wireless generation has traditionally been discovered through public fitness troubles, and the arrival of 5G is not an exception. As this advanced wi-fi network stays rolled out globally, it's essential to cautiously look at the scientific evidence and address the lingering uncertainties surrounding the ability health implications of 5G exposure. This paper will try to provide a balanced and comprehensive evaluation, imparting policy guidelines and hints for destiny research to ensure the accountable and safe deployment of the 5G era

1.1 Background of Wireless Technology

Wireless generation started with the primary technology (1G), which was brought within the Eighties and marked a giant milestone in cell communique. This analog system inspired modern cellular networks, frequently designed for voice calls. It only permitted voice calls within one U.S.A. The 1G era encompassed systems such as the Mobile Telephone System (MTS), Advanced Mobile Telephone System (AMTS), Improved Mobile Telephone System (IMTS), and Push To Talk (PTT). The frequency band for 1G changed into 824-894 MHz, with Frequency Division Duplexing used to split the uplink and downlink signals. The uplink (mobile to base station) operated on 824-849 MHz, while the downlink (base station to mobile) functioned on 869-894 MHz Despite its pioneering role, 1G faced numerous demanding situations, including negative voice, lack of security, high strength consumption, and great fees, ultimately bringing about its alternative through 2G technology (Chávez-Santiago et al., 2015).

1991, the second era (2G) generation was added, marking a considerable advancement over its predecessor. This era is characterized by digital smartphone calls and messaging services, leveraging the Global System for Mobile Communication (GSM) era. GSM hired virtual modulation to beautify voice nicely, supplying clearer and extra dependable verbal exchange. 2G generation may be divided into TDMA-primarily based (Time Division Multiple Access) and CDMA-primarily based (Code Division Multiple Access) requirements. TDMA separated signals into time slots, allowing a couple of users to share the equal channel by assigning precise time slots to each consumer. CDMA, on the other hand, enabled a couple of users to share the identical channel and frequency allocation within a selected band by using particular spreading codes. Operating within 850 to 1900 MHz frequencies, 2G greatly improved service first-rate and security phrases. However, it still confronted troubles with inconsistent insurance and slower information transfer fees, particularly in rural and faraway areas (Matrouk, et al., 2023).

The launch of the third technology (3G) generation in 2000 marked another leap forward, bringing excessive-velocity net connectivity, video calling capabilities, and quicker communication to cell customers. This technology introduced 3 unique kinds of cellular entry to generation: TD-SCDMA (Time Division Synchronous Division Multiple Access), WCDMA (Wideband Code Division Multiple Access), and CDMA2000 (Code Division Multiple Access). These technologies allowed for shipping various online services and multimedia programs, making smartphones a vital part of regular existence. The frequency bandwidth for 3G ranged from 850 MHz to 2100 MHz, permitting higher insurance and faster statistics quotes than preceding generations (Chiaraviglio, et al., 2021).

In 2009, the fourth (4G) era was introduced, constructing the advancements of 3G with extra offerings, including Multimedia Messaging Services (MMS), digital television, and diverse leisure alternatives. 4G utilized Multiple Input Multiple Output (MIMO) and Orthogonal Frequency Division Multiplexing (OFDM) technologies to beautify information transmission efficiency and community potential. The frequency bandwidth for 4G spanned from 700 MHz to 2600 MHz, imparting even faster net speeds and greater dependable connectivity for a huge range of packages, which includes high-definition video streaming and superior gaming (Karipidis, et al.,2021).

The 5th technology (5G), the latest wireless era development, was launched in 2019. 5G represents a giant evolution over 4G, imparting extraordinary speeds, decreased latency, and the capability to attach various devices concurrently. It is based on OFDM and Massive MIMO technology, with a greater variety of antennas than 4G. The frequency bandwidth variety for 5G consists of sub-6 GHz frequencies (six hundred MHz-3700 MHz) and millimeter wave frequencies (24,000 MHz-47,000 MHz). These higher frequencies allow 5G to help information-in-depth applications, which include augmented truth (AR), virtual fact (VR), and the Internet of Things (IoT), riding innovation throughout diverse sectors (Djomadji et al., 2023).

As the wi-fi generation has evolved from 1G to 5G, each era has extended the frequency range and delivered new talents, reworking how we communicate and interact with the sector. While each technology has brought large upgrades in terms of velocity, capability, and high-quality carrier, the progression from analog to virtual, and now to excessive-frequency millimeter waves, has also raised new challenges and possibilities (Yarrabothu, R. S., 2018). The deployment of 5G mainly holds the promise of revolutionizing industries and society. However, it additionally necessitates cautious attention to the ability of health effects and the development of robust regulatory frameworks to ensure safe and equitable access to this transformative technology.

Table 1. Details of network generations

Generation	Year	Technology	Frequency Bandwidth Range
1G	1980	Analog Signal	824-894MHz
2G	1991	GSM, CDMA, TDMA	850-1900 MHz
3G	2000	CDMA200, WCMA(UMTS), TD-SCDMA	850-2100 MHz
4G	2010	OFDM, MIMO	700-2600 MHz
5G	2019	OFDM, Massive MIMO	● Sub-6-GHz Frequencies: 600MHz-3700MHz. ● Millimeter Wave Frequencies: 24,000 MHz-47,000 MHz.

1.2 Comparative Analysis of Different Frequency Band Exposure

Figure 1 illustrates the frequency range of wireless technology from 1G to 5G. The minimum and maximum frequencies in the graph show how the frequency band has expanded over time with each generation of wireless technology. One of the key distinguishing features of 5G technology is its use of different frequency groups, including the higher-frequency millimeter wave range. While the potential health impacts of these higher frequencies have been the subject of continuous scientific investigation, the conclusions drawn from different studies have been mixed.

Some studies have found no solid proof to recommend that 5G or other radio frequency communication signals will cause or contribute to adverse health impacts. However, other investigations have demonstrated that 5G exposure may have adverse systemic effects, not just on the skin and eyes as commonly believed. These contrasting findings highlight the need for a more comprehensive understanding of the complex relationship between 5G technology and its potential health implications. To accommodate these differing viewpoints, it is essential to carefully examine the test conditions and research methodologies utilized in these studies and to compare them with the real-world settings in which 5G hardware and infrastructure operate (Rao & Prasad, 2018). Factors such as exposure duration, signal strength, and usage patterns can all influence the potential health effects of 5G technology. By critically analyzing the data and considering the context in which it was obtained, researchers and policymakers can gain a more nuanced and balanced perspective on the potential health impacts of 5G, ultimately informing evidence-based decision-making and policy development.

Figure 1. Frequency range across mobile generation

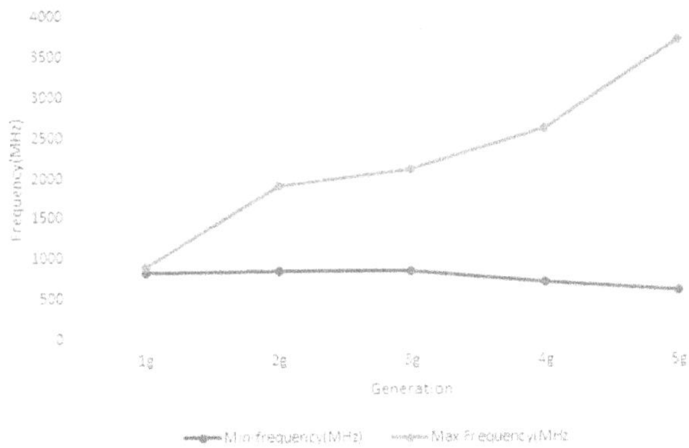

1.3 Historical Context of Wireless Technologies and Health Concerns

The advancement of wireless communication technologies has been accompanied by discussions and debates about their potential health impacts. Since the presentation of earlier generations of wireless technologies, similar to 2G, 3G, and 4G, broad concerns have been raised about the potential health risks related to the radiofrequency electromagnetic areas transmitted by these systems. These concerns have been boosted as wireless networks have become progressively ubiquitous in our day-to-day lives, directing more prominent public investigations and requests for a comprehensive assessment of the scientific evidence (Akyildiz et al., 2016). The history of public health concerns surrounding wireless technologies can be followed back several decades. As early as the 1950s, there were reports of potential biological impacts from exposure to radiofrequency radiation, especially on the central nervous system. All through the following decades, as mobile phone networks became broader, a developing body of research examined the possible links between RF exposure and different health results, including cancer and neurological disorders. This continuous scientific dialogue has fueled public debates and policy discussions about understanding and moderating potential health risks (Ishtiaq et al., 2021).

The International Agency for Research on Cancer, a specialized cancer research agency of the World Health Organization, has classified radiofrequency electromagnetic zones as possibly carcinogenic to people. This category demonstrates that there's limited evidence of a causal relationship between introduction to these areas and certain types of cancer, similar to head and neck cancers. The IARC's assessment has further fueled the continuous debate and open concern around the potential health risks of wireless technologies involving the developing 5G network. These concerns have persisted and evolved with the presentation of each modern era of wireless technology. With the rollout of 5G networks, there are renewed fears about the potential health impacts of the higher-frequency millimeter wave spectrum utilized by 5G. Whereas a few have found no dependable proof of adverse health impacts from 5G exposure, other research proposes the possibility of systemic effects beyond just the skin and eyes. This continuous scientific uncertainty has led to a continued public investigation and the requirement for further research to completely understand the health implications of 5G technology. As the arrangement of 5G networks advances, there is a developing emphasis on developing technological solutions and optimization techniques to relieve potential health risks and address the public's concerns.

1.4 Factors Influencing 5G Deployment and Adoption

The hit deployment and adoption of the 5G generation hinge on a couple of interrelated factors that span technological, economic, regulatory, social, and industry dimensions. Technologically, the transition to 5G demands full-size improvements to present infrastructure and the installation of new base stations, small cells, and advanced technologies including Massive MIMO and beamforming.

The availability and allocation of spectrum are crucial, as 5G operates throughout low, mid, and excessive bands, every with awesome characteristics (Di Ciaula, 2018). The manufacturing and affordability of 5G-compatible devices additionally power purchaser adoption costs. Economically, big monetary investments are required from telecom operators for infrastructure upgrades, spectrum acquisition, and generation improvement. Governments and private investors play a pivotal position in offering investment support, even as telecom operators should evaluate the return on investment through ability sales from new offerings and packages. Consumer demand for superior connectivity, in addition, drives adoption, stimulated by using increasing facts intake and the want for reliable network overall performance (Russell, 2018).

Regulatory frameworks considerably form 5G deployment, with government guidelines and pointers governing spectrum allocation, network deployment, and protection requirements. Harmonizing spectrum allocation and regulatory standards globally is important for seamless interoperability. Addressing public health issues through proof-primarily based regulations and transparent conversation is essential for gaining public agreement and recognition. Social and cultural factors, such as public perception and the capability to bridge the digital divide, impact 5G adoption (Simkó & Mattsson, 2019). Effective public schooling campaigns and tasks to increase 5G blessings to rural and underserved regions are critical for promoting digital inclusion. Cultural attitudes towards generation adoption range across areas, necessitating tailor-made deployment strategies. Industry collaboration and innovation are key to using 5G deployment. Partnerships across the telecom environment, ongoing research and development efforts, and standardization projects contribute to advancing 5G technology and addressing emerging demanding situations. Collaborative efforts foster innovation, lessen prices, and accelerate community rollout. The deployment and adoption of the 5G era are influenced by a complex interplay of factors that require a collaborative and multi-faceted method. By information and addressing these effects, stakeholders can ensure the successful and sustainable rollout of 5G networks, unlocking the transformative potential of this subsequent-generation generation (Latif et al., 2017).

1.5 Economic and Social Impact of 5G Deployment

The deployment of 5G technology can bring big financial and social effects, both at the macro and micro tiers. At the macroeconomic degree, the supply of immoderate-speed, low-latency, and dependable 5G networks can power productivity gains and permit new corporation models at some point in numerous industries. For example, mixing 5G with rising technologies, the Internet of Things, artificial intelligence, and robotics, can revolutionize production, healthcare, and transportation (Vornoli et al., 2019). In manufacturing, 5G-enabled IoT devices can display device fitness and optimize production strategies, drastically lowering downtime and enhancing output. 5G can facilitate a way of surgical methods and real-time affected man or woman monitoring, enhancing affected individual outcomes and decreasing healthcare costs. Improved connectivity and records transmission abilities provided through the manner of 5G can also facilitate the development of clever cities, improving the performance of town infrastructure, public offerings, and useful resource management. Smart towns use IoT sensors and facts analytics to optimize urban infrastructure, public services, and useful resource management. For example, clever visitor management structures can reduce congestion and improve air quality by dynamically adjusting visitor's signals based totally on real-time visitor situations. This can create a stronger, more pleasant existence for residents and foster sustainable development (Bailey et al., 2020).

At the microeconomic degree, 5G can allow new sales streams and enterprise possibilities for each firm and people. The low latency and high throughput of 5G can allow the deployment of innovative applications in areas that include augmented and digital facts, far-off healthcare, and commercial automation. For instance, virtual truth reports can become extra immersive and responsive, far-off healthcare services can provide real-time consultations and tracking, and commercial automation can become extra particular and efficient. Furthermore, the improved connectivity and access to virtual services can bridge the cellular virtual divide, especially in underserved rural and suburban areas. This can empower individuals and small agencies, growing new avenues for entrepreneurship and financial growth (Chiaraviglio et al., 2022). Beyond the direct economic influences, the large adoption of the 5G generation can also have tremendous social implications. Improved connectivity and proper entry to virtual services can decorate social inclusion, permitting marginalized communities to participate more completely in the virtual monetary system. Additionally, the mixture of 5G with eras like telemedicine and remote schooling can beautify the right of access to healthcare and academic opportunities, especially in areas with restricted infrastructure. For example, far-off education powered with the resource of way of 5G can offer extraordinary, interactive reading reviews to college university college students in rural areas, and telemedicine can offer specialized healthcare offerings to sufferers in foreign places (McCredden et al., 2023; Zaitseva et al., 2020). Ultimately, the capability of the 5G era to strain financial boom, foster innovation, and enhance social welfare is massive. By improving productivity, permitting new industrial company fashions, and improving the right of entry to digital offerings and opportunities, 5G has the electricity to transform how we stay, work, and engage.

However, the perception of these advantages could require considerate and accountable deployment strategies prioritizing the desires and concerns of all stakeholders, including residents, agencies, and policymakers. This includes addressing capability health concerns, ensuring equitable access to 5G services, and developing guidelines that aid innovation while safeguarding public interests.

2. EVALUATION OF CANCER AND NEUROLOGICAL DISORDER RISKS

The medical network has diligently studied the capacity connections between publicity to 5G and the risks of cancer and neurological diseases. The World Health Organization's International Agency for Research on Cancer (IARC) has categorized radio-frequency electromagnetic fields, which embody 5G frequencies, as possibly carcinogenic to humans. This classification stems from proof suggesting a potential causal dating between exposure to these fields and certain kinds of cancer, in particular, brain tumors. Such findings have caused heightened public challenges and calls for further research to make clear the fitness implications of 5G technology (Marchetti, 2017).

However, the study's findings on the capability fitness influences of 5G exposure stay blended and inconclusive. While some studies mean a probable hyperlink between 5G exposure and prolonged risks of maximum cancers and neurological problems, together with mind tumors and neurodegenerative situations, other rigorously designed research has not discovered large evidence to aid those claims. This clinical uncertainty underscores the complicated and multifactorial nature of those health worries. Various synergistic factors, which incorporate individual vulnerability, genetic predisposition, and exceptional environmental exposures, may additionally affect the capability effects of 5G publicity. Additionally, the perfect 5G deployment strategies, including antenna placement, electricity output, and insurance areas, can substantially impact the actual publicity degrees skilled with the resources of community populations. This variability necessitates a nuanced, region-precise assessment of the capability fitness risks related to 5G generation (Pennanen et al., 2024; Aslam et al., 2024).

Ongoing studies and collaboration with the scientific community, regulatory bodies, and enterprise stakeholders are crucial to elucidate the capacity fitness effects of 5G and develop powerful mitigation techniques. A comprehensive, proof-primarily based technique that prioritizes public fitness and protection is important for the responsible and sustainable deployment of 5G networks. Such a method should include non-prevent monitoring of exposure levels, prolonged-term epidemiological studies, and improving hints that reflect modern-day medical evidence (Barry et al., 2024). It is essential to word that the regulatory panorama governing the deployment of 5G and one-of-a-kind wireless generation is constantly evolving. International pointers and countrywide policies, developed through organizations incorporating the International Commission on Non-Ionizing Radiation Protection (ICNIRP) and the International Telecommunication Union (ITU), provide a framework for ensuring the safety of 5G and extraordinary wireless era. However, implementing and interpreting these pointers can vary within the path of one in every type of country and region, requiring a thorough knowledge of the regulatory landscape in particular geographic areas. Some global places may adopt stricter exposure limits and rigorous tracking protocols, even as others might prioritize speedy deployment and technological development. As the deployment of the 5G era continues to progress globally, each stakeholder—the scientific community, enterprise, and policymakers—must paint collaboratively to deal with capacity fitness concerns, enforce appropriate regulatory measures, and foster a balanced and nicely informed public discourse. Transparency and effective verbal exchange are key to preserving public acceptance as true and making sure that the benefits of the 5G generation are found out without compromising public health. By accomplishing open talk, sharing research findings, and growing policies that stabilize innovation with safety, stakeholders can navigate the challenges associated with 5G deployment and pave the way for a future where technological development and public fitness coexist harmoniously (Rathore et al., 2012; Qureshi et al., 2022).

2.1. International Guidelines and National Regulations

The assessment of 5G fitness dangers wants to maintain in thought the worldwide pointers and national rules governing the appearance of electromagnetic radiation. These tips, created through groups with the International Commission on Non-Ionizing Radiation Protection (ICNIRP) and the International Telecommunication Union (ITU), provide a whole framework for ensuring the safety of 5G and precise wireless generation. The ICNIRP, an independent clinical business organization recognized through manner of the World Health Organization (WHO), has installed pointers that set limits on exposure to electromagnetic fields to guard human fitness. These tips are primarily based on big studies and are periodically updated to mirror the current scientific information. Similarly, the United Nations specialized Organization for Information and Communication Technologies, the ITU, has advanced requirements and tips for consistently deploying 5G and one-of-a-kind wi-fi networks (Lodato et al., 2024).

However, noting variations in implementing and interpreting these hints in numerous countries and areas is crucial. An in-depth understanding of those regulatory differences is crucial in assessing the capability health risks of 5G deployments, especially in geographic regions. For example, some global locations may adopt more stringent exposure limits than others, even as the enforcement and tracking of compliance with these guidelines may also vary. Additionally, unique 5G deployment strategies, including antenna placement, strength output, and insurance regions, can extensively affect the actual publicity ranges professionals by the close populace. This necessitates a nuanced, region-unique assessment of fitness risks, considering the 5G network infrastructure's particular tendencies and surroundings (Farmer et al., 2013).

Furthermore, the regulatory panorama is usually evolving as new clinical evidence emerges and policymakers try to balance the blessings of the 5G generation with capacity health issues. Ongoing collaboration and dialogue between the medical community, regulatory bodies, and industry stakeholders are essential in ensuring that sturdy safety measures are carried out and that there may be an obvious conversation with the general public concerning the deployment of 5G networks. By staying abreast of the ultra-modern studies and regulatory trends, stakeholders could make knowledgeable selections that prioritize public health, even facilitating the technological advancements delivered approximately by way of 5G.

This collaborative approach can assist in coping with public issues, foster acceptance as true, and ensure that the deployment of the 5G generation proceeds safely and accountable. Additionally, one must remember the evolving nature of the 5G era in assessing health risks. As 5G networks make bigger and include new frequencies and technologies, continuous monitoring and studies are necessary to apprehend their lengthy-term fitness implications. This requires investment in clinical studies, public health surveillance, and the improvement of the latest methodologies for assessing exposure and health results. Policymakers must remain flexible and aware of new proof, adjusting regulations and recommendations to guard public fitness. The assessment of 5G fitness risks is a complex and dynamic technique that consists of detailed know-how of world guidelines, countrywide suggestions, and specific deployment techniques. It calls for a multi-faceted technique that considers the present-day scientific studies, regulatory requirements, and the right traits of each geographic vicinity. Through non-forestall collaboration, apparent verbal exchange, and willpower toward public fitness, stakeholders can ensure that the deployment of the 5G era is both stable and useful for society (Shi, 2022).

2.2 Ethical Considerations

As 5G technology is developed and deployed worldwide, scientists, industry, and policymakers must work responsibly and transparently to address potential health concerns. Ensuring that 5G networks are deployed in a way that prioritizes human safety should be a priority (Jeschke et al., 2022). This requires a fine-grained and site-specific approach that considers the specific characteristics of the 5G infrastructure, regulatory framework, and environment. Ongoing research and stakeholder collaboration are essential to develop effective mitigation strategies and inform evidence-based policies. Furthermore, it is the moral obligation of all concerned events to talk about the study's findings and implications transparently to the general public. Given the ever-present and regularly non-consensual nature of wi-fi radiation exposures, even a small variety of great bioeffects require thorough research and transparent verbal exchange. This is necessary to save you from any fallacious conclusions or assumptions of protection that would jeopardize public and environmental health. As the deployment of the 5G era continues to progress globally, it's imperative that all stakeholders - the medical community, industry, and policymakers - work in unison to deal with potential fitness worries responsibly and transparently. This collaboration has to be aware of developing proof-based totally policies and regulatory frameworks that strike a cautious balance between the blessings of 5G technology and its ability to fitness impacts. At the same time, the public ought to be engaged and knowledgeable all through this method. A transparent communique of study findings, capability dangers, and mitigation techniques is important to building public belief and ensuring the responsible deployment of 5G. Without open dialogue and transparency, the general public may broaden unwarranted fears or misconceptions about 5G, which can bog down its sizeable adoption and limit the belief in its societal advantages.

In addition to obvious communication, the clinical network, agencies, and policymakers must moreover paint collectively to cope with functionality health troubles through revolutionary technological answers proactively.

This also can embody optimizing 5G radio frequency designs, implementing superior manipulation algorithms to lessen exposure, and strategically deploying 5G infrastructure to restrict human contact. By leveraging the stylish improvements in communications engineering, stakeholders can ensure that 5G is deployed in a manner that prioritizes public fitness and protection. To ensure the moral deployment of the 5G generation, it's far more important for worldwide requirements and regulatory frameworks to be installed and adhered to. Organizations, which include the International Commission on Non-Ionizing Radiation Protection (ICNIRP) and the Institute of Electrical and Electronics Engineers (IEEE), play vital roles in placing guidelines and standards for safe publicity stages and technological practices. Harmonizing those standards throughout borders can assist in mitigating discrepancies in protection regulations and provide a unified method for managing capability health risks related to 5G (Zhang et al., 2021). Moreover, stakeholder engagement wants to be no longer constrained to immediate worries; however, it needs to include lengthy-term tracking and assessment. The evolving nature of 5G technology necessitates a non-stop assessment of its effects on human fitness and the environment. Implementing periodic evaluations and updates to protection suggestions based on new study findings can be important in adapting to any rising issues. This iterative method will guide the accountable advancement of the 5G generation while ensuring that public protection stays a top priority. Ultimately, fostering surroundings of collaborative research, transparent verbal exchange, and rigorous standardization may be instrumental in addressing moral issues surrounding

5G deployment. By taking one's measures, stakeholders can collectively work toward a future wherein technological innovation and public health coexist harmoniously (Campbell et al., 2017).

2.3 Public Perception and Acceptance

The public perception and popularity of the 5G era are vital elements in its fulfillment deployment. As with any new technological development, 5G has been met with numerous public reactions, from pleasure to skepticism. On the splendid side, many humans view 5G as a transformative innovation that ensures bringing in a brand-new generation of hyper-connectivity, quicker statistics speeds, and a huge style of groundbreaking programs. The functionality of 5G to allow smart towns, enhance far-off healthcare, and decorate employer automation has generated sizeable enthusiasm among early adopters and technology lovers. As society increasingly depends on digital infrastructure, the predicted societal and economic advantages of 5G were emphasized as key drivers for its big adoption (Abu-Taieh et al., 2020). However, the public belief in 5G has been marked by using issues concerning capability fitness dangers associated with prolonged exposure to electromagnetic radiation. Despite assurances from regulatory bodies that 5G networks perform internal setup protection pointers, some population segments live skeptically. This skepticism is partly because of restricted public statistics on the technical elements of 5G and the proliferation of wrong records and unsubstantiated claims about its health consequences. The non-ionizing nature of 5G radio frequencies, which lack the electricity to ionize atoms or molecules, and the absence of conclusive proof linking them to dangerous fitness effects have not fully allayed public fears. Addressing one's troubles requires a complete and multi-faceted approach.

First and foremost, obvious and ongoing communication from businesses, government businesses, and scientific establishments is essential. The public wants to be provided with clear, correct, and without issues accessible records of approximately 5G technology, its capability advantages, and scientific proof regarding its protection. This verbal exchange must also address the unique annoying situations posed via 5G, collectively with its use of better frequency bands and the resulting want for a denser network infrastructure. It is important to explain how these factors are cautiously managed to mitigate capability health risks. In parallel, the medical network must behavior rigorous, independent studies to research the capacity fitness influences of 5G. This research should embody not only the biological consequences of exposure to 5G frequencies but also discover progressive technological answers that could, in addition, reduce exposure and beautify the safety of 5G deployments. By working closely with enterprise stakeholders and policymakers, the scientific community can help form evidence-based guidelines and deployment techniques that prioritize public health and safety.

Community engagement and close-by-degree discussions are also critical additives of this method. Addressing the worries of unique groups and incorporating their comments into 5G rollout plans can help construct take delivery as real and alleviate fears. Engaging with the overall public through several channels, collectively with social media, network conferences, and public forums, can facilitate open talk and make sure that the voices of worried citizens are heard and addressed. By adopting the one's techniques, stakeholders can correctly manipulate public troubles, sell more knowledgeable information about the 5G era, and foster a supportive environment for its deployment (Gupta et al., 2019).

2.4. Technological Advancement to Minimize Risk

With the rollout of 5G networks shifting forward, there may be a growing number of technological solutions to mitigate fitness concerns. This development is essential for ensuring that the 5G generation may be released correctly and sustainably, addressing public issues regarding its potential fitness results. One of the key areas of attention entails optimizing the 5G radio frequency layout. Researchers and engineers are substantially reading how to improve the directionality and beam-shaping abilities of 5G antennas (Kim & Nasim, 2020).

Enhanced directionality permits a greater unique focus on signal transport, probably decreasing the publicity experienced by people, specifically in densely populated urban areas. By excellent tuning radio frequency configuration and beam patterns, 5G systems can be designed to direct signal power precisely wherein it is wished maximum while concurrently minimizing spillover and exposure in adjacent areas (Rokade & Mishra, 2024). Another essential aspect of technological advancement in 5G is superior management mechanisms. 5G networks depend upon dynamic supervision to keep sign first-class and control transmission manipulation constantly. This dynamic adjustment considers different factors, consisting of user region, tool traits, and network conditions. The 5G gadget can optimize power efficiency using state-of-the-art control algorithms, lowering pointless exposure ranges. This method now complements strength efficiency and mitigates potential health risks related to immoderate electromagnetic radiation (Routray et al., 2018; Emara et al., 2014; Emara, 2016; Emara, 2019).

The strategic placement of 5G cellular websites is critical in minimizing publicity degrees. Network directors must cautiously select locations to ensure that the 5G infrastructure is deployed, limiting human publicity while supplying essential coverage and capability. For example, placing cell websites away from densely populated regions can significantly lessen publicity. Additionally, prioritizing rooftops or accelerated systems for cell site placement can limit human contact with electromagnetic fields. Natural or synthetic limitations can also direct indicators far from touchy regions, imparting a delivered layer of protection (Yang & Jeon, 2020).

As technological enhancements continue, studies in communications engineering are increasingly centered on expertise in the interaction between the 5G era and human fitness. Researchers are utilizing superior modeling, simulation, and estimation strategies to create more statistics on how electromagnetic fields from 5G networks propagate and are absorbed via human tissues. This expertise is treasured in shaping coverage selections, improving more significant regular 5G requirements and tips, and ultimately assisting in this groundbreaking technology's accountable and sustainable deployment (Zhang et al., 2020).

In addition to technological and engineering solutions, public schooling and transparent communication are essential in addressing fitness issues related to 5G. Providing clean, evidence-based information about the safety measures and regulations in the region can help alleviate public anxiety. Engaging with groups, paying attention to their worries, and addressing them via genuine, technology-based discussions can foster extra reputation and belief in the 5G generation (Silva & Guerreiro, 2020).

Figure 2. Reduction in Exposure to Electromagnetic Fields

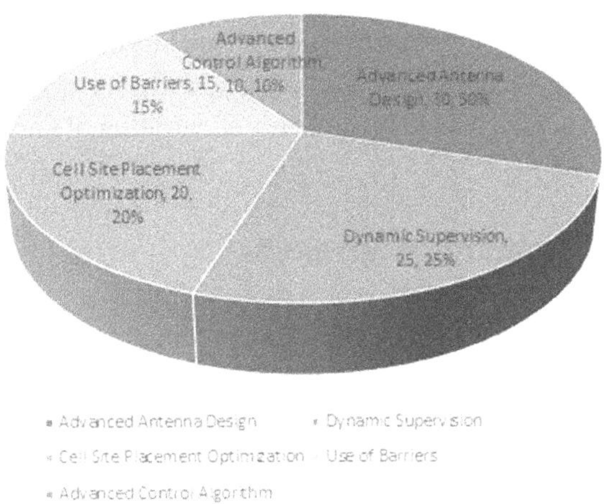

Moreover, ongoing collaboration among the scientific network, regulatory bodies, and industry stakeholders is critical. Such collaboration guarantees that the modern-day medical findings are integrated into regulatory frameworks and industry practices, selling the improvement and deployment of 5G technology that prioritizes public fitness and safety. As research maintains and new statistics emerge, it's vital to remain adaptable and responsive, updating recommendations and practices to mirror the most present-day expertise of 5G's ability fitness influences. The deployment of the 5G generation provides each opportunity and challenge. Through technological innovations, strategic making plans, ongoing studies, and transparent conversation, it's far viable to cope with fitness concerns and understand the whole capability of 5G. By prioritizing protection and sustainability, the rollout of 5G can continue in a manner that blesses society while minimizing capacity risks.

3. CASE STUDY OF 5G DEPLOYMENT IN VARIOUS COUNTRIES

The rapid development of telecommunication technology has ushered in the arrival of 5G, a transformative wireless trend that guarantees terrific speeds, low latency, and superior connectivity. As global locations round the sector race to install 5G infrastructure, policymakers and regulators are grappling with the complicated challenges that stand up from this technological shift. A key component of 5G deployment is the regulatory framework that governs its implementation. Policy-makers play an important function in shaping the panorama for the 5G rollout, as their decisions can notably impact the rate, success, and typical effectiveness of its adoption. With streamlined approval procedures, spectrum allocation, and incentives for infrastructure funding, regulatory support can substantially facilitate the well-timed and large deployment of 5G. Additionally, harmonizing spectrum allocation throughout borders is essential for seamlessly integrating 5G networks, permitting go-border connectivity and roaming.

United States: Proactive Approach

In the USA, the Federal Communications Commission (FCC) has proactively facilitated the deployment of 5G structures. By simplifying regulatory methods and making extra spectrum to be had, the FCC has enabled a swift rollout of 5G infrastructure. This regulatory help, blended with the robust presence of essential telecommunications corporations, has pushed speedy 5G deployment. However, public issues regarding capacity fitness risks from expanded radiofrequency (RF) exposure have persevered, necessitating ongoing threat checks and transparent communique to cope with these problems. The FCC's projects have covered public consultations and the guide of RF exposure suggestions based on cutting-edge scientific research to make certain public self-belief and protection (Benzaid & Taleb, 2020).

Europe: Cautious and Regulated

European international locations have generally followed an extra cautious method to 5G deployment. Nations like Switzerland and Belgium have imposed strict limits on electromagnetic radiation publicity, reflecting heightened public sensitivity to fitness concerns. This conservative stance emphasizes the importance of ensuring the safety of 5G networks before tremendous adoption. European regulatory bodies prioritize complete danger checks and proof-based guidelines to address capacity fitness implications, balancing technological development with public health concerns. For example, the European Commission has established stringent requirements for RF publicity and mandated big environmental and fitness effect assessments before 5G community approval. These measures are designed to reassure the general public and make sure that 5G generation is deployed safely (Mohanta et al., 2019)

Asia: Diverse Strategies

In Asia, the 5G panorama is diverse. South Korea and China were at the forefront of 5G deployment, leveraging their technological abilities and favorable regulatory environments to boost the rollout. South Korea's aggressive funding in 5G infrastructure and China's country-driven method have positioned these countries as worldwide leaders in the 5G era. Both nations have made full-size investments in research and improvement, infrastructure, and regulatory frameworks to help the speedy deployment of 5G. Conversely, India has taken a greater deliberate approach, cautiously weighing the potential advantages and dangers before committing to huge scale 5G infrastructure investments. This measured method reflects India's awareness of ensuring the protection and sustainability of 5G deployment, with full-size trials and checks to apprehend the results of 5G generation in various environmental and populace settings (Wu et al., 2021)

Comparative Analysis: The nearby variations in 5G deployment methodologies highlight the complex interplay between technological development, public fitness worries, and regulatory structures. The proactive approach within the US contrasts with the cautious stance in Europe, while Asia presents a mixture of aggressive and measured techniques. Understanding those factors is important in assessing the capacity health influences of 5G, as the implementation and monitoring of 5G systems vary considerably throughout distinct areas. Policymakers and enterprise stakeholders should navigate those challenges via effective verbal exchange, robust research, and collaborative efforts to ensure the safe and successful deployment of the 5G generation (Pradhan et al., 2023).

Figure 3. Comparison of 5G Deployment Factors Across Regions

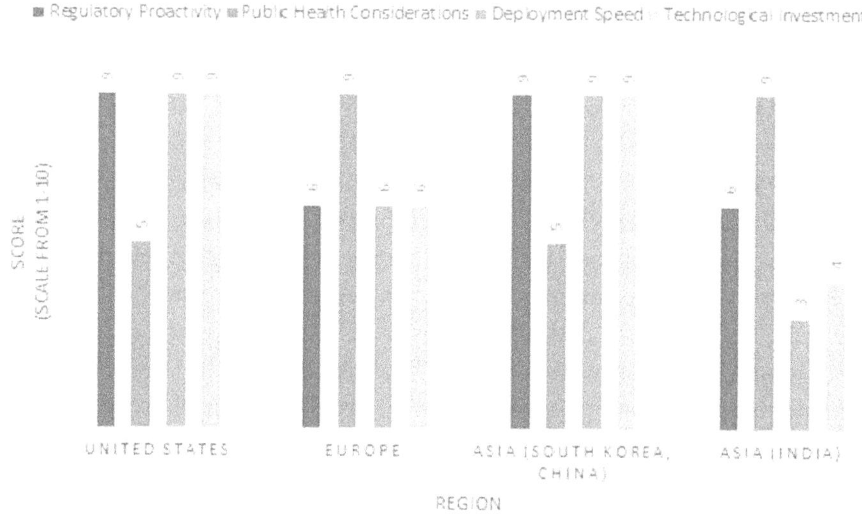

4. POLICY RECOMMENDATIONS

Based on the current state of analytic understanding and nonsupervisory landscape, some policy suggestions can be made to address the concerns around the health impacts of 5G:

Intensify and expand research efforts: There is a critical need for continued and multiplied research funding to understand better the long-term health influences of 5G exposure, such as capability synergistic effects with other environmental factors. This research must encompass huge-scale epidemiological research, in-intensity animal and cellular-based experiments, and superior computational modeling and simulation techniques. The aim is to expand extra complete information on how 5G radiofrequency signals interact with the human frame and probably contribute to destructive health consequences. This research should look at numerous publicity eventualities, including extended and intermittent publicity, and consider vulnerable populations, youngsters, the aged, and people with preexisting health situations. Additionally, studies have to explore the biological mechanisms underlying any found fitness consequences, imparting a clearer photograph of how 5G exposure might affect cell techniques and contribute to most cancers and neurological disorders.

Establish harmonized international standards and guidelines: Developing a harmonized international framework for placing 5G exposure limits and protection standards is important to ensure consistency, transparency, and evidence-based total policymaking throughout specific nations and areas. This framework has to be advanced through collaboration among worldwide health companies, regulatory bodies, and the clinical community. A unified set of pointers for deploying and operating 5G networks will foster public belief and ensure the safety of public fitness. These requirements must be regularly reviewed and updated in light of the latest clinical evidence, ensuring they stay applicable and powerful. Harmonized standards may also facilitate the worldwide deployment of the 5G era, providing clean and steady regulations for enterprise players to comply with, thereby decreasing regulatory uncertainties and selling international cooperation (West, 2016).

Facilitate collaborative efforts between stakeholders: Advancing significant collaboration between policymakers, the scientific network, and the telecommunications enterprise is crucial to ensure that 5G deployment is carried out responsibly and transparently, with a number one consciousness on protecting public health and safety. This collaborative technique must involve open conversation, records sharing, and joint chance evaluation efforts to align the interests of all stakeholders and develop solutions that balance technological development and public well-being. Policymakers must interact with scientists and industry experts to apprehend state-of-the-art research findings and technological developments. At the same time, the medical community must work carefully with regulators to ensure their studies address policy-relevant questions. The telecommunications enterprise has to participate actively in these discussions, sharing technical information and insights while committing to transparency and public safety. By fostering a way of life of collaboration and mutual recognition, stakeholders can paint collectively to address public issues, beautify the protection of the 5G generation, and promote its blessings (Miller et al., 2019).

Explore innovative solutions to minimize exposure: Researchers and engineers should explore modern technological solutions to decrease exposure levels and reduce capacity fitness risks associated with 5G. This includes developing advanced antenna designs with progressed directionality and beam-shaping capabilities for greater focused sign transport. By enhancing the precision of signal transmission, those technologies can reduce the quantity of electromagnetic radiation to which humans are exposed, especially in densely populated urban regions. Strategic planning for mobile site locations is also essential. Network operators must prioritize setting antennas in places that minimize human exposure, consisting of rooftops or extended systems, and use natural or synthetic obstacles to guard sensitive regions. Furthermore, state-of-the-art manipulation algorithms must be hired to dynamically alter signal electricity and power use based on actual-time situations, optimizing each network's performance and safety. These algorithms can assist in balancing the need for high-velocity connectivity to minimize unnecessary publicity to electromagnetic fields (Ullah et al., 2019).

Enhance public education and communication: Public education and obvious communique are essential in addressing fitness issues related to 5G. Providing clear, proof-based information about the safety measures and guidelines in the area can help alleviate public anxiety. Efforts should be made to engage with groups, listen to their concerns, and cope with them via genuine, science-based discussions. Public training campaigns must explain the benefits and risks of the 5G era, the measures taken to ensure safety, and the continued studies efforts. Transparent communication from regulatory bodies, the scientific network, and the telecommunications industry can construct public agreement with and foster a more informed and balanced talk about 5G. By addressing these vital policy recommendations and expanding the scientific expertise of 5G technology, we will work toward an extra knowledgeable and balanced method to deploy this transformative era. This will ultimately guard the well-being of individuals and groups, even knowing the massive benefits that 5G offers.

5. FUTURE DIRECTION:

As the world continues to embrace the transformative capabilities of 5G, the next generation of wireless technology, 6G, is already on the horizon. Expected to provide remarkable speeds, surprisingly low latency, and advanced connectivity, 6G guarantees to revolutionize industries and societal infrastructures further. The present-day state of research and development suggests that 6G will not only decorate present packages but also allow new ones, which incorporate an alternative state-of-the-art AI system, immersive virtual and augmented truth opinions, and more interconnected smart cities. 6G is expected to deliver faster speeds and connections, supporting several current applications and services. These improvements can transform sectors such as artificial intelligence (AI), the Internet of Things (IoT), and smart cities that will improve productivity and normal functions. Moreover, the 6G era could enable real-time holographic communications, advanced telemedicine, and incredible futuristic applications beyond what we can presently achieve.

However, with those improvements come new fitness issues. The better frequencies and novel generation utilized in 6G may additionally introduce new organic effects that are not understood. Continuous research is important to discover those capability fitness consequences, focusing on prolonged-term publicity, synergistic outcomes with different environmental elements, and new biological effects. This would require big-scale epidemiological research, in-intensity cell-based total experiments, and superior computational modeling and simulation techniques. Regulatory frameworks will even need to adapt to address the specific demanding situations posed with the aid of 6G. Developing harmonized worldwide requirements and guidelines will ensure consistency, transparency, and evidence-based policymaking throughout unique international locations and areas. International cooperation will be key in setting these standards and ensuring the safe and responsible deployment of the 6G technology.

6. CONCLUSION

The deployment of 5G technology has raised valid worries about its ability and effects on human health, specifically regarding the dangers of cancer and neurological problems. The current scientific research presents blended findings, and the communications engineering attitude proposes that the real-world settings of 5G equipment might also differ from the controlled test situations in this research. As the rollout of 5G continues, the clinical network, administrative bodies, and enterprise stakeholders must paint collaboratively to thoroughly examine the health influences of 5G exposure. This collaborative attempt should involve comprehensive studies, open communication, and developing processes and guidelines prioritizing public safety. Only via a complete and independent evaluation of the to-be-had evidence can we make certain that the benefits of the 5G era are realized even as successfully mitigating any capacity dangers to human fitness. Every stakeholder must work together to balance technological progress and public well-being, ensuring that the deployment of 5G is completed responsibly and with the maximum subject for the health and safety of individuals and communities.

Looking ahead, the appearance of 6G technology guarantees even more advancements in speed, connectivity, and applications, similarly revolutionizing industries and societal infrastructures. However, 6G can even introduce new health concerns associated with higher frequencies and novel technologies that are not fully understood. Continuous studies are crucial to explore these ability health consequences that specialize in long-term publicity, synergistic results with other environmental elements, and new biological impacts. Developing harmonized global requirements and tips might be essential for the safe and responsible deployment of the 6G technology. By fostering interdisciplinary collaboration, investing in complete studies, and evolving regulatory frameworks, we can harness the total capacity of 6G while minimizing its dangers. This balanced approach will ensure that the next wireless era will benefit society, even safeguarding public fitness.

REFERENCES:

Abu-Taieh, E. H., Al Hadid, I., & Zolait, A. (2020). *5G road map to communication revolution*. Cyberspace., DOI: 10.5772/intechopen.92319

Akyildiz, I. F., Nie, S., Lin, S., & Chandrasekaran, M. (2016). 5G roadmap: 10 key enabling technologies. *Computer Networks*, *106*, 17–48. DOI: 10.1016/j.comnet.2016.06.010

Aslam, A. B., Iqbal, F., Talpur, U., Syed, Z. S., & Shaikh, F. K. (2024). Artificial intelligence-enabled 6G mobile systems. *Signals and Communication Technology*, 49-79. https://doi.org/DOI: 10.1007/978-3-031-58527-2_3

Bailey, W. H., Cotts, B. R., & Dopart, P. J. (2020). Wireless 5G Radiofrequency technology — An overview of small cell exposures, standards and science. *IEEE Access : Practical Innovations, Open Solutions*, *8*, 140792–140797. DOI: 10.1109/ACCESS.2020.3010677

Barry, M., Haque, A. A., & Jan, M. T. (2024). Mobile commerce adoption in Malaysia: A conceptual framework. *Open Journal of Economics and Commerce, 5*(1), 04-12. https://doi.org/DOI: 10.22259/2638-549X.0501002

Benzaid, C., & Taleb, T. (2020). AI-driven zero touch network and service management in 5G and beyond: Challenges and research directions. *IEEE Network*, *34*(2), 186–194. DOI: 10.1109/MNET.001.1900252

Campbell, K., Diffley, J., Flanagan, B., Morelli, B., O'Neil, B., & Sideco, F. (2017). The 5G economy: How 5G technology will contribute to the global economy.

Chávez-Santiago, R., Szydełko, M., Kliks, A., Foukalas, F., Haddad, Y., Nolan, K. E., Kelly, M. Y., Masonta, M. T., & Balasingham, I. (2015). 5G: The convergence of wireless communications. *Wireless Personal Communications*, *83*(3), 1617–1642. DOI: 10.1007/s11277-015-2467-2 PMID: 27076701

Chiaraviglio, L., Elzanaty, A., & Alouini, M. (2021). Health risks associated with 5G exposure: A view from the communications engineering perspective. *IEEE Open Journal of the Communications Society*, *2*, 2131–2179. DOI: 10.1109/OJCOMS.2021.3106052

Chiaraviglio, L., Lodovisi, C., Franci, D., Grillo, E., Pavoncello, S., Aureli, T., Blefari-Melazzi, N., & Alouini, M. (2022). How much exposure from 5G towers is radiated over children, teenagers, schools and hospitals? *IEEE Open Journal of the Communications Society*, *3*, 1592–1614. DOI: 10.1109/OJCOMS.2022.3208938

Di Ciaula, A. (2018). Towards 5G communication systems: Are there health implications? *International Journal of Hygiene and Environmental Health, 221*(3), 367–375. DOI: 10.1016/j.ijheh.2018.01.011 PMID: 29402696

Djomadji, E. M., Garga, M., Fouba, B. A., & Bouetou, T. B. (2024). Genetic algorithm for mode selection in device-to-device (D2D) communication for 5G cellular networks. *American Journal of Networks and Communications, 13*(1), 30–43. DOI: 10.11648/j.ajnc.20241301.13

Emara, T. (2019). Adaptive Power-Saving Mechanism for VoIP Over WiMAX Based on Artificial Neural Network. In *Algorithms, Methods, and Applications in Mobile Computing and Communications* (pp. 158–177). IGI Global. https://doi.org/DOI: 10.4018/978-1-5225-5693-0.ch007

Emara, T. Z. (2016). Maximizing power saving for voip over wimax systems. [IJMCMC]. *International Journal of Mobile Computing and Multimedia Communications, 7*(1), 32–40. DOI: 10.4018/IJMCMC.2016010103

Emara, T. Z., Saleh, A. I., & Arafat, H. (2014). Power saving mechanism for VoIP services over WiMAX systems. *Wireless Networks, 20*(5), 975–985. DOI: 10.1007/s11276-013-0650-5

Farmer, P., Kim, J. Y., Kleinman, A., & Basilico, M. (2013). *Reimagining global health: an introduction* (Vol. 26). Univ of California Press.

Gupta, P., Malviya, L., & Charhate, S. V. (2019). 5G multi-element/port antenna design for wireless applications:a review. *International Journal of Microwave and Wireless Technologies, 11*(9), 918–938. DOI: 10.1017/S1759078719000382

Ishtiaq, M., Saeed, N., & Khan, M. A. (2021). Edge computing in IoT: A 6G perspective. https://doi.org/DOI: 10.36227/techrxiv.17031665

Jeschke, P., Alteköster, C., Hansson Mild, K., Israel, M., Ivanova, M., Schiessl, K., Shalamanova, T., Soyka, F., Stam, R., & Wilén, J. (2022). Protection of workers exposed to Radiofrequency electromagnetic fields: A perspective on open questions in the context of the new ICNIRP 2020 guidelines. *Frontiers in Public Health, 10*, 875946. Advance online publication. DOI: 10.3389/fpubh.2022.875946 PMID: 35757616

Karipidis, K., Mate, R., Urban, D., Tinker, R., & Wood, A. (2021). 5G mobile networks and health—A state-of-the-science review of the research into low-level RF fields above 6 GHz. *Journal of Exposure Science & Environmental Epidemiology, 31*(4), 585–605. DOI: 10.1038/s41370-021-00297-6 PMID: 33727687

Kim, S., & Nasim, I. (2020). Human electromagnetic Field exposure in 5G at 28 GHz. *IEEE Consumer Electronics Magazine*, *9*(6), 41–48. DOI: 10.1109/MCE.2019.2956223

Latif, S., Qadir, J., Farooq, S., & Imran, M. (2017). How 5G wireless (and concomitant technologies) will revolutionize healthcare? *Future Internet*, *9*(4), 93. DOI: 10.3390/fi9040093

Lodato, F., Garzia, A., Valbonesi, S., Ruello, G., Iodice, A., Matera, F., Salvo, P., & Massa, R. (2024). Ray tracing tools assessment for the evaluation of EMF levels generated by 5G NR systems: An overview. *2024 IEEE International Symposium on Measurements & Networking (M&N), 120*, 1-6. https://doi.org/DOI: 10.1109/MN60932.2024.10615802

Marchetti, N. (2017). Towards the 5th generation of wireless communication systems. *arXiv preprint arXiv:1702.00370.*

Matrouk, K., Trabelsi, Y., Gomathy, V., Arun Kumar, U., Rathish, C., & Parthasarathy, P. (2023). Energy efficient data transmission in intelligent transportation system (ITS): Millimeter (Mm wave) based routing algorithm for connected vehicles. *Optik (Stuttgart)*, *273*, 170374. DOI: 10.1016/j.ijleo.2022.170374

McCredden, J. E., Weller, S., & Leach, V. (2023). The assumption of safety is being used to justify the rollout of 5G technologies. *Frontiers in Public Health*, *11*, 1058454. Advance online publication. DOI: 10.3389/fpubh.2023.1058454 PMID: 36815158

Miller, A. B., Sears, M. E., Morgan, L. L., Davis, D. L., Hardell, L., Oremus, M., & Soskolne, C. L. (2019). Risks to health and well-being from radio-frequency radiation emitted by cell phones and other wireless devices. *Frontiers in Public Health*, *7*, 223. Advance online publication. DOI: 10.3389/fpubh.2019.00223 PMID: 31457001

Mohanta, B., Das, P., & Patnaik, S. (2019). Healthcare 5.0: A paradigm shift in digital healthcare system using artificial intelligence, IOT and 5G communication. *2019 International Conference on Applied Machine Learning (ICAML)*. https://doi.org/DOI: 10.1109/ICAML48257.2019.00044

Pennanen, H., Hänninen, T., Tervo, O., Tölli, A., & Latva-aho, M. (2024). 6G: The Intelligent Network of Everything—A Comprehensive Vision, Survey, and Tutorial. *arXiv preprint arXiv:2407.09398.*

Pradhan, B., Das, S., Roy, D. S., Routray, S., Benedetto, F., & Jhaveri, R. H. (2023). An AI-assisted smart healthcare system using 5G communication. *IEEE Access : Practical Innovations, Open Solutions*, *11*, 108339–108355. DOI: 10.1109/ACCESS.2023.3317174

Qureshi, H. N., Manalastas, M., Ijaz, A., Imran, A., Liu, Y., & Al Kalaa, M. O. (2022). Communication requirements in 5G-Enabled healthcare applications: Review and considerations. *Health Care, 10*(2), 293. DOI: 10.3390/healthcare10020293 PMID: 35206907

Rao, S. K., & Prasad, R. (2018). Impact of 5G technologies on industry 4.0. *Wireless Personal Communications, 100*(1), 145–159. DOI: 10.1007/s11277-018-5615-7

Rathore, A. K., Chaurasia, R. K., Mishra, R., & Kumar, H. (2012). Road map and challenges in 4G wireless system. *Journal of Electrical & Electronics, 01*(03). Advance online publication. DOI: 10.4172/2167-101X.1000104

Rokade, S., & Mishra, N. (2024). Intelligent health: Empowering illness and prognosis with machine learning-enabled electronic health records on blockchain. *2024 International Conference on Advancements in Smart, Secure and Intelligent Computing (ASSIC), 23*, 1-5. https://doi.org/DOI: 10.1109/ASSIC60049.2024.10508019

Routray, S. K., Jha, M. K., Sharma, L., Sarkar, S., Javali, A., & Tengshe, R. (2018). Energy consumption aspects of 5G waveforms. *2018 International Conference on Wireless Communications, Signal Processing and Networking (WiSPNET)*, 1-5. https://doi.org/DOI: 10.1109/WiSPNET.2018.8538674

Russell, C. L. (2018). 5 G wireless telecommunications expansion: Public health and environmental implications. *Environmental Research, 165*, 484–495. DOI: 10.1016/j.envres.2018.01.016 PMID: 29655646

Shi, L. (2022). Analysis of the security of 5G technology from the network level. *SHS Web of Conferences, 144*, 02007. https://doi.org/DOI: 10.1051/shsconf/202214402007

Silva, M. M. D., & Guerreiro, J. (2020). On the 5G and Beyond. *Applied Sciences (Basel, Switzerland), 10*(20), 7091. DOI: 10.3390/app10207091

Simkó, M., & Mattsson, M. (2019). 5G wireless communication and health effects—A pragmatic review based on available studies regarding 6 to 100 GHz. *International Journal of Environmental Research and Public Health, 16*(18), 3406. DOI: 10.3390/ijerph16183406 PMID: 31540320

Ullah, H., Gopalakrishnan Nair, N., Moore, A., Nugent, C., Muschamp, P., & Cuevas, M. (2019). 5G communication: An overview of vehicle-to-Everything, drones, and healthcare use-cases. *IEEE Access : Practical Innovations, Open Solutions, 7*, 37251–37268. DOI: 10.1109/ACCESS.2019.2905347

Vornoli, A., Falcioni, L., Mandrioli, D., Bua, L., & Belpoggi, F. (2019). The contribution of in vivo mammalian studies to the knowledge of adverse effects of Radiofrequency radiation on human health. *International Journal of Environmental Research and Public Health, 16*(18), 3379. DOI: 10.3390/ijerph16183379 PMID: 31547363

West, D. M. (2016). How 5G technology enables the health internet of things. *Brookings Center for Technology Innovation, 3*(1), 20.

Wu, Q., Xu, J., Zeng, Y., Ng, D. W. K., Al-Dhahir, N., Schober, R., & Swindlehurst, A. L. (2021). A comprehensive overview on 5G-and-beyond networks with UAVs: From communications to sensing and intelligence. *IEEE Journal on Selected Areas in Communications, 39*(10), 2912–2945. DOI: 10.1109/JSAC.2021.3088681

Yang, W., & Jeon, J. Y. (2020). Design strategies and elements of building envelope for urban acoustic environment. *Building and Environment, 182*, 107121. DOI: 10.1016/j.buildenv.2020.107121

Zaitseva, N., Shur, P., Chetverkina, K., & Khasanova, A. (2020). Developing methodical approaches to substantiating average annual maximum permissible concentrations of hazardous substances in ambient air in settlements as per acceptable health risk. *Health Risk Analysis*, (3), 39–48. DOI: 10.21668/health.risk/2020.3.05.eng

Zhang, C., Ueng, Y., Studer, C., & Burg, A. (2020). Artificial intelligence for 5G and beyond 5G: Implementations, algorithms, and optimizations. *IEEE Journal on Emerging and Selected Topics in Circuits and Systems, 10*(2), 145–148. DOI: 10.1109/JETCAS.2020.2999944

Zhang, D., Rodrigues, J. J., Zhai, Y., & Sato, T. (2021). Design and implementation of 5G e-Health systems: Technologies, use cases, and future challenges. *IEEE Communications Magazine, 59*(9), 80–85. DOI: 10.1109/MCOM.001.2100035

Chapter 13
AI-Driven Laboratory Automation:
Transforming Experimental Phases of Drug Discovery

Amena Mahmoud
Department of Computer Science, Faculty of Computers and Information, Kafrelsheikh University, Kafr Elsheikh, Egypt

Wael A. Awad
Department of Computer Science, Faculty of Computers and Artificial Intelligence, Damietta University, Damietta, Egypt

ABSTRACT

The integration of artificial intelligence (AI) into laboratory automation systems is revolutionizing the drug discovery process by enhancing efficiency and accuracy in experimental phases. By leveraging machine learning algorithms and robotic systems, researchers can achieve higher throughput in compound screening, optimize experimental designs, and reduce human error. A case study was discussed that demonstrated successful applications of AI in laboratory settings, highlighting advancements in high-throughput screening, data analysis, and predictive modeling. Additionally, we address the challenges associated with implementing AI in laboratory automation, including data integration, system interoperability, and the need for skilled personnel.

DOI: 10.4018/979-8-3693-9208-9.ch013

1. INTRODUCTION

The landscape of drug development is constantly evolving, at least partly in response to technological advancements. Innovations in biophysical platforms for target validation, disease mobilization, lead identification, and preclinical development are also snowballing. Drug developers are in hot pursuit of innovation, while the demand for efficiency, accuracy, and cost reduction is mounting higher (Liu et al., 2024). In silico re-evaluation of the existing literature and data brings out compelling de novo drug candidates. Consequently, practical and competitive pressures impose the integration of AI and laboratory systems. AI-driven laboratory automation in drug discovery has garnered critical attention in recent years for refining the methodologies and infrastructures. The rapid application of this powerful integrating technology and the reformation of the traditional drug discovery process is significant to stress upon in these lines, fig 1 presents the role of AI in drug discovery and healthcare (Mac et al, 2023).

Figure 1. The Role of AI in Drug Discovery and Healthcare

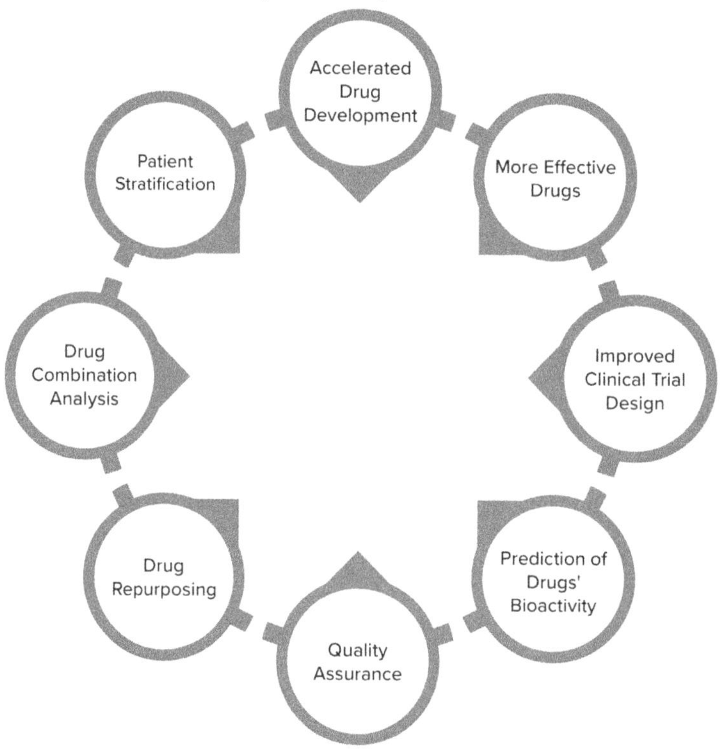

The possible innovation in the combinations and settings is aimed at understanding how effective we can be in the identification of drug candidates and the integration of both AI and laboratory robots. New drugs come to market faster and with less later-stage attrition, i.e., saving money. It is also possible that the artificial intelligence that drives such robots can become very proficient at discovering accurate drugs with low toxicity, which historically would be very expensive (Patel et al., 2022). In terms of GV20 value, shorter timelines and high success rates make drug development more accessible to the consumer. In general, it means we can create a social and economic structure where it makes sense to develop new medicines regularly, with lower profit margins in an effort to serve underserved populations around the world. This chapter lays the groundwork for this paradigm shift, with a particular focus on essential drug targets and clinical trial addresses, and how new technological developments can lead us to these new ways of drug discovery (Tran et al., 2023).

1.1. Background and Significance

The search for innovative medications used to be rooted in serendipitous discovery. With time, innovations in fields such as molecular and cell biology allowed more rational hypotheses to be generated for novel ligands. Biomolecular screening against potential targets has largely replaced serendipity. Innovative genomics and in-silico methods have accelerated mechanistic hypothesis generation, but there is general scientific consensus that the drug compound attrition rate seen with traditional target-based approaches can be significantly lower (Kang et al., 2024). Limitations within biomolecular drug interactions and cell-based systems have been primarily addressed by using more sophisticated high-content screening assays. Some notable exceptions do exist, such as the search for antibiotics, which have continued to rely on growth inhibition assays to have a better chance of selecting hit compounds with greater chances of being active later on. Pharma companies continue to utilize in-vitro culturing assays, albeit now typically using immortalized cell lines. Large libraries of compounds tested in these assays require automated screening systems (Soice et al., 2021).

There is an increased demand for machine learning methodology supported by an on-chip approach in the drug market. This is both to service lead optimization and to satisfy compound libraries intended for screening. Furthermore, the global drug market is on an exponential increase, at 5% per year, and this trajectory is likely to stabilize soon. It is suggested that companies have a vested interest in innovation, as demonstrated by a steadily increasing number of new chemical entities approved per year. There is marked growth in traditional in-vitro and in-vivo assessments including assays and imaging modalities (Rome et al., 2022). This ongoing reliance on traditional techniques means recent scientific progress, which would otherwise make significant contributions to drug discovery, is often bypassed. Such conventional approaches show marked limitations for speed, cost, and use as a reliable predictive maintenance strategy. It is anticipated that innovation within medicinal in-vitro testing approaches will significantly accelerate drug discovery and reduce the 12-year time horizon for developing approved drugs. There otherwise seems no impetus or motivation for advancing. The inability to design and bring to market new chemical entities swiftly is a health risk to global populations. Attention should then be focused not solely on discovering therapeutic compounds but also on designing fundamental non-traditional approaches to significantly accelerate ADMET assays and ADME testing strategies with new velocity and reduce costs (Alnowaiser et al., 2024).

2. FOUNDATIONS OF AI IN DRUG DISCOVERY

In general, when we talk about the application of AI in drug discovery, it is important that we have some foundational understanding of how AI works. In a traditional sense, AI can be divided into three branches: machine learning, deep learning, and natural language processing. Machine learning is at the heart of AI, where models learn from data, find relevant patterns, and make predictions. Deep learning is an advanced form of machine learning, where the pattern recognition algorithm is more complex. It is particularly good for image and speech recognition. Natural language processing, on the other hand, is for language processing, fig 2 illustrates the use of AI to accelerate drug discovery.

Figure 2. Using AI To Accelerate Drug Discovery

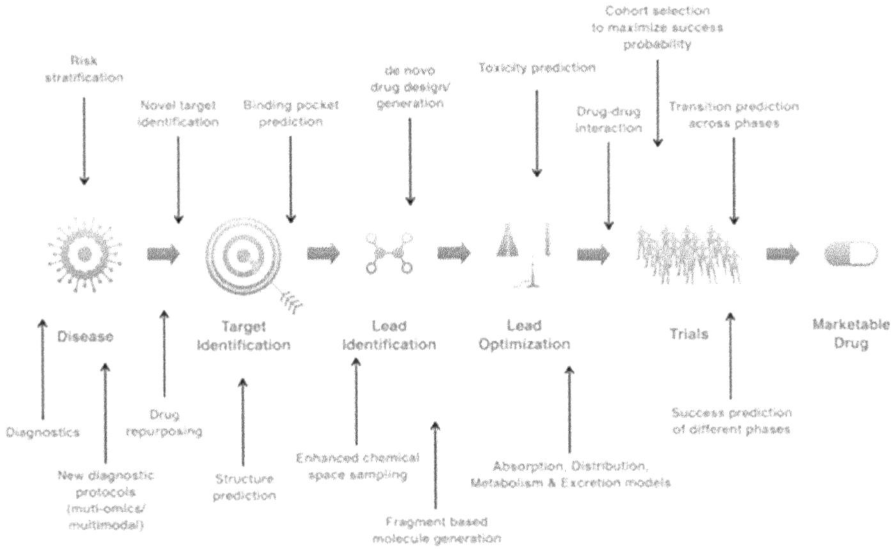

In drug discovery, AI-machine learning algorithms can help in various facets—discovering lead molecules, developing general medicines, or researching personalized drugs. Supervised and unsupervised learning are the core AI-machine learning algorithm types that we can apply in resolving pharmaceutical research issues like predicting compound properties, elucidating structure-activity relationships, simplifying ligand-based modeling, assuming protein folding, handling high-content imaging in phenotypic screening, and many more (González et al., 2022).

The idea of AI/machine learning in drug discovery is to train computers to learn from the data, recognize patterns in the data, and make recognition or identification depending on what they have learned from the data. The concept is detailed in a four-step plan to extract insights from the drug discovery data, composed of different facets like data, artificial intelligence, drug discovery data, data science, momentum analytics, insights, impact, and value. This plays a crucial role in the development of an AI model, the quality of the data, and the data's quantitative value. Generally, AI is not self-sufficient; it needs a major part of data science with human intervention and knowledge for the discovery of drugs or any other commercially useful inputs. It incorporates different branches like molecular informatics, cheminformatics, bioinformatics, computational biology, as well as chemistry (González et al., 2022).

2.1. Machine Learning and Deep Learning

In the field of AI, two important components exist, namely, machine learning (ML) and deep learning. In the traditional sense, machine learning includes two types of algorithms: supervised learning and unsupervised learning. The algorithms used are related to the analysis of a huge amount of biological omics data obtained. Deep learning, on the other hand, consists of powerful neural network models that can be used to discover intricate structures within big data, which typically consist of multiple layers of very simple computational nodes. Traditional computational methods generally use predefined features of molecular interactions and bypass experimental and costly lab experiments, while deep learning is state of the art in computational biology for uncovering intramolecular and intermolecular interactions, band gap energy, or any interesting molecular to systemic properties when the 3D structures and the related activity data are available. Importantly, deep learning can identify fingerprints of features including images and texts that can automatically learn more useful features from molecular data (Soleymani et al., 2022).

A critical aspect of deep learning is feature selection, which involves altering models to make them automatically select features of interest. As its name indicates, feature selection is a process of choosing input layer features considered as predictors that have the greatest impact in determining the target layer or classification. Owing to the extreme computational demand and requirement for big data, feature dimension reduction is applied to the gene/chemical approach. There still exist major challenges in deep learning systems, such as overfitting, generalization, and reproducibility of data. Despite these challenges, these AI techniques have been integrated into laboratory automation, resulting in improved accuracy and speed in experimental biology applications, particularly in mechanistic interpretation and drug discovery. Deep learning has been employed in the cardiology field to detect complex patterns in echocardiography videos, enabling the prediction of heart failure from the videos (Mahmoud et al., 2024).

3. CURRENT CHALLENGES IN DRUG DISCOVERY

The drug discovery process is expensive, with costs incurred at every step required to bring a new drug to market. An often-cited figure is that each new drug requires around $2.5 billion to be successfully brought to market, of which attrition in clinical testing accounts for around 60% of the costs. Over 90% of agents entering clinical trials eventually fail, leading to high drug development costs that are now recognized as unsustainable, and as a result, investing in R&D by the pharmaceutical industry is decreasing (Bhutani et al., 2021).

Traditional dose-response assays used to validate hits and leads are time-consuming and resource-intensive. Moreover, the biological complexity and variability of human biology mean it is impossible to predict all drug responses in the development process. Consequently, any one of many drug responses could mean a compound is abandoned in development, raising everyday costs. Over the years, the process of bringing a molecule to the clinic has become increasingly onerous due to the requirement for increasingly stringent testing for patient safety and efficacy by regulators. Failure costs are now growing due to the removal and expense of drug candidate compounds from clinical studies. Collectively, procedures throughout the original drug discovery process decrease the likelihood of safe and effective drugs being brought to patients due to high attrition rates (Barrett et al., 2022).

Phenotypic screening is a more relevant approach to discover new drugs because it refers to pathway mechanisms intrinsic in human-relevant whole cells to identify leads that may make it through development processes. High-content imaging is alongside high-throughput electrophysiology and gene sequence analysis to define more closely the responses of molecules with the disease, all gaining traction in facilitating this approach. Phenotypic assays, however, are complex and increase the technology expertise required, which is beyond traditional target-based assay experience and can become time-consuming for those inexperienced in drug discovery. A detailed review on the use of phenotypic assays utilizing high-content imaging towards cell culture studies was recently published, again whose approach does not take into account the regulatory element of clinical trials during drug discovery (Mahmoud et al., 2023).

3.1. High Costs and Time-Consuming Processes

Generating novel therapeutics comes with high costs. Generally, drug discovery and development can be a decade-long or even two-decade-long journey from the early research phase to the final marketing application when it is approved by the regulatory agencies. In particular, clinical trials are among the more expensive phases of a drug's development, and they might account for up to 70% of a project's final costs. Even if the overall time needed for the scientific development and regulatory approval of a new pharmaceutical has decreased in the past half-century, the industry is reporting peaks in premarketing drug development times. This delay is instead the result of the increased time needed to gather safety and efficacy data. Particularly in the drug discovery space, fermenting this environment is the novelty of molecules entered in clinical experiments, as well as the specificity of biological pathways being considered for therapeutic intervention. However, it is worth noting that more work does not always pay off: a project that has required an extensive biological research phase can gain speed with a rapid medicinal chemistry effort, combining the screening results from several projects and focusing on derivatization and lead optimization (Pandey et al., 2022).

Automation, in the form of chemical reaction automation and storage facilities, has promisingly revolutionized several aspects of the medicinal chemist's profession, also exerting a highly impactful effect on the initial stages of the drug discovery process: the increasing levels of compound availability have shortened the time needed to produce sufficient data to prioritize the compounds and to decide whether to continue a research program or to stop it. Bringing even a fresh drug to success is a dangerous horse race, and the quick abandonment of early-development compounds, characterized only partially, is fundamental to avoid spending the early budgets up to profitability and existential success in the sector. This represents one additional reason why the medicinal chemist, in the very beginning, should have at hand compounds in multiples as if cars in the emission line are controlled by the traffic lights to increase the overall appeal for the trip (Farghali et al., 2021).

In that race for the area under the survival curve, the investment in much bigger databases, where information literally grows at the pace of zeros and ones, and the decision to prioritize the compounds is also of utmost relevance. Even more importantly, machine learning and artificial intelligence driven computational tools now pervade several aspects of biological screening, with the capability to expedite the discovery of innovative therapeutics. In the end, it is only by reducing the long isolation of time required for the generation of a hypothesis and the performance of its verification that we will fully realize the potential for time and cost savings made available by modern applications (Farghali et al., 2021).

4. BENEFITS OF AI-DRIVEN AUTOMATION

Drug discovery can be improved due to AI-driven automation in many ways. It can make data analysis 100 times faster than human labor; it allows us to see our data in totality. For instance, humans are unable to close a detailed clinical trial, as it produces terabytes of data, and a human is unable to account for its totality. Common benefits of AI-driven automation include increased efficiency, higher accuracy, the ability to analyze larger datasets, and the ability to analyze more complex datasets. AI can use predictive algorithms to also increase the accuracy in predicting viable drug candidates. Cutting down on human error can also add greater accuracy to the discoveries. Additionally, AI's ability to analyze is estimated to be 100 times faster than any human, making the decision process much more time efficient. By cutting down on human error and helping to maximize the chances of choosing a viable drug, AI helps cut some costs as well. By quickly and thoroughly analyzing enormous datasets, AI helps cut down on time and the amount of work that a human would have been able to do in order to retrieve the same quantities of data. This is a marketing point, albeit used often, that revolves around the potential for great savings. Taking only the high estimate of the median pharma return on investment, AI in drug development can roughly double and up to almost triple the current ROI of around 7% for existing drugs (Hassani et al., 2020).

4.1. Increased Efficiency and Accuracy

One of the primary benefits of AI-driven laboratory automation in drug discovery is increased efficiency and accuracy. By deploying intelligent systems for this purpose, the aim is to streamline routine, manual tasks. This is significant because, at present, the vast majority of drug discovery workflows involve running a large number of mostly mundane, repetitive experiments. By automating these routine activities, AI means that we can handle high-throughput screening at a scale that would be impractical for a human lab to implement. AI systems can not only collect vast amounts of data very quickly but can also manipulate this data immediately, on the fly if required. Instead of a bottleneck being created by too much data with insufficient people to analyze it, AI automation means that more data can be dealt with in real time, inspiring more and more questions around drug development (Pun et al., 2023).

The time and cost saved by automation can also be a significant productivity tool, allowing human researchers to focus on the more intellectually demanding and creative aspects of the work. Making informed decisions is further enhanced by AI's predictive capabilities. In drug discovery, there are many promising molecules to choose from, but that will fail to transform a patient's life because they aren't effective or safe enough. AI can use developed, curated, rich datasets combined with deep learning models to predict likely outcomes in humans from tests done in laboratories. This will allow drug developers to perform fewer but higher impact clinical studies, mitigating the risk of human involvement. In addition, carrying out certain studies under controlled environments through automation allows researchers to access more accurate results compared to manually executed experiments. The accuracy of drug compound effectiveness prediction by AI is crucial because decisions made during pre-clinical trials can lead to significant financial commitments later in drug development. Increasingly, AI is already being used to progress drugs and get more accurate results in clinical trials for the best chance of success in humans (Sharma et al., 2022).

5. ETHICAL CONSIDERATIONS IN AI-DRIVEN DRUG DISCOVERY

One of the main considerations when implementing AI-based models in drug development and discovery is the inclusion of patient data in the datasets used to train algorithms. In this sense, transparency in machine learning methods, particularly when working with patient data, is critical. Moreover, in the area of drug development and discovery, where human lives are directly impacted, technologies that could be labeled as "black boxes" are less likely to be considered viable and appropriate. Together with the issue of transparency, considerations of responsibility become increasingly relevant. In other words, how can responsibility in this context be attributed? As in most disciplines, AI in drug development may be perceived as a means of automating complex tasks rather than replacing the human element, and how to define human and AI responsibility in this context is still a matter of debate. One can argue for a governance strategy in drug development that is based on the desirable values that we expect to be evident and properly accounted for in the technology, or we appeal to the expected functionality of the technology by referring to ethical frameworks. In the case of advanced machine learning models that promise to automate complex decision-making, the question of how such systems should be regulated is closely related to these ethical considerations (Nagi et al., 2022).

Another potential problem in machine learning techniques is the introduction of bias. AI algorithms are essentially an expression of the data that is used to train them. As a result, a critical assessment should be made with respect to the re-interpretation of the existing bias that is already present in the algorithm and consideration of the inclusion of new procedures in drug development concerning the development of research activity, results, and findings. In other words, baseline information that artificially prejudices results and impacts the outcome is treated as a representation of objective fact, thus contributing to creating a profile that humans must depict and accept (Araby et al., 2022).

However, with an AI algorithm, the unintended consequences of failing to interpret the data might lead to an accumulation that unwittingly conceals a potentially significant side effect or outcome, thereby hampering the implications of the newly developed drug. Likewise, these biases have the potential to eventually contribute to a lack of diversity in AI-driven drug development and discovery. The potential to question certain areas in decision-making can be argued, but it can also harness unintended consequences in every part of the work from being described in the process to the re-assessment procedure of results, including improvements or alterations. For example, a misinterpretation of the data repurposing for new identification of drugs with prior approval for on-the-market drugs could avoid examining targeted genes or diseases that may potentially be candidates for effective new drug treatments.

Regulatory approval of new AI tools for drug repurposing in drug development and discovery must ensure compulsory and standardized data governance. Data governance can be defined as a set of procedures, policies, and processes that regulate and standardize the acquisition, storage, access, use, and disposal of datasets. Each procedure postulated for data curation and standardized governance procedures must be in compliance with relevant and contemporary regional and international legislation regarding privacy and protected health information (Cheng et al., 2021).

5.1. Data Privacy and Security

Enabling rapid integration of AI-assisted technologies into drug discovery while ensuring data privacy and maximum security of the data will be crucial. Sensitive patient information may potentially be exposed through the use of big data in conjunction with AI. Research data should be as open as possible, especially when it is intended for the public good, but as closed as necessary to ensure the legitimate interests of the subjects are protected. Scientific researchers who use AI must comply with regulations and be ethical, especially with information from human data. New regulations in combination with AI will help significantly advance research and application, but opposition and challenges will remain. The coexistence of big data and AI confronts all stakeholders with new challenges (Mahmoud et al., 2020).

Using big data merely as a source and means of information for AI ignores the right of everyone who has generated or is concerned with their personality, including genetic information, to be treated as a human being and as a person who, while giving confidential information, expects it to be respected in line with human rights and ethical norms. Patient data misuse, data privacy violations, and security failures can have drastic implications for the reputation and perception of a data subject, as well as for the public's trust in a certain field of science or in drug development in general. We would like to emphasize the potential consequences of AI-driven misuse of patient data and suggest that the need for the highest level of encryption and adaptive access control becomes evident when the consequences of misuse are considered. Data integrity on all levels can only be commensurate with the rapid AI-driven expansion in both drug discovery and sales. Protecting your data is protecting your future (White et al., 2022).

6. CASE STUDIES IN AI-DRIVEN LABORATORY AUTOMATION

AI has proven to transform drug discovery programs and has recently offered numerous opportunities to pharmaceutical industries. Many experts in public and private organizations have invested a great amount of time and effort to develop and establish various AI tools and technologies in a myriad of experimental and computational tasks in the pharmaceutical industry. In this section, we present several use cases showing AI-driven laboratory automation in practice.

In the first case on polymer drug formulation, a workflow optimization approach based on generative models and active learning can meaningfully increase the efficiency of a complex high-throughput form-evaluate-print platform. The second case on the identification of chemical artifacts shows how AI can assist the characterization of automated workflows: deep learning-based, "smart experiment" designs do not just predict success or failure but also generalize the reasons behind failed experiments. The third case shows how an AI-driven approach can predict the outcomes of pharmaceutical amendment filings, delivering results in minutes. The fourth case presents a strategy based on reinforcement learning that reduces the amount of laboratory work required to define the critical quality attributes and critical process parameters for active pharmaceutical ingredient formulations (Hassan et al., 2024). The final case discusses the use of Gaussian Process models to inform the design of "multi-omics" experiments in the field of personalized medicine. In all cases, a diverse range of application areas is addressed, and the methodologies exhibited are equally diverse. Moreover, the case studies also offer insights into the specific challenges faced, for example, with regard to data availability and data quality, and the solutions implemented thus far. There is ample evidence of the success and efficiency of laboratory workflows through AI-driven automation (Bender et al., 2021).

6.1. Application in High-Throughput Screening

High-throughput screening (HTS) assesses the presence or biological activity of large numbers of potential compounds. It hovers over the middle section of the funnel because of its efficiency in revealing a simple cause-and-effect linkage, separating compound hit identification from lead optimization. In general, the thousands-to-millions scale of compounds are used to assess interactions against one or more target enzymes, proteins, cell lines, or tissues. The resultant data is used to predict the potential usefulness, or not, of a given compound for further structure-activity relationship studies. AI can speed up and automate large-scale assessment activities, thus critically expanding the base of the funnel and forging a deeper moat on the hit identification side (Aldewachi et al., 2021).

Application in High-Throughput Screening HTS, when recently shed of some of its historical and overly large trappings, has seen a resurgence in application because of the scope and the scale of assays it can solve and the power of cheminformatics to allocate resources toward new avenues of research. Combining increasingly specialized pharmacological assays with the machine learning required to interpret the ever-growing number of data points and squeeze out biologically interesting structure-activity relationships has allowed the application of HTS-derived compound libraries to expand as a result of increased scalability of the assessment, thus expanding the bottom of the funnel on the hit identification side. These scaling operations that are increasingly used to handle HTS bottlenecks occur typically alongside optimization of lead characteristics to prepare the candidate compound for a development path. Predictions about compound behavior are derived from algorithms such as support vector machines, deep learning, and other neural networks, random forest, Bayesian models, k-means clustering, and many others, and new algorithms are always being sought to increase predictivity. Challenges in interpretation of HTS data motivate planning and analysis of complex experiments that use orthogonal assays in response to the most predictive and most relevant algorithms. Applications of AI in HTS enable us to discover and equip potential leads growing in already SAR'd libraries simultaneous with eligibility of implementation in candidate move-ahead research programs. In theory, AI's automation of HTS alone should produce new winners in the drug development space at a 7-fold rate over random screening (Cao et al., 2024).

7. FUTURE DIRECTIONS AND EMERGING TECHNOLOGIES

Artificial intelligence can be symptomatically divided into four main waves of progress. Each wave brought about significant improvements in its impact and uptake throughout the pharmaceutical industry. Finally, we are observing a gradual transition into a mature, potent AI regime. As of now, we expect that this change is going to accelerate in pace, driven by developed technology. The use of advanced computational technologies promises systematic impacts across various aspects of the pharmaceutical field. Some of the most relevant components behind an overall successful digitalization push are incremental improvements in industry-wide access, technologies for data generation, and lower hardware costs. In the pharmaceutical sector, the first three waves, while still in evolutionary stages, enabled insightful decision-making support systems for scientists. The next waves, however, stand out for the shift from impactful tools toward operational optimization occurring because of the novel hybrid between AI and robotics (Sadybekov et al., 2023).

The most anticipated and immediately applicable innovation is likely to be a hybrid of traditional AI tools with robotics on the laboratory floor. Scaling up existing operational efficiencies, via a non-trivial increase in automation, has the potential to move the more complex processes of drug discovery into a rethinking phase. It will give the industry flexibility in applying traditional methods and bring to life newer perspectives from the hundreds of millions of small molecules. Machine learning approaches that will make more efficient use of experimental time can make broader screening more practicable and lead the way to applying a plethora of innovative techniques well beyond molecular discovery, such as the production of personalized medicine or tumor-targeting therapeutics (Marescotti et al., 2022).

For all those purposes, the drop in genome sequencing costs and the emergence of single-cell biology have the potential to open new opportunities for drug discovery. Potential game changers include, for example, fast data analysis tools, mechanistic models for the response to compound therapies, or advances in solid-phase synthesis. It is also clear in the case of artificial intelligence that achieving the full potential will require a focused effort based on interdisciplinary collaboration. It's about biochemists working closely with techies, data scientists, and AI experts to deploy all relevant human and machine intelligence. Looking further into its evolution, we anticipate the future AI-driven laboratory operations and the management of large molecules to be scalable, efficient, and cloud-based. The ability to process and interpret assays in real time is unproven but can be a transformative feature of the drug discovery process. In the same way that integrated circuits revolutionized computer operations—technology we can't even build processors without—deep learning techniques for data interpretation have the potential to revolutionize how decisions are made in drug discovery (Padyana et al., 2023).

7.1. Integration of Robotics and AI

Recent technological developments in areas such as robotics allow for the automation of many tasks that previously had to be executed by lab technicians. The most general advantage of robotics is the potential for running automated and consistent processes that can lead to higher quality when evaluating the outcomes of large experiments. In this context, robotic systems can meet the needs of AI systems, which happen to be much better at analyzing experimental outcomes. In drug discovery, robotics can automate certain tasks such as sample handling, sample transportation and loading, sample processing, compound handling, reagent-based processing of all kinds, plate repair, and many other tasks that do not need human intervention in a laboratory. Technically, robotic systems are commonly built to run repeated processes with high-throughput applications in large facilities. In order to advance robotics, the development of sample-handling devices, movement and transport of microplates, plate handling, liquids, and robotics needs to be integrated. There is a range of commercial systems available. Another issue relates to the laboratory infrastructure in which the robotics are placed. Many constraints exist in terms of space, facility design, the need for clean rooms, and many other factors that have to be taken into account. In a sense, robotics can replace technicians as well as protect them from risks, increase quality, generate new production paradigms and environments, and respond to and sustain human production skills. They also interact with

AI and provide responses from these systems automatically, which increases the scope of data generation and hence prediction for a given system. On the negative side, they retain costs and constrain the experimental settings. In addition, it may be difficult to install these systems as autonomous units in a laboratory where older instruments are warehoused (Saber et al, 2023).

8. CONCLUSION AND IMPLICATIONS

Both successful case studies of AI-driven laboratory automation and critical examination of the key enablers highlight the opportunities and challenges in this new and emerging field. The massive investment in AI and AI-driven drug discovery makes it likely that we will see the first clinical candidates with significant AI contributions and that at least one key drug discovery bottleneck will be addressed exclusively by automated AI approaches. The application and success of AI and ML in drug discovery depend on the development of efficient and cost-sensitive laboratory automation and increased flexibility in process design. Many people do indeed view the new and emerging field as an invitation to create more and better technological applications.

The future of drug discovery might be an area ripe for innovation, where cutting-edge technology could become a key differentiator between companies. Efficient and commercially attractive small molecule active ingredient discovery and development, coupled with an understanding of their interplay with human biology, will allow the pharma and biotech industry to develop truly effective treatments. Progress will depend in part on society's ability to address complex ethical issues related to data integrity and the use of non-human and human data in AI and ML learning systems.

Finally, further validation of AI-driven automation in industrial and non-industrial complex processes requires ongoing research implementation and the sharing of data and compatible protocols to facilitate real-world AI applications. AI technologies as game changers in drug discovery are, in line with their objectives, a daring proposition for us to choose, rather than something that has been simply handed down to us. The development and optimization of AI-driven laboratory automation can ultimately drive a fundamental transformation in the paradigm for drug discovery: from systematic screening to rationalized drug design. The future of drug discovery is as much a human choice as it is a given to be embraced. In conclusion, AI-driven automation in the digital laboratory is the next step in this process. Enabled by powerful and vast resources, the accurate prediction of properties and active ingredients, as well as the properties and actions of proteins, DNAs, RNAs, and other molecules, is now becoming optimizable, more applicable, and then usable in real-world tests.

REFERENCES

Aldewachi, H., Al-Zidan, R. N., Conner, M. T., & Salman, M. M. (2021). High-throughput screening platforms in the discovery of novel drugs for neurodegenerative diseases. *Bioengineering, 8*(1), 3.

Alnowaiser, K., Saber, A., Hassan, E., & Awad, W. A. (2024). An optimized model based on adaptive convolutional neural network and grey wolf algorithm for breast cancer diagnosis. *PloS One, 19*(8), e0304868.

Bender, A., & Cortés-Ciriano, I. (2021). Artificial intelligence in drug discovery: What is realistic, what are illusions? Part 1: Ways to make an impact, and why we are not there yet. *Drug Discovery Today, 26*(1), 511-524.

Bhutani, P., Joshi, G., Raja, N., & Bachhav, N. (2021). US FDA approved drugs from 2015–June 2020: A perspective. *Journal of Medicinal Chemistry, 64*(15), 10644-10684.

Cao, Y., Wu, B., Xu, Y., Wang, M., Wu, X., & Liang, X. (2024). Discovery of GPX4 inhibitors through FP-based high-throughput screening. *European Journal of Drugs, 15*(3), 128-137.

Cheng, L., Varshney, K. R., & Liu, H. (2021). Socially responsible AI algorithms: Issues, purposes, and challenges. *Journal of Artificial Intelligence Research, 70*, 377-395.

Farghali, H., Canová, N. K., & Arora, M. (2021). The potential applications of artificial intelligence in drug discovery and development. *Physiological Research, 70*(1), 19-30.

González, M. G., Janssen, A. P. A., & IJzerman, A. P. (2022). Oncological drug discovery: AI meets structure-based computational research. *Drug Discovery Today, 27*(9), 2404-2418.

Hassan, E., Saber, A., & Elbedwehy, S. (2024). Knowledge distillation model for acute lymphoblastic leukemia detection: Exploring the impact of Nesterov-accelerated adaptive moment estimation optimizer. *Biomedical Signal Processing and Control, 94*, 106246.

Hassani, H., Silva, E. S., Unger, S., & TajMazinani, M. (2020). Artificial intelligence (AI) or intelligence augmentation (IA): What is the future? *AI, 1*(2), 143-155.

Kang, S., Chen, E. C., Cifuentes, H., Cole, G., & Graham, J. (2024). Complex in vitro models positioned for impact to drug testing in pharma: A review. *Drug Testing Reports, 12*(4), 455-470.

Liu, J., Du, H., Huang, L., Xie, W., & Liu, K. (2024). AI-powered microfluidics: Shaping the future of phenotypic drug discovery. *ACS Applied Materials & Interfaces, 16*(12), 2204-2212.

Mahmoud, A., Alhussein, M., Aurangzeb, K., & Takaoka, E. (2024). Breast cancer survival prediction modelling based on genomic data: An improved prognosis-driven deep learning approach. *IEEE Access, 12*(6), 455-470.

Mahmoud, A., Awad, N. A., Alsubaie, N., Ansarullah, S. I., Alqahtani, M. S., Abbas, M., Usman, M., Soufiene, B. O., & Saber, A. (2023). Advanced deep learning approaches for accurate brain tumor classification in medical imaging. *Symmetry, 15*(3), 571.

Mahmoud, A., Hamza, T., & Rashad, M. Z. (2020). Prediction of chemical toxicity for drug design using AIRS algorithms and hybrid classifiers. *International Journal of Applied Mathematics & Information Sciences, 14*(2), 133-144.

Mak, K. K., Wong, Y. H., & Pichika, M. R. (2023). Artificial intelligence in drug discovery and development. In *Drug Discovery and Evaluation [Springer.]. *Safety (Basel, Switzerland)*, •••, 345–378.

Marescotti, D., Narayanamoorthy, C., Bonjour, F., Kuwae, K., et al. (2022). AI-driven laboratory workflows enable operation in the age of social distancing. *SLAS Technology, 27*(3), 129-138.

Padyana, U. K., Rai, H. P., Ogeti, P., & Fadnavis, N. S. (2023). AI and machine learning in cloud-based internet of things (IoT) solutions: A comprehensive review and analysis. *Integrated Journal for AI Solutions, 15*(7), 45-62.

Pandey, S. S., Kovaleva, M., & Barelle, C. J. (2022). Overview, generation, and significance of variable new antigen receptors (VNARs) as a platform for drug and diagnostic development. In *Single-Domain Antibodies* (pp. 223–240). Springer. DOI: 10.1007/978-1-0716-2075-5_2

Patel, V., & Shah, M. (2022). Artificial intelligence and machine learning in drug discovery and development. *Intelligent Medicine, 3*(4), 295-309.

Pun, F. W., Ozerov, I. V., & Zhavoronkov, A. (2023). AI-powered therapeutic target discovery. *Trends in Pharmacological Sciences, 44*(2), 166-177.

Rome, B. N., Egilman, A. C., & Kesselheim, A. S. (2022). Trends in prescription drug launch prices, 2008-2021. *JAMA, 327*(21), 2145-2150.

Saber, A., Hussien, A. G., Awad, W. A., & Mahmoud, A. (2023). Adapting the pre-trained convolutional neural networks to improve the anomaly detection and classification in mammographic images. *Scientific Reports, 13*(1), 14877.

Sadybekov, A. V., & Katritch, V. (2023). Computational approaches streamlining drug discovery. *Nature, 621*(5), 243-250.

Sharma, S., Gupta, S., Gupta, D., Juneja, S., Mahmoud, A., El-sappagh, S., & Kwak, K. S. (2022). Transfer learning based modified inception model for diagnosis of Alzheimer's disease. *Frontiers in Computational Neuroscience, 16*(119), 1-13.

Soice, E., & Johnston, J. (2021). Immortalizing cells for human consumption. *International Journal of Molecular Sciences, 22*(5), 2315.

Soleymani, F., Paquet, E., Viktor, H., & Michalowski, W. (2022). Protein–protein interaction prediction with deep learning: A comprehensive review. *Computational Biology, 37*(4), 123-137.

Tran, T. T. V., Wibowo, A. S., & Tayara, H. (2023). Artificial intelligence in drug toxicity prediction: Recent advances, challenges, and future perspectives. *Journal of Chemical Biology, 56*(2), 411-429.

White, T., Blok, E., & Calhoun, V. D. (2022). Data sharing and privacy issues in neuroimaging research: Opportunities, obstacles, challenges, and monsters under the bed. *Human Brain Mapping, 43*(6), 1830-1850.

Chapter 14

The Intersection of AI, Cloud Computing, and Healthcare: A Comprehensive Overview

Wael A. Awad
Department of Computer Science, Faculty of Computers and Artificial Intelligence, Damietta University, Damietta, Egypt

Amena Mahmoud
Department of Computer Science, Faculty of Computers and Information, Kafrelsheikh University, Kafr Elsheikh, Egypt

ABSTRACT

The convergence of Artificial Intelligence (AI) and Cloud Computing has ushered in a new era of innovation across various industries, including healthcare. AI, with its ability to analyze vast datasets, identify patterns, and make intelligent decisions, offers transformative potential for improving patient outcomes and enhancing healthcare efficiency. Cloud Computing, on the other hand, provides a scalable and flexible infrastructure for storing, processing, and accessing healthcare data, enabling seamless collaboration among healthcare professionals and the development of innovative applications. This overview presents a comprehensive overview of the intersection of AI and Cloud Computing in healthcare, exploring their applications, benefits, challenges, and ethical considerations. The survey will delve into various aspects of cloud computing in healthcare, including cloud adoption and usage, challenges and opportunities, future trends, and expectations.

DOI: 10.4018/979-8-3693-9208-9.ch014

1. INTRODUCTION

AI and Cloud Computing have emerged as transformative technologies with the potential to revolutionize various industries. In the realm of healthcare, these technologies are poised to redefine the way medical services are delivered, improving patient outcomes, and enhancing overall healthcare efficiency. AI offers advanced capabilities such as machine learning (ML), natural language processing, and computer vision, enabling healthcare providers to analyze vast amounts of medical data, identify patterns, and make more accurate diagnoses. Cloud Computing provides a scalable and flexible infrastructure for storing, processing, and accessing healthcare data, facilitating collaboration among healthcare professionals, and enabling the development of innovative applications (El-Attar et al., 2021).

Cloud computing is a "model for delivering IT resources" that provides "a novel concept for maintaining and providing services over the web" (Egger et al., 2022; Milletari et al., 2018). Cloud computing, as defined by leading industry providers, is "the on-demand availability of computing resources (such as storage and infrastructure), as services over the internet" or "the delivery of computing services—including servers, storage, databases, networking, software, analytics, and intelligence—over the internet" (Zhou et al., 2021). The advantages of cloud computing include pay-as-you-go access to numerous services, from hardware to software, on-demand storage, availability, and scalability (Emara et al., 2023). Figure 1 illustrates the role of AI-based cloud computing service in medical diagnosing (Hwang et al., 2019).

Figure 1. AI-based cloud computing service for medical diagnosing

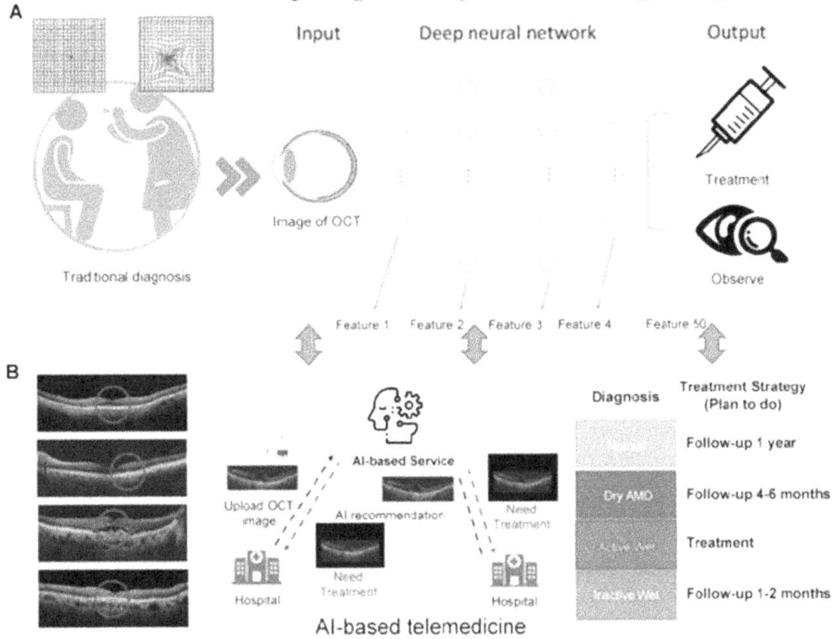

(Hwang et al., 2019)

Healthcare systems' evolution aims to take advantage of cloud services. Cloud computing may considerably improve information technology and information-rich services like medical imaging. To share medical data and pictures, hospitals and other medical organizations must work together. Virtual archives make it simple for various healthcare providers to access patient data, which facilitates data sharing and drastically lowers the need for local storage. Privacy concerns can arise when cloud services store private information. However, there are a lot of benefits to using cloud computing services when interpreting complex clinical cases. Global experts in various medical specialties can confer on the diagnosis (Nasser et al., 2020; El-Attar et al., 2022).

Although the number of medical imaging investigations is growing at a pace of about 3% to 5% per year, this growth rate is not particularly noteworthy. The amount of medical imaging studies, particularly CT and MRI, is expanding at a significantly quicker rate than before, though, and this has resulted in a 10% annual increase in storage requirements. The cost of cloud storage has been declining more quickly than that of enterprise storage, and this trend is probably going to accelerate the use of the cloud for medical image storage (Alnowaiser et al., 2024; Saber et al., 2023; Ahmed et al., 2022; Saber et al., 2021; Sakr et al., 2020).

Healthcare systems' evolution aims to take advantage of cloud services. Cloud computing may considerably improve information technology and information-rich services like medical imaging. Medical data and photos must be shared collaboratively between hospitals and medical organizations. Virtual archives make it simple for various healthcare providers to access patient data, which facilitates data sharing and drastically lowers the need for local storage (El-Matary et al., 2019; Hassan et al., 2024).

This article will delve into the intersection of AI and cloud computing in healthcare, exploring their applications, benefits, and challenges. We will discuss cloud computing and AI adoption and usage and how these technologies are being employed to improve various aspects of medical care. Additionally, we will examine the ethical considerations and potential implications of AI and cloud computing in healthcare challenges and opportunities. Finally, the future trends and expectations are discussed.

2. CLOUD COMPUTING AND AI ADOPTION AND USAGE

2.1 Cloud Computing and AI for Medical Image Analysis

Studierfenster, is a free, non-commercial science client-server framework developed for biomedical image analysis. It offers a range of capabilities, including visualization of medical data in 2D and 3D space, calculation of medical metrics, manual slice-by-slice outlining of structures, placement of landmarks in medical images, visualization of medical data in virtual reality (VR), and facial reconstruction and registration for augmented reality (AR). The framework also includes features like automatic cranial implant design, inpainting of aortic dissections, and CNN for automatic aortic landmark detection in CT angiography images. The study demonstrates the potential of Studierfenster for various fields, including medical applications for humans, veterinary clinics, optical measuring techniques, astronomy, and archaeology (Egger et al., 2022).

Milletari et al. (2018) applied a deep learning (DL) technique to various computer vision and medical image analysis problems, but many methods are not easily accessible or used. This paper proposes an open-source framework called TOMAAT, which provides AI-enabled medical image analysis through a network. TOMAAT consists of an announcement service, a public registry of distributed server nodes, and client software with simple interfaces. It is used for tasks like segmentation, diffeomorphic deformable atlas registration, landmark localization, and workflow integration.

Zhou et al. (2021) developed an open-access, high-performance medical intelligence cloud computing platform (XCloud-pFISTA) to reconstruct MRI images from undersampled k-space data using ML and AI. The platform successfully implements two state-of-the-art pFISTA approaches, potentially benefiting integrated reconstruction and online diagnosis systems.

A cloud computing platform for medical imaging, addressing scaling issues and introducing volume rendering. It discusses a method for accelerating CPU-based volume rendering and scaling to a software-as-a-service solution. The platform also discusses universal data access, security, and an analytic engine. The design decisions and results are discussed (Parsonson et al., 2019).

2.2 Cloud Computing for medical imaging Storage and Access

Cloud computing has fundamentally transformed the field of medical imaging by providing robust solutions for data storage and access. Traditional imaging methods often grappled with significant challenges, including limited data storage capacity, accessibility issues, and security concerns. As the volume of imaging data continues to grow exponentially, cloud computing emerges as a scalable and efficient alternative that addresses these critical needs (Emara & Huang, 2020).

One of the primary advantages of cloud computing in medical imaging is its ability to offer virtually unlimited storage capacity. Traditional systems often require substantial investments in physical infrastructure, such as servers and storage devices, which can be both costly and cumbersome to manage. In contrast, cloud-based solutions allow healthcare providers to store vast amounts of imaging data without the constraints of physical hardware. This flexibility not only reduces costs but also enables medical facilities to scale their storage needs effortlessly as imaging volumes increase.

Accessibility is another key benefit of cloud computing. In a traditional setup, accessing imaging data often involves navigating complex networks or waiting for data transfers, which can delay critical patient care. Cloud platforms enable healthcare professionals to access images and related data from any location with an internet connection. This capability is particularly valuable for remote consultations and telemedicine, allowing specialists to review and analyze images in real-time, regardless of their geographical location. As a result, patients receive faster diagnoses and treatment, enhancing the overall quality of care.

Moreover, cloud computing enhances data security in medical imaging. Traditional systems are vulnerable to various security threats, including data breaches and unauthorized access. Cloud providers typically implement advanced security measures, such as encryption, access controls, and regular security audits, to protect sensitive patient information. These robust security frameworks help ensure compliance with regulations like HIPAA, instilling confidence in healthcare providers and patients alike. Collaboration among healthcare professionals is also improved through cloud-based imaging solutions. Multiple clinicians can access the same imaging data simultaneously, facilitating discussions and consultations that can lead to more informed decision-making. This collaborative environment supports multidisciplinary approaches to patient care, ensuring that all relevant specialists can contribute their expertise without the barriers posed by traditional data access methods.

Additionally, cloud computing can enhance the efficiency of medical imaging workflows. Automated backup processes and centralized data management reduce the administrative burden on healthcare staff, allowing them to focus more on patient care rather than data handling. Integrating AI and ML within cloud platforms can further streamline processes, such as image analysis and diagnostic support, leading to improved outcomes.

Viana-Ferreira (2017) proposed an intelligent Cloud storage gateway for medical imaging repositories to optimize data access times through a new cache architecture. The system combines static rules and pattern recognition for eviction and prefetching, reducing image retrieval times by over 60%. The system can achieve cache hit ratios of around 80%, even with small cache sizes, and can adjust to specific workflows of different institutions.

2.3 Cloud Computing for Smart Healthcare

Cloud computing is a game-changer in the healthcare sector, driving the development of smart healthcare systems that enhance patient care, streamline operations, and facilitate innovative solutions. By leveraging the cloud, healthcare organizations can access a range of advanced technologies that support personalized medicine, real-time data analytics, and improved collaboration among providers.

Nasser et al. (2023) proposed an intelligent healthcare system integrating IoT-cloud technologies, using smart connectivity sensors and deep learning for thoughtful decision-making. The system tracks patient status in real-time, delivering reliable, timely, and high-quality healthcare facilities at a low cost. The system uses a sensor to record, transfer, and track healthcare data, and a cognitive module stores CT scan images in the cloud. The system uses a state-of-the-art classification algorithm based on DL, ResNet50, to detect and classify patients as normal or infected by COVID-19. The system's performance is validated using two benchmark datasets, demonstrating high accuracy, specificity, sensitivity, and F1-score. This system will be helpful in medical diagnosis research and healthcare systems, supporting medical experts in COVID-19 screening.

Shukla et al. (2023) revolutionized cloud computing and artificial intelligence (CC-AI) in the medical field, providing automated processes and efficient analysis of patient databases. This innovative solution aims to provide superior surveillance and disease detection through IoT wearable sensor data. The proposed AI-based solution uses whale optimization and fuzzy neural networks (WO-FNN) to evaluate the accuracy, sensitivity, specificity, and computation time of these techniques, enhancing the overall healthcare experience.

Faruqui et al. (2023) introduced Healthcare-As-A-Service (HAAS), a cloud-compatible CNN that can improve lung cancer diagnosis. It has an accuracy rate of 96.07% and is globally accessible via the Internet. HAASNet, a cloud-compatible CNN, integrates predictions with physio-symptomatic data from the Internet of Medical Things (IoMT) to generate accurate and reliable reports. The service achieves average precision, recall, and f1-scores of 96.47%, 95.39%, and 94.81%, respectively.

3. THE INTERSECTION OF AI AND CLOUD COMPUTING IN HEALTHCARE: CHALLENGES AND OPPORTUNITIES

The rapid advancement of technology has transformed numerous sectors, with healthcare being one of the most significantly impacted. Among the leading innovations in this space are AI and cloud computing. Together, they present a powerful combination that has the potential to revolutionize healthcare delivery, enhance patient outcomes, and streamline operational efficiency. However, the integration of these technologies also brings forth a myriad of challenges that must be addressed to fully realize their potential. This exploration delves into the opportunities and challenges at the intersection of AI and cloud computing in healthcare, shedding light on how they can synergistically improve health services while navigating the complexities involved (Shah et al., 2022; Cuimei et al., 2019).

A. Challenges

- **Data Privacy and Security:** Healthcare data is highly sensitive and subject to stringent regulations. Ensuring the privacy and security of patient data in cloud environments is a major concern. Data breaches and unauthorized access can have severe consequences for both patients and healthcare providers (El-Attar et al., 2016; El-Morshedy et al., 2023).
- **Ethical Considerations:** The use of AI in healthcare raises ethical questions regarding bias, accountability, and transparency. Algorithms may inadvertently perpetuate existing biases present in the data, leading to discriminato-

ry outcomes. Additionally, determining who is responsible for the decisions made by AI systems can be complex (Morley et al., 2020).

- **Interoperability:** Integrating AI and cloud-based solutions with existing healthcare systems can be challenging due to differences in standards and protocols. Ensuring seamless interoperability is crucial for effective data sharing and collaboration.
- **Cost and Infrastructure:** Implementing AI and cloud-based solutions can be expensive, requiring significant investments in hardware, software, and skilled personnel. Additionally, organizations may need to upgrade their existing infrastructure to accommodate these technologies.
- **Regulatory Compliance:** Healthcare providers must comply with various regulations, such as HIPAA in the United States and GDPR in the European Union. Ensuring that AI and cloud-based solutions adhere to these regulations is essential to avoid legal and financial penalties.

B. Opportunities

- **Improved Diagnosis and Treatment:** AI can analyze medical images, electronic health records, and other data to assist in diagnosis and treatment planning. This can lead to more accurate and timely interventions.
- **Personalized Medicine:** AI can help identify personalized treatment plans based on a patient's individual characteristics and genetic makeup. This can improve treatment outcomes and reduce adverse effects.
- **Drug Discovery and Development:** AI can accelerate drug discovery by analyzing vast amounts of data to identify potential drug targets and optimize drug development processes.
- **Remote Patient Monitoring:** Cloud-based platforms can enable remote patient monitoring, allowing healthcare providers to track patients' vital signs and symptoms from a distance. This can improve patient care, especially for those living in rural or underserved areas.
- **Administrative Efficiency:** AI and cloud computing can automate administrative tasks, such as scheduling appointments, billing, and claims processing, freeing up healthcare professionals to focus on patient care (Mistry et al., 2024).

By carefully addressing the challenges and leveraging the opportunities presented by AI and cloud computing, healthcare organizations can improve patient outcomes, enhance efficiency, and drive innovation in the industry. The summary of cloud computing's benefits and challenges in medical imaging is included in Table. 1.

Table 1. Cloud computing's benefits and challenges in medical imaging

Benefits	Challenges
Cost Efficiency: Healthcare facilities can save IT expenditures associated with hardware maintenance and updates by eliminating the need for large on-premises infrastructure. Cloud services often operate on a pay-per-use model, allowing enterprises to expand their resources according to demand.	**Data Security and Privacy:** While cloud companies employ security safeguards, worries regarding data privacy and regulatory compliance remain high. Healthcare businesses must select reliable providers and understand the security processes in place.
Enhanced Security: Cloud companies frequently use strong security measures to protect sensitive medical data. This can reduce data loss or breach risks, which are more common in local server setups.	**Integration Issues:** Moving from old systems to cloud-based solutions may necessitate sophisticated integration processes. Ensuring compatibility with existing systems and procedures is critical for a successful transition.
Improved Patient Care: Better access to imaging data can lead to faster diagnoses and treatment decisions, eventually improving patient outcomes. Prioritizing key patients and streamlining workflows reduces physician burnout and improves care delivery.	**Dependence on Internet Connectivity:** Cloud-based solutions require consistent Internet access, which can be challenging in places with weak connectivity. This reliance can impact the availability of imaging data over important times.

4. FUTURE TRENDS AND EXPECTATIONS FOR AI AND CLOUD COMPUTING IN HEALTHCARE

4.1 Expansion of AI and ML Integration

The rapid advancement of AI and ML technologies is poised to revolutionize healthcare, especially when integrated with cloud computing. As these technologies continue to evolve, cloud platforms will increasingly serve as the backbone for AI-driven tools that enhance diagnostics, predictive analytics, and personalized treatment plans. This synergy promises to improve the overall quality of patient care and streamline various healthcare processes (Saber et al., 2021).

One of the most significant impacts of AI and ML in healthcare is their ability to improve the accuracy of disease prediction and risk assessment. Advanced algorithms can analyze vast amounts of data from diverse sources, including electronic health records (EHRs), medical imaging, and genomic information. By leveraging cloud computing's scalable storage and processing capabilities, these algorithms can identify patterns and correlations that might elude human analysts. For instance, AI can sift through thousands of patient histories to pinpoint risk factors associated with specific conditions, enabling healthcare providers to intervene earlier and tailor preventive measures to individual patients (El-Attar et al., 2022).

Furthermore, AI-powered solutions enhance decision-making in clinical settings. By integrating real-time data analysis into clinical workflows, healthcare professionals can receive evidence-based recommendations that inform their treatment choices. For example, AI systems can analyze a patient's unique health data and suggest personalized treatment plans that consider not only their medical history but also the latest research and clinical guidelines. This level of customization helps ensure that patients receive the most effective therapies, ultimately improving outcomes and reducing trial-and-error approaches in treatment.

In addition to diagnostics and treatment, AI and ML are transforming the management of medical imaging. Traditional methods of analytical images often require significant time and human expertise. However, AI algorithms can rapidly analyse imaging data—such as X-rays, MRIs, and CT scans—detecting anomalies with remarkable precision. Cloud computing enables the storage and sharing of these images across platforms, allowing for collaboration among radiologists and specialists. This not only speeds up diagnosis but also enhances the accuracy of interpretations, as multiple experts can contribute their insights seamlessly.

Natural language processing (NLP), another subset of AI, is proving invaluable for optimizing EHR management. By analyzing unstructured data within EHRs, NLP can extract relevant information, summarize patient histories, and flag potential issues. This capability reduces the administrative burden on healthcare providers, allowing them to focus more on patient care rather than paperwork. Moreover, with cloud computing, these NLP tools can be continually updated and improved, ensuring that they adapt to the evolving language of medicine and patient interactions (Kaswan et al., 2021).

Automation of administrative tasks is yet another area where AI and cloud computing intersect to enhance efficiency in healthcare settings. Routine tasks, such as appointment scheduling, billing, and claims processing, can be automated using AI-driven systems. This not only minimizes errors but also frees up staff to engage in more critical patient-facing activities. By streamlining these processes, healthcare facilities can improve their operational efficiency and enhance the overall patient experience.

4.2 Growth of Telemedicine and Remote Patient Monitoring

The adoption of telemedicine and remote patient monitoring is rapidly expanding, largely fueled by the capabilities of cloud computing. As healthcare continues to evolve, cloud-based platforms are emerging as crucial enablers of seamless virtual consultations, real-time health monitoring, and secure sharing of patient data. This transformation is particularly significant for improving access to healthcare services for patients in remote or underserved areas, where traditional healthcare delivery can be limited (Tang et al., 2022).

One of the most immediate benefits of cloud-based telemedicine is its ability to facilitate virtual consultations. Patients can connect with healthcare providers from the comfort of their homes, reducing the need for travel and minimizing wait times. This convenience is especially vital for individuals with mobility issues or those living in rural regions where healthcare facilities may be far away. Cloud computing supports high-quality video and audio connections, ensuring that consultations are effective and that patients receive the same level of care they would in a face-to-face appointment (Jin et al., 2015).

In addition to virtual visits, cloud computing enables real-time health monitoring through remote patient monitoring (RPM) technologies. Wearable devices and mobile health applications can collect and transmit health data, such as heart rate, blood pressure, and glucose levels, directly to healthcare providers. This continuous stream of information allows clinicians to monitor patients' conditions proactively, identifying potential issues before they escalate. For example, patients with chronic conditions like diabetes or hypertension can benefit significantly from regular monitoring, allowing for timely interventions and adjustments to treatment plans (Hariharanm et al., 2021).

The integration of wearable devices with advanced cloud infrastructure enhances the overall patient experience and engagement. Patients can easily track their health metrics, set reminders for medication, and receive educational resources—all in one place. This level of engagement empowers individuals to take charge of their health, leading to better adherence to treatment plans and improved health outcomes. Moreover, the ability to share this data securely with healthcare providers fosters a collaborative approach to managing chronic diseases, where patients and providers work together to optimize care.

Another critical aspect of cloud-based telemedicine is the secure sharing of patient data. Privacy and security concerns are paramount in healthcare, and cloud platforms are designed with robust encryption and compliance with regulations such as HIPAA. This ensures that patient information remains confidential while allowing healthcare providers to access the data they need for informed decision-making. Secure data sharing not only protects patients but also enhances the quality of care by enabling providers to have a complete view of a patient's health history, facilitating better diagnostics and personalized treatment plans.

The expansion of telemedicine and remote monitoring also addresses healthcare disparities. Many patients in underserved areas face barriers to accessing care due to geographical., economic, or social factors. Telemedicine breaks down these barriers, allowing individuals to connect with specialists who may not be available in their localities. This democratization of healthcare access can lead to improved health equity, as more patients can receive timely and appropriate medical attention.

As cloud technology continues to advance, the future of telemedicine looks promising. The integration of AI and ML with cloud platforms will further enhance the capabilities of telemedicine, enabling predictive analytics and personalized care plans based on real-time data. Additionally, as more healthcare providers adopt these technologies, the potential for collaborative care networks will grow, creating a more interconnected healthcare ecosystem.

4.3 Enhanced Interoperability and Data Integration

The future of cloud solutions in healthcare is poised to significantly enhance interoperability between diverse healthcare systems and data sources. As the demand for integrated care grows, the development of standardized data formats and protocols will become essential for seamless integration and sharing of patient information across various platforms. This evolution is crucial in ensuring that healthcare providers can access and utilize patient data effectively, leading to improved care coordination and clinical outcomes.

Interoperability is a critical challenge in today's healthcare landscape, where various systems—ranging from electronic health records (EHRs) to laboratory information systems—often operate in silos. This fragmentation can hinder the timely exchange of vital patient information, resulting in duplicated tests, fragmented care, and increased healthcare costs. Future cloud solutions will address these issues by fostering the creation of health information exchanges (HIEs) that prioritize interoperability. These cloud-based HIEs will facilitate the seamless flow of patient data among healthcare providers, regardless of the specific systems they use.

The establishment of standardized data formats and protocols is key to achieving this interoperability. By adopting widely accepted standards, healthcare organizations can ensure that data from different sources can be easily integrated and interpreted. For instance, the Fast Healthcare Interoperability Resources (FHIR) standard is gaining traction as a method for sharing electronic health information. FHIR's structured data formats allow for more efficient data exchange, making it easier for providers to access comprehensive patient records without the need for complex mapping processes.

Cloud-based interoperable EHR systems will also play a pivotal role in enhancing care coordination. These systems will enable healthcare providers to view a patient's complete medical history, including diagnoses, medications, and lab results, all in one place. By having access to a holistic view of a patient's health, clinicians can make more informed decisions, reducing the likelihood of errors and improving treatment efficacy. This integration will be especially beneficial in managing chronic conditions, where coordinated care from multiple providers is often necessary.

Moreover, greater data integration will streamline workflows within healthcare organizations. By reducing the need for repetitive data entry and facilitating automatic updates, providers can focus more on patient care rather than administrative tasks. This efficiency will not only save time but also minimize the potential for human error, ultimately leading to better patient safety and outcomes.

In addition to improving care coordination and workflow efficiency, enhanced interoperability will support more robust clinical decision-making. Access to comprehensive and up-to-date patient information allows healthcare providers to employ data-driven approaches to diagnosis and treatment. For example, clinicians can utilize integrated data to identify patterns that inform personalized treatment plans, improving the likelihood of successful outcomes. Additionally, the ability to analyze aggregated data from various sources can facilitate population health management, enabling providers to identify trends and implement preventative measures.

The future of cloud solutions also holds promise for empowering patients. With interoperable systems, individuals will have greater access to their health information across various platforms. This transparency can enhance patient engagement, allowing individuals to take an active role in their care and fostering better communication between patients and providers.

4.4 Advancements in Data Security and Privacy Measures

As the volume of sensitive health data stored and processed in the cloud continues to surge, the emphasis on advancing data security and privacy measures will become increasingly critical. The healthcare sector is particularly vulnerable to data breaches and cyber threats due to the highly sensitive nature of patient information. As a result, organizations must prioritize the development and implementation of comprehensive security strategies to safeguard this data (Chikhaoui et al., 2017).

Future trends in cloud security will likely focus on the adoption of more sophisticated encryption techniques. Traditional encryption methods, while effective, may not be sufficient to combat the evolving landscape of cyber threats. Emerging technologies, such as homomorphic encryption and advanced algorithms, will allow data to be processed in its encrypted form, reducing the risk of exposure during storage and transmission. By ensuring that even if data is intercepted, it remains unreadable to unauthorized parties, these techniques will provide an additional layer of protection (Radwan et al., 2017).

Another significant trend will be the implementation of zero-trust security models. In a zero-trust architecture, no one—whether inside or outside the organization—is automatically trusted. Every user, device, and application must be verified before being granted access to sensitive data. This approach not only mitigates the risks associated with insider threats but also strengthens overall data security. By continuously monitoring and validating user access, organizations can respond swiftly to any anomalies or suspicious activities.

Advanced threat detection systems will also play a crucial role in enhancing cloud security for healthcare data. ML and AI can be leveraged to identify patterns and anomalies in data access and usage. These systems can analyze vast amounts of data in real-time, enabling proactive identification of potential threats before they escalate into significant breaches. By incorporating these advanced technologies, healthcare organizations can significantly improve their incident response times and minimize the impact of any security incidents (El-Morshedy et al., 2020).

Cloud providers will be compelled to invest in robust security frameworks to protect patient information and ensure compliance with ever-evolving regulations. As regulatory bodies introduce stricter guidelines concerning data privacy and security—such as the Health Insurance Portability and Accountability Act (HIPAA) and the General Data Protection Regulation (GDPR)—cloud service providers must adapt their practices to meet these requirements. Compliance not only helps avoid legal repercussions but also fosters trust among patients and healthcare professionals, who need assurance that their data is secure. Also, several works investigated isolating the storage level of the data from the analysis level to enhance data privacy (Emara & Huang. 2019a; Emara & Huang. 2019b).

Moreover, enhanced security measures will directly address concerns related to data breaches, unauthorized access, and cyber threats. As cybercriminals develop increasingly sophisticated techniques to infiltrate systems, organizations must stay ahead of the curve. Regular security audits, penetration testing, and vulnerability assessments will be essential to identify potential weaknesses and reinforce defenses. Additionally, organizations must cultivate a culture of security awareness among employees, providing training on best practices for data handling and recognizing phishing attempts.

Collaboration between cloud providers and healthcare organizations will also be essential in establishing effective security protocols. By sharing threat intelligence and best practices, both parties can strengthen their defenses against emerging threats. This collaborative approach not only enhances security but also fosters innovation in the development of new security solutions tailored to the unique challenges faced by the healthcare sector.

4.5 Increased Use of Blockchain for Data Integrity in Cloud

Blockchain technology is poised to significantly impact the integrity and security of health data stored in the cloud. By providing a decentralized, immutable ledger for recording transactions and data exchanges, blockchain enhances transparency and reduces the risk of data tampering, making it a compelling solution for the challenges faced in healthcare data management (Wei et al., 2020).

One of the primary advantages of blockchain technology is its ability to create an unchangeable record of data transactions. Each entry on the blockchain is encrypted and linked to the previous one, forming a chain that cannot be altered without consensus from the network participants. This characteristic ensures that once health data is recorded, it remains secure and tamper-proof. In an era where data breaches and unauthorized access to sensitive health information are prevalent, the implementation of blockchain could serve as a robust defence mechanism, bolstering the overall security of patient data stored in the cloud.

In healthcare, one of the most promising applications of blockchain is secure patient consent management. Currently, managing patient consent for data sharing and treatment can be cumbersome and lacks standardization. Blockchain can streamline this process by providing patients with a secure platform to grant and revoke consent for their health data to be used or shared. Each consent transaction would be recorded on the blockchain, ensuring a transparent and auditable trail that enhances accountability. This not only empowers patients but also helps healthcare organizations comply with regulations regarding data privacy and consent, such as HIPAA and GDPR.

Additionally, blockchain can revolutionize the recording of clinical trial data. Historically, the integrity of clinical trial results has been a concern, with instances of data manipulation undermining trust in medical research. By utilizing blockchain for clinical trial data recording, researchers can ensure that all data points are securely stored and easily traceable. This transparency can foster greater trust among stakeholders, including patients, healthcare providers, and regulatory bodies. Moreover, immutable records can facilitate real-time monitoring of trial progress, enabling timely interventions if issues arise.

Another critical application of blockchain in healthcare is the improved tracking of medical supply chains. The global pharmaceutical supply chain is complex and often plagued by inefficiencies and vulnerabilities, including counterfeit drugs and lack of traceability. Blockchain technology can create a transparent supply chain where each transaction—from manufacturer to distributor to pharmacy—is recorded on a decentralized ledger. This level of traceability ensures that all parties can verify the authenticity of medical products, reducing the likelihood of counterfeit drugs entering the market and enhancing patient safety.

Furthermore, the use of blockchain can support interoperability between disparate healthcare systems. By establishing a standardized framework for data exchange, blockchain can enable various healthcare providers and organizations to share information seamlessly. This interoperability is essential for coordinated patient care, as it allows healthcare professionals to access comprehensive patient histories without facing the barriers often associated with traditional data-sharing methods.

As the healthcare industry continues to embrace digital transformation, the integration of blockchain technology will likely become more prevalent. Its potential to enhance data security, improve patient consent management, ensure the integrity of clinical trial data, and streamline supply chain tracking positions blockchain as a vital tool in the future of healthcare. By adopting this technology, healthcare organizations can not only protect sensitive patient information but also build trust among patients, providers, and regulatory bodies (Wei et al., 2020).

4.6 Evolution of Cloud-Native Health Applications

Cloud-native applications designed specifically for the healthcare industry will become more prevalent. These applications will leverage the flexibility and scalability of cloud infrastructure to deliver innovative solutions for patient management, clinical workflows, and administrative functions. Cloud-native health applications will be designed to easily integrate with existing systems, offer real-time analytics, and support scalable deployments to accommodate varying demands (Deng et al., 2024).

As the healthcare landscape evolves, cloud-native applications specifically designed for the industry are set to become increasingly prevalent. These applications harness the inherent flexibility and scalability of cloud infrastructure, paving the way for innovative solutions that enhance patient management, optimize clinical workflows, and streamline administrative functions. Unlike traditional applications, which often struggle with integration and scalability, cloud-native solutions are built from the ground up to leverage cloud capabilities, making them ideal for the dynamic and complex environment of healthcare. This shift is not merely about technology; it represents a fundamental change in how healthcare providers deliver care, manage resources, and engage with patients.

One of the most significant advantages of cloud-native health applications is their ability to seamlessly integrate with existing systems. Healthcare organizations often rely on a multitude of software solutions for electronic health records, practice management, and billing. However, many of these legacy systems can be cumbersome and difficult to integrate, leading to data silos that hinder effective communication and collaboration among providers. Cloud-native applications, on the other hand, are designed with interoperability in mind. By utilizing open standards and application programming interfaces, these applications can easily connect with existing healthcare technologies. This interoperability ensures that patient data flows freely between systems, allowing for a comprehensive view of each patient's health status and enabling more coordinated care.

Moreover, cloud-native applications offer real-time analytics capabilities that are crucial for informed decision-making in healthcare settings. Traditional systems often rely on batch processing of data, which can result in delays in accessing critical information. In contrast, cloud-native solutions can process and analyze data in real time, providing healthcare providers with immediate insights into patient conditions, resource utilization, and operational efficiency. For instance, a cloud-native application could analyze patient vitals from monitoring devices and alert clinicians to any significant changes that require immediate attention. This proactive approach not only enhances patient safety but also improves the overall quality of care. By making data-driven decisions in real time, healthcare providers can respond more swiftly to patient needs and optimize treatment plans based on the most current information available.

The scalability of cloud-native applications is another key factor driving their adoption in healthcare. As patient populations grow and healthcare demands fluctuate, organizations need solutions that can easily scale to meet changing requirements. Cloud infrastructure allows for the rapid deployment of additional resources, whether for handling increased patient loads during a seasonal surge or for accommodating new services as a healthcare organization expands. This flexibility is particularly important in today's healthcare environment, where unexpected challenges—such as a public health crisis—can quickly change demand patterns. By utilizing cloud-native applications, healthcare providers can quickly adapt their operations, ensuring that they are always equipped to deliver high-quality care to their patients.

In addition to these operational benefits, cloud-native applications support innovative patient engagement strategies. As healthcare becomes more patient-centric, providers are increasingly focusing on ways to empower individuals to take an active role in their healthcare journeys. Cloud-native solutions can facilitate enhanced patient engagement through user-friendly portals and mobile applications that provide access to health information, appointment scheduling, medication reminders, and educational resources. These tools not only foster better communication between patients and providers but also encourage adherence to treatment plans and preventive care measures. When patients are more informed and involved in their healthcare, the overall effectiveness of care improves, leading to better health outcomes.

Another crucial aspect of cloud-native applications is their ability to support data security and compliance. In an industry where patient data security is paramount, cloud-native applications can offer advanced security features such as encryption, access controls, and continuous monitoring. Healthcare organizations are required to comply with regulations like HIPAA, and cloud service providers are often equipped with tools and protocols designed to ensure compliance. By leveraging cloud-native solutions, healthcare organizations can focus on delivering care while entrusting the security of their data to specialized providers. This partnership allows healthcare organizations to maintain a high standard of data protection without diverting resources away from patient care.

4.7 Expansion of Personalized and Precision Medicine

Cloud computing is set to revolutionize the field of personalized and precision medicine by providing the infrastructure necessary for the storage, processing, and analysis of vast amounts of genomic and omics data. As the volume of biological data continues to surge, the traditional methods of data management are becoming increasingly inadequate. Cloud computing offers a scalable, flexible, and cost-effective solution that not only meets the needs of researchers and healthcare providers but also enhances the overall effectiveness of medical interventions.

One of the most significant advantages of cloud computing is its ability to facilitate the storage and integration of diverse datasets, including genomic sequences, proteomic profiles, metabolomic information, and clinical data. This integration is critical for precision medicine, which aims to tailor treatment plans to the unique genetic makeup of each patient. By harnessing the power of cloud-based platforms, researchers can easily access and analyze data from various sources, enabling them to uncover patterns and correlations that might otherwise remain hidden. This holistic approach is essential for identifying genetic markers associated with specific diseases and treatment responses.

Advanced analytics tools available in cloud environments play a crucial role in transforming raw data into actionable insights. ML algorithms and artificial intelligence can be deployed to sift through large datasets, identifying genetic variations that may influence patient outcomes. These tools can predict how individuals are likely to respond to specific treatments, thereby guiding clinicians in selecting the most effective therapies. For example, by analyzing the genomic data of cancer patients, cloud-based systems can help identify which individuals are likely to benefit from targeted therapies, leading to more personalized treatment plans that enhance efficacy while minimizing adverse effects.

Moreover, the collaborative nature of cloud computing fosters innovation in precision medicine. Researchers across institutions can share data and insights in real-time, facilitating interdisciplinary collaborations that are essential for advancing medical knowledge. This collaborative framework allows for the rapid dissemination of findings and encourages the development of new methodologies that can be integrated into clinical practice. As researchers work together, they can aggregate data from diverse populations, leading to more robust conclusions and a deeper understanding of how genetic factors influence health and disease.

The speed at which cloud computing can process and analyze data significantly accelerates research and development in precision medicine. Traditional computational methods can be time-consuming and resource-intensive, often delaying crucial discoveries and advancements. In contrast, cloud-based analytics can handle large-scale computations efficiently, allowing researchers to test hypotheses and validate findings much more quickly. This agility is particularly important in a rapidly evolving field like precision medicine, where new discoveries can change the landscape of treatment options almost overnight.

Additionally, cloud computing enhances the security and accessibility of sensitive medical data. With robust encryption and compliance with regulatory standards, cloud platforms can protect patient information while allowing authorized researchers and healthcare providers to access the data they need. This balance between security and accessibility is vital for fostering trust among patients and encouraging participation in genomic studies. As more individuals contribute their genomic data to research, the breadth and depth of knowledge available to scientists will expand, ultimately leading to better-targeted therapies.

Furthermore, cloud computing enables real-time monitoring and feedback in clinical settings, which is essential for implementing precision medicine strategies effectively. Wearable devices and mobile health applications can continuously collect health data, which can be analyzed in the cloud to provide insights into a patient's condition. This information can then be used to adjust treatment plans dynamically, ensuring that interventions are responsive to a patient's changing health status. Such proactive approaches not only improve patient outcomes but also optimize resource utilization within healthcare systems (Hassan et al., 2024; Dalal et al., 2023a; Dalal et al., 2023b).

4.8 Development of Smart Health Ecosystems

The future of cloud computing in healthcare is poised to usher in the era of smart health ecosystems, where various technologies and services converge to create a cohesive network for managing patient care. Central to this evolution is the integration of cloud-based platforms with Internet of Things (IoT) devices, AI tools, and health applications. By leveraging the scalability and accessibility of cloud computing, these ecosystems will facilitate seamless data flow across multiple healthcare settings, enabling providers to access and analyze patient information in real time. This integrated approach not only enhances the efficiency of healthcare delivery but also empowers clinicians to make informed decisions swiftly, ultimately leading to improved patient outcomes. The incorporation of diverse technologies ensures that healthcare systems can adapt to the evolving landscape, accommodating the growing volume of data and the increasing complexity of patient needs (Pourqasem, 2018).

One of the most significant benefits of smart health ecosystems is their ability to support real-time monitoring and proactive patient management. IoT devices, such as wearable health monitors and remote patient monitoring tools, will continuously collect vital signs and health metrics, transmitting this data to cloud platforms for analysis. By utilizing AI algorithms, healthcare providers can gain insights into patient health trends, enabling them to identify potential issues before they escalate into serious conditions. For instance, a patient with chronic heart disease wearing a smart monitor could have their heart rate and blood pressure tracked in real time. If the data indicates an anomaly, healthcare providers can intervene early, adjusting treatment plans or providing additional support as needed. This proactive approach not only enhances patient safety but also reduces the burden on emergency services and hospitals, leading to more efficient use of healthcare resources.

Moreover, the seamless integration of these technologies within smart health ecosystems promotes coordinated care across various healthcare settings. With patient data accessible in real time, healthcare providers can collaborate more effectively, whether they are working within the same facility or across different institutions. For example, when a patient transitions from a hospital to outpatient care, their health data can be easily transferred to their primary care physician, specialists, and home health providers. This continuity of care ensures that all team members are informed about the patient's medical history, treatment plans, and any recent developments. Additionally, patient engagement applications can empower individuals to take an active role in their healthcare journey. These applications can provide reminders for medications, facilitate communication with providers, and offer personalized health tips based on the patient's unique data. As patients become more involved in their own care, health literacy and adherence to treatment plans are likely to improve, leading to better health outcomes and patient satisfaction.

4.9 Emphasis on Patient-Centric Solutions

Cloud computing will enable the development of more patient-centric solutions, focusing on enhancing the patient experience and engagement. Personalized health dashboards, secure patient portals, and mobile apps will empower patients to access their health information, schedule appointments, and communicate with healthcare providers. The cloud will facilitate the delivery of tailored health content and education, supporting patients in managing their health and making informed decisions (Mahmoud et al., 2023; Elbedwehy et al., 2024).

In summary, the future of cloud computing in healthcare promises significant advancements and innovations. As technologies evolve, cloud solutions will drive improvements in AI and ML integration, telemedicine, data interoperability, security, and personalized medicine. By embracing these trends, healthcare organizations can enhance patient care, streamline operations, and contribute to a more efficient and effective healthcare system (Nagim et al., 2022).

5. CONCLUSION:

The survey highlighted a strong consensus among respondents that the adoption of AI and cloud computing can significantly enhance patient experiences, lower healthcare costs, and improve access to care, particularly in underserved areas. However, several challenges must be addressed to realize these benefits fully. Key concerns include ensuring interoperability between different systems, navigating regulatory hurdles, and investing in the necessary infrastructure and skilled workforce.

To achieve successful implementation of AI and cloud computing in healthcare, a collaborative approach is essential. Healthcare providers, technology companies, and policymakers must work together to tackle these challenges and leverage the opportunities presented by these advanced technologies. By fostering partnerships and open communication, stakeholders can develop solutions that promote seamless integration and innovation. Moreover, addressing issues such as data privacy and security will be crucial to gaining the trust of patients and providers alike. Prioritizing these elements will not only enhance the efficacy of AI and cloud computing but also contribute to a more patient-centric and equitable healthcare system.

In conclusion, by overcoming the obstacles associated with AI and cloud computing, we can create a healthcare landscape that is more efficient, accessible, and responsive to patient needs. Embracing these technologies holds the potential to revolutionize healthcare delivery, ensuring that quality care is available to all, regardless of their circumstances.

REFERENCES

Ahmed, A. S., Keshk, A. E. M., Abo-Seida, O., & Sakr, M. (2022). Tumor detection and classification in breast mammography based on fine-tuned convolutional neural networks. IJCI. *International Journal of Computers and Information*, *9*(1), 74–84.

Alnowaiser, K., Saber, A., Hassan, E., & Awad, W. A. (2024). An optimized model based on adaptive convolutional neural network and grey wolf algorithm for breast cancer diagnosis. *PLoS One*, *19*(8), e0304868. DOI: 10.1371/journal.pone.0304868 PMID: 39159151

Chikhaoui, E., Sarabdeen, J., & Parveen, R. (2017). Privacy and Security Issues in the Use of Clouds in e-Health in the Kingdom of Saudi Arabia. *Communications of the IBIMA*, *2017*, 1–18. DOI: 10.5171/2017.369309

Cuimei, A. D. K. S. B., & Boafoh, K. B. (2019). The Emergence of AI and IoT on Cloud Computing: Evolution, Technology, Future Research and Challenges. *Emergence (Mahwah, N.J.)*, *10*(7).

Dalal, S., Goel, P., Onyema, E. M., Alharbi, A., Mahmoud, A., Algarni, M. A., & Awal, H. (2023b). Application of machine learning for cardiovascular disease risk prediction. *Computational Intelligence and Neuroscience*, *2023*(1), 9418666. DOI: 10.1155/2023/9418666

Dalal, S., Poongodi, M., Lilhore, U. K., Dahan, F., Vaiyapuri, T., Keshta, I., Aldossary, S. M., Mahmoud, A., & Simaiya, S. (2023a). Optimized LightGBM model for security and privacy issues in cyber-physical systems. *Transactions on Emerging Telecommunications Technologies*, *34*(6), e4771. DOI: 10.1002/ett.4771

Deng, S., Zhao, H., Huang, B., Zhang, C., Chen, F., Deng, Y., & Zomaya, A. Y. (2024). Cloud-native computing: A survey from the perspective of services. *Proceedings of the IEEE*. DOI: 10.1109/JPROC.2024.3353855

Egger, J., Wild, D., Weber, M., Bedoya, C. A. R., Karner, F., Prutsch, A., Schmied, M., Dionysio, C., Krobath, D., Jin, Y., Gsaxner, C., Li, J., & Pepe, A. (2022). Studierfenster: An open science cloud-based medical imaging analysis platform. *Journal of Digital Imaging*, *35*(2), 340–355. DOI: 10.1007/s10278-021-00574-8 PMID: 35064372

El-Attar, N. E., Awad, W. A., & Omara, F. A. (2016, December). Empirical assessment for security risk and availability in public cloud frameworks. In 2016 11th International Conference on Computer Engineering & Systems (ICCES) (pp. 17-25). IEEE. DOI: 10.1109/ICCES.2016.7821969

El-Attar, N. E., El-Morshedy, D. S., & Awad, W. A. (2021). A new hybrid automated security framework to cloud storage system. *Cryptography*, *5*(4), 37. DOI: 10.3390/cryptography5040037

El-Attar, N. E., Moustafa, B. M., & Awad, W. A. (2022). Deep Learning Model to Detect Diabetes Mellitus Based on DNA Sequence. *Intelligent Automation & Soft Computing*, *31*(1). Advance online publication. DOI: 10.32604/iasc.2022.019970

El-Attar, N. E., Sabbeh, S. F., Fasihuddin, H., & Awad, W. A. (2022). An Improved DeepNN with Feature Ranking for Covid-19 Detection. *Computers, Materials & Continua*, *71*(2). Advance online publication. DOI: 10.32604/cmc.2022.022673

El-Matary, D. M., El-Attar, N. E., Awad, W. A., & Hanafy, I. M. (2019, February). Automated negotiation framework based on intelligent agents for cloud computing. In *2019 International Conference on Innovative Trends in Computer Engineering (ITCE)* (pp. 156-161). IEEE. DOI: 10.1109/ITCE.2019.8646481

El-Morshedy, D. S., El-Attar, N. E., Awad, W. A., & Hanafy, I. M. (2020). Trustworthy Self-protection for Data Auditing in Cloud Computing Environment. In Internet of Things—Applications and Future [Springer Singapore.]. *Proceedings of ITAF*, *2019*, 23–39.

El-Morshedy, D. S., El-Attar, N. E., Hanafy, I. M., & Awad, W. A. (2023). Cryptographic Algorithms for Enhancing Security in Cloud Computing. *Alfarama Journal of Basic & Applied Sciences*, *4*(3), 433–455.

Elbedwehy, S., Hassan, E., Saber, A., & Elmonier, R. (2024). Integrating neural networks with advanced optimization techniques for accurate kidney disease diagnosis. *Scientific Reports*, *14*(1), 21740. DOI: 10.1038/s41598-024-71410-6 PMID: 39289394

Elmasry, S. A., & Awad, W. A., & Abd El-hafeez, S. A. (2020). Review of different image fusion techniques: Comparative study. In Internet of Things—Applications and Future: Proceedings of ITAF 2019 (pp. 41-51). Springer Singapore.

Emara, T. Z., & Huang, J. Z. (2019a). A distributed data management system to support large-scale data analysis. *Journal of Systems and Software*, *148*, 105–115. DOI: 10.1016/j.jss.2018.11.007

Emara, T. Z., & Huang, J. Z. (2019b). RRPlib: A spark library for representing HDFS blocks as a set of random sample data blocks. *Science of Computer Programming*, *184*, 102301. DOI: 10.1016/j.scico.2019.102301

Emara, T. Z., & Huang, J. Z. (2020). Distributed Data Strategies to Support Large-Scale Data Analysis Across Geo-Distributed Data Centers. *IEEE Access : Practical Innovations, Open Solutions*, *8*, 178526–178538. DOI: 10.1109/ACCESS.2020.3027675

Emara, T. Z., Trinh, T., & Huang, J. Z. (2023). Geographically distributed data management to support large-scale data analysis. *Scientific Reports*, *13*(1), 17783. DOI: 10.1038/s41598-023-44789-x PMID: 37853092

Faruqui, F., Yousuf, M. A., Kateb, F. A., Abdul Hamid, M., & Monowar, M. M. (2023). Healthcare As a Service (HAAS): CNN-based cloud computing model for ubiquitous access to lung cancer diagnosis. *Heliyon*, *9*(11), e21520. DOI: 10.1016/j.heliyon.2023.e21520 PMID: 37942151

Hariharan, U., Rajkumar, K., Akilan, T., & Jeyavel, J. (2021). Smart Wearable Devices for Remote Patient Monitoring in Healthcare 4.0. Internet of Medical Things: Remote Healthcare Systems and Applications, 117-135.

Hassan, E., Saber, A., & Elbedwehy, S. (2024). Knowledge distillation model for Acute Lymphoblastic Leukemia Detection: Exploring the impact of nesterov-accelerated adaptive moment estimation optimizer. *Biomedical Signal Processing and Control*, *94*, 106246. DOI: 10.1016/j.bspc.2024.106246

Hwang, D. K., Hsu, C. C., Chang, K. J., Chao, D., Sun, C. H., Jheng, Y. C., Yarmishyn, A. A., Wu, J.-C., Tsai, C.-Y., Wang, M.-L., Peng, C.-H., Chien, K.-H., Kao, C.-L., Lin, T.-C., Woung, L.-C., Chen, S.-J., & Chiou, S. H. (2019). Artificial intelligence-based decision-making for age-related macular degeneration. *Theranostics*, *9*(1), 232–245. DOI: 10.7150/thno.28447 PMID: 30662564

Jin, Z., & Chen, Y. (2015). Telemedicine in the cloud era: Prospects and challenges. *IEEE Pervasive Computing*, *14*(1), 54–61. DOI: 10.1109/MPRV.2015.19

Kaswan, K. S., Gaur, L., Dhatterwal, J. S., & Kumar, R. (2021). AI-based natural language processing for the generation of meaningful information electronic health record (EHR) data. In Advanced AI techniques and applications in bioinformatics (pp. 41-86). CRC Press.

Mahmoud, A., Awad, N. A., Alsubaie, N., Ansarullah, S. I., Alqahtani, M. S., Abbas, M., Usman, M., Soufiene, B. O., & Saber, A. (2023). Advanced deep learning approaches for accurate brain tumor classification in medical imaging. *Symmetry*, *15*(3), 571. DOI: 10.3390/sym15030571

Milletari, F., Frei, J., Aboulatta, M., Vivar, G., & Ahmadi, S. A. (2018). Cloud deployment of high-resolution medical image analysis with TOMAAT. *IEEE Journal of Biomedical and Health Informatics*, *23*(3), 969–977. DOI: 10.1109/JBHI.2018.2885214 PMID: 30530377

Mistry, H. K., Mavani, C., Goswami, A., & Patel, R. (2024). The Impact Of Cloud Computing And Ai On Industry Dynamics And Competition. *Educational Administration: Theory and Practice*, *30*(7), 797–804.

Morley, J., Machado, C. C., Burr, C., Cowls, J., Joshi, I., Taddeo, M., & Floridi, L. (2020). The ethics of AI in health care: A mapping review. *Social Science & Medicine*, *260*, 113172. DOI: 10.1016/j.socscimed.2020.113172 PMID: 32702587

Nagi, A. T., Awan, M. J., Mohammed, M. A., Mahmoud, A., Majumdar, A., & Thinnukool, O. (2022). Performance analysis for COVID-19 diagnosis using custom and state-of-the-art deep learning models. *Applied Sciences (Basel, Switzerland)*, *12*(13), 6364. DOI: 10.3390/app12136364

Nasser, N., Emad-ul-Haq, Q., Imran, M., Ali, A., Razzak, I., & Al-Helali, A. (2023). A smart healthcare framework for detection and monitoring of COVID-19 using IoT and cloud computing. *Neural Computing & Applications*, *35*(19), 1–15. DOI: 10.1007/s00521-021-06396-7 PMID: 34522068

Pourqasem, J. (2018). Cloud-based IoT: Integration cloud computing with Internet of things. *International Journal of Research in Industrial Engineering*, *7*(4), 482–494.

Radwan, T., Azer, M. A., & Abdelbaki, N. (2017). Cloud computing security: Challenges and future trends. *International Journal of Computer Applications in Technology*, *55*(2), 158–172. DOI: 10.1504/IJCAT.2017.082865

Saber, A., Hussien, A. G., Awad, W. A., Mahmoud, A., & Allakany, A. (2023). Adapting the pre-trained convolutional neural networks to improve the anomaly detection and classification in mammographic images. *Scientific Reports*, *13*(1), 14877. DOI: 10.1038/s41598-023-41633-0 PMID: 37689757

Saber, A., Sakr, M., Abo-Seida, O. M., & Keshk, A. (2021, May). Automated Breast Cancer Detection and Classification Techniques–A survey. In 2021 International Mobile, Intelligent, and Ubiquitous Computing Conference (MIUCC) (pp. 200-207). IEEE.

Saber, A., Sakr, M., Abou-Seida, O., & Keshk, A. (2021). A novel transfer-learning model for automatic detection and classification of breast cancer based deep CNN. *Kafrelsheikh Journal of Information Sciences*, *2*(1), 1–9. DOI: 10.21608/kjis.2021.192207

Sakr, M., Saber, A. M., Abo-Seida, O., & Keshk, A. (2020). Machine learning for breast cancer classification using k-star algorithm. *Applied Mathematics & Information Sciences*, *14*(5), 855–863. DOI: 10.18576/amis/140513

Shah, V., & Konda, S. R. (2022). Cloud Computing in Healthcare: Opportunities, Risks, and Compliance. *Revista Española de Documentación Cientifica*, *16*(3), 50–71.

Shukla, A. K., & Kumar, V. S. (2023). Cloud computing with artificial intelligence techniques for effective disease detection. *International Journal of Data Informatics and Intelligent Computing*, *2*(1), 32–41. DOI: 10.59461/ijdiic.v2i1.45

Tang, M., Mehrotra, A., & Stern, A. D. (2022). Rapid Growth Of Remote Patient Monitoring Is Driven By A Small Number Of Primary Care Providers: Study examines the growth of remote patient monitoring. *Health Affairs*, *41*(9), 1248–1254. DOI: 10.1377/hlthaff.2021.02026 PMID: 36067430

Viana-Ferreira, C., Guerra, A., Silva, J. F., Matos, S., & Costa, C. (2017). An intelligent cloud storage gateway for medical imaging. *Journal of Medical Systems*, *41*(9), 141. DOI: 10.1007/s10916-017-0790-8 PMID: 28780714

Wei, P., Wang, D., Zhao, Y., Tyagi, S. K. S., & Kumar, N. (2020). Blockchain data-based cloud data integrity protection mechanism. *Future Generation Computer Systems*, *102*, 902–911. DOI: 10.1016/j.future.2019.09.028

Zhou, Y., Qian, C., Guo, Y., Wang, Z., Wang, J., Qu, B., . . . Qu, X. (2021, November). XCloud-pFISTA: A medical intelligence cloud for accelerated MRI. In 2021 43rd Annual International Conference of the IEEE Engineering in Medicine & Biology Society (EMBC) (pp. 3289-3292). IEEE. DOI: 10.1109/EMBC46164.2021.9630813

Chapter 15

Integration of Brain-Computer Interfaces in Modern Healthcare

Manish Kumar
Chitkara University, India

Hakam Singh
Chitkara University, India

Nilamadhab Mishra
https://orcid.org/0000-0002-1330-4869
VIT Bhopal University, India

Ramamani Tripathy
Chitkara University, India

Rudra Kalyan Nayak
https://orcid.org/0000-0003-4447-8391
VIT Bhopal University, India

Amit Thakur
https://orcid.org/0009-0007-7550-4990
VIT Bhopal University, India

ABSTRACT

The world of Brain-Computer Interface (BCI) is kind of like a big deal in healthcare. This research paper dives into the many sides of BCI, checking out how it started, works, and integrates with Artificial Intelligence (AI) in healthcare. Starting with a look back in time, the paper follows the path of BCIs from the beginning to now, sharing some cool moments. It talks a lot about why BCIs are important in healthcare and how they could change how we care for patients and improve their lives. Then, it gets into the details of how BCIs work – turning brain signals into things we can do. It's a bit tricky, but the paper breaks it down. It also chats about different ways we can use brain signals for real-world stuff in healthcare, and there's a bit about making special datasets for healthcare. This research paper describes special steps for a conceptual device that converts brain signals into audio signals, which can be useful for mute individuals.

DOI: 10.4018/979-8-3693-9208-9.ch015

1. INTRODUCTION

A Brain-Computer Interface (BCI) is a new and promising area of study and tech. It is more relatable to Biotechnology. It lets your brain directly talk and interact with computers or prosthetic devices like artificial arms and legs without needing nerves or muscles (Wolpaw et al., 2000). BCI interfaces try to understand and use your brain's signals to control outside things. It is a big thing because it can change how we interact with machines, especially when the usual methods might not work properly (Wolpaw et al., 2002). The main idea behind BCIs is to capture, Figureure out, and try to turn brain activity into commands we can use. These commands generally are binary signals (Turnip & Hong, 2012). We use sensors like electroencephalography (EEG) or functional magnetic resonance imaging (fMRI) to measure brain signals. Since these signals are very complex, we use complex techniques like machine learning to understand them better. So, it's like a mix of brain science, signal tricks, and smart computer learning that builds the foundation for BCIs (Khan & Hong, 2015).

1.1 Types of BCIs

a) **Electroencephalography (EEG) BCIs:** EEG is a gentle way of checking out the electric activity on your head with tiny sensors. People often use it for brain training, games, and simple chatting (Moghimi et al., 2013). The most common type of EEG is scalp-based EEG. They Quantify the currents that occur in the postsynaptic membranes' summated activity. In this paper, we are using this as a base for our concept because any action in a transmitting neuron initiates the discharge of neurotransmitters toward the synapse. The chemical messenger attaches to the receiving neuron's dendrite receptors and diffuses across the synaptic cleft (Naseer & Hong, 2015). The activity of various receptors leads to the movement of ions into or out of the dendrites, which causes compensatory electrical currents in the surrounding extracellular space. This current generates the EEG voltages. The EEG does not pick the action that travels along axons (Weyand et al., 2015). Instead, it measures different types of brainwave patterns on the scalp. These patterns vary in their spatial layout and frequency and how they relate to different states of brain activity, like whether someone is sleeping or active. These oscillations are responsible for representing activities that are synchronized over a neuron network. In the neuronal network, certain brainwave patterns are more understood, like the thalamus-cerebral synchronization responsible for generating sleep-related bursts. However, many other brainwave systems remain elusive, such as the one responsible for generating the posterior basic rhythm, which is still challenging to understand.

Even EEG signals come from post-synaptic potentials, which means you cannot see what each neuron or dendrite is doing just by an EEG of the scalp. Instead, the surface EEG shows the combined action of thousands of neurons lined up the same way. The EEG does not detect currents that do not matter to the scalp (Hyder et al., 2001). Like this, the EEG Advantages with Asynchronous are the vertical alignment of the top branches of neurons in the brain's outer layer. Due to this, Electrical fields Reduce in strength proportional to the fourth power of the radius. Detecting signals from deeper areas is very challenging compared to detecting currents close to the surface of the skull.

b) **Intercortical Electrodes:** We put tiny electrodes inside the brain or in certain brain spots. These electrodes are like intercortical ones that go into the brain for precise signal catching. People mostly link this kind of BCI with medical uses.
c) **Functional Near-Infrared Spectroscopy (FNIRS) BCIs:** These focus on changes in how much oxygen is in your blood, specifically in the brain. It's like a low-impact way to see what is happening in your brain, and it works well when you need to keep observing things regularly. It is like Figureuring out how hard your brain is working.
d) **Magnetoencephalography (MEG) BCIs**: MEG checks out the magnetic fields made by brain activity (Shulman et al., 2007). It's not as usual as EEG. It shows us things in more detail. People use it in studies and hospitals to understand how the brain works.
e) **Intracranial BCIs:** Intracranial BCIs mean sticking electrodes inside the skull but not in the brain. People do this a lot to monitor epilepsy and get lots of detailed info about the brain.
f) **Hybrid BCIs**: Hybrid BCIs use different brain-scan methods together to work better. For instance, putting EEG and other imaging tricks together makes it easier to see things in the brain more clearly and in a better timeframe.
g) **Motor Imagery BCIs:** These BCIs work by trying to understand brain signals when you're thinking about moving, not moving. You can make things happen outside by thinking about doing something with your muscles. People find these useful, especially for getting better at moving after an injury.
h) **P300-Based BCIs:** These P300-based Brain-Computer Interfaces (BCIs) pick up on a brain signal called the P300 in EEG signals. You can choose things by just thinking about the option you want. They're handy for talking or messaging stuff (Rebsamen et al., 2010).

i) **Intraneural BCIs**: Intraneural BCIs are like putting tiny electrodes inside nerves. People are Figureuring this out for things like prosthetics and trying to bring back feelings. It's still a work in progress.
j) **Neural Dust BCIs:** These are tiny wireless sensors known as neural dust; these are tiny sensors that get put into the brain to keep an eye on things (Li et al., 2010). These little sensors can send brain signals without any wires, and it can change how we keep track of various brain signals and activities happening in the brain for a long time.
k) **Cortical Brain-Machine Interfaces (BMI):** The BMI, which is Cortical, directly links the brain and gadgets outside the body. They are handy in periprosthetics, allowing people to guide robot arms with their thoughts.

1.2 Evolution of Brain-Computer Interface Technology

The research on BCIs was started in 1970 by Jacques Vidal. After that, the journey of BCIs has seen some very big changes, going from basic experiments to becoming useful in the real world. In the beginning, people were checking if it was possible to turn brain signals into actions (Aghalya et al., 2024). As time passed, BCIs got better and started helping people with major issues, making helpful things like neuroprosthetics and devices that assist us. BCIs also combined with Artificial Intelligence (AI), mainly smart algorithms like machine learning models (Donchin et al., 2000). These algorithms learn from how a person's brain works, improving over time to do their job more accurately. When BCIs and AI work together, it opens up new possibilities for doing unimaginable things, like making our brains work better, helping with disabilities, and maybe even making us better at things. There are many possibilities, such as brain-to-brain communication, accessing servers directly and downloading their data, producing new outcomes based on that data, providing contactless control over computer devices, etc.

1.3 Brain-Computer Interface Technology in Healthcare

BCIs have a very high potential in healthcare. They can help the person dealing with neurological issues or spinal cord injuries that make it difficult for a person to move or sometimes talk. The main idea behind using your brain to control things directly can change how we care for patients. Also, when BCIs are integrated with AI, they can become smarter by making interfaces that suit each person's unique brain activity (Castermans et al., 2011). However, some big questions arise when thinking about using BCIs. There are things like privacy concerns and how they might be used in the wrong way. This makes it super important to develop and use BCIs responsibly. As BCIs get better, we are still Figureuring out new ways they can connect with us, better solutions for understanding signals, and the right ways to use them without causing problems. It looks like BCIs might shape the future, where our brains can smoothly talk to machines (Ma et al., 2011; Khosrowabadi et al., 2013). But it's uncertain because we have not achieved that level of accuracy that ensures the errorless or lossless extraction of brain signals.

1.4 Importance of Brain-Computer Interface in Healthcare

a) **Neurological Rehabilitation:** Neurological Rehabilitation: BCIs open up new possibilities for helping people with conditions like paralysis or difficulty moving. By connecting directly with the brain, BCIs can control artificial limbs, wheelchairs, or other helpful gadgets. This can help restore movement and independence, although it's uncertain how well it will work for everyone.

b) **Communication Augmentation:** In situations where, regular talking methods are hard, like for people with locked-in syndrome or tough speech issues, BCIs offer a direct way to communicate (Liu et al., 2014). People can share what they're thinking using brain signals, helping them express what they want and need, even when talking is tough.

c) **Treatment of Neurological Disorders:** BCIs might help with different brain problems like epilepsy or chronic pain. The fancy ones can watch your brain in real-time, catching weird patterns early on. This could mean we could do something about it sooner or try treatments that fit you better.

d) **Cognitive Enhancement**: BCIs improve our thinking skills in brain health. Scientists are looking into ways BCIs can help with memory, attention, and staying focused (Panicker et al., 2011). This could be helpful for people facing memory problems, brain injuries, or diseases that affect the brain. But, you know, it's still uncertain.

e) **Objective Assessment of Mental Health:** BCIs might help determine how our minds are doing by looking at brain activity patterns linked to feeling down, anxious, or stressed. This could help doctors understand our mental health better, making it easier for them to give personalized help and treatments.

f) **Facilitation of Prosthetics and Exoskeletons:** if someone is missing a limb or has trouble moving, BCIs can make prosthetic limbs and exoskeletons work better. They read the signals from your brain, making moving these things feel more natural and easier. It's like trying to make these helpful gadgets work in a way that feels right for each person.

g) **Research and Understanding**: BCIs help us learn more about the brain and its workings. By studying and collecting data with BCIs, we get useful info about how the brain functions, what goes wrong with it, and how we might help fix issues.
h) **Adaptive Healthcare Interfaces:** When Brain-Computer Interfaces (BCIs) team up with Artificial Intelligence (AI), they create clever systems that try to learn and change based on how each person's brain works. This personal touch makes using healthcare gadgets better for each person, ensuring they fit their needs and situations. So, having BCIs in healthcare is like a big change towards making things more about the patient, adapting to them, and being more effective. We get new options for personal and precise healthcare solutions by directly connecting to the brain's signals. BCIs have become an important and sometimes necessary part of how we do things in healthcare (Meng et al., 2011; Rokade & Mishra, 2024).

2. METHODOLOGY

2.1 The Fundamental Working of BCIs (Figure 1) Involves the Following Steps

a) **Signal Acquisition**: BCIs grab signals your brain makes. The usual way is with electroencephalography (EEG), like recording the electricity in your head. There are other ways, too, like sticking stuff inside your brain, using fancy light (Functional Near-Infrared Spectroscopy - fNIRS), or checking out magnetic fields.
b) **Magnetoencephalography – MEG**: It depends on how invasive it is and what we want to do with it.
c) **Signal Processing:** We take the brain signals we get and do some things to them. We use techniques like filtering and amplifying to improve the signals and pick out the important stuff connected to what the person wants to do.
d) **Feature Extraction:** We take things like imagining moving, feeling stuff, or certain brainwave patterns from the signals we've looked at. These things are like the starting point to determine what the person wants or is telling the system to do (Wei et al., 2018).
e) **Decoding:** They use computer programs or smart techniques to determine what they find in the brain signals. The system learns to connect these signal patterns with what the person wants to do. It tries to understand thoughts and turn them into things you can do.

f) **Command Execution:** once we Figureure out what the brain is saying, we can use that info to do stuff with gadgets or computer programs. It could be as basic as moving a mouse on a screen or typing, or it might get fancier, like making a robot arm move or playing around in a virtual world.

g) **Feedback Loop:** In some cases, this Brain-Computer Interface (BCI) use a loop to give users information about what happens when they do something (any activity) like pictures, sounds, or feelings that help the user to understand how their actions connect with the system when they do certain different kind of things. This makes it easier for people to learn how to use the system effectively.

Figure 1. Structure of BCI

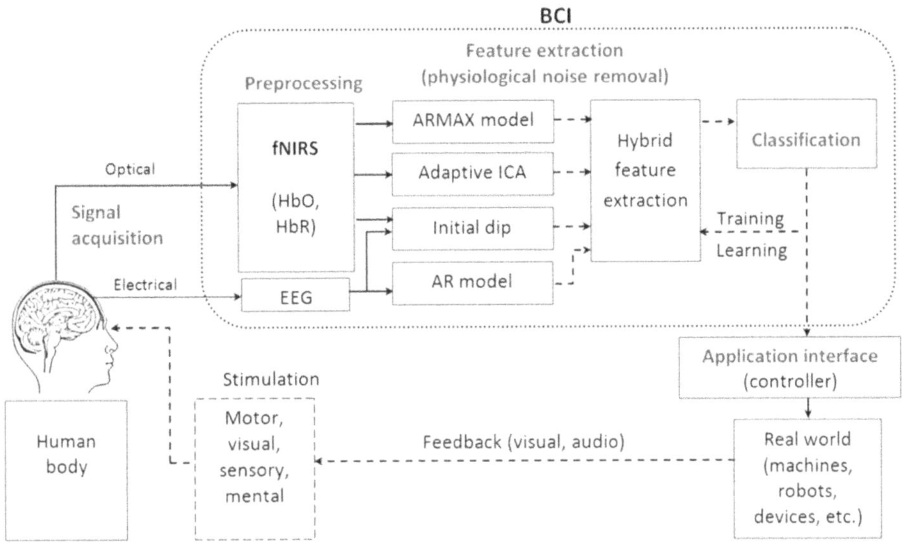

(AR: auto-regressive, ARMAX: auto-regressive moving average with exogenous signals, ICA: independent component analysis)

2.2 Training Datasets for Brain-Computer Interfaces

It is important to make and train datasets to teach Brain-Computer Interfaces (BCIs) because we have to integrate BCIs with AI, as it's very hard to understand brain signals, and there are numerous combinations or possibilities. These datasets help the model to learn and understand the brain signals better. Datasets are like the core of BCIs because we use them to train smart computer programs and AI models. The process involves gathering and marking data from people doing activities or feeling different emotions. We use different methods like (EEG) Electroencephalography or (fNIRS) Functional Near-Infrared Spectroscopy to record brain signals from human and animal brains. People do specific tasks while we record their brain signals, so creating a dataset that shows a wide range of brain activity is beneficial. It is important to have a diverse dataset that includes different ages, genders, and thinking states to make BCIs work well for everyone. Labels or annotations are like tags that tell us what the brain signals mean or what actions they are related to. For example, if the dataset contains signals, when someone moves their left hand or right hand, then each part of that data must be marked properly. These labels help the computer program to learn how to connect specific patterns in brain signals with specific actions or states of a person. In this case, the dataset's quality matters for how well the BCI works (Combaz & Van Hulle, 2015; Yin et al., 2013). A well-organized dataset helps the program learn the details, adapt to differences between people, and work for new users or tasks (Trinh et al., 2021; Emara et al., 2023). There should be some data cleanup steps like filtering out unwanted stuff and extracting important features to improve the dataset. Sharing and making datasets easy to understand is crucial for improving BCI research. Open-access datasets let researchers compare their computer programs, share results, and work together (VuThi et al., 2024; Emara & Huang, 2019). Projects like the BCI Competition series are a big concern because they give us standard datasets and ways to measure how well BCIs are doing. The training of datasets for BCIs involves a careful process of collecting, marking, and organizing data. Tweaking and growing these datasets helps BCI technology improve by making brain-computer interfaces stronger, more accurate, and useful for many different things, as depicted in Figure 2.

Figure 2. Decoding attention from the user.

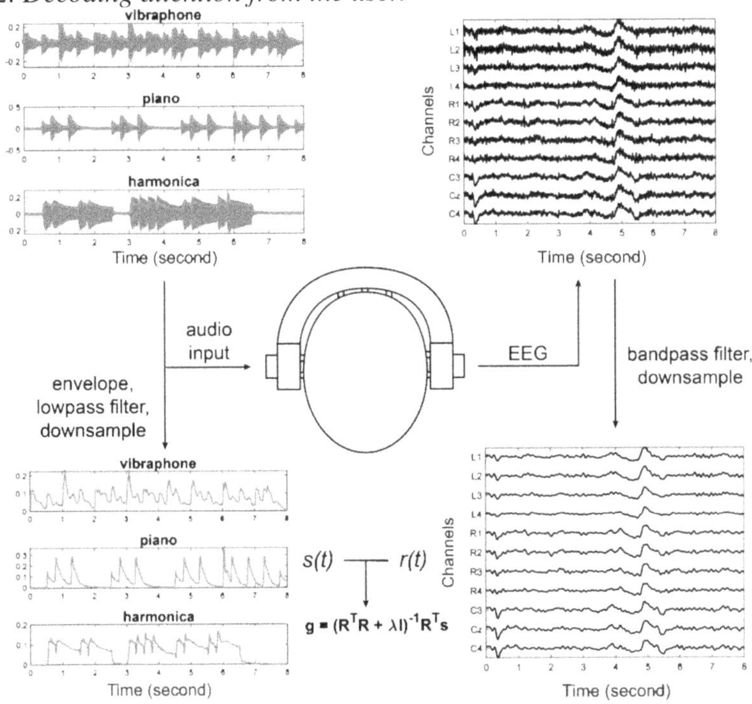

3. ANTICIPATED BCI TOPOGRAPHIES

In this research work, we look closely at how brain-computer interfaces can help people who can't speak. We discuss a simple idea for a device that turns a person's brain signals into sound.

Signal Acquisition: We use electrodes to understand brain function during speech planning. These electrodes are carefully put on the person's head in spots related to speech planning. The goal is to place them where the brain's signals for speech are strongest. These electrodes are like sensitive sensors that pick up the electrical activity made by the brain. When we think about speaking, specific brain parts get busy, and these electrodes catch tiny electrical signals from the brain cells in those areas. It can be easily understood using an analogy in which the brain talks using electricity, and these electrodes act like special sensors that catch those signals. A good understanding of the brain's workings is needed to finalize the position or part where these electrodes should be placed. Different parts of the brain do different things for speech, from thinking about anything to how to say it. So, we put the electrodes in those specific brain parts to get a detailed look at the brain's activity when someone speaks. Detecting the electrical signals when someone tries to say is a quick and real-time job. As people think about what to say, their brain makes electrical patterns showing these thoughts. The electrodes quickly pick up these patterns and create a kind of snapshot of what is happening in the person's brain related to speech. This gathered electrical signals, often shown as brainwaves. These brainwaves are then sent to a Brain-Computer Interface (BCI) system for more processing. Advanced computer technologies like signal processing and machine learning help understand and translate these signals into commands for synthesized speech. Putting the electrodes in the right places is important to ensure the BCI system gets it right. Scientists want to ensure that the signals they are interested in are clear and not mixed up with other activities happening in the brain. This careful connection between electrodes and brain activity shows how amazing BCIs are, giving people who can't speak a way to express themselves through computer-generated speech. Improving the technology of these electrodes and how we place them keeps pushing forward BCI research, making it even better at understanding the complex signals linked to speech intentions (Figure 3). We are talking about showing the signals from our brains with math; we write it like V(t). This V(t) thing depends on time and includes stuff like what our brain cells are doing and how they're connected (neuronal activity and synapses) (Beverina et al., 2003; Thulasidas et al., 2006). But here's the tricky part: some random stuff (noise) is mixed in. To make sense of all this, we use math equations, like differential equations, that describe how the brain cell coverings (membrane potential) change over time. It's like decoding a secret message from our brain but with math.

Figure 3. The EEG system detects brain signals through the use of electrodes.

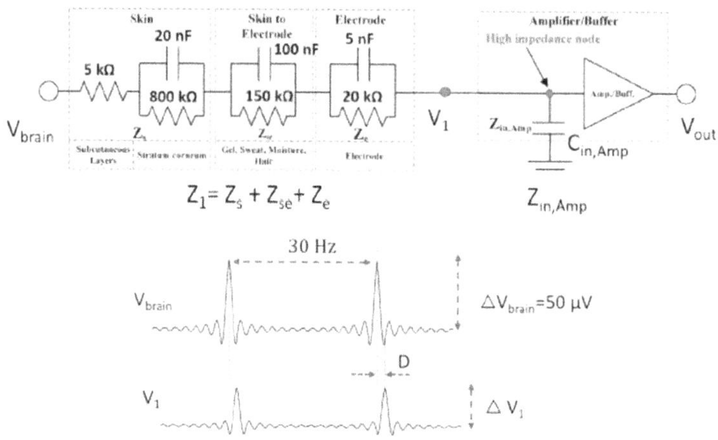

The EEG uses electrodes to track brain signals. It's like making a connection with electrical circuits. We need to figure out something called AV_1 when the brain signal is at 30 Hz and also find the delay (D) from V_am to V_1 at the same frequency. It's a bit uncertain, but that's what we're trying to sort out.

Signal Processing: In Brain-Computer Interface (BCI), dealing with the brain's signals is a bit like a combination of smart signal tricks and clever computer learning. It's a big deal in turning what our brains do into things we can do. So, when the little sensors on your head pick up signals from your brain while you're planning to speak, the data goes through some complicated steps. First, we use advanced signal tricks to clean up and understand the brain signals. This helps make the signals better by getting rid of any extra stuff that might mess things up. The tricks also look at when things happen in the brain and what kind of signals show up. All of this helps figure out the different plans your brain has for speaking. Then, we bring in computer learning to make the system better at understanding what you want to do. It is similar to training a computer to recognize the patterns in the signals that match what the person is thinking to say. The more the system interacts with you, the smarter it gets by learning from how your brain works. There is a thing called reinforcement learning that helps the system by giving feedback when it makes good predictions and improving things inside when it needs to get better. So, it is a process where the system learns and improves over time. But it's not always sure about everything. Brain-computer interface (BCI) systems which are dealing with the brain's signals require good computational power and clever computer learning. It sounds easy but it's a complex task in turning what our brains do into things we can do. So, when the little sensors on your head pick up signals from your brain while you're planning to speak, the data goes through some complicated steps. First, it uses advanced signal filtering and amplifying to clean up and understand the brain signals. This improves the signals by eliminating extra data that might mess with things. The tricks also examine when things happen in the brain and what signals appear. All of this helps determine your brain's different plans for speaking. Then, we bring in computer learning to make the system better at understanding what you want to do. It's like training a computer to recognize the patterns in the signals that match what you're thinking about saying. The more the system hangs out with you, the smarter it gets by learning how your brain works. There's a good thing called reinforcement learning that helps by giving the system a reward on the back when it makes good predictions and tweaking things inside when it needs to get better. So, it's a bit like human behavior, where the system learns and improves over time, but it's not always sure about everything (Huang et al., 2015).

EEG Signal - Preprocessing and Decoding: From a computational perspective, raw EEG signals are essentially discrete, multivariate time-series data. Every unit in the chronological data series relates to an example taken from multiple EEG streams, with the size of every element determined by the number of channels. The total quantity of elements in the temporal sequence is influenced by the duration of recording and the frequency of sampling (for instance, 270 samples per second). Raw EEG signals typically include direct current shifts, fluctuations, electromagnetic interference, and distortions. Necessitating initial signal processing to filter out these unwanted components. This preprocessing step is crucial, and extensive research has focused on addressing noise and artifact issues in EEG signals. For further details, refer to our post on EEG distortions and signal processing methods.

After cleaning the signals, the next step to improve and identify cognitive patterns and neural connections of relevance. Often, these brain processes are confined to specific frequency bands, for instance, the N200 elicited reaction in the alpha frequency range (8-12 Hz) or the alteration of mu oscillations in the 8-15 Hz range. Simple frequency filters, such as high-cut or band-limited filters, are used to separate these frequency ranges and eliminate irrelevant frequencies. Post-preprocessing involves extracting significant attributes derived from the processed EEG information before the arrival of deep learning, attribute extraction depended on improvised techniques tailored to specific mind processes, covering a spectrum from custom-designed attributes to complex approaches that involve linear and nonlinear spatial enhancement methods. Techniques for this include principal component analysis, independent component analysis, CSPs for power characteristics, and X-Dawn for time-based attributes. Figure 3above shows a basic feature extraction method that subsamples EEG signals in the temporal or Spectrum of oscillations. The extracted features are application-specific, aiding in tasks like distinguishing between experimental conditions, predicting behavior, or identifying anomalies (Lenhardt et al., 2008).

EEG Signals Data Decoding: After feature extraction, they are used to decode EEG data automatically, typically through supervised learning. This approach employs a training dataset to build a model capable of classifying, predicting, or recognizing EEG patterns from the features. Common methods include linear classifiers (LDA), multiple linear regression, SVMs, random forests, neural networks, and sophisticated combinations of these techniques. Supervised learning requires a labeled training dataset for model training and evaluation, usually through cross-validation. EEG decoders face unique challenges due to EEG's non-stationary and subject-specific nature. Features extracted for one individual at a certain time may not be suitable later or for another individual. Consequently, models need retraining with updated datasets. Initially, decoders were specific to participants and sessions, requiring a dedicated training set for each, which is labor-intensive and impractical outside research environments. Adjusting every individual is time-consuming and costly. To mitigate this, modern techniques aim to reduce calibration and develop solid ways that function throughout the duration and among various individuals. A critical aspect is determining the exact time of relevant EEG information. While some applications (e.g., EEG speller) have clear time points, many BCI and neuro tech applications, such as home seizure detection or neurorehabilitation, require asynchronous processing. This adds complexity, as distinguishing patterns of interest must also account for background EEG. To address this, asynchronous decoding often uses a sliding window approach, computing outputs for each window independently. Throughout the training process, the background EEG is categorized as "rest," and the beginning of motion is detected through calibration methods such as EMG signals or buttons. Subsequently, identical supervised learning techniques can be utilized to educate the decoder, which is used in a sliding window to provide continuous decoding (Hoffmann et al., 2005).

Speech Synthesis: After figuring out the tricky neural signals in a Brain-Computer Interface (BCI) model, the next big step is kind of uncertain. It involves connecting these brain patterns to a system that talks out loud, like making actual speech. This step is super important because it links the brain signals about talking with creating understandable speech. The processed brain signals, now mixed with info from smart signal tricks and computer learning, hold the details of what the person wants to say. We use a clever speech system to turn this brain code into meaningful speech. This system acts like a translator between the complicated brain patterns and the spoken words that match what the person wants to say. There are different ways to do this, and one common method is turning the brain patterns into written words first. It is similar to creating a language plan for what the person wants to say. This written plan or pattern goes through a speech synthesis engine, turning words into spoken signals. This engine can mimic how people speak naturally by considering tone, rhythm, and pronunciation. It aims to produce speech that sounds identical to how the person wants to express themselves. The speech system is a flexible AI modal that can work with different languages, accents, and speech styles, making communication more inclusive and personal. This flexibility is crucial to make sure the spoken words match the person's talking accent and how they like to talk. Also, using artificial intelligence (AI) in the speech system adds more possibilities for dynamic and context-based communication. The AI can pick up the clues from the brain signals and adjust the speech to express feelings, intentions, or meanings. Ultimately, the mix of processed brain signals and the speech system finishes the journey from thoughts in the brain to spoken words (Figure 4). This whole process shows how awesome BCIs are in giving people a way to speak out loud, opening up new ways for communication and empowerment, especially for those who might find regular speech a bit challenging (Cooney et al., 2018).

Figure 4. Conversion of neuro signals to auditory feedback.

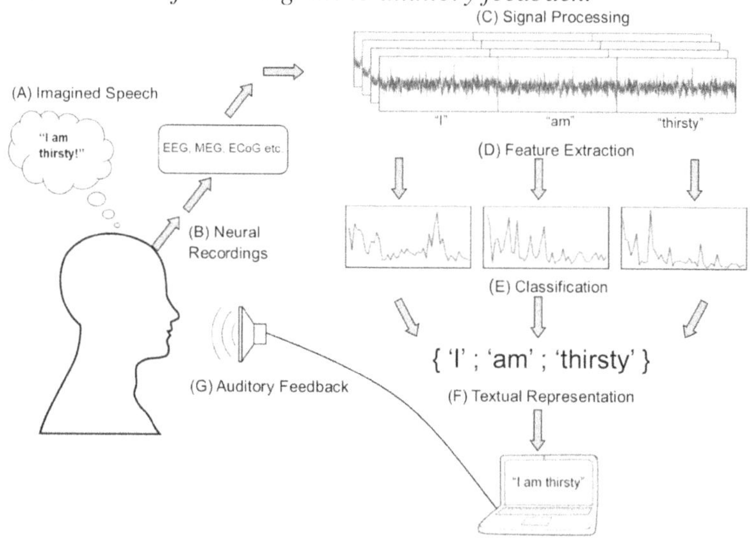

Audio Output: Once the Brain-Computer Interface (BCI) figure out how to turn thoughts into speech, the next big thing is getting that speech out to the world through a speaker. This speaker is like a buddy connected to the BCI device, turning electronic signals into actual sounds so people nearby can hear what the user wants to say. The speech, carefully made by the BCI system, holds all the words and sounds the user is meant to use. Now, this electronic speech is all set to go from the brain's world to the real world of hearing. The BCI device's and the speaker's link is super important for this sound journey. The electronic signals, showing the synthesized speech, travel smoothly from the BCI device to the speaker through a friendly connection. This connection ensures the information travels well, keeping the speech as intended during the transmission. When the signals reach the speaker, they undergo a crucial change for the whole talking process. With its ability, the speaker turns these electronic signals into sound waves you can hear. This change follows the steps of the rule, converting the electrical signal into radio waves by the BCI device and transmitting that wirelessly to the speaker, which can be attached to the BCI device. The sound coming from the speaker is the person's thoughts turned into audio output. The external speaker behaves like the person is speaking himself, turning the synthesized speech from brain signals into shared communication that everyone can hear. This smooth way of sending a speech to a speaker is a big deal in tech that mixes brain science, signal conversion, and hearing tech to let people talk even if natural speaking is hard. This way of communicating through BCI opens up new possibilities for people with different communication needs by giving them a powerful way to express their thoughts to other people.

Real-time Feedback: Real-time feedback in a brain-computer interface (BCI) system reduces latency. It improves the user experience and helps improve speech intentions over time. This feedback loop lets the user and the BCI talk both ways, not just send signals from the brain. In speech, the user thinks about what they want to say, and the BCI system catches and processes those brain signals linked to those thoughts. The system uses smart algorithms to figure out these brain patterns and turn them into speech that we can hear. However, here is the twist: the BCI system does not stop there. It gives back this speech to the user in real-time. This loop is like a mirror for the users, thoughts and letting them hear and judge how well their intended speech is coming out. Instant feedback allows the user to check if what they are thinking matches the speech they are hearing. This feedback system has many benefits. It helps the user check how accurate and clear the speech is. Hearing their thoughts immediately lets users make quick changes in what they think, possibly making their speech more specific. This loop is a continuous learning experience for the user and the BCI system. With each interaction, the user gets familiar with how the BCI talks and changes how they think to make speech better. At the same time, the BCI system, using smart machine learning models, learns user patterns and preferences and improves over time. This improvement cycle is important for dealing with challenges and changes in brain signals, ensuring the BCI can always communicate well. This real-time feedback loop, which shares lots of information, strengthens the teamwork between human thinking and technology and helps make BCI applications better in speech and easier to use.

Customization and Training: It is important to ensure that Brain-Computer Interface (BCI) gadgets benefit everyone, especially people with different mental conditions. People have different skills, so creating BCIs that work for everyone, regardless of physical or mental abilities, is a big deal. For this, we cannot just stick to the normal ways of using BCI devices with buttons or making gestures. We need to think beyond that. Things like hand gestures are still important, but we can make BCIs more versatile by adding other ways to control them. One effective way is using eye tracking. This technology watches how your eyes move and helps you do things on the computer by looking. It is super helpful for people who cannot move much, as they can use their eyes to click and move around. Another way is using facial expressions. Modern systems can understand the face movements and turn them into commands. This helps people with motor challenges and lets BCIs work more naturally, like how we express ourselves through our faces. Bringing these different ways of controlling BCIs is about making them flexible and easy to use. This recognizes that people have different abilities and tries to give them personalized and effective ways to use BCIs. Also, using various control methods, BCIs can work well for people with different movement challenges and make technology more inclusive. It is not just about adapting but changing how we design BCIs. By understanding and. using what people can naturally do, regardless of their motor skills, BCIs become more user-friendly and open to everyone. This approach improves using BCIs and positively impacts society by making advanced tech available to everyone (Karthika et al., 2023).

Natural language processing: The NLP is a method of calculating for systematic examination and illustration of human speech, where distributed semantics represents the definitions of terms and phrases utilizing vector representations and stored in vector databases that handle multidimensional databases. When we read words and sentences, our brain processes language and creates processing data, which is cognitive, like the EEG signals. Modern NLP models are the foundation of Word embeddings. High-dimensional vectors in semantic space represent words, phrases, and sentences. These vectors are obtained by training large language models on large Datasets to encode contextual information. Each vector represents the meaning of a word. Evaluation and comparison of the quality of different word embeddings is a known, majorly open challenge. Traditionally, word embeddings are analyzed using internal or external methods (Hollenstein et al., 2019). External evaluation analyses the quality of the embedding based on their performance on downstream NLP tasks, such as question-answering or recognizing entities. But we can train and fine-tune embeddings for specific tasks. Automated language conversion is an essential component of natural language processing that seeks to develop a dependable encoding-decoding framework. An entirely connected, feed-forward architecture cannot leverage the fact that analogous features frequently emerge at various positions in the ECoG data sequence. For instance, articulating a specific phoneme typically exhibits a comparable pattern at a designated electrode, independent of its timing. To effectively recognize such trends, the architecture consistently utilizes the same short-term filter throughout the whole input sequence. Implementing a stride larger than one sample effectively reduces the size of the resulting output sequence. The architecture acquires a collection of these filters, generating filtered sequences down-sampled to 16 Hz. A trained recurrent neural network (RNN) is intended to transform a sequence of neural activity representing sentences into an abstract form. It then decodes this form precisely into English sentences, achieving a word error rate of 3-4% (Makin et al., 2019; Emara, 2019).

Integration with Communication Devices: Connecting Brain-Computer Interface (BCI) systems with devices like phones is a big step forward. It opens up many possibilities for how people interact with BCIs and how the signal is processed in an external device rather than the BCI device itself. By integrating with everyday devices like smartphones and tablets, BCIs can go beyond the traditional ways we interact, specifically for people with different communication needs (Emara et al., 2014; Emara, 2016). One key is making a good link between the BCI and the user's phone or tablet. This link lets communication go both ways. The BCI gets the user's commands and sends info to other devices. Having a strong and quick way to communicate is crucial for real-time interaction. It makes sure users can smoothly interact with their surroundings. In real life, users can use their BCIs to do things on their phones or tablets. They can write messages, make calls, or use apps by using their BCIs only using their eye movements or just thinking. The BCI understands what the user's brain is trying to say, turns it into commands, and makes the connected device do what is needed. The connection is not just about words or sounds users can use their BCIs to move things around on their phones or tablets. This is super helpful for people who might have trouble moving around. We can also integrate IOT to help physically challenged people by reducing their physical movement in daily home activities. It gives them an easy and direct way to handle digital things. This effective connection with gadgets boosts how users express themselves and helps them be more independent by working with common tools like phones; BCIs make it easier for people to socialize, work, and be part of the digital world. Users can send emails and join video calls all through their BCI-connected devices. The way BCIs work with smartphones and tablets makes them an important part of this technology. Ultimately, the smooth connection between BCI systems and communication gadgets is revolutionary. It breaks down the communication barriers and creates digital spaces where everyone feels included and connected (Shad et al., 2020).

4. DEEP LEARNING FOR EEG-BASED BCI

Using Deep Learning with EEG-Based BCIs can be very beneficial. Deep learning is a subpart of particular machine learning methods. These algorithms can be utilized to study the EEG trends (Mattei et al., 2024). These algorithms refer to the deep Neural Networks. The deep learning system's major characteristic is that it combines many neurons embedded in multiple layers with multiple combinations that enable them to acquire features closely similar to human cognitive abilities. The Ordinary systems for Deep Learning include recurrent neural networks (RNN), convolutional neural networks (CNN), and generative adversarial networks (GAN). For the EEG-based BCI applications, Deep Learning is widely used. It helps extract EEG features and combines them with the existing machine learning to achieve a high regression or classification work level. In this Portion, Impulsive EEG-based BCI applications with the CNN model, the use of GANs and the process and applications of RNN are discussed. Additionally, this discussion will cover Long Short-Term Memory (LSTM) and the integration of CNN with LSTM (Samal & Hashmi, 2024).

Convolutional Neural Networks: They are neural networks that work on feedforwarding. They are specially designed for processing data with grid-like structures, e.g., vectors and matrices, and as we know, matrices can represent sound signals very well; in this, the EEG signals flow through mainly three layers. These layers are the pooling, convolutional, and fully connected layers. In the convolutional layer, convolve operation is performed on the tensor and filters; this results in the formation of the feature map, and then the pooling layer helps in reducing the feature dimensions, which results in a reduction of the feature map in terms of height and width and also reduces the calculations. Lastly, the fully connected layer contains a neural network whose input layer is connected with the last feature maps of each dimension. This layer is responsible for classifying the desired output. It applies and adjusts the weights for the prediction of the correct labels. The CNN applies filters so that the EEG patterns will be identified. The continuous progress in CNN, mainly in classification purposes, is important and aimed at advancing BCI technology, like the enhancement of EEG-based fatigue evaluation. For instance, a new EEG-based technique using a specific convolutional neural network (CNN) was introduced (Kumari et al., 2024). In this approach, the CNN captures temporal patterns and combines them with thick layers of neurons to process and classify spatial-temporal EEG data; in the results, the model successfully achieves 96-97% accuracy in identifying different fatigue levels and examined exhaustion degrees in a graphical EEG study. That study introduced a cross-level CNN that employs a combined spatial-temporal-spectral characteristics retrieval technique for analyzing raw EEG signals, implementing a layered approach similar to inception.

Recurrent neural networks are used for sequential and time series-based data. The RNN uses a chronological order to capture the memory. In natural language, each word has meaning and relation with its previous and next words, so this is a kind of sequential data. As we know, EEG signals are time series-based data that change over time. RNNs are designed to recognize patterns in data sequences, which means they are good at temporal Pattern Recognition. RNNs can remember information about different time steps. This is beneficial because brain signals keep changing over time. RNNs can learn and adapt to new brain signal patterns; this can help in personalized training as each person can have slightly different brain signals for the same thing. This can also be helpful in processing and extracting the words and semantics of what the person is trying to say (Vansteensel et al., 2023).

Generative adversarial networks: In a few cases, real-world EEG data is not present in high amounts for training the Machine Leering model. Developing some Machine Learning or Deep Learning models can produce many challenges. To confront this situation, the GAN includes two concurrently trained Neural Networks: generator networks and discriminator networks. The generator is used to create or train the model, but the discriminator is used to refine the model by feedback. The accuracy of the model is increased with the help of a discriminator (Wahab et al., 2024).

Research has recently been introduced, mainly focused on data upgrade purposes. These delayers fabricate new data instances to mirror real-world information by implementing generative models (Maiseli et al., 2023). Another better GAN model was newly published (Umar & Mishra, 2024) that understood the EEG features and expanded the dataset's dimensions by producing artificial samples to increase the classification performance. Another new framework (Chevallier et al., 2024) based on EEG, Conditional Wasserstein GAN, was introduced to refine Sentiment detection by producing superior-grade fabricated EEG data from SEED and DEAP datasets.

5. CONCLUSIONS

The research begins with exploring the history and evolution of BCIs and discusses their role in acquiring and translating brain signals into actionable commands and audio output. This paper explains various types of BCIs, from EEG-based systems to advanced technologies like Neural Dust BCIs and Cortical Brain Machine Interfaces. The importance of BCIs in healthcare is very promising, focusing on their potential applications in communication augmentation, treatment of neurological disorders, cognitive enhancement, objective mental health assessment, prosthetics and exoskeletons functioning, research, and the development of adaptive healthcare interfaces. Integrating BCIs with Artificial Intelligence opens new possibilities for personalized and precise healthcare solutions. The proposed work in the paper introduces a conceptual device aimed to cover brain signals into audio signals, which is particularly beneficial for mute people. The methodology section in the research paper focuses on the workings of BCIs, including signal acquisition, processing, feature extraction, decoding, command execution, and the importance of real-time feedback. This research signifies the training of datasets for BCIs, highlighting their role in teaching smart computer programs and AI models to understand better and interpret brain signals. The proposed device's functionality outlines the steps of signal acquisition, processing, feature extraction, decoding, command execution, and the important aspect of real-time feedback for improved user experience.

This paper also explains the customized training of BCIs for individuals with various mental and physical conditions, including technologies such as eye tracking and facial expressions for more control. Integrating BCIs with communication devices, like smartphones and tablets, will provide users with enhanced interaction and connectivity in their daily lives. As tech keeps getting better. Combining BCI, healthcare, and AI brings many new ideas. It is like a puzzle where everything fits perfectly. This integration helps patients and takes us to a future where smart solutions, like our devices, change how people and machines work together. The journey of BCIs from the past till now has given the best results with AI. This shows a really good path. We are focused on making tech better for healthcare, especially for those who find it hard to communicate. In the future, we will dive even deeper into our device, making it real implications.

REFERENCES

Aghalya, S., Sarasu, R., Mishra, N., Arunachalam, G., Gobinath, D., & Naganathan, S. B. T. (2024, April). Tele-Rheumatology-Advanced Neural Network Model for Remote Assessment and Management of Rheumatic Conditions. *In 2024 10th International Conference on Communication and Signal Processing* (ICCSP) (pp. 453-458). IEEE.

Beverina, F., Palmas, G., Silvoni, S., Piccione, F., & Giove, S. (2003). User adaptive BCIs: SSVEP and P300 based interfaces. *PsychNology Journal, 1*(4), 331–354.

Castermans, T., Duvinage, M., Petieau, M., Hoellinger, T., Saedeleer, C. D., Seetharaman, K., Bengoetxea, A., Cheron, G., & Dutoit, T. (2011). Optimizing the Performances of a P300-Based Brain–Computer Interface in Ambulatory Conditions. *IEEE Journal on Emerging and Selected Topics in Circuits and Systems, 1*(4), 566–577. DOI: 10.1109/JETCAS.2011.2179421

Chevallier, S., Carrara, I., Aristimunha, B., Guetschel, P., Sedlar, S., Lopes, B., Velut, S., Khazem, S., & Moreau, T. (2024). The largest EEG-based BCI reproducibility study for open science: the MOABB benchmark. *ArXiv, abs/2404.15319.*

Combaz, A., & Van Hulle, M. M. (2015). Simultaneous detection of P300 and steady-state visually evoked potentials for hybrid brain-computer interface. *PLoS One, 10*(3), e0121481. DOI: 10.1371/journal.pone.0121481 PMID: 25815815

Cooney, C., Folli, R., & Coyle, D. (2018). Neurolinguistics research advancing development of a direct-speech brain-computer interface. *iScience, 8*, 103–125. DOI: 10.1016/j.isci.2018.09.016 PMID: 30296666

Cruz, M. V., Jamal, S., & Sethuraman, S. C. (2024). A Comprehensive Survey of Brain-Computer Interface Technology in Healthcare: Research Perspectives.

Donchin, E., Spencer, K. M., & Wijesinghe, R. (2000). The mental prosthesis: Assessing the speed of a P300-based brain-computer interface. *IEEE Transactions on Rehabilitation Engineering, 8*(2), 174–179. DOI: 10.1109/86.847808 PMID: 10896179

Emara, T. (2019). Adaptive Power-Saving Mechanism for VoIP Over WiMAX Based on Artificial Neural Network. In *Algorithms, Methods, and Applications in Mobile Computing and Communications* (pp. 158–177). IGI Global. DOI: 10.4018/978-1-5225-5693-0.ch007

Emara, T. Z. (2016). Maximizing power saving for voip over wimax systems. [IJMC-MC]. *International Journal of Mobile Computing and Multimedia Communications*, *7*(1), 32–40. DOI: 10.4018/IJMCMC.2016010103

Emara, T. Z., & Huang, J. Z. (2019). RRPlib: A spark library for representing HDFS blocks as a set of random sample data blocks. *Science of Computer Programming*, *184*, 102301. DOI: 10.1016/j.scico.2019.102301

Emara, T. Z., Saleh, A. I., & Arafat, H. (2014). Power saving mechanism for VoIP services over WiMAX systems. *Wireless Networks*, *20*(5), 975–985. DOI: 10.1007/s11276-013-0650-5

Emara, T. Z., Trinh, T., & Huang, J. Z. (2023). Geographically distributed data management to support large-scale data analysis. *Scientific Reports*, *13*(1), 17783. DOI: 10.1038/s41598-023-44789-x PMID: 37853092

Hoffmann, U., Garcia, G., Vesin, J. M., Diserens, K., & Ebrahimi, T. (2005, March). A boosting approach to P300 detection with application to brain-computer interfaces. *In Conference Proceedings. 2nd* [IEEE.]. *International IEEE/EMBS Conference on Neural Engineering*, *2005*, 97–100.

Hollenstein, N., de la Torre, A., Langer, N., & Zhang, C. (2019, November). CogniVal: A Framework for Cognitive Word Embedding Evaluation. In Bansal, M., & Villavicencio, A. (Eds.), *Proceedings of the 23rd Conference on Computational Natural Language Learning (CoNLL)* (pp. 538–549). DOI: 10.18653/v1/K19-1050

Huang, H. P., Liu, Y. H., Lin, W. Z., Kang, Z. H., Cheng, C. A., & Huang, T. H. (2015). Development of a P300 brain–machine interface and design of an elastic mechanism for a rehabilitation robot. *International Journal of Automation and Smart Technology*, *5*(2), 123–145.

Hyder, F., Kida, I., Behar, K. L., Kennan, R. P., Maciejewski, P. K., & Rothman, D. L. (2001). Quantitative functional imaging of the brain: Towards mapping neuronal activity by BOLD fMRI. NMR in Biomedicine. *NMR in Biomedicine*, *14*(7-8), 413–431. DOI: 10.1002/nbm.733 PMID: 11746934

Karthika, K., Dhanalakshmi, S., Murthy, S. M., Mishra, N., Sasikala, S., & Murugan, S. (2023, October). Raspberry Pi-enabled Wearable Sensors for Personal Health Tracking and Analysis. *In 2023 International Conference on Self Sustainable Artificial Intelligence Systems* (ICSSAS) (pp. 1254-1259). IEEE.

Khan, M. J., & Hong, K. S. (2015). Passive BCI based on drowsiness detection: An fNIRS study. *Biomedical Optics Express*, *6*(10), 4063–4078. DOI: 10.1364/BOE.6.004063 PMID: 26504654

Khosrowabadi, R., Quek, C., Ang, K. K., & Wahab, A. (2013). ERNN: A biologically inspired feedforward neural network to discriminate emotion from EEG signal. *IEEE Transactions on Neural Networks and Learning Systems, 25*(3), 609–620. DOI: 10.1109/TNNLS.2013.2280271 PMID: 24807454

Kumari, A., Edla, D. R., Reddy, R. R., Jannu, S., Vidyarthi, A., Alkhayyat, A., & de Marin, M. S. G. (2024). EEG-based motor imagery channel selection and classification using hybrid optimization and two-tier deep learning. *Journal of Neuroscience Methods, 409*, 110215. DOI: 10.1016/j.jneumeth.2024.110215 PMID: 38968976

Lenhardt, A., Kaper, M., & Ritter, H. J. (2008). An adaptive P300-based online brain–computer interface. *IEEE Transactions on Neural Systems and Rehabilitation Engineering, 16*(2), 121–130. DOI: 10.1109/TNSRE.2007.912816 PMID: 18403280

Li, Y., Long, J., Yu, T., Yu, Z., Wang, C., Zhang, H., & Guan, C. (2010). An EEG-based BCI system for 2-D cursor control by combining Mu/Beta rhythm and P300 potential. *IEEE Transactions on Biomedical Engineering, 57*(10), 2495–2505. DOI: 10.1109/TBME.2010.2055564 PMID: 20615806

Liu, Y. H., Wu, C. T., Cheng, W. T., Hsiao, Y. T., Chen, P. M., & Teng, J. T. (2014). Emotion recognition from single-trial EEG based on kernel Fisher's emotion pattern and imbalanced quasiconformal kernel support vector machine. *Sensors (Basel), 14*(8), 13361–13388. DOI: 10.3390/s140813361 PMID: 25061837

Ma, R., Aghasadeghi, N., Jarzebowski, J., Bretl, T., & Coleman, T. P. (2011). A stochastic control approach to optimally designing hierarchical flash sets in P300 communication prostheses. *IEEE Transactions on Neural Systems and Rehabilitation Engineering, 20*(1), 102–112. DOI: 10.1109/TNSRE.2011.2179560 PMID: 22203722

Maiseli, B. J., Abdalla, A. T., Massawe, L. V., Mbise, M., Mkocha, K., Nassor, N. A., Ismail, M., Michael, J., & Kimambo, S. (2023). Brain–computer interface: Trend, challenges, and threats. *Brain Informatics, 10*(1), 20. DOI: 10.1186/s40708-023-00199-3 PMID: 37540385

Makin, J. G., Moses, D. A., & Chang, E. F. (2019). Machine translation of cortical activity to text with an encoder–decoder framework. *Nature Neuroscience, 23*(4), 575–582. DOI: 10.1038/s41593-020-0608-8 PMID: 32231340

Mattei, E., Lozzi, D., Matteo, A. D., Polsinelli, M., Manes, C., Mignosi, F., & Placidi, G. (2024). Deep Learning Architecture analysis for EEG-Based BCI Classification under Motor Execution. *2024 IEEE 37th International Symposium on Computer-Based Medical Systems (CBMS)*, 549-555.

Meng, L., Jin, J., & Wang, X. (2011, October). A comparison of three electrode channels selection methods applied to SSVEP BCI. *In 2011 4th international conference on biomedical engineering and informatics* (BMEI) (Vol. 1, pp. 584-587). IEEE.

Moghimi, S., Kushki, A., Marie Guerguerian, A., & Chau, T. (2013). A review of EEG-based brain-computer interfaces as access pathways for individuals with severe disabilities. *Assistive Technology, 25*(2), 99–110. DOI: 10.1080/10400435.2012.723298 PMID: 23923692

Naseer, N., & Hong, K. S. (2015). fNIRS-based brain-computer interfaces: A review. *Frontiers in Human Neuroscience, 9*, 3. PMID: 25674060

Panicker, R. C., Puthusserypady, S., & Sun, Y. (2011). An asynchronous P300 BCI with SSVEP-based control state detection. *IEEE Transactions on Biomedical Engineering, 58*(6), 1781–1788. DOI: 10.1109/TBME.2011.2116018 PMID: 21335304

Rebsamen, B., Guan, C., Zhang, H., Wang, C., Teo, C., Ang, M. H., & Burdet, E. (2010). A brain controlled wheelchair to navigate in familiar environments. *IEEE Transactions on Neural Systems and Rehabilitation Engineering, 18*(6), 590–598. DOI: 10.1109/TNSRE.2010.2049862 PMID: 20460212

Rokade, S., & Mishra, N. (2024, January). Intelligent Health: Empowering Illness and Prognosis with Machine Learning-Enabled Electronic Health Records on Blockchain. *In 2024 International Conference on Advancements in Smart, Secure and Intelligent Computing* (ASSIC) (pp. 1-5). IEEE.

Samal, P., & Hashmi, M. F. (2024). Role of machine learning and deep learning techniques in EEG-based BCI emotion recognition system: A review. *Artificial Intelligence Review, 57*(3), 50. DOI: 10.1007/s10462-023-10690-2

Shad, E. H. T., Molinas, M., & Ytterdal, T. (2020). Impedance and noise of passive and active dry EEG electrodes: A review. *IEEE Sensors Journal, 20*(24), 14565–14577. DOI: 10.1109/JSEN.2020.3012394

Shulman, R. G., Rothman, D. L., & Hyder, F. (2007). A BOLD search for baseline. *NeuroImage, 36*(2), 277–281. DOI: 10.1016/j.neuroimage.2006.11.035 PMID: 17223362

Thulasidas, M., Guan, C., & Wu, J. (2006). Robust classification of EEG signal for brain-computer interface. *IEEE Transactions on Neural Systems and Rehabilitation Engineering, 14*(1), 24–29. DOI: 10.1109/TNSRE.2005.862695 PMID: 16562628

Trinh, T., Duc, L. P., Tran, C. T., Duy, T. T., & Emara, T. Z. (2021). A new stratified block model to process large-scale data for a small cluster. *International Conference on Artificial Intelligence and Big Data in Digital Era*, 253–263. Springer International Publishing Cham. https://doi.org/DOI: 10.1007/978-3-030-97610-1_21

Turnip, A., & Hong, K. S. (2012). Classifying mental activities from EEG-P300 signals using adaptive neural network. Int. *J. Innov. Comp. Inf. Control*, *8*(7), 5839–5850.

Umar, Y. R. M., & Mishra, N. (2024, January). GeneOntology's Disease Gene Identification Assessment by DisGeNET. *In 2024 International Conference on Advancements in Smart, Secure and Intelligent Computing (ASSIC)* (pp. 1-7). IEEE.

Vansteensel, M. J., Klein, E., van Thiel, G., Gaytant, M., Simmons, Z., Wolpaw, J. R., & Vaughan, T. M. (2023). Towards clinical application of implantable brain–computer interfaces for people with late-stage ALS: Medical and ethical considerations. *Journal of Neurology*, *270*(3), 1323–1336. DOI: 10.1007/s00415-022-11464-6 PMID: 36450968

VuThi., K., VuongThi, N., NguyenXuan, H., Emara, T. Z., & Trinh, T. (2024). Feature Selection Based on Top Frequency. *The International Conference on Intelligent Systems & Networks*, 58–65. Springer Nature Singapore. https://doi.org/ DOI: 10.1007/978-981-97-5504-2_7

Wahab, A., Khan, U. S., Nawaz, T., Akbar, H., Shah, S. T., Khalid, A., Ansari, A. R., & Nawaz, R. (2024). Improved Accuracy for Subject-Dependent and Subject-Independent Deep Learning-Based SSVEP BCI Classification: A User-Friendly Approach. *IEEE Access : Practical Innovations, Open Solutions*, *12*, 115935–115950. DOI: 10.1109/ACCESS.2024.3442235

Wei, C., Salloum, S., Emara, T. Z., Zhang, X., Huang, J. Z., & He, Y. (2018). A Two-Stage Data Processing Algorithm to Generate Random Sample Partitions for Big Data Analysis. In M. Luo & L.-J. Zhang (Eds.), *Cloud Computing -- CLOUD 2018* (Vol. 10967, Issue 2018060, pp. 347–364). Springer International Publishing. DOI: 10.1007/978-3-319-94295-7_24

Weyand, S., Takehara-Nishiuchi, K., & Chau, T. (2015). Weaning off mental tasks to achieve voluntary self-regulatory control of a near-infrared spectroscopy brain-computer interface. *IEEE Transactions on Neural Systems and Rehabilitation Engineering*, *23*(4), 548–561. DOI: 10.1109/TNSRE.2015.2399392 PMID: 25667356

Wolpaw, J. R., Birbaumer, N., Heetderks, W. J., McFarland, D. J., Peckham, P. H., Schalk, G., Donchin, E., Quatrano, L. A., Robinson, C. J., & Vaughan, T. M. (2000). Brain-computer interface technology: A review of the first international meeting. *IEEE Transactions on Rehabilitation Engineering*, *8*(2), 164–173. DOI: 10.1109/TRE.2000.847807 PMID: 10896178

Wolpaw, J. R., Birbaumer, N., McFarland, D. J., Pfurtscheller, G., & Vaughan, T. M. (2002). Brain–computer interfaces for communication and control. *Clinical Neurophysiology*, *113*(6), 767–791. DOI: 10.1016/S1388-2457(02)00057-3 PMID: 12048038

Yin, E., Zhou, Z., Jiang, J., Chen, F., Liu, Y., & Hu, D. (2013). A speedy hybrid BCI spelling approach combining P300 and SSVEP. *IEEE Transactions on Biomedical Engineering*, *61*(2), 473–483. PMID: 24058009

Chapter 16

AI-Powered Smart Tourism Reimagining Travel

Mohamed Badouch
https://orcid.org/0000-0003-0495-3784
Faculty of Sciences, Ibn Zohr University, Agadir, Morocco

Mehdi Boutaounte
https://orcid.org/0009-0001-0459-7532
National School of Commerce and Management, Ibn Zohr University, Dakhla, Morocco

Hasna Mahmoud
https://orcid.org/0009-0006-5457-3594
Faculty of Sciences, Ibn Zohr University, Agadir, Morocco

Nagwa Elmobark
https://orcid.org/0000-0002-6757-088X
Mansoura University, Mansoura, Egypt

ABSTRACT

The tourism industry is undergoing a transformative shift through the integration of artificial intelligence (AI) and applied life sciences. This chapter explores AI's multifaceted applications, from personalized travel planning and AI-powered chat-bots to optimizing resource management and promoting sustainability. AI-driven predictive analytics forecast tourist demand, while AR/VR enhances the travel experience. Additionally, AI facilitates environmental monitoring and promotes traveler well-being. However, ethical considerations regarding data privacy and responsible AI development are crucial. This chapter examines the potential of AI

DOI: 10.4018/979-8-3693-9208-9.ch016

to revolutionize tourism, fostering a future where travel is personalized, sustainable, and ethically sound.

INTRODUCTION

The tourism industry, a vibrant force in global economies and a catalyst for cultural exchange, is undergoing a dramatic transformation fueled by the rapid advancements in artificial intelligence (AI). This technological revolution, often referred to as "smart tourism," is not merely a trend; it's a fundamental shift redefining how we travel, experience destinations, and interact with the tourism ecosystem (Alsahafi, R. et al., 2023). AI-powered systems are poised to revolutionize the industry, offering personalized experiences, optimized resource allocation, and a deeper understanding of traveler needs, ultimately fostering a more responsible and impactful tourism sector. This transformation promises to enhance the traveler experience, improve efficiency, and contribute to a more sustainable future for tourism. However, this technological advancement comes with a critical responsibility, demanding careful consideration of its potential drawbacks and ethical implications.

The integration of AI into tourism creates a complex landscape with numerous potential benefits. For example, AI-powered recommendation systems analyze vast amounts of data to provide personalized itineraries, suggesting attractions, activities, and accommodations tailored to individual preferences. Imagine a system that suggests a unique boutique hotel based on a traveler's interest in art and local culture, or a guided tour that seamlessly integrates historical information with interactive experiences. Moreover, AI-powered chatbots can provide instant assistance with booking flights, handling reservations, and addressing inquiries, improving the overall traveler experience, and reducing stress (Sharma, I. and Aggarwal, A., 2024). These examples demonstrate the transformative potential of AI to personalize travel and enhance visitor satisfaction.

However, this technological advancement also raises crucial questions about data privacy, algorithmic biases, and the potential for unintended consequences. The collection and analysis of vast amounts of traveler data raise concerns about privacy and security, requiring robust safeguards to protect personal information. Furthermore, the algorithms driving these systems may perpetuate existing biases, leading to unfair or discriminatory recommendations (Lukita, C., et al. 2023). These ethical considerations require careful attention to ensure that AI-driven tourism is not only innovative but also equitable and responsible. This chapter addresses these critical needs by providing a comprehensive analysis of the potential and challenges of AI-driven tourism.

We explore the key contributions of AI in enhancing the traveler experience, optimizing resource allocation, and fostering a more sustainable tourism sector, while also analyzing the challenges associated with its implementation. Specifically, we examine the ethical implications of AI in tourism, including data privacy concerns, algorithmic biases, and the potential for unintended consequences. Our analysis provides valuable insights for researchers, practitioners, and policymakers, guiding responsible innovation and shaping the future of AI in tourism (Bokolo, A. J. 2021). The significance of this exploration lies in its potential to inform the development of ethical and sustainable AI-driven tourism solutions, fostering a more responsible and impactful tourism sector that benefits both travelers and destinations. By understanding the potential and challenges of AI in tourism, we can pave the way for a future where technology enhances the travel experience while promoting responsible and sustainable practices.

1. ENHANCING TOURIST EXPERIENCES THROUGH AI-POWERED PERSONALIZATION

1.1. Personalized Itinerary Generation

The shift towards personalized travel planning is driven by the increasing use of AI algorithms that analyze vast datasets to understand individual traveler preferences and tailor travel plans accordingly (Badouch, M., and Boutaounte, M. 2024). These datasets encompass a variety of information, including:

Explicit Preferences

- "I want to experience a cultural immersion trip to Japan, focusing on traditional arts and cuisine." *(This is a clear preference articulated by a user.)*
- "My budget for this trip is under $1000." *(This defines a constraint on the travel plan.)*
- "I prefer to stay in charming boutique hotels rather than large resorts." *(This indicates a preference for a particular type of accommodation.)*

Implicit Preferences

- Booking History: A traveler who has repeatedly booked flights to European cities, stayed in historic hotels, and booked tours focused on art and culture suggests an interest in these types of experiences (Indaryanto, A., et al. 2023).

- Accommodation History: If a user frequently selects hotels with swimming pools or has booked accommodations near beaches, this reveals a preference for relaxation and recreation.
- Transportation Preferences: A traveler who consistently chooses public transportation over rental cars suggests an interest in exploring cities by foot or taking trains, reflecting a preference for urban experiences.

Social Media Activity

- "Exploring the beauty of Santorini." *(A travel post indicating interest in picturesque landscapes.)*
- "Loved the traditional cuisine in Marrakech!" *(Expressing interest in local food and cultural experiences.)*
- "Looking for recommendations for a hiking trip in the Alps." *(A search query indicating interest in outdoor activities and adventurous travel.)*

Real-Life Examples

To illustrate this process, consider the real-world applications of AI-driven personalization in popular travel platforms:

- **Expedia**: Utilizes machine learning algorithms to analyze user preferences and past travel behavior to recommend suitable destinations and activities. For instance, a traveler who frequently opts for budget-friendly accommodations and enjoys outdoor activities might be recommended hiking trails, national parks, or affordable camping options in their chosen destination. This data-driven approach helps create a more relevant and tailored travel plan.
- **Booking.com**: Customizes lodging suggestions based on criteria like budget, preferred facilities, and proximity to popular sites. For example, a traveler seeking a beachfront hotel with a family-friendly atmosphere might receive recommendations for hotels with children's play areas and proximity to a sandy beach. This demonstrates how AI can be utilized to efficiently find suitable accommodations based on user preferences.

Figure 1. Architecture for tourism recommender system

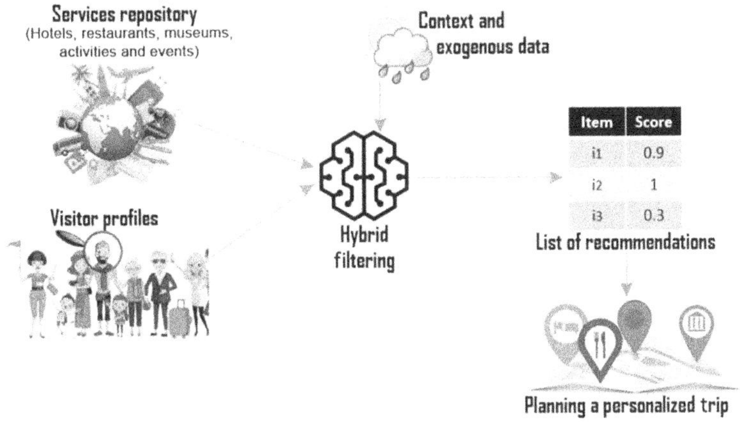

1.1.1. Dataset Description and Examples

The integration of artificial intelligence (AI) into the travel industry is ushering in a new era of personalized experiences, powered by sophisticated data analysis and predictive algorithms. This transformation, while promising a more tailored and engaging travel experience, raises critical questions about the ethical implications of data collection and utilization. AI-powered platforms are increasingly relying on vast and diverse datasets, often proprietary to the companies involved, to understand traveler preferences and predict their needs. This data, collected through user interactions, website activity, social media engagement, and even device location tracking, encompasses a wide range of information, including user demographics, travel history, preferences, and even browsing behavior on travel websites. Platforms like Expedia, for instance, might track user preferences for specific types of accommodations (hotels, vacation rentals, resorts), activities (sightseeing, dining, shopping), and destinations (Mbunge, E., and Muchemwa, B. 2022). This data forms the foundation for creating detailed user profiles, allowing platforms to tailor travel recommendations and personalize experiences for each traveler, creating a seemingly more efficient and enjoyable experience. However, this data-driven approach also raises concerns about data privacy, algorithmic biases, and potential unintended consequences. Let's delve deeper with some real-world examples:

Personalized Recommendations: Imagine booking a flight on a platform like Kayak or Skyscanner. The platform, utilizing your past travel history and preferences, might suggest flights that align with your budget, preferred airlines, and travel times. This might include showcasing flights with shorter layovers, direct flights, or even highlighting airlines with known amenities that match your preferences. This personalized approach can be beneficial, but raises concerns about the potential for bias. If the algorithm prioritizes flights based on past travel patterns, it might inadvertently exclude destinations that might be more relevant to the user but aren't within their typical travel radius.

Targeted Advertising: As you browse travel websites or social media, you might encounter targeted advertisements for specific destinations or travel experiences. These ads, based on your browsing history and user profile, might showcase luxury resorts in the Caribbean if you recently viewed articles about beach vacations or suggest budget-friendly hostels in Southeast Asia if you've expressed interest in backpacking. This approach, while effective for advertisers, raises concerns about data privacy. What data is being used to target these ads? Are users aware of the extent to which their browsing activity is being used to tailor advertising?

Location-based Recommendations (Trinh, et al., 2019): Imagine using a travel app like Google Maps or Tripadvisor to find local restaurants or attractions. These apps, utilizing your location data, might recommend restaurants based on your proximity, past reviews, and even your dietary restrictions. While this can be convenient, it raises concerns about the potential for location tracking to be used for more invasive purposes. What safeguards are in place to ensure that your location data isn't being misused or shared without your explicit consent?

These examples illustrate the potential benefits and challenges associated with AI-driven personalization in tourism. While AI can offer tailored experiences, it's crucial to ensure that data is collected and utilized ethically, with transparency and user control, to avoid potential biases and unintended consequences. The future of travel relies on finding the right balance between personalization and ethical considerations, ensuring that technology enhances the travel experience while safeguarding user privacy and promoting inclusivity.

1.1.2. Data Collection Ethics, Methods, and Sources

The ethical considerations surrounding data collection in AI-powered tourism are paramount, demanding a delicate balance between leveraging data for enhanced user experience and respecting individual privacy. Platforms, driven by the promise of personalized travel experiences, gather vast amounts of user data through a multitude of channels, including user interactions, website activity, social media engagement, and even device location tracking. This data trove, encompassing user demographics, travel history, preferences, and even browsing behavior on travel websites, forms the foundation for sophisticated user profiles, enabling platforms to tailor recommendations and personalize experiences.

Obtaining explicit user consent is a fundamental ethical requirement in data collection. This involves providing clear and concise information about what data is being collected, how it will be used, and the user's options for managing their data. Platforms often employ clear privacy policies and opt-in mechanisms to ensure that users actively consent to data collection. For example, a travel booking platform might display a pop-up message when a user first accesses the site, requesting their consent to collect data related to their browsing activity, travel preferences, and location data. This process, however, requires careful attention to ensure that consent is truly informed and that users understand the implications of sharing their data. Furthermore, platforms must implement robust security measures to protect user data from unauthorized access, breaches, and misuse. This includes adopting comprehensive data encryption protocols, implementing strict access controls, and leveraging advanced technologies like blockchain for secure data storage and management. Imagine a travel app using blockchain technology to securely store user data, ensuring that only authorized individuals have access to it and preventing unauthorized access or data breaches.

Once collected, data is processed using a variety of machine learning techniques to generate tailored recommendations and personalize travel experiences. One common approach is collaborative filtering, which analyzes patterns in the preferences of similar users to predict what a given user might enjoy. For example, if a user has expressed interest in historical sites, collaborative filtering might recommend similar destinations based on the preferences of other users who have enjoyed similar experiences. Imagine a travel app using collaborative filtering to recommend a visit to the Colosseum in Rome based on the fact that other users who enjoyed visiting the Acropolis in Athens also enjoyed visiting the Colosseum. Another approach, content-based filtering, matches user preferences to specific destinations and activities based on their content. For example, if a user has previously searched for information about hiking trails, content-based filtering might recommend destinations with similar hiking trails. Imagine a travel app using content-based filtering to recommend hiking trails in the Swiss Alps after a user searches for information about hiking trails in the Himalayas. More sophisticated platforms often employ hybrid approaches, combining both collaborative filtering and content-based filtering to generate more nuanced and accurate recommendations.

The ethical implications of this data collection and processing are far-reaching. The collection and use of vast amounts of user data raise concerns about data privacy and security, necessitating robust safeguards to protect personal information. Furthermore, the algorithms driving these systems may perpetuate existing biases, leading to unfair or discriminatory recommendations. For instance, an algorithm might disproportionately recommend destinations based on a user's demographic data, potentially reinforcing existing travel patterns and overlooking diverse travel options. This necessitates a focus on algorithm transparency, fairness, and accountability to mitigate potential biases and promote equitable travel experiences. Imagine a travel app using an algorithm that inadvertently favors destinations with a certain demographic profile, potentially excluding diverse options and perpetuating travel patterns that might not be representative of the wider population.

Addressing these ethical challenges is paramount to ensuring the responsible and ethical development of AI-driven tourism. Platforms must prioritize transparency and provide clear and concise information about the data they collect, how it is used, and the options available to users for managing their privacy settings. The industry should establish ethical guidelines for the development and deployment of AI-powered travel tools, addressing issues like data privacy, algorithmic bias, and potential unintended consequences. Continued research and development are crucial for exploring solutions for data privacy, algorithmic fairness, and the promotion of diverse and authentic travel experiences.

1.1.3. Ethical Implications and Challenges

The collection and analysis of vast amounts of traveler data, a cornerstone of AI-driven personalization, raises significant concerns about privacy and security. The potential for misuse, breaches, or unauthorized access to sensitive personal information necessitates robust safeguards to protect user data. This requires a commitment to transparency, providing users with clear and concise information about the data being collected, its purpose, and their options for managing their privacy settings. Implementing comprehensive data encryption protocols, strict access controls, and secure data storage practices are crucial to mitigate risks and ensure data security.

Furthermore, the algorithms driving these AI-powered travel systems may perpetuate existing biases, leading to unfair or discriminatory recommendations. Algorithms trained on historical data may inadvertently reflect societal biases, potentially excluding certain demographics from travel opportunities or recommending experiences that reinforce existing inequalities. For example, an algorithm might recommend destinations based on historical travel patterns, leading to a limited range of suggestions that might not adequately represent the diverse interests and needs of all travelers. This could lead to a situation where a travel app primarily recommends beach vacations to users with certain demographic profiles, while overlooking other interests like cultural exploration or adventure travel. To mitigate these biases, ongoing efforts to develop fair and unbiased algorithms, coupled with rigorous testing and monitoring, are essential.

The integration of AI in tourism also raises concerns about potential unintended consequences. The pursuit of personalization could lead to the homogenization of travel experiences, eroding the unique character and authenticity of destinations. Imagine a scenario where an AI-powered platform predominantly recommends the same popular tourist attractions, resulting in a less diverse and authentic travel experience for everyone. Additionally, relying heavily on AI-driven recommendations could potentially limit exploration and discourage travelers from venturing beyond their comfort zones, leading to a less diverse and enriching travel experience. Navigating this ethical landscape requires a proactive and comprehensive approach, prioritizing user privacy, mitigating algorithmic biases, and fostering responsible innovation. Several key strategies can be implemented to ensure ethical and responsible AI-driven tourism:

Transparency and User Control: Platforms should provide users with clear and concise information about the data they collect, how it is used, and the options available to users for managing their privacy settings. This ensures that users are informed about the data collection process and have the ability to control their data.

Algorithmic Fairness and Accountability (Elahi, et al., 2021): Developing fair and unbiased algorithms requires rigorous testing and monitoring, ensuring that they do not perpetuate existing biases and are inclusive of all travelers. This can be achieved through ongoing research and development, collaborating with experts in ethical AI, and establishing mechanisms for holding algorithms accountable. This could involve using techniques like differential privacy to protect individual user data while ensuring the fairness of algorithms, or implementing explainable AI to understand and address potential biases within the algorithms.

Fostering Diverse and Authentic Experiences: The pursuit of personalized travel experiences should not come at the cost of diversity and authenticity. Platforms should encourage exploration, expose travelers to diverse experiences, and promote the unique character of destinations, ensuring that the benefits of AI are realized without compromising the richness and variety of the travel landscape. This might involve incorporating features that recommend lesser-known attractions, promote local culture, and offer diverse experiences that cater to different interests and backgrounds.

1.1.4. Advantage and Disadvantages of IA In Tourism

This advancement also brings with it a range of challenges and ethical considerations that necessitate careful attention. One critical gap lies around data privacy and security. The collection and analysis of vast amounts of traveler data raise concerns about potential breaches, misuse of information, and the risk of profiling and discrimination. Moreover, AI algorithms can be susceptible to biases inherent in the data they are trained on, leading to unfair or discriminatory recommendations and resource allocation. This creates a need for greater transparency and accountability in AI-driven tourism systems.

Advantages of AI in Tourism

- **Enhanced Traveler Experience:** AI can personalize travel plans, offering tailored recommendations based on individual preferences and past behavior. This can lead to a more engaging and fulfilling travel experience.
- **Optimized Resource Allocation:** By analyzing data and predicting demand, AI can optimize resource allocation in tourism destinations, leading to reduced waste and improved efficiency in staffing, transportation, and infrastructure management.
- **Improved Customer Service:** AI-powered chatbots and virtual assistants can provide instant assistance, answer queries, and resolve issues, improving customer satisfaction and streamlining travel experiences.
- **Sustainable Tourism:** AI can help to reduce the environmental impact of tourism by optimizing resource usage, predicting demand, and informing sustainable practices, such as promoting responsible travel and supporting eco-tourism initiatives (Loureiro, S. M. C., and Nascimento, J., 2021).
- **Increased Accessibility:** AI can facilitate access to tourism information for individuals with disabilities and those who do not speak the local language, making travel more inclusive and accessible to a wider range of travelers.

Disadvantages of AI in Tourism

- **Data Privacy and Security:** The collection and analysis of vast amounts of personal data raise concerns about privacy and security (Alyasiri, et al., 2024). There's a risk of data breaches, misuse of information, and the potential for profiling and discrimination.
- **Algorithmic Bias:** AI algorithms trained on biased data may perpetuate existing social inequalities and lead to discriminatory practices. This can include unfair recommendations, limited access to certain destinations or experiences, and unequal distribution of tourism resources.
- **Job Displacement:** The automation of tasks traditionally performed by human workers, such as customer service and travel planning, could lead to job displacement and economic disruption.
- **Lack of Transparency:** The complex workings of AI algorithms can be difficult to understand, hindering transparency and accountability. This can create challenges in addressing potential biases, ensuring fairness, and mitigating potential harms.

- **Dependence on Technology:** Excessive reliance on AI-driven systems can lead to a loss of human connection and a diminished appreciation for local culture and experiences.

1.2. AI-Powered Chatbots and Virtual Assistants for Enhanced Communication

The application of AI-driven chatbots and virtual assistants is transforming communication in the tourism sector, offering travelers a higher level of accessibility and convenience. NLP algorithms are essential for the functionality of these smart systems, allowing them to comprehend and address various user inquiries, promoting a more human-like interaction with travelers (Álvarez-Carmona, M. Á., et al. 2022).

Creating a connection between humans and machines: NLP algorithms allow chatbots to analyze and comprehend human language, recognizing the purpose of user questions, even if they are expressed differently (Abro, W. A., et al. 2022). These algorithms use methods such as tokenization, part-of-speech tagging, and named entity recognition to break down user input and identify important details. For instance, a chatbot can understand that a request such as "I need assistance booking a flight to Paris for the month of June" indicates a user's desire to make a flight reservation and identify important information like the destination (Paris) and travel date (June). Advantages for the Tourism Sector: AI-driven chatbots provide numerous benefits for the tourism industry, such as:

- Enhanced Customer Service: Chatbots provide 24/7 availability, offering instant responses to traveler queries and resolving issues promptly, thereby improving customer satisfaction (Ahmad, M., et al. 2024).
- Streamlined Booking Assistance: Chatbots can guide users through the booking process, providing information about available options, managing reservations, and handling cancellations, leading to increased efficiency and a smoother booking experience.
- Seamless Information Provision: Chatbots can access and deliver relevant information about destinations, attractions, and travel guidelines, acting as readily available knowledge resources for travelers.

Various leading tourism companies have effectively incorporated AI-driven chatbots into their businesses. KLM Royal Dutch Airlines utilizes a chatbot called "BB" to communicate with passengers and assist them with flight reservations, luggage details, and travel questions (BR, S. R. 2022). Marriott International uses "Ask Marriott," a chatbot created to respond to guest inquiries, offer hotel details, and assist with booking requests. These demonstrations display the real-world advantages of chatbots powered by AI in enhancing communication and enhancing the travel experience for travelers.

1.3. Immersive Experiences through AR/VR

The convergence of artificial intelligence (AI), augmented reality (AR), and virtual reality (VR) is not just reshaping travel, it's revolutionizing the very way we perceive and interact with the world (Sharma, M., et al. 2024). This essay delves into the scientific keys and cutting-edge technologies driving this transformation, exploring how algorithms, computer vision, and immersive experiences are poised to redefine the future of travel (Bretos, M. A. et al., 2024).

At the core of this revolution lies the power of immersion. Augmented reality, fueled by computer vision algorithms, overlays digital information onto the real world. Imagine strolling through the ancient Roman Forum, your smartphone's camera revealing digital overlays identifying key structures like the Temple of Saturn or the Basilica Julia. These overlays could provide historical context, explaining the purpose and significance of these buildings, or even offer interactive 3D models that allow you to virtually explore the structures in their original grandeur. This is powered by sophisticated image recognition algorithms that analyze the camera feed, identifying objects and landmarks, and using location data to provide relevant context. These algorithms are often based on deep learning models like Convolutional Neural Networks (CNNs), trained on vast datasets of images and location data, enabling them to identify objects with remarkable accuracy.

Virtual reality, on the other hand, takes immersion to a whole new level, transporting us to entirely simulated environments. Imagine "walking" through the ancient city of Pompeii, experiencing the bustling streets, shops, and homes as if you were transported back in time. This is made possible through algorithms that generate 3D models of environments, often using techniques like Structure from Motion (SfM) and Multi-View Stereo (MVS) (Kholil, et al., 2021). These techniques, often powered by deep learning models, process multiple images or videos of a scene to create a 3D representation, capturing intricate details and textures. With VR, you can explore the ruins of the Colosseum in Rome, walk through the Great Wall of China, or even journey to the bottom of the ocean, encountering diverse marine life in a realistic virtual environment.

AI further enhances these immersive experiences through its ability to understand the complex relationships within a scene. Graph Neural Networks (GNNs) are being employed to model the connections between objects within a virtual environment, allowing for more natural and interactive experiences. Imagine a virtual tour of the Forbidden City in Beijing, where the AI understands the flow of foot traffic, the interactions between virtual visitors, and even the impact of weather conditions on the environment, resulting in a truly dynamic and believable experience.

But the potential goes far beyond just visual experiences. AI can create personalized travel experiences tailored to individual preferences. Imagine an AI-powered travel app that analyzes your past travel history, browsing habits, and social media data to recommend destinations, activities, and experiences that perfectly align with your interests (Liu, et al., 2020). If you're a history buff, the app might suggest a trip to the ancient ruins of Machu Picchu in Peru, while an art enthusiast could be recommended a visit to the Louvre Museum in Paris. This personalization is achieved through recommendation algorithms, often based on collaborative filtering or matrix factorization techniques, which identify patterns in user data to suggest items or experiences that similar users have enjoyed.

The impact of this convergence on the travel industry is profound. VR allows potential travelers to "visit" destinations virtually, providing a more informed decision-making process and potentially increasing booking rates. Travelers can experience the atmosphere of a destination before committing to a trip, making more confident decisions and potentially leading to higher booking rates. AR enhances the actual travel experience, providing real-time information, interactive guides, and personalized recommendations, enriching the overall journey. Imagine using your phone to point at a historical landmark in Florence, and instantly accessing information about its history, architecture, and significance, or using AR to find nearby restaurants with reviews and menus tailored to your dietary preferences (Zhou, Y., et al. 2022).

However, significant challenges remain. Achieving truly realistic and immersive VR experiences requires further advances in real-time rendering, ensuring seamless interactions between users and the virtual environment. This involves optimizing algorithms to process vast amounts of data in real time, accurately simulating lighting effects, and creating dynamic environments that respond to user actions.

Furthermore, ensuring widespread adoption requires addressing affordability and accessibility, particularly for VR/AR devices and platforms in developing countries. Providing affordable and accessible technology will be crucial in realizing the full potential of these technologies for a truly inclusive future of travel.

- **3D Reconstruction**: This involves other methods such as Structure from Motion-SfM and Multi-View Stereo-MVS that are generally enhanced by the models of DL, like CNNs. These process several 2D images or videos taken from multiple perspectives to develop a 3D model of any given scene (Fitzgibbon, A., and Zisserman, A. 1998). It affords the general offer of realistic representations of an environment, building, and landmark in a virtual domain.

- **Semantic segmentation**: AI algorithms segment images and videos to identify and classify different objects or elements of the environment, such as buildings, trees, and people, to improve understanding of the composition of the scene.
- **Scene Understanding**: AI models, such as GNNs (Zhou, J., et al. 2020), recognize the relations involving objects and other elements of the scene that will help develop more interactive or less interactive, yet contextually aware, VR/AR experiences.
- **Real-time rendering** is a very challenging methodology, highly complex with advanced AI-based frameworks ensuring smooth and interactive experiences in virtual worlds. Ray tracing and path tracing techniques enable the simulation of all real-world lighting and its reflection in making the renderings of virtual surroundings more realistic.

The convergence of AI, AR, and VR represents a paradigm shift in travel, offering unprecedented opportunities for exploration, understanding, and connection (Shukla, V., et al. 2024). As these technologies continue to evolve, they promise to redefine our relationship with the world, creating a more informed, accessible, and immersive travel experience. The future of travel is undoubtedly linked to the continued development and integration of these transformative technologies, and the journey ahead is full of exciting possibilities.

Figure 2. The different parts in an intelligent tourist recommendation system

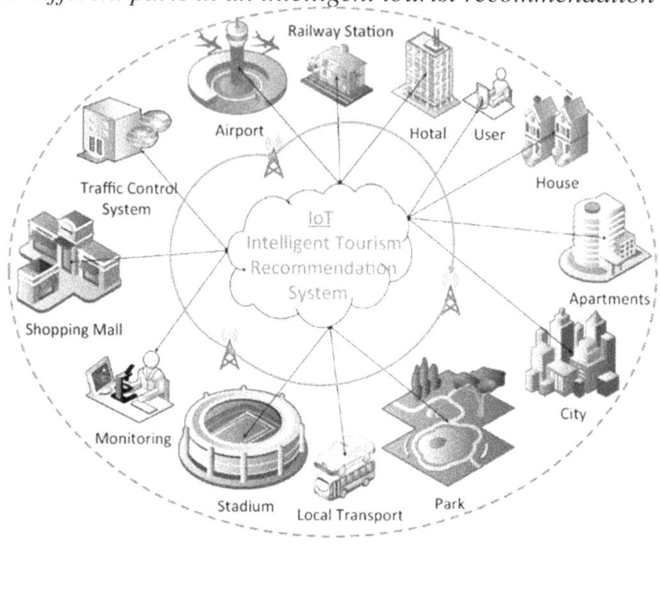

2. OPTIMIZING RESOURCES AND FOSTERING SUSTAINABILITY THROUGH AI

2.1. Predictive Analytics for Forecasting Tourist Demand

Predictive analytics plays a pivotal role in optimizing tourism operations, enabling destinations and businesses to anticipate demand, allocate resources effectively, and enhance visitor experiences (Trinh, T., et al. 2019). By analyzing historical data and identifying patterns, AI algorithms can forecast future tourist arrivals, allowing for proactive planning and resource management.

Implementation Tools and Environment: The development and deployment of AI-driven predictive analytics systems in tourism necessitates a robust technological infrastructure. This typically involves the following:

- **Statistical Computing Software:** Python Programming (Abril-Pla, O. et al., 2023) environments provide the foundation for data analysis (Emara, et al., 2023), model building, and algorithm implementation. This platform is equipped with libraries for time series analysis, regression modeling, and machine learning, enabling the construction of sophisticated predictive models.
- **Cloud Computing Platforms** (Naseer, I., 2023): Cloud computing services (AWS) offer scalable resources and storage solutions to handle large datasets and complex algorithms. These platforms allow for the deployment of AI models for real-time prediction and analysis, facilitating data processing and model execution at scale.
- **Data Visualization Tools:** Power BI enable the visualization of data patterns and insights. This allows for the clear interpretation of model results, the identification of trends, and the informed understanding of the factors driving tourist demand. Visualizations can facilitate data exploration and provide a clear picture of the patterns and trends within the data.
- **Specialized Forecasting Software:** "Prophet" (developed by Facebook) provide advanced time series forecasting capabilities, adept at incorporating seasonality and trends in the data. This specialized software is well-suited for tourism forecasting, effectively capturing the cyclical nature of tourist arrivals and incorporating key factors like weather, economic conditions, and events (Garlapati, et al., 2021).

Programming Languages and Environments: To implement these systems, researchers often use Python, utilizing libraries such as scikit-learn, TensorFlow, and PyTorch for machine learning. These libraries offer a range of algorithms for diverse types of predictive modeling, including time series analysis and regression modeling. These tools provide the necessary functionalities to build and deploy predictive models for tourist demand forecasting.

Data Sources and Methods: The foundation of predictive analytics in tourism rests on the availability and quality of data. To train predictive models, researchers often draw upon the following:

- **Tourist Arrival Data:** Historical records of tourist arrivals, categorized by season, origin, and demographic characteristics, provide a comprehensive understanding of overall trends and fluctuations.
- **Booking Data:** Information on booking patterns, including hotel reservations, flight bookings, and tour bookings, allows for the analysis of seasonal fluctuations, potential surges in demand, and changes in traveler behavior.
- **External Factors:** Data related to external factors impacting tourism demand, such as weather patterns, economic conditions, and events (festivals, sporting events, political events) These data sources enable the identification of factors that influence fluctuations in tourist arrivals and provide crucial context for accurate prediction.

Predictive Models

- **Time Series Analysis:** The Autoregressive Integrated Moving Average (ARIMA) model effectively captures the time-dependent patterns in tourist arrivals (Yao, L., et al. 2021). The Prophet model, developed by Facebook, provides a user-friendly solution for forecasting time series data, handling seasonality and trends. Both models leverage historical data to forecast future arrivals and provide valuable insights for resource allocation and operational planning.
- **Regression Models:** Linear regression and support vector regression models can be used to establish relationships between tourist demand and various influential variables (Vanegas, et al. 2020). These models are trained on historical data and enable the prediction of future demand based on forecasted values of key factors like weather conditions, economic circumstances, and marketing drives.
- **Machine Learning Classifiers:** Techniques like Random Forest or Gradient Boosting Machines classify tourists based on their travel preferences, enabling the identification of different tourist segments. This classification is crucial for tailoring resource allocation to meet the diverse needs of different groups of travelers.

- **Staffing Optimization:** Predictive analytics helps businesses like hotels, restaurants, and attractions optimize staffing levels by accurately forecasting demand (Richard, et al., 2017). By utilizing historical data and external factors, AI models can provide valuable insights into peak seasons, helping businesses ensure efficient and effective service delivery with the appropriate number of employees.
- **Transportation Management:** Predictive analytics enables the optimization of transportation services by anticipating and responding to anticipated demand. For instance, transportation authorities can adjust bus routes, frequency (VuThi, et al., 2024), and capacity based on forecasted tourism dynamics, improving efficiency, reducing congestion, and ensuring sufficient transportation capacity for travelers.
- **Infrastructure Development:** Long-term projections derived from predictive analytics assist in planning and developing infrastructure to meet future tourism demand (Melnychenko, et al., 2020). By analyzing historical data and trends, destinations can anticipate future growth and implement infrastructure development strategies for accommodation, transportation, and waste management. This enables proactive planning and resource allocation to ensure adequate capacity and mitigate potential issues related to overtourism.

This section presents a more detailed overview of the technological landscape for developing and deploying AI-driven predictive analytics systems in tourism. By focusing on specific tools, programming languages, and data sources, it provides a clearer picture of how these systems function. This analysis provides a deeper understanding of the capabilities of predictive analytics in tourism, highlighting its practical applications and its role in achieving sustainable and responsible tourism.

2.2. AI-Driven Crowd Management for Smoother Experiences

Managing large crowds effectively is paramount for tourism destinations, particularly during peak seasons or major events. Congestion can lead to negative visitor experiences, including long queues, delays, and increased stress, potentially impacting satisfaction and even deterring future visits. AI-driven crowd management systems offer a powerful solution to optimize infrastructure and enhance the overall visitor experience by leveraging real-time data and predictive analytics. These systems employ computer vision algorithms, often powered by Convolutional Neural Networks (CNNs), to analyze data from CCTV cameras and other sensor networks, estimating crowd density, movement patterns, and identifying potential congestion hotspots (Singh, V., et al. 2020). Additionally, IoT sensors deployed throughout a destination capture data on pedestrian traffic, environmental conditions, and other relevant factors, providing a comprehensive view of crowd behavior and infrastructure utilization (Al-Nabhan, N., et al. 2021). Social media analytics, utilizing Natural Language Processing (NLP) techniques (Fanni, S. C. et al., 2023), further contribute by gauging crowd sentiment and identifying emerging issues from real-time social media posts and comments (Saquete, E., et al. 2020).

By analyzing historical data and integrating real-time sensor data, AI models can predict crowd behavior and potential congestion points. Techniques like agent-based modeling simulate individual tourist behavior within a virtual environment to anticipate crowd flow and identify potential bottlenecks. Kalman filtering algorithms track and predict the movement of individuals or groups, providing real-time estimates of crowd density and movement patterns (Li, B., et al. 2018). Furthermore, reinforcement learning (RL) algorithms dynamically adjust factors like traffic light timings, pedestrian routing, and resource allocation in response to changing crowd conditions, optimizing resource utilization, and minimizing congestion through iterative learning and optimization.

AI-driven wayfinding systems are revolutionizing the tourist experience by providing dynamic and personalized guidance to navigate crowded areas and optimize resource utilization. By leveraging sophisticated algorithms based on graph theory and network analysis (Zhang, et al., 2017), these systems offer alternative routes, real-time congestion information, and personalized recommendations, effectively guiding travelers towards less-visited areas and contributing to a more sustainable and enjoyable travel experience. This evolution in wayfinding involves the integration of crowd sensing technology to provide real-time data on visitor density, enabling travelers to make informed choices and avoid peak hours. Moreover, AI-powered predictive modeling techniques analyze historical data, including visitor arrival patterns, weather conditions, and special events, to anticipate demand and optimize resource allocation for services like staffing, cleaning, and waste management. This ensures that resources are available at the right time and place, mitigating the challenges associated with overtourism and improving operational efficiency.

The implementation of AI in wayfinding and resource management is evident in leading tourist destinations like Changi Airport in Singapore and the city of Barcelona. At Changi Airport, AI-powered wayfinding systems provide travelers with real-time information on gate locations, baggage claim areas, and transit routes, leading to improved passenger flow and reduced wait times. Similarly, in Barcelona, AI is being used to manage crowds and optimize resource allocation, resulting in improved efficiency in cleaning, security, and public transportation services. This proactive approach has led to a more sustainable and enjoyable tourist experience. The use of smart tourist apps that integrate AI-driven wayfinding features also plays a vital role in providing personalized travel guidance and enabling travelers to explore destinations more effectively. These apps leverage deep learning and natural language processing to offer alternative routes, real-time updates, and suggestions for less-crowded areas.

The impact of AI-driven wayfinding and resource optimization extends beyond improving the traveler experience. By minimizing congestion, reducing reliance on private transportation, and guiding travelers towards less-visited areas, AI can contribute to a more sustainable tourism industry. It also provides valuable insights into traveler behavior, allowing destination managers to optimize resources, make informed decisions, and create a more welcoming and enjoyable experience for visitors. The integration of AI in tourism has the potential to drive economic growth and sustainability, making travel experiences more enjoyable for all. This innovative approach to wayfinding and resource management, fueled by the power of deep learning, natural language processing, and crowd sensing, represents a crucial step towards creating a more vibrant and responsible future for the tourism sector.

2.3. Environmental Practices Enabled by AI-Powered Sensors and Drones

It tries to explain the role of sensors and drones based on Artificial Intelligence in revolutionizing environmental stewardship in the tourism industry through higher-order information/decision-making processes relevant to sustainable tourism development. It is fitted with high-resolution cameras, multispectral sensors, LiDAR technology (Roriz, R., et al. 2021), and hyperspectral cameras, thus having the capability of acquiring highly detailed aerial imagery for information on the health and dynamics of large landscapes. On the other, it feeds AI algorithms, in particular deep learning models such as CNNs. It automatically detects feature objects within imagery—for example, trees, buildings, or wildlife—and classifies them to deliver monitoring of deforestation, illegal construction, or changes in biodiversity. The semantic segmentation of land by CNNs (Hassan, E., 2024) contributes to the analysis of changes in land cover and the detection of places under environmental stress.

These AI change detection algorithms will keep observing sites with time, detecting altered vegetation health, water quality, or coastal erosion to provide insight on its impacts due to tourist activity. In this line, an integrated network of IoT sensors shall observe and collect data regarding air and water quality, noise pollution, and waste generation in real-time at destinations to act accordingly with environmental mitigation measures. This will be intelligent bins with sensors that can help in the routing of waste collection to optimize the process for better management of waste. Machine-learning-based predictive modeling methodologies applied to this rich dataset will be able to provide better forecasts regarding the potential environmental bottlenecks and inform effective implementation of prevention strategies, which improve resource allocation in waste management, water conservation, and energy efficiency (Zhong, S., et al. 2021). The platform can relate tourist behaviors and appraisals to have interpretations regarding possible directions for sustainability efforts that could trigger more environmentally sensitive travel choices, commensurate with the growth of interest in sustainable tourism. For instance, Booking.com estimated recently that 70% of all travelers would opt for accommodation which applies a positive environmental policy. Integration of AI-driven monitoring and management systems will cement this three-dimensional shift toward sustainability with minimal ecological impact, essentially preserving natural resources to the maximum for future generations. In such a manner, the change will be three-dimensional—one that makes good use of the strategic policy initiatives designed with artificial intelligence insights, one that is actively engaged with stakeholders, and one that continues to further the environmental sustainability agenda of the tourism sector in the long run (Birendra, K. C., et al. 2021).

Figure 3. The different data sources using sensors

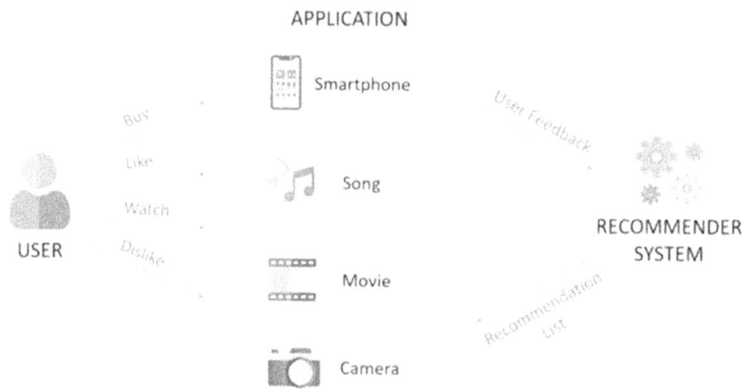

3. ETHICAL CONSIDERATIONS AND SHAPING THE FUTURE OF AI-DRIVEN TOURISM

3.1. Data Privacy in Tourism Platforms

The burgeoning use of AI in the tourism industry necessitates a deep dive into the ethical considerations surrounding data privacy (Gong, Y. and Ashley, S. 2022). With the increasing collection and analysis of traveler data, ensuring responsible and ethical data handling becomes paramount to building trust and fostering a sustainable future for smart tourism.

The Risks: AI-driven platforms often gather a wealth of personal data, including explicit and implicit preferences, travel history, and even online behavior. This information, if not handled responsibly, can expose individuals to privacy risks. Imagine a scenario where a traveler's browsing history on a travel website is used to profile their interests and suggest destinations based on these preferences. While this may seem beneficial for personalization, it can also raise concerns about the potential for misuse of this data, including targeted advertising, discriminatory practices, and even financial fraud.

Exploring the Challenges: Beyond these general risks, AI-driven tourism systems pose specific challenges related to data privacy:

- **Data Aggregation and Profiling:** AI platforms often combine data from various sources, creating comprehensive profiles of individuals that can be used for tracking their travel patterns and preferences over time. This can lead to a loss of anonymity and raise concerns about surveillance and the potential for misuse of this information.
- **Data Retention and Disposal:** The length of time that platforms retain user data is a crucial factor in safeguarding privacy. Storing data for extended periods without clear purpose and proper safeguards can pose a significant risk, especially in light of the potential for data breaches or unauthorized access.
- **Data Sharing and Cross-border Transfers:** AI-driven tourism often involves data sharing between different platforms and countries. This raises complex legal and ethical challenges related to international data transfer regulations and the need for strong privacy protections across borders.
- **Algorithmic Bias and Discrimination:** The algorithms that power AI systems can be influenced by existing biases in the data they are trained on. This can result in unfair recommendations, discriminatory practices, and unequal access to tourism resources based on factors like race, religion, or socio-economic status.

Safeguarding Privacy through Data Handling Practices: To mitigate these risks, a multi-faceted approach to data privacy is necessary:

- **Data Minimization:** This fundamental principle emphasizes collecting only the necessary data for the intended purpose. Travel platforms should clearly define their data collection policies, specify the type of data collected, and minimize the amount of data stored.
- **Data Anonymization and Pseudonymization:** To ensure individual privacy, AI systems can employ techniques to mask personal identifiers, such as replacing names with pseudonyms or removing personally identifiable information (PII). However, it is crucial to ensure that these techniques do not create new identifiers or inadvertently facilitate re-identification.
- **Strong Encryption:** Data transmission and storage should be protected by robust encryption methods to prevent unauthorized access and data breaches. This requires the use of strong encryption algorithms, regular security updates, and secure data handling practices.
- **Access Control and Data Security:** Access to user data should be strictly controlled. Only authorized personnel should have access to sensitive data, and stringent security measures should be in place to prevent unauthorized access, disclosure, or manipulation.

Real-world Examples and Analysis:

- **The Use of AI-Powered Recommendation Engines:** Platforms like Booking.com and Expedia (Londhe, et al., 2024) use AI to generate customized travel itineraries. These platforms collect data on traveler preferences, booking history, and social media activity. They then use this data to provide personalized recommendations for accommodations, activities, and destinations. This type of personalization requires careful consideration of data privacy and the potential for profiling.
  - **Analysis:** While AI-powered recommendations can enhance the traveler experience, they raise questions about the extent to which platforms can analyze and use data for profiling and targeting individuals.
- **The Role of Facial Recognition Technology:** Facial recognition technology can be used in tourism to enhance security, provide personalized services, and improve access to information. However, it can also raise significant privacy concerns, as it involves collecting and analyzing facial images. This technology must be carefully implemented and deployed with user consent and proper safeguards to avoid unwarranted surveillance.

- o **Analysis:** Facial recognition presents complex ethical issues that require careful balancing of potential benefits with individual privacy rights.
- **The Use of Location Data for Personalized Experiences:** AI can use location data to personalize travel experiences by providing real-time recommendations based on the traveler's current location. For instance, AI can recommend nearby restaurants or attractions based on the traveler's preferences and location. This can be highly convenient but raises concerns about privacy.
  - o **Analysis:** Location data can offer valuable insights for enhancing the travel experience. However, it necessitates strong data privacy protections to ensure that it is used responsibly.

Data privacy regulations like the General Data Protection Regulation (GDPR) play a crucial role in protecting user privacy in the tourism sector. These regulations establish a set of standards for the collection, use, and storage of personal data. However, as AI technology evolves, there is a need for continued dialogue and collaboration between researchers, practitioners, and regulators to develop best practices and ensure the responsible use of data in AI-driven tourism. The financial and reputational consequences of data breaches, as evidenced by IBM's Data Breach Report, highlight the importance of placing data privacy at the forefront of AI development in the tourism sector.

By prioritizing data privacy, security, and compliance with relevant regulations, the tourism industry can establish trust with users, ensure ethical practices, and foster a more sustainable and impactful future for smart tourism.

This section explores a wide range of challenges and considerations related to data privacy in the context of AI-driven tourism. It analyzes the risks associated with data aggregation and profiling, the potential for algorithmic bias, and the importance of data minimization and encryption. It further underscores the significance of relevant legal frameworks like the GDPR and emphasizes the ongoing need for ethical considerations in shaping the future of AI in tourism. By acknowledging these issues, this exploration highlights the need for continuous dialogue and collaboration between researchers, practitioners, and policymakers to ensure the responsible and ethical use of AI in tourism.

3.2. AI to Ensure Equitable Tourism Experiences

The burgeoning field of AI in tourism presents a complex interplay between technological advancement and societal impact. While offering immense potential to enhance travel experiences, AI systems risk amplifying existing societal biases if not carefully designed and implemented. This section explores the multifaceted nature of bias in AI-driven tourism, highlighting the challenges, innovative solutions, and the need for ongoing vigilance to foster a truly equitable and inclusive tourism landscape.

AI systems learn from the data they are trained on, and if this data reflects existing societal biases, these biases can be unknowingly propagated in the AI's outputs. This phenomenon, known as algorithmic bias, poses significant ethical challenges in tourism. For instance, consider an AI-powered hotel reservation system trained on historical data that reveals a preference for luxury hotels among affluent travelers. Without careful design and intervention, this system may disproportionately recommend luxury accommodations to users with similar profiles, potentially excluding travelers with different budgets and needs. This can lead to reinforcement of socioeconomic inequalities and the exclusion of diverse traveler segments.

- **Fairness in Destination Recommendations:** AI-powered recommendation systems that analyze past travel patterns and user preferences could unintentionally perpetuate historical biases in the selection of destinations. If the data reflects a historical preference for certain destinations among specific demographics, the AI might perpetuate this bias in its recommendations. This could lead to the exclusion of lesser-known destinations or experiences that offer unique opportunities for cultural exchange and diversity.
- **Bias in Pricing and Access:** AI algorithms employed in pricing models and access control systems can be vulnerable to biases that stem from historical data (Schwartz, et al., 2022). For example, if historical pricing data reflects discriminatory practices based on race, gender, or socio-economic status, these biases might be incorporated into the AI model's pricing and access recommendations, leading to unequal distribution of resources and limited access to tourism experiences for marginalized groups.
- **Automated Customer Service:** AI-powered chatbots used for customer service in tourism can inadvertently reinforce existing biases. If the chatbot is trained on data reflecting stereotypes or prejudices, it might exhibit unfair or discriminatory behavior in its interactions with travelers. This could result in a negative customer experience and further marginalization of specific user groups.

To combat the pitfalls of algorithmic bias and ensure fairness in AI-driven tourism, researchers and practitioners are exploring a range of innovative approaches:

- **Fairness-Aware Machine Learning:** This field of study investigates techniques for designing algorithms that are inherently fair and minimize the impact of bias. These methods can incorporate fairness constraints into the learning process, encouraging equitable outcomes by penalizing biases and promoting fairness across different demographics (Le Quy, et al. 2022).

- **Adversarial Debiasing Techniques:** These innovative methods leverage adversarial training to detect and mitigate biases by introducing opposing forces to counter existing patterns and encourage fair representation in the model's outputs. This technique can be applied to various AI tasks in tourism, such as recommendation systems and pricing algorithms, to ensure fairness in outcomes.
- **Explainable AI (XAI) for Bias Detection:** XAI tools provide insights into the decision-making processes behind AI algorithms, enabling a more transparent understanding of how biases may be influencing recommendations and other outputs. By analyzing the factors that influence the model's predictions (Arrieta, A. B., et al. 2020), XAI helps to identify and address discriminatory practices. This is particularly relevant in the tourism sector, where AI plays a critical role in personalizing travel experiences.
- **Diversity in Data and Representation:** Training AI models on diverse and representative datasets is crucial to minimizing bias. By incorporating data from various communities, socioeconomic backgrounds, and cultural perspectives, these models can learn more accurately and provide fairer recommendations. This requires intentional efforts to collect data that reflects the real-world complexity of travel experiences.

The ethical implications of AI in tourism extend beyond mitigating bias. Key considerations include:

- **Transparency and Accountability:** The complex workings of AI algorithms necessitate greater transparency to ensure accountability and encourage responsible use. This means clearly explaining how algorithms make decisions, identifying potential biases, and allowing for user input and oversight.
- **Data Privacy and Security:** Ensuring robust data privacy protections is paramount. This involves implementing secure data handling practices, obtaining informed consent, and adhering to relevant regulations like the GDPR (Ke, T. T., 2023).
- **Job Displacement and Workforce Transition:** As AI automates certain tasks in tourism, there's a need to proactively address potential job displacement and support workforce transition. This includes training programs, retraining initiatives, and the exploration of new job roles in the AI-powered tourism ecosystem.

3.3. Responsible AI Development

In any case, responsible AI development within the tourism sector needs to carry a wider set of ethical and social connotations regarding the use of such powerful tools. It is important to ensure that the techniques of model interpretability are being applied to the development and deployment processes of AI models so that transparency enables users to obtain insights into how AI-driven recommendations and services are made. For instance, an AI-powered hotel recommendation engine could recommend a hotel based on factors like past booking behavior and proximity to a set of favored destinations and budget. Such a move would go a long way in ushering transparency and trust in the deal. Further, this confirms findings from Accenture, which pointed out that 83% of consumers opined organizations should be transparent concerning their use of AI. The process should have incorporated principles of human-centered design (Dimitriadis, et al., 2021), wherein development mainly focused on addressing human needs and values and further should mitigate potential risks and biases, with techniques such as fairness-aware machine learning. Ethics and consideration towards the society's well-being mean that continuous assessment and monitoring frameworks must be applied continuously, putting into consideration unforeseen outcomes or bias developing within the AI system. This includes adversarial training and identification of model drift. Collaboration among AI researchers, developers, policy thinkers, and others would be essential in coming up with holistic ethical guidelines and best practices about how to operate AI in tourism; there should be coherence in understanding responsible deployment of AI in terms of development and application. Clear accountability frameworks regarding the operation of AI in tourism foster trust, transparency, and responsible innovation.

Strong ethics and regulations, directed by the FAIR framework in principles of fairness, accountability, and transparency, are meant to ensure that AI technologies have a positive contribution to the sustainability, inclusivity, and equity of different categories of stakeholders within the tourism industry (Dangi, T. B., and Petrick, J. F. 2021). Companies that focus on ethical AI development, including taking such drastic steps as federated learning to achieve privacy-centric data sharing, obviously gain trust from customers and stakeholders much better. In turn, long-term sustainability will have better reputations. An Edelman study concluded that 80% of its respondents agreed that companies have a responsibility to use technology ethically. It embraces the responsible culture of AI development and deployment in the tourism sector to unleash transformational powers in protection of ethical and societal values within the prosperity framework or the sustainable tourism framework. Such foresight shall ensure that AI remains the harbinger of positive change to contribute toward an innovative, efficient, fair, inclusive, and environmentally conscious future in travel and tourism.

CONCLUSION

The journey through the realm of AI-powered smart tourism has unveiled a future brimming with possibilities. This chapter has explored the transformative power of AI to reshape the tourism sector, emphasizing the interplay between technological advancements and ethical considerations. We have witnessed how AI is revolutionizing travel experiences, offering personalized itineraries, optimized resource allocation, and a deeper understanding of traveler needs.

From personalized travel plans tailored to individual preferences and immersive AR/VR experiences that transport travelers to distant lands to sophisticated demand forecasting and intelligent crowd management systems that ensure smoother and more sustainable journeys (Rane, N., et al. 2023), AI is proving to be a powerful tool in enhancing the tourism experience. The applications of AI extend beyond the digital realm, influencing environmental monitoring and sustainable practices, further supporting informed decision-making and promoting responsible tourism development.

However, this rapid evolution necessitates a vigilant approach to ethical considerations. The increasing reliance on data privacy, the potential for algorithmic bias, and the need for responsible development and deployment of AI technologies raise crucial concerns. Ensuring user privacy, mitigating bias in algorithms, and promoting transparency are essential for building trust and fostering a sustainable future for smart tourism.

A human-centered approach, prioritizing the needs and well-being of travelers, is paramount. This requires transparency in how AI systems make decisions, empowering travelers with control over their data, and ensuring that AI-driven solutions promote equity and inclusivity. It emphasizes a commitment to ethical AI practices, safeguarding against potential risks and maximizing the benefits of this technology for all stakeholders.

In conclusion, the future of AI-powered smart tourism rests on a commitment to responsible innovation. By prioritizing ethical considerations, embracing a human-centered approach, and fostering collaboration between researchers, practitioners, and policymakers, the tourism industry can unlock the full potential of AI to deliver a travel experience that is personalized, efficient, sustainable, and ethical. This chapter underscores the importance of ongoing dialogue, research, and development to ensure that AI continues to enhance the journey, creating a more vibrant, inclusive, and sustainable tourism landscape for all. The future of travel, in its most impactful form, rests on embracing the power of AI while prioritizing ethical considerations and human values.

REFERENCES

Abril-Pla, O., Andreani, V., Carroll, C., Dong, L., Fonnesbeck, C. J., Kochurov, M., Kumar, R., Lao, J., Luhmann, C. C., Martin, O. A., & Zinkov, R. (2023). PyMC: A modern, and comprehensive probabilistic programming framework in Python. *PeerJ. Computer Science*, *9*, e1516. DOI: 10.7717/peerj-cs.1516 PMID: 37705656

Abro, W. A., Aicher, A., Rach, N., Ultes, S., Minker, W., & Qi, G. (2022). Natural language understanding for argumentative dialogue systems in the opinion building domain. *Knowledge-Based Systems*, *242*, 108318. DOI: 10.1016/j.knosys.2022.108318

Ahmad, M., Naeem, M. K. H., Mobo, F. D., Tahir, M. W., & Akram, M. (2024). Navigating the Journey: How Chatbots Are Transforming Tourism and Hospitality. In Design and Development of Emerging Chatbot Technology (pp. 236-255). IGI Global. https://doi.org/DOI: 10.4018/979-8-3693-1830-0.ch014

Al-Nabhan, N., Alenazi, S., Alquwaifili, S., Alzamzami, S., Altwayan, L., Alaloula, N., Alowaini, R., & Al Islam, A. A. (2021). An intelligent IoT approach for analyzing and managing crowds. *IEEE Access : Practical Innovations, Open Solutions*, *9*, 104874–104886. DOI: 10.1109/ACCESS.2021.3099531

Alsahafi, R., Alzahrani, A., & Mehmood, R. (2023). Smarter sustainable tourism: Data-driven multi-perspective parameter discovery for autonomous design and operations. *Sustainability (Basel)*, *15*(5), 4166. DOI: 10.3390/su15054166

Álvarez-Carmona, M. Á., Aranda, R., Rodríguez-Gonzalez, A. Y., Fajardo-Delgado, D., Sánchez, M. G., Pérez-Espinosa, H., & Díaz-Pacheco, Á. (2022). Natural language processing applied to tourism research: A systematic review and future research directions. *Journal of King Saud University. Computer and Information Sciences*, *34*(10), 10125–10144. DOI: 10.1016/j.jksuci.2022.10.010

Alyasiri, O. M., Selvaraj, K., Younis, H. A., Sahib, T. M., Almasoodi, M. F., & Hayder, I. M. (2024). A Survey on the Potential of Artificial Intelligence Tools in Tourism Information Services. *Babylonian Journal of Artificial Intelligence*, *2024*, 1–8. DOI: 10.58496/BJAI/2024/001

Arrieta, A. B., Díaz-Rodríguez, N., Del Ser, J., Bennetot, A., Tabik, S., Barbado, A., García, S., & Herrera, F. (2020). Explainable Artificial Intelligence (XAI): Concepts, taxonomies, opportunities and challenges toward responsible AI. *Information Fusion*, *58*, 82–115. DOI: 10.1016/j.inffus.2019.12.012

Badouch, M., & Boutaounte, M. (2024). Design and Implementation of a Hotel Recommendation System Using Deep Learning. In *AI and Data Engineering Solutions for Effective Marketing* (pp. 388–408). IGI Global., DOI: 10.4018/979-8-3693-3172-9.ch019

Birendra, K. C., Dhungana, A., & Dangi, T. B. (2021). Tourism and the sustainable development goals: Stakeholders' perspectives from Nepal. *Tourism Management Perspectives*, *38*, 100822. DOI: 10.1016/j.tmp.2021.100822

Bokolo, A. J. (2021). Application of telemedicine and eHealth technology for clinical services in response to COVID-19 pandemic. *Health and Technology*, *11*(2), 359–366. DOI: 10.1007/s12553-020-00516-4 PMID: 33469474

BR. S. R. (2022). Information and Communication Technology Application in the Asian Tourism Industry. In Handbook of Technology Application in Tourism in Asia (pp. 19-48). Singapore: Springer Nature Singapore.

Bretos, M. A., Ibáñez-Sánchez, S., & Orús, C. (2024). Applying virtual reality and augmented reality to the tourism experience: A comparative literature review. *Spanish Journal of Marketing - ESIC*, *28*(3), 287–309. DOI: 10.1108/SJME-03-2023-0052

Dangi, T. B., & Petrick, J. F. (2021). Augmenting the role of tourism governance in addressing destination justice, ethics, and equity for sustainable community-based tourism. *Tourism and Hospitality*, *2*(1), 15–42. DOI: 10.3390/tourhosp2010002

Dimitriadis, Y., Martínez-Maldonado, R., & Wiley, K. (2021). Human-centered design principles for actionable learning analytics. Research on E-learning and ICT in education: Technological, pedagogical and instructional perspectives, 277-296. https://doi.org/DOI: 10.1007/978-3-030-64363-8_15

Elahi, M., Abdollahpouri, H., Mansoury, M., & Torkamaan, H. (2021, June). Beyond algorithmic fairness in recommender systems. In *Adjunct proceedings of the 29th ACM conference on user modeling, adaptation and personalization* (pp. 41-46). https://doi.org/DOI: 10.1145/3450614.3461685

Emara, T. Z., Trinh, T., & Huang, J. Z. (2023). Geographically distributed data management to support large-scale data analysis. *Scientific Reports*, *13*(1), 17783. DOI: 10.1038/s41598-023-44789-x PMID: 37853092

Fanni, S. C., Febi, M., Aghakhanyan, G., & Neri, E. (2023). Natural language processing. In *Introduction to Artificial Intelligence* (pp. 87–99). Springer International Publishing., DOI: 10.1007/978-3-031-25928-9_5

Fitzgibbon, A., & Zisserman, A. (1998, September). Automatic 3D model acquisition and generation of new images from video sequences. In *9th European Signal Processing Conference (EUSIPCO 1998)* (pp. 1-8). IEEE.

Garlapati, A., Krishna, D. R., Garlapati, K., Rahul, U., & Narayanan, G. (2021, April). Stock price prediction using Facebook Prophet and Arima models. In 2021 6th International Conference for Convergence in Technology (I2CT) (pp. 1-7). IEEE. https://doi.org/DOI: 10.1109/I2CT51068.2021.9418057

Gong, Y., & Schroeder, A. (2022). A systematic literature review of data privacy and security research on smart tourism. *Tourism Management Perspectives*, *44*, 101019. DOI: 10.1016/j.tmp.2022.101019

Hassan, E. (2024). Enhancing coffee bean classification: A comparative analysis of pre-trained deep learning models. *Neural Computing & Applications*, *36*(16), 9023–9052. DOI: 10.1007/s00521-024-09623-z

Indaryanto, A., Harijadi, B. D., & Sinaga, E. (2023). The growing use and impact of artificial intelligence technologies in the tourism industry. *Sustainable Engineering and Innovation*, *5*(2), 189–204. DOI: 10.37868/sei.v5i2.id238

Ke, T. T., & Sudhir, K. (2023). Privacy rights and data security: GDPR and personal data markets. *Management Science*, *69*(8), 4389–4412. DOI: 10.1287/mnsc.2022.4614

Kholil, M., Ismanto, I., & Fu'Ad, M. N. (2021, February). 3D reconstruction using structure from motion (SFM) algorithm and multi view stereo (MVS) based on computer vision. In *IOP Conference Series: Materials Science and Engineering* (Vol. 1073, No. 1, p. 012066). IOP Publishing. https://doi.org/DOI: 10.1088/1757-899X/1073/1/012066

Le Quy, T., Roy, A., Iosifidis, V., Zhang, W., & Ntoutsi, E. (2022). A survey on datasets for fairness-aware machine learning. *Wiley Interdisciplinary Reviews. Data Mining and Knowledge Discovery*, *12*(3), e1452. DOI: 10.1002/widm.1452

Li, B., Han, X., & Wu, D. (2018, January). Real-time crowd density estimation based on convolutional neural networks. In *2018 International Conference on Intelligent Transportation, Big Data & Smart City (ICITBS)* (pp. 690-694). IEEE. https://doi.org/DOI: 10.1109/ICITBS.2018.00179

Liu, X., Mehraliyev, F., Liu, C., & Schuckert, M. (2020). The roles of social media in tourists' choices of travel components. *Tourist Studies*, *20*(1), 27–48. DOI: 10.1177/1468797619873107

Londhe, K., Dharmadhikari, N., Zaveri, P., & Sakoglu, U. (2024). Enhanced Travel Experience using Artificial Intelligence: A Data-driven Approach. *Procedia Computer Science*, *235*, 1920–1928. DOI: 10.1016/j.procs.2024.04.182

Loureiro, S. M. C., & Nascimento, J. (2021). Shaping a view on the influence of technologies on sustainable tourism. *Sustainability (Basel)*, *13*(22), 12691. DOI: 10.3390/su132212691

Lukita, C., Pangilinan, G. A., Chakim, M. H. R., & Saputra, D. B. (2023). Examining the impact of artificial intelligence and internet of things on smart tourism destinations: A comprehensive study. [ATT]. *Aptisi Transactions on Technopreneurship*, *5*(2sp), 135–145. DOI: 10.34306/att.v5i2sp.332

Mbunge, E., & Muchemwa, B. (2022). Deep learning and machine learning techniques for analyzing travelers' online reviews: a review. Optimizing digital solutions for hyper-personalization in tourism and hospitality, 20-39. https://doi.org/DOI: 10.4018/978-1-7998-8306-7.ch002

Melnychenko, S., Boiko, M., Okhrimenko, A., Bosovska, M., & Mazaraki, N. (2020). Foresight technologies of economic systems: Evidence from the tourism sector of Ukraine. *Problems and Perspectives in Management*, *18*(4), 303–318. DOI: 10.21511/ppm.18(4).2020.25

Naseer, I. (2023). AWS Cloud Computing Solutions: Optimizing Implementation for Businesses. Statistics. *Computing and Interdisciplinary Research*, *5*(2), 121–132. DOI: 10.52700/scir.v5i2.138

Rane, N., Choudhary, S., & Rane, J. (2023). Sustainable tourism development using leading-edge Artificial Intelligence (AI), Blockchain, Internet of Things (IoT), Augmented Reality (AR) and Virtual Reality (VR) technologies. Blockchain, Internet of Things (IoT), Augmented Reality (AR) and Virtual Reality (VR) technologies (October 31, 2023). http://dx.doi.org/DOI: 10.2139/ssrn.4642605

Richard, B. (2017). Hotel chains: Survival strategies for a dynamic future. *Journal of Tourism Futures*, *3*(1), 56–65. DOI: 10.1108/JTF-06-2016-0018

Roriz, R., Cabral, J., & Gomes, T. (2021). Automotive LiDAR Technology: A Survey. *IEEE Transactions on Intelligent Transportation Systems*, *23*(7), 6282–6297. DOI: 10.1109/TITS.2021.3086804

Saquete, E., Tomás, D., Moreda, P., Martínez-Barco, P., & Palomar, M. (2020). Fighting post-truth using natural language processing: A review and open challenges. *Expert Systems with Applications*, *141*, 112943. DOI: 10.1016/j.eswa.2019.112943

Schwartz, R., Schwartz, R., Vassilev, A., Greene, K., Perine, L., Burt, A., & Hall, P. (2022). Towards a standard for identifying and managing bias in artificial intelligence (Vol. 3, p. 00). US Department of Commerce, National Institute of Standards and Technology. https://doi.org/DOI: 10.6028/NIST.SP.1270

Sharma, I., & Aggarwal, A. (2024). Revolutionizing Creative Tourism: Integrating AI in Marketing for Immersive Travel Experiences. In Manohar, S., Mittal, A., Raju, S., & Nair, A. (Eds.), *Innovative Technologies for Increasing Service Productivity* (pp. 179–197). IGI Global., DOI: 10.4018/979-8-3693-2019-8.ch011

Sharma, M., Mahajan, Y., & Saber, A. (2024). Using Augmented Reality (AR) and the Internet of Things (IoT) to Improve Water Management Maintenance and Training. In Innovations in Machine Learning and IoT for Water Management (pp. 252-270). IGI Global. https://doi.org/DOI: 10.4018/979-8-3693-1194-3.ch013

Shukla, V., Rana, S., & Prashar, S. (2024). Examining the potential of virtual and augmented reality in enhancing tourism experiences. *The Bottom Line (New York, N.Y.)*. Advance online publication. DOI: 10.1108/BL-05-2023-0139

Singh, V., Singh, S., & Gupta, P. (2020). Real-time anomaly recognition through CCTV using neural networks. *Procedia Computer Science*, *173*, 254–263. DOI: 10.1016/j.procs.2020.06.030

Trinh, T., Nguyen, N.-T., Wu, D., Huang, J. Z., & Emara, T. Z. (2019). A new location-based topic model for event attendees recommendation. *2019 IEEE-RIVF International Conference on Computing and Communication Technologies (RIVF)*, 1–6. IEEE. https://doi.org/DOI: 10.1109/RIVF.2019.8713716

Vanegas, J., Valencia, M., Restrepo, J., & Muñeton, G. (2020). Modeling determinants of tourism demand in Colombia. *Tourism and Hospitality Management*, *26*(1), 49–67. DOI: 10.20867/thm.26.1.4

Yao, L., Ma, R., & Wang, H. (2021). Baidu index-based forecast of daily tourist arrivals through rescaled range analysis, support vector regression, and autoregressive integrated moving average. *Alexandria Engineering Journal*, *60*(1), 365–372. DOI: 10.1016/j.aej.2020.08.037

Zhang, W., Chien, J., Yong, J., & Kuang, R. (2017). Network-based machine learning and graph theory algorithms for precision oncology. *NPJ Precision Oncology*, *1*(1), 25. DOI: 10.1038/s41698-017-0029-7 PMID: 29872707

Zhong, S., Zhang, K., Bagheri, M., Burken, J. G., Gu, A., Li, B., Ma, X., Marrone, B. L., Ren, Z. J., Schrier, J., Shi, W., Tan, H., Wang, T., Wang, X., Wong, B. M., Xiao, X., Yu, X., Zhu, J.-J., & Zhang, H. (2021). Machine learning: New ideas and tools in environmental science and engineering. *Environmental Science & Technology*, *55*(19), 12741–12754. DOI: 10.1021/acs.est.1c01339 PMID: 34403250

Compilation of References

Aagaard, A., & Tucci, C. (2024). AI-Driven Business Model Innovation: Pioneering New Frontiers in Value Creation. In *Business Model Innovation: Game Changers and Contemporary Issues* (pp. 295–328). Springer International Publishing. DOI: 10.1007/978-3-031-57511-2_10

Abade, A., Ferreira, P. A., & De Barros Vidal, F. (2021). Plant diseases recognition on images using convolutional neural networks: A systematic review. *Computers and Electronics in Agriculture*, *185*, 106125. DOI: 10.1016/j.compag.2021.106125

Abdullah Al, W., Cha, W., & Yun, I. D. (2022). Reinforcement Learning Based Vocal Fold Localization in Preoperative Neck CT for Injection Laryngoplasty. *Applied Sciences (Basel, Switzerland)*, *13*(1), 262. DOI: 10.3390/app13010262

Abhilasha, A., Sreenivasulu, A., & Manimozhi, T. KUMAR, P. S., Selvakumar, P., & Singh, P. (2022, April). The Model of Smart Sensing Device For Sensitive Nano-clusters Modification in Sensing Properties. In 2022 2nd International Conference on Advance Computing and Innovative Technologies in Engineering (ICACITE) (pp. 1043-1047). IEEE.

Abril-Pla, O., Andreani, V., Carroll, C., Dong, L., Fonnesbeck, C. J., Kochurov, M., Kumar, R., Lao, J., Luhmann, C. C., Martin, O. A., & Zinkov, R. (2023). PyMC: A modern, and comprehensive probabilistic programming framework in Python. *PeerJ. Computer Science*, *9*, e1516. DOI: 10.7717/peerj-cs.1516 PMID: 37705656

Abro, W. A., Aicher, A., Rach, N., Ultes, S., Minker, W., & Qi, G. (2022). Natural language understanding for argumentative dialogue systems in the opinion building domain. *Knowledge-Based Systems*, *242*, 108318. DOI: 10.1016/j.knosys.2022.108318

Abu-Taieh, E. H., Al Hadid, I., & Zolait, A. (2020). *5G road map to communication revolution*. Cyberspace., DOI: 10.5772/intechopen.92319

Academy of Royal Medical Colleges. (2019) Artificial Intelligence in Healthcare. https://www.aomrc.org.uk/wpcontent/uploads/2019/01/Artificial_intelligence_in_healthcare_0119.pdf

Adewusi, A. O., Asuzu, O. F., Olorunsogo, T., Iwuanyanwu, C., Adaga, E., & Daraojimba, D. O. (2024). AI in precision agriculture: A review of technologies for sustainable farming practices. *World Journal of Advanced Research and Reviews*, *21*(1), 2276–2285. DOI: 10.30574/wjarr.2024.21.1.0314

AeoLogic. (2023, April 21). *Drone Technology in Agriculture: Benefits and Challenges for Rural India*. Retrieved from AeoLogic: https://www.aeologic.com/blog/drone-technology-in-agriculture-benefits-and-challenges-for-rural-india/

Aghalya, S., Sarasu, R., Mishra, N., Arunachalam, G., Gobinath, D., & Naganathan, S. B. T. (2024, April). Tele-Rheumatology-Advanced Neural Network Model for Remote Assessment and Management of Rheumatic Conditions. *In 2024 10th International Conference on Communication and Signal Processing* (ICCSP) (pp. 453-458). IEEE.

Agrawal, K., & Nargund, N. (2024, January). Deep Learning in Industry 4.0: Transforming Manufacturing Through Data-Driven Innovation. In *International Conference on Distributed Computing and Intelligent Technology* (pp. 222-236). Cham: Springer Nature Switzerland. DOI: 10.1007/978-3-031-50583-6_15

Ahmad, M. (2024, June 11). Top 6 applications of AI in aquaculture: 2023. PanaceaLogics -. https://panacealogics.com/blogs/top-6-applications-of-ai-in-aquaculture-2023/

Ahmad, M., Naeem, M. K. H., Mobo, F. D., Tahir, M. W., & Akram, M. (2024). Navigating the Journey: How Chatbots Are Transforming Tourism and Hospitality. In Design and Development of Emerging Chatbot Technology (pp. 236-255). IGI Global. https://doi.org/DOI: 10.4018/979-8-3693-1830-0.ch014

Ahmadi, A. (2023). The AI Food Revolution: Reshaping Food Sciences through Artificial Intelligence. [IJBLS]. *International Journal of BioLife Sciences*, *2*(1), 62–71.

Ahmed, A. A. M., Deo, R. C., Raj, N., Ghahramani, A., Feng, Q., Yin, Z., & Yang, L. (2021). Deep Learning Forecasts of Soil Moisture: Convolutional Neural Network and Gated Recurrent Unit Models Coupled with Satellite-Derived MODIS, Observations and Synoptic-Scale Climate Index Data. *Remote Sensing (Basel)*, *13*(4), 554. DOI: 10.3390/rs13040554

Ahmed, A. S., Keshk, A. E. M., Abo-Seida, O., & Sakr, M. (2022). Tumor detection and classification in breast mammography based on fine-tuned convolutional neural networks. *IJCI.International Journal of Computers and Information*, *9*(1), 74–84.

AI in fish growth and health. (2024, October 26). PanaceaLogics -. https://panacealogics.com/blogs/role-of-artificial-intelligence-ai-in-fish-growth-and-health/

Akintuyi, O. B. (2024). Adaptive AI in precision agriculture: a review: investigating the use of self-learning algorithms in optimizing farm operations based on real-time data. *Research Journal of Multidisciplinary Studies, 7*(02), 016-030.

Akintuyi, O. B. (2024). AI in agriculture: A comparative review of developments in the USA and Africa. *Research Journal of Science and Engineering, 10*(02), 060-070.

Akkem, Y., Biswas, S. K., & Varanasi, A. (2024). Streamlit-based enhancing crop recommendation systems with advanced explainable artificial intelligence for smart farming. *Neural Computing & Applications*, *36*(32), 20011–20025. DOI: 10.1007/s00521-024-10208-z

Akram, A., Jamjoom, A. A., Innab, N., Almujally, N. A., Umer, M., Alsubai, S., & Fimiani, G. (2024). SkinMarkNet: An automated approach for prediction of monkeyPox using image data augmentation with deep ensemble learning models. *Multimedia Tools and Applications*. Advance online publication. DOI: 10.1007/s11042-024-19862-w

Akter, M. S. (2024). AI for sustainability: leveraging technology to address global environmental. *Journal of Artificial Intelligence General science (JAIGS) ISSN: 3006-4023, 3*(1), 40-48.

Akter, M. S. (2024). Harnessing technology for environmental sustainability: utilizing AI to tackle global ecological challenge. *Journal of Artificial Intelligence General science (JAIGS) ISSN: 3006-4023, 2*(1), 61-70.

Akyildiz, I. F., Nie, S., Lin, S., & Chandrasekaran, M. (2016). 5G roadmap: 10 key enabling technologies. *Computer Networks*, *106*, 17–48. DOI: 10.1016/j.comnet.2016.06.010

Al Kuwaiti, A., Nazer, K., Al-Reedy, A., Al-Shehri, S., Al-Muhanna, A., Subbarayalu, A. V., Al Muhanna, D., & Al-Muhanna, F. A. (2023). A Review of the Role of Artificial Intelligence in Healthcare. *Journal of Personalized Medicine*, *13*(6), 951. DOI: 10.3390/jpm13060951 PMID: 37373940

Aldewachi, H., Al-Zidan, R. N., Conner, M. T., & Salman, M. M. (2021). High-throughput screening platforms in the discovery of novel drugs for neurodegenerative diseases. *Bioengineering, 8*(1), 3.

Aldoseri, A., Al-Khalifa, K. N., & Hamouda, A. M. (2024). AI-Powered Innovation in Digital Transformation: Key Pillars and Industry Impact. *Sustainability (Basel), 16*(5), 1790. DOI: 10.3390/su16051790

Alibabaei, K., Gaspar, P. D., Assunção, E., Alirezazadeh, S., & Lima, T. M. (2022). Irrigation optimization with a deep reinforcement learning model: Case study on a site in Portugal. *Agricultural Water Management, 263*, 107480. DOI: 10.1016/j.agwat.2022.107480

Ali, H., Muzammil, M. A., Dahiya, D. S., Ali, F., Yasin, S., Hanif, W., & Al-Haddad, M. (2024). Artificial intelligence in gastrointestinal endoscopy: A comprehensive review. *Annals of Gastroenterology, 37*(2), 133. DOI: 10.20524/aog.2024.0861 PMID: 38481787

Alkan, M. (2024, October 18). *Soil Health and Plant Nutrient Analysis with Agricultural Drones*. Retrieved from Cropim.com: https://cropim.com/soil-health-plant-nutrient-analysis-agricultural-drones/#:~:text=Drones%20equipped%20with%20multispectral%20cameras%20and%20sensors%20can,allowing%20them%20to%20optimize%20fertilization%20and%20irrigation%20strategies

Al-Nabhan, N., Alenazi, S., Alquwaifili, S., Alzamzami, S., Altwayan, L., Alaloula, N., Alowaini, R., & Al Islam, A. A. (2021). An intelligent IoT approach for analyzing and managing crowds. *IEEE Access : Practical Innovations, Open Solutions, 9*, 104874–104886. DOI: 10.1109/ACCESS.2021.3099531

Alnowaiser, K., Saber, A., Hassan, E., & Awad, W. A. (2024). An optimized model based on adaptive convolutional neural network and grey wolf algorithm for breast cancer diagnosis. *PloS One, 19*(8), e0304868.

Alnowaiser, K., Saber, A., Hassan, E., & Awad, W. A. (2024). An optimized model based on adaptive convolutional neural network and grey wolf algorithm for breast cancer diagnosis. *PLoS One, 19*(8), e0304868. DOI: 10.1371/journal.pone.0304868 PMID: 39159151

Alrowais, F., Jamjoom, A. A., Karamti, H., Umer, M., Alsubai, S., Abate, A. F., & Ashraf, I. (2024). Automated approach to predict cerebral stroke based on fuzzy inference and convolutional neural network. *Multimedia Tools and Applications*. Advance online publication. DOI: 10.1007/s11042-024-19254-0

Alsahafi, R., Alzahrani, A., & Mehmood, R. (2023). Smarter sustainable tourism: Data-driven multi-perspective parameter discovery for autonomous design and operations. *Sustainability (Basel), 15*(5), 4166. DOI: 10.3390/su15054166

Altayeb, J. M., Eleyan, H., Wishah, N. D., Elmahmoum, A. E., Khalil, A. J., Abu-Nasser, B. S., & Abu-Naser, S. S. (2024). AI-Driven Innovations in Agriculture: Transforming Farming Practices and Outcomes.

Álvarez-Carmona, M. Á., Aranda, R., Rodríguez-Gonzalez, A. Y., Fajardo-Delgado, D., Sánchez, M. G., Pérez-Espinosa, H., & Díaz-Pacheco, Á. (2022). Natural language processing applied to tourism research: A systematic review and future research directions. *Journal of King Saud University. Computer and Information Sciences*, *34*(10), 10125–10144. DOI: 10.1016/j.jksuci.2022.10.010

Alyasiri, O. M., Selvaraj, K., Younis, H. A., Sahib, T. M., Almasoodi, M. F., & Hayder, I. M. (2024). A Survey on the Potential of Artificial Intelligence Tools in Tourism Information Services. *Babylonian Journal of Artificial Intelligence*, *2024*, 1–8. DOI: 10.58496/BJAI/2024/001

Alzamily, J. Y., Bakeer, H., Almadhoun, H., Abunasser, B. S., & Abu-Naser, S. S. (2024). Artificial Intelligence in Healthcare: Transforming Patient Care and Medical Practices.

Anderson, D. (2019). Artificial Intelligence and Applications in PM&R. *American Journal of Physical Medicine & Rehabilitation*, *98*(11), e128–e129. DOI: 10.1097/PHM.0000000000001171 PMID: 30839314

Anjum, M. N., Cheema, M. J. M., Hussain, F., & Wu, R. S. (2023). Precision irrigation: Challenges and opportunities. *Precision Agriculture*, •••, 85–101.

Antonakoudis, A., Barbosa, R., Kotidis, P., & Kontoravdi, C. (2020). The era of big data: Genome-scale modelling meets machine learning. *Computational and Structural Biotechnology Journal*, *18*, 3287–3300. DOI: 10.1016/j.csbj.2020.10.011 PMID: 33240470

Anwar, H., Anwar, T., & Mahmood, G. (2023). Nourishing the Future: AI-Driven Optimization of Farm-to-Consumer Food Supply Chain for Enhanced Business Performance. *Innovative Computing Review, 3*(2).

AquaAdapt. (2024, November 4). Aquaculture and artificial intelligence: Challenges and opportunities. Retrieved from https://aquadapt.org/2024/11/04/aquaculture-artificial-intelligence

Aquaculture artificial intelligence: A history for the curious. (2023, September 12). Fish n' bits. https://blog.manolinaqua.com/en/aquaculture-artificial-intelligence-history

Araújo, S. O., Peres, R. S., Ramalho, J. C., Lidon, F., & Barata, J. (2023). Machine Learning Applications in Agriculture: Current Trends, Challenges, and Future Perspectives. *Agronomy (Basel)*, *13*(12), 2976. DOI: 10.3390/agronomy13122976

Arora, R., Basu, A., Mianjy, P., & Mukherjee, A. (2018). *Understanding Deep Neural Networks with Rectified Linear Units* (arXiv:1611.01491). arXiv. https://doi.org//arXiv.1611.01491DOI: 10.48550

Arrieta, A. B., Díaz-Rodríguez, N., Del Ser, J., Bennetot, A., Tabik, S., Barbado, A., García, S., & Herrera, F. (2020). Explainable Artificial Intelligence (XAI): Concepts, taxonomies, opportunities and challenges toward responsible AI. *Information Fusion*, *58*, 82–115. DOI: 10.1016/j.inffus.2019.12.012

Arthy, J., & Raja, K. (2024, April). A Study on Design, Development and Deployment of Web Crawler Algorithms and Their Metrics. In *2024 International Conference on Advances in Data Engineering and Intelligent Computing Systems (ADICS)* (pp. 1-6). IEEE.

Artificial intelligence and its aquaculture applications. (2025, February 4). Ocean News & Technology. https://ecomagazine.com/news/fisheries-aquaculture/artificial-intelligence-and-its-aquaculture-applications/

Ashraf Rather, M., Ahmad, I., Shah, A., Ahmad Hajam, Y., Amin, A., Khursheed, S., Ahmad, I., & Rasool, S. (2024). Exploring opportunities of Artificial Intelligence in aquaculture to meet increasing food demand. *Food Chemistry: X*, *22*(101309), 101309. DOI: 10.1016/j.fochx.2024.101309 PMID: 38550881

Aslam, A. B., Iqbal, F., Talpur, U., Syed, Z. S., & Shaikh, F. K. (2024). Artificial intelligence-enabled 6G mobile systems. *Signals and Communication Technology*, 49-79. https://doi.org/DOI: 10.1007/978-3-031-58527-2_3

AsteriaAerospace. (2024, December 12). *Drone Technology in Agriculture: Benefits and Challenges*. Retrieved from asteria.co.in: https://asteria.co.in/blog/drone-technology-in-agriculture-benefits-and-challenges

Asvial, M., Zagloel, T. Y. M., Fitri, I. R., Kusrini, E., & Whulanza, Y. (2023). Resolving Engineering, Industrial and Healthcare Challenges through AI-Driven Applications. *International Journal of Technology*, *14*(6), 1177. DOI: 10.14716/ijtech.v14i6.6767

Aung, T., Razak, R., & Nor, A. (2024). Artificial intelligence methods used in various aquaculture applications: A systematic literature review. *Journal of the World Aquaculture Society*. Advance online publication. DOI: 10.1111/jwas.13107

Ayyaz, M. S., Lali, M. I. U., Hussain, M., Rauf, H. T., Alouffi, B., Alyami, H., & Wasti, S. (2021). Hybrid deep learning model for endoscopic lesion detection and classification using endoscopy videos. *Diagnostics (Basel)*, *12*(1), 43. DOI: 10.3390/diagnostics12010043 PMID: 35054210

Azam, M. A., Sampieri, C., Ioppi, A., Africano, S., Vallin, A., Mocellin, D., & Peretti, G. (2022). Deep learning applied to white light and narrow band imaging videolaryngoscopy: Toward real-time laryngeal cancer detection. *The Laryngoscope*, *132*(9), 1798–1806. DOI: 10.1002/lary.29960 PMID: 34821396

Badji, A., Machida, L., Kwemoi, D. B., Kumi, F., Okii, D., Mwila, N., Agbahoungba, S., Ibanda, A., Bararyenya, A., Nghituwamhata, S. N., Odong, T., Wasswa, P., Otim, M., Ochwo-Ssemakula, M., Talwana, H., Asea, G., Kyamanywa, S., & Rubaihayo, P. (2020). Factors Influencing Genomic Prediction Accuracies of Tropical Maize Resistance to Fall Armyworm and Weevils. *Plants*, *10*(1), 29. DOI: 10.3390/plants10010029 PMID: 33374402

Badouch, M., & Boutaounte, M. (2024). Design and Implementation of a Hotel Recommendation System Using Deep Learning. In *AI and Data Engineering Solutions for Effective Marketing* (pp. 388–408). IGI Global., DOI: 10.4018/979-8-3693-3172-9.ch019

Baig, M. M., GholamHosseini, H., Moqeem, A. A., Mirza, F., & Lindén, M. (2017). A Systematic Review of Wearable Patient Monitoring Systems–Current Challenges and Opportunities for Clinical Adoption. *Journal of Medical Systems*, *41*(7), 115. DOI: 10.1007/s10916-017-0760-1 PMID: 28631139

Bailey, W. H., Cotts, B. R., & Dopart, P. J. (2020). Wireless 5G Radiofrequency technology — An overview of small cell exposures, standards and science. *IEEE Access : Practical Innovations, Open Solutions*, *8*, 140792–140797. DOI: 10.1109/ACCESS.2020.3010677

Bainomugisha, E., Warigo, P. A., Daka, F. B., Nshimye, A., Birungi, M., & Okure, D. (2024). AI-driven environmental sensor networks and digital platforms for urban air pollution monitoring and modelling. *Societal Impacts*, *3*, 100044. DOI: 10.1016/j.socimp.2024.100044

Balasubramanian, A., & Elangeswaran, S. V. J. (2024). A novel power aware smart agriculture management system based on RNN-LSTM. *Electrical Engineering*. Advance online publication. DOI: 10.1007/s00202-024-02640-0

Bałazy, K., Struski, Ł., Śmieja, M., & Tabor, J. (2023). r-softmax: Generalized Softmax with Controllable Sparsity Rate. In J. Mikyška, C. De Mulatier, M. Paszynski, V. V. Krzhizhanovskaya, J. J. Dongarra, & P. M. A. Sloot (Eds.), *Computational Science – ICCS 2023* (Vol. 14074, pp. 137–145). Springer Nature Switzerland. DOI: 10.1007/978-3-031-36021-3_11

Baldini, C., Azam, M. A., Sampieri, C., Ioppi, A., Ruiz-Sevilla, L., Vilaseca, I., & Mattos, L. S. (2024). An automated approach for real-time informative frames classification in laryngeal endoscopy using deep learning. *European Archives of Oto-Rhino-Laryngology*, *281*(8), 1–10. DOI: 10.1007/s00405-024-08676-z PMID: 38698163

Ballocchi, A. (2024, February 11). Artificial intelligence in aquaculture: Enhancing a time-honoured practice with AI, robots, and drones. Tech4Future. https://tech4future.info/en/aquaculture-artificial-inteligence/

Barry, M., Haque, A. A., & Jan, M. T. (2024). Mobile commerce adoption in Malaysia: A conceptual framework. *Open Journal of Economics and Commerce, 5*(1), 04-12. https://doi.org/DOI: 10.22259/2638-549X.0501002

Basha, S. H. S., Dubey, S. R., Pulabaigari, V., & Mukherjee, S. (2020). Impact of fully connected layers on performance of convolutional neural networks for image classification. *Neurocomputing*, *378*, 112–119. DOI: 10.1016/j.neucom.2019.10.008

Batra, R., Arora, S., Sharma, M. M., Rana, S., Raheja, K., Saber, A., & Shah, M. A. (2024). Integration of LSTM networks with gradient boosting machines (GBM) for assessing heating and cooling load requirements in building energy efficiency. *Energy Exploration & Exploitation*, *42*(6), 01445987241268075. DOI: 10.1177/01445987241268075

Bauer, P., Dueben, P., Chantry, M., Doblas-Reyes, F., Hoefler, T., McGovern, A., & Stevens, B. (2023). Deep learning and a changing economy in weather and climate prediction. *Nature Reviews. Earth & Environment*, *4*(8), 507–509. DOI: 10.1038/s43017-023-00468-z

Baxevanis, A. D., & Ouellette, B. F. F. (Eds.). (2004). *Bioinformatics: A Practical Guide to the Analysis of Genes and Proteins* (3rd ed.). Wiley-Interscience.

Bedi, G., Carrillo, F., Cecchi. (2015). Automated analysis of free speech predicts psychosis onset in high-risk youths. *npj Schizophr 1, 15030. https://doi.org/.*DOI: 10.1038/npjschz.2015.30

Behmann, J., Mahlein, A.-K., Rumpf, T., Römer, C., & Plümer, L. (2015). A review of advanced machine learning methods for the detection of biotic stress in precision crop protection. *Precision Agriculture, 16*(3), 239–260. DOI: 10.1007/s11119-014-9372-7

Ben Ayed, R., & Hanana, M. (2021). Artificial intelligence to improve the food and agriculture sector. *Journal of Food Quality, 2021*(1), 5584754.

Bender, A., & Cortés-Ciriano, I. (2021). Artificial intelligence in drug discovery: What is realistic, what are illusions? Part 1: Ways to make an impact, and why we are not there yet. *Drug Discovery Today, 26*(1), 511-524.

Benos, L., Tagarakis, A. C., Dolias, G., Berruto, R., Kateris, D., & Bochtis, D. (2021). Machine Learning in Agriculture: A Comprehensive Updated Review. *Sensors (Basel), 21*(11), 3758. DOI: 10.3390/s21113758 PMID: 34071553

Benzaid, C., & Taleb, T. (2020). AI-driven zero touch network and service management in 5G and beyond: Challenges and research directions. *IEEE Network, 34*(2), 186–194. DOI: 10.1109/MNET.001.1900252

Berg, S. (2018). Nudge Theory Explored to Boost Medication Adherence. American Medical Association., Available online www.ama-assn.org/delivering-care/patient-support-advocacy/nudge-theory-exploredboost-medication-adherence, Retrieved January 1, 2023, from.

Beverina, F., Palmas, G., Silvoni, S., Piccione, F., & Giove, S. (2003). User adaptive BCIs: SSVEP and P300 based interfaces. *PsychNology Journal, 1*(4), 331–354.

Bhandari, R. (2023, June 26). *Himachal Pradesh Embraces Drones for Efficient Agriculture and Horticulture Operations*. Retrieved from The News Himachal: https://thenewshimachal.com/2023/06/himachal-pradesh-embraces-drones-for-efficient-agriculture-and-horticulture-operations/

Bhardwaj, A., Kishore, S., & Pandey, D. K. (2022). Artificial intelligence in biological sciences. *Life (Chicago, Ill.), 12*(9), 1430. PMID: 36143468

Bhati, N., Duggar, R., & Saber, A. (2024). Empowering Safety by Embracing IoT for Leak Detection Excellence. In Innovations in Machine Learning and IoT for Water Management (pp. 231-251). IGI Global.

Bhat, S., Thakur, P., & Bhatt, A. (2024). A novel fish classification system using deep learning algorithms. *AIP Conference Proceedings*, 050013. DOI: 10.1063/5.0227890

Bhutani, P., Joshi, G., Raja, N., & Bachhav, N. (2021). US FDA approved drugs from 2015–June 2020: A perspective. *Journal of Medicinal Chemistry, 64*(15), 10644-10684.

Bian, J., Kang, M., Xi, Y., Wang, Y., & Zi, W. (2024). A Review of the Measurement and Control Technologies for the Critical Parameters of Microwave Drying Processes: Temperature and Humidity. *Food and Bioprocess Technology*, •••, 1–24. DOI: 10.1007/s11947-024-03614-7

Bibel, W. (2014). Artificial Intelligence in a historical perspective. *AI Communications*, *27*(1), 87–102. DOI: 10.3233/AIC-130576

Birendra, K. C., Dhungana, A., & Dangi, T. B. (2021). Tourism and the sustainable development goals: Stakeholders' perspectives from Nepal. *Tourism Management Perspectives*, *38*, 100822. DOI: 10.1016/j.tmp.2021.100822

Biswas, P. (2023, September 3). *Maharashtra's Agri-Entrepreneurs Take to the Skies: How Drone are Being Used Tackle Farm Labour Shortage*. Retrieved from IndianExpress.com: https://indianexpress.com/article/cities/pune/drones-the-new-tool-for-farmers-in-maharashtra-8921730/

Biswas, A., Kumari, A., Gaikwad, D. S., & Pandey, D. K. (2023). Revolutionizing biological science: The synergy of genomics in health, bioinformatics, agriculture, and artificial intelligence. *OMICS: A Journal of Integrative Biology*, *27*(12), 550–569. DOI: 10.1089/omi.2023.0197 PMID: 38100404

Bohara, K., Joshi, P., Acharya, K. P., & Ramena, G. (2023). Emerging technologies revolutionising disease diagnosis and monitoring in aquatic animal health. *Reviews in Aquaculture*, *16*(2), 836–854. DOI: 10.1111/raq.12870

Bokolo, A. J. (2021). Application of telemedicine and eHealth technology for clinical services in response to COVID-19 pandemic. *Health and Technology*, *11*(2), 359–366. DOI: 10.1007/s12553-020-00516-4 PMID: 33469474

Boquin Instrument. (2024). The impact of turbidity on fish health in aquaculture systems. Retrieved from https://www.boquinstrument.com/the-impact-of-turbidity-on-fish-health-in-aquaculture-systems

BR. S. R. (2022). Information and Communication Technology Application in the Asian Tourism Industry. In Handbook of Technology Application in Tourism in Asia (pp. 19-48). Singapore: Springer Nature Singapore.

Bradford, E., Schweidtmann, A. M., Zhang, D., Jing, K., & del Rio-Chanona, E. A. (2018). Dynamic modeling and optimization of sustainable algal production with uncertainty using multivariate Gaussian processes. *Computers & Chemical Engineering, 118*, 143–158. DOI: 10.1016/j.compchemeng.2018.07.015

Bretos, M. A., Ibáñez-Sánchez, S., & Orús, C. (2024). Applying virtual reality and augmented reality to the tourism experience: A comparative literature review. *Spanish Journal of Marketing - ESIC, 28*(3), 287–309. DOI: 10.1108/SJME-03-2023-0052

Brian, K., & Bommu, R. (2022). Revolutionizing Healthcare IT through AI and Microfluidics: From Drug Screening to Precision Livestock Farming. *Unique Endeavor in Business & Social Sciences, 1*(1), 84–99.

Buchanan, B. G., & Shortliffe, E. H. (1984). *Rule-Based Expert Systems: The MYCIN Experiments of the Stanford Heuristic Programming Project*. Addison-Wesley.

Bwambale, E., Abagale, F. K., & Anornu, G. K. (2022). Smart irrigation monitoring and control strategies for improving water use efficiency in precision agriculture: A review. *Agricultural Water Management, 260*, 107324. DOI: 10.1016/j.agwat.2021.107324

Camacho, D. M. (2018). Machine learning in biological networks: Predicting interactions and responses. *Nature Reviews. Genetics, 19*(7), 321–332. DOI: 10.1038/s41576-018-0002-3

Campbell, K., Diffley, J., Flanagan, B., Morelli, B., O'Neil, B., & Sideco, F. (2017). The 5G economy: How 5G technology will contribute to the global economy.

Cao, Y., Wu, B., Xu, Y., Wang, M., Wu, X., & Liang, X. (2024). Discovery of GPX4 inhibitors through FP-based high-throughput screening. *European Journal of Drugs, 15*(3), 128-137.

Cao, L. AI in Finance: A Review (July 10, 2020). Available at *SSRN*: https://ssrn.com/abstract=3647625 or http://dx.doi.org/DOI: 10.2139/ssrn.3647625

Carbonell, P., Radivojevic, T., & García Martín, H. (2019). Opportunities at the Intersection of Synthetic Biology, Machine Learning, and Automation. *ACS Synthetic Biology, 8*(7), 1474–1477. DOI: 10.1021/acssynbio.8b00540 PMID: 31319671

Castermans, T., Duvinage, M., Petieau, M., Hoellinger, T., Saedeleer, C. D., Seetharaman, K., Bengoetxea, A., Cheron, G., & Dutoit, T. (2011). Optimizing the Performances of a P300-Based Brain–Computer Interface in Ambulatory Conditions. *IEEE Journal on Emerging and Selected Topics in Circuits and Systems, 1*(4), 566–577. DOI: 10.1109/JETCAS.2011.2179421

Cervantes, J., Garcia-Lamont, F., Rodríguez-Mazahua, L., & Lopez, A. (2020). A comprehensive survey on support vector machine classification: Applications, challenges and trends. *Neurocomputing*, *408*, 189–215. DOI: 10.1016/j.neucom.2019.10.118

Chakhvashvili, E. M., Machwitz, M., Antala, M., Rozenstein, O., Prikaziuk, E., Schlerf, M., Naethe, P., Wan, Q., Komárek, J., Klouek, T., Wieneke, S., Siegmann, B., Kefauver, S., Kycko, M., Balde, H., Paz, V. S., Jimenez-Berni, J. A., Buddenbaum, H., Hänchen, L., & Rascher, U. (2024). Crop Stress Detection from UAVs: Best Practices and Lessons Learned for Exploiting Sensor Synergies. *Precision Agriculture*, *25*(5), 2614–2642. DOI: 10.1007/s11119-024-10168-3

Chang, C., Wang, J., Wu, J., Hsieh, Y., Wu, T., Cheng, S., Chang, C., Juang, J., Liou, C., Hsu, T., Huang, Y., Huang, C., Lin, C., Peng, Y., Huang, R., Jhang, J., Liao, Y., & Lin, C. (2021). Applying Artificial Intelligence Techniques to Implement a Practical Smart Cage Aquaculture Management System. *Journal of Medical and Biological Engineering*, *41*, 652–658. DOI: 10.1007/s40846-021-00621-3

Charbuty B, Abdulazeez A.(2021)."Classification Based on Decision Tree Algorithm for Machine Learning". J. Appl. Sci. Technol. Trends.,2022

Chaterji, S., Naghizadeh, P., Alam, M. A., Bagchi, S., Chiang, M., Corman, D., Henz, B., Jana, S., Li, N., Mou, S., Oishi, M., Peng, C., Rompf, T., Sabharwal, A., Sundaram, S., Weimer, J., & Weller, J. (2020). *Resilient Cyberphysical Systems and their Application Drivers: A Technology Roadmap* (Version 1). arXiv. DOI: 10.48550/ARXIV.2001.00090

Chaudhry, M., Shafi, I., Mahnoor, M., Vargas, D. L. R., Thompson, E. B., & Ashraf, I. (2023). A Systematic Literature Review on Identifying Patterns Using Unsupervised Clustering Algorithms: A Data Mining Perspective. *Symmetry*, *15*(9), 1679. DOI: 10.3390/sym15091679

Chávez-Santiago, R., Szydełko, M., Kliks, A., Foukalas, F., Haddad, Y., Nolan, K. E., Kelly, M. Y., Masonta, M. T., & Balasingham, I. (2015). 5G: The convergence of wireless communications. *Wireless Personal Communications*, *83*(3), 1617–1642. DOI: 10.1007/s11277-015-2467-2 PMID: 27076701

Cheng, L., Varshney, K. R., & Liu, H. (2021). Socially responsible AI algorithms: Issues, purposes, and challenges. *Journal of Artificial Intelligence Research, 70*, 377-395.

Chen, H., Shen, Z., Wang, L., Qi, C., & Tian, Y. (2022). Prediction of undrained failure envelopes of skirted circular foundations using gradient boosting machine algorithm. *Ocean Engineering*, *258*, 111767. DOI: 10.1016/j.oceaneng.2022.111767

Chen, S.-F., & Kuo, Y.-F. (2022). *Artificial Intelligence for Image Processing in Agriculture*. Sensing, Data Managing, and Control Technologies for Agricultural Systems., DOI: 10.1201/9781003214892

Chevallier, S., Carrara, I., Aristimunha, B., Guetschel, P., Sedlar, S., Lopes, B., Velut, S., Khazem, S., & Moreau, T. (2024). The largest EEG-based BCI reproducibility study for open science: the MOABB benchmark. *ArXiv, abs/2404.15319.*

Chiaraviglio, L., Elzanaty, A., & Alouini, M. (2021). Health risks associated with 5G exposure: A view from the communications engineering perspective. *IEEE Open Journal of the Communications Society*, *2*, 2131–2179. DOI: 10.1109/OJCOMS.2021.3106052

Chiaraviglio, L., Lodovisi, C., Franci, D., Grillo, E., Pavoncello, S., Aureli, T., Blefari-Melazzi, N., & Alouini, M. (2022). How much exposure from 5G towers is radiated over children, teenagers, schools and hospitals? *IEEE Open Journal of the Communications Society*, *3*, 1592–1614. DOI: 10.1109/OJCOMS.2022.3208938

Chikhaoui, E., Sarabdeen, J., & Parveen, R. (2017). Privacy and Security Issues in the Use of Clouds in e-Health in the Kingdom of Saudi Arabia. *Communications of the IBIMA*, *2017*, 1–18. DOI: 10.5171/2017.369309

Ching, T., Himmelstein, D. S., Beaulieu-Jones, B. K., Kalinin, A. A., Do, B. T., Way, G. P., Ferrero, E., Agapow, P. M., Zietz, M., Hoffman, M. M., Xie, W., Rosen, G. L., Lengerich, B. J., Israeli, J., Lanchantin, J., Woloszynek, S., Carpenter, A. E., Shrikumar, A., Xu, J., & Greene, C. S. (2018). Opportunities and obstacles for deep learning in biology and medicine. *Journal of the Royal Society, Interface*, *15*(141), 20170387. DOI: 10.1098/rsif.2017.0387 PMID: 29618526

Chisom, O. N., Biu, P. W., Umoh, A. A., Obaedo, B. O., Adegbite, A. O., & Abatan, A. (2024). Reviewing the role of AI in environmental monitoring and conservation: A data-driven revolution for our planet. *World Journal of Advanced Research and Reviews*, *21*(1), 161–171. DOI: 10.30574/wjarr.2024.21.1.2720

Chlingaryan, A., Sukkarieh, S., & Whelan, B. (2018a). Machine learning approaches for crop yield prediction and nitrogen status estimation in precision agriculture: A review. *Computers and Electronics in Agriculture*, *151*, 61–69. DOI: 10.1016/j.compag.2018.05.012

Choi, J., Shin, K., Jung, J., Bae, H. J., Kim, D. H., Byeon, J. S., & Kim, N. (2020). Convolutional neural network technology in endoscopic imaging: Artificial intelligence for endoscopy. *Clinical Endoscopy*, *53*(2), 117–126. DOI: 10.5946/ce.2020.054 PMID: 32252504

Choudhury, S.. (2022) Reconstructing Kinetic Models for Dynamical Studies of Metabolism using Generative Adversarial Networks. *bioRxiv*.DOI: 10.1101/2022.01.06.475020

Cho, W. K., & Choi, S.-H. (2022). Comparison of convolutional neural network models for determination of vocal fold normality in laryngoscopic images. *Journal of Voice*, *36*(5), 590–598. DOI: 10.1016/j.jvoice.2020.08.003 PMID: 32873430

Collins, F. S., & Fink, L. (1995). The Human Genome Project. *Alcohol Health and Research World*, *19*(3), 190–195. PMID: 31798046

Colmerauer, A., & Roussel, P. (1993). *The birth of Prolog*. The Second ACM SIGPLAN Conference on History of Programming Languages. Presented at the HOPL93: History of Programing Languages 1993, Cambridge Massachusetts USA. DOI: 10.1145/154766.155362

Combaz, A., & Van Hulle, M. M. (2015). Simultaneous detection of P300 and steady-state visually evoked potentials for hybrid brain-computer interface. *PLoS One*, *10*(3), e0121481. DOI: 10.1371/journal.pone.0121481 PMID: 25815815

Confalonieri, R., Coba, L., Wagner, B., & Besold, T. R. (2021). A historical perspective of explainable Artificial Intelligence. *Wiley Interdisciplinary Reviews. Data Mining and Knowledge Discovery*, *11*(1), e1391. Advance online publication. DOI: 10.1002/widm.1391

Cooney, C., Folli, R., & Coyle, D. (2018). Neurolinguistics research advancing development of a direct-speech brain-computer interface. *iScience*, *8*, 103–125. DOI: 10.1016/j.isci.2018.09.016 PMID: 30296666

Coşgun, A., Günay, M. E., & Yıldırım, R. (2022). Analysis of lipid production from Yarrowia lipolytica for renewable fuel production by machine learning. *Fuel*, *315*, 122817. Advance online publication. DOI: 10.1016/j.fuel.2021.122817

Coulibaly, A. (2024). *Smart Farming: Computer Simulation and Predictive Model for Cassava*. DOI: 10.20381/RUOR-30362

Cruz, M. V., Jamal, S., & Sethuraman, S. C. (2024). A Comprehensive Survey of Brain-Computer Interface Technology in Healthcare: Research Perspectives.

Cuimei, A. D. K. S. B., & Boafoh, K. B. (2019). The Emergence of AI and IoT on Cloud Computing: Evolution, Technology, Future Research and Challenges. *Emergence (Mahwah, N.J.)*, *10*(7).

Dalal, S., Goel, P., Onyema, E. M., Alharbi, A., Mahmoud, A., Algarni, M. A., & Awal, H. (2023b). Application of machine learning for cardiovascular disease risk prediction. *Computational Intelligence and Neuroscience*, *2023*(1), 9418666. DOI: 10.1155/2023/9418666

Dalal, S., Poongodi, M., Lilhore, U. K., Dahan, F., Vaiyapuri, T., Keshta, I., Aldossary, S. M., Mahmoud, A., & Simaiya, S. (2023a). Optimized LightGBM model for security and privacy issues in cyber-physical systems. *Transactions on Emerging Telecommunications Technologies*, *34*(6), e4771. DOI: 10.1002/ett.4771

Dangi, T. B., & Petrick, J. F. (2021). Augmenting the role of tourism governance in addressing destination justice, ethics, and equity for sustainable community-based tourism. *Tourism and Hospitality*, *2*(1), 15–42. DOI: 10.3390/tourhosp2010002

Davenport, T., & Kalakota, R. (2019). The potential for artificial intelligence in healthcare. *Future Healthcare Journal*, *6*(2), 94–98. DOI: 10.7861/futurehosp.6-2-94 PMID: 31363513

De Zarzà, I., De Curtò, J., Hernández-Orallo, E., & Calafate, C. T. (2023). Cascading and Ensemble Techniques in Deep Learning. *Electronics (Basel)*, *12*(15), 3354. DOI: 10.3390/electronics12153354

Debnath, J., Kumar, K., Roy, K., Choudhury, R. D., & U, A. K. P. (2024). Precision Agriculture: A Review of AI Vision and Machine Learning in Soil, Water, and Conservation Practice. *International Journal for Research in Applied Science and Engineering Technology*, *12*(12), 2130–2141. DOI: 10.22214/ijraset.2024.66166

Del Rio-Chanona, E. A., Manirafasha, E., Zhang, D., Yue, Q., & Jing, K. (2016). Dynamic modeling and optimization of cyanobacterial C-phycocyanin production process by artificial neural network. *Algal Research*, *13*, 7–15. DOI: 10.1016/j.algal.2015.11.004

Deloitte. (2020). *Transforming Agriculture through Digital Technologies*. Deloitte.

Deng, S., Zhao, H., Huang, B., Zhang, C., Chen, F., Deng, Y., & Zomaya, A. Y. (2024). Cloud-native computing: A survey from the perspective of services. *Proceedings of the IEEE*. DOI: 10.1109/JPROC.2024.3353855

Detection, B., & Classification, S. (2023). Using YOLOv5 and Deep Transfer Learning Models Hoang-Tu Vo, Nhon Nguyen Thien, Kheo Chau Mui Software Engineering Department, FPT University,Cantho city, Vietnam. *IJACSA-International Journal of Advanced Computer Science and Applications*, *14*(7), 2023.

Devarajan, G. G., & Nagarajan, S. M., T.V., R., T., V., Ghosh, U., & Alnumay, W. (. (2023). DDNSAS: Deep reinforcement learning based deep Q-learning network for smart agriculture system. *Sustainable Computing : Informatics and Systems*, *39*, 100890. DOI: 10.1016/j.suscom.2023.100890

Dhanwe, S. S., Abhangrao, C. M., & Liyakat, K. K. S. (2024). AI-driven IoT in Robotics: A Review. *Journal of Mechanisms and Robotics*, *9*(1), 41–48.

Di Ciaula, A. (2018). Towards 5G communication systems: Are there health implications? *International Journal of Hygiene and Environmental Health*, *221*(3), 367–375. DOI: 10.1016/j.ijheh.2018.01.011 PMID: 29402696

Dimitriadis, Y., Martínez-Maldonado, R., & Wiley, K. (2021). Human-centered design principles for actionable learning analytics. Research on E-learning and ICT in education: Technological, pedagogical and instructional perspectives, 277-296. https://doi.org/DOI: 10.1007/978-3-030-64363-8_15

Ding, W., Abdel-Basset, M., Alrashdi, I., & Hawash, H. (2024). Next generation of computer vision for plant disease monitoring in precision agriculture: A contemporary survey, taxonomy, experiments, and future direction. *Information Sciences*, *665*, 120338. DOI: 10.1016/j.ins.2024.120338

Ding, X., & Du, W. (2024). Optimizing Irrigation Efficiency using Deep Reinforcement Learning in the Field. *ACM Transactions on Sensor Networks*, *20*(4), 1–34. DOI: 10.1145/3662182

Djomadji, E. M., Garga, M., Fouba, B. A., & Bouetou, T. B. (2024). Genetic algorithm for mode selection in device-to-device (D2D) communication for 5G cellular networks. *American Journal of Networks and Communications*, *13*(1), 30–43. DOI: 10.11648/j.ajnc.20241301.13

Donchin, E., Spencer, K. M., & Wijesinghe, R. (2000). The mental prosthesis: Assessing the speed of a P300-based brain-computer interface. *IEEE Transactions on Rehabilitation Engineering*, *8*(2), 174–179. DOI: 10.1109/86.847808 PMID: 10896179

Dronepilotgroundschool. (2025, February 7). *Drone Applications & Uses: A Comprehensive Guide for Drone Pilots*. Retrieved from Dronepilotgroundschool.com: https://www.dronepilotgroundschool.com/drone-uses/#:~:text=Drones%20equipped%20with%20advanced%20spraying%20systems%20can%20precisely,usage%2C%20and%20minimizes%20human%20exposure%20to%20hazardous%20substances

DroneXperts. (2025, February 17). *How are Drones Redefining Sustainable Agriculture?* Retrieved from DroneXperts.com: https://www.dronexperts.com/en/article/how-are-drones-redefining-sustainable-agriculture/

Duarte, P. M., Adesola, R. O., Priyadarsini, S., Singh, R., Shaheen, M. N., Ogundijo, O. A., Gulumbe, B. H., Lounis, M., Samir, M., Govindan, K., Adebiyi, O. S., Scott, G. Y., Ahmadi, P., Mahmoodi, V., Chogan, H., Gholami, S., Shirazi, O., Moghadam, S. K., Jafari, N., & Tazerji, S. S. (2024). Unveiling the global surge of Mpox (Monkeypox): A comprehensive review of current evidence. *The Microbe*, *4*, 100141. DOI: 10.1016/j.microb.2024.100141

Dubashi, N. F. (2023, May 10). *How AgTech is Poised to Transform India into a Farming Powerhouse*. Retrieved from McKinsey.com: https://www.mckinsey.com/industries/agriculture/our-insights/how-agtech-is-poised-to-transform-india-into-a-farming-powerhouse

Ebrahimpour, A., Rahman, R. N. Z. R. A., Ean Ch'ng, D. H., Basri, M., & Salleh, A. B. (2008). A modeling study by response surface methodology and artificial neural network on culture parameters optimization for thermostable lipase production from a newly isolated thermophilic Geobacillus sp. strain ARM. *BMC Biotechnology*, *8*(1), 96. DOI: 10.1186/1472-6750-8-96 PMID: 19105837

Egger, J., Wild, D., Weber, M., Bedoya, C. A. R., Karner, F., Prutsch, A., Schmied, M., Dionysio, C., Krobath, D., Jin, Y., Gsaxner, C., Li, J., & Pepe, A. (2022). Studierfenster: An open science cloud-based medical imaging analysis platform. *Journal of Digital Imaging*, *35*(2), 340–355. DOI: 10.1007/s10278-021-00574-8 PMID: 35064372

Elahi, M., Abdollahpouri, H., Mansoury, M., & Torkamaan, H. (2021, June). Beyond algorithmic fairness in recommender systems. In *Adjunct proceedings of the 29th ACM conference on user modeling, adaptation and personalization* (pp. 41-46). https://doi.org/DOI: 10.1145/3450614.3461685

El-Attar, N. E., Awad, W. A., & Omara, F. A. (2016, December). Empirical assessment for security risk and availability in public cloud frameworks. In 2016 11th International Conference on Computer Engineering & Systems (ICCES) (pp. 17-25). IEEE. DOI: 10.1109/ICCES.2016.7821969

El-Attar, N. E., El-Morshedy, D. S., & Awad, W. A. (2021). A new hybrid automated security framework to cloud storage system. *Cryptography*, *5*(4), 37. DOI: 10.3390/cryptography5040037

El-Attar, N. E., Moustafa, B. M., & Awad, W. A. (2022). Deep Learning Model to Detect Diabetes Mellitus Based on DNA Sequence. *Intelligent Automation & Soft Computing*, *31*(1). Advance online publication. DOI: 10.32604/iasc.2022.019970

El-Attar, N. E., Sabbeh, S. F., Fasihuddin, H., & Awad, W. A. (2022). An Improved DeepNN with Feature Ranking for Covid-19 Detection. *Computers, Materials & Continua*, *71*(2). Advance online publication. DOI: 10.32604/cmc.2022.022673

Elavarasan, D., Vincent, D. R., Sharma, V., Zomaya, A. Y., & Srinivasan, K. (2018). Forecasting yield by integrating agrarian factors and machine learning models: A survey. *Computers and Electronics in Agriculture*, *155*, 257–282. DOI: 10.1016/j.compag.2018.10.024

Elbasi, E., Zaki, C., Topcu, A. E., Abdelbaki, W., Zreikat, A. I., Cina, E., Shdefat, A., & Saker, L. (2023). Crop Prediction Model Using Machine Learning Algorithms. *Applied Sciences (Basel, Switzerland)*, *13*(16), 9288. DOI: 10.3390/app13169288

Elbaz, N. F., & Abd Al Fatah, M. E. (2024). Bacterial diseases outbreaks in some freshwater fish farms in Kafr El-Sheikh, Egypt. *Journal of Applied Aquaculture*, *36*(1), 1–23. DOI: 10.1080/10454438.2022.2105673

Elbaz, N., Abd Al Fatah, M., Gaafar, A., & Soliman, M. (2023). *Investigations on virulence genes of some bacterial infections in Nile tilapia in Kafr El-Sheikh farms*. Damanhour Journal of Veterinary Sciences. DOI: 10.21608/djvs.2023.200798.1112

Elbedwehy, S., Hassan, E., Saber, A., & Elmonier, R. (2024). Integrating neural networks with advanced optimization techniques for accurate kidney disease diagnosis. *Scientific Reports*, *14*(1), 21740. Advance online publication. DOI: 10.1038/s41598-024-71410-6 PMID: 39289394

Elmasry, S. A., & Awad, W. A., & Abd El-hafeez, S. A. (2020). Review of different image fusion techniques: Comparative study. In Internet of Things—Applications and Future: Proceedings of ITAF 2019 (pp. 41-51). Springer Singapore.

El-Matary, D. M., El-Attar, N. E., Awad, W. A., & Hanafy, I. M. (2019, February). Automated negotiation framework based on intelligent agents for cloud computing. In *2019 International Conference on Innovative Trends in Computer Engineering (ITCE)* (pp. 156-161). IEEE. DOI: 10.1109/ITCE.2019.8646481

El-Morshedy, D. S., El-Attar, N. E., Awad, W. A., & Hanafy, I. M. (2020). Trustworthy Self-protection for Data Auditing in Cloud Computing Environment. In Internet of Things—Applications and Future [Springer Singapore.]. *Proceedings of ITAF*, *2019*, 23–39.

El-Morshedy, D. S., El-Attar, N. E., Hanafy, I. M., & Awad, W. A. (2023). Cryptographic Algorithms for Enhancing Security in Cloud Computing. *Alfarama Journal of Basic & Applied Sciences*, *4*(3), 433–455.

Elmuogy, S., Hikal, N. A., & Hassan, E. (2021). An efficient technique for CT scan images classification of COVID-19. *Journal of Intelligent & Fuzzy Systems*, *40*(3), 5225–5238. DOI: 10.3233/JIFS-201985

Emara, T. (2019). Adaptive Power-Saving Mechanism for VoIP Over WiMAX Based on Artificial Neural Network. In *Algorithms, Methods, and Applications in Mobile Computing and Communications* (pp. 158–177). IGI Global. https://doi.org/DOI: 10.4018/978-1-5225-5693-0.ch007

Emara, T. Z. (2016). Maximizing power saving for voip over wimax systems. [IJMCMC]. *International Journal of Mobile Computing and Multimedia Communications*, *7*(1), 32–40. DOI: 10.4018/IJMCMC.2016010103

Emara, T. Z., & Huang, J. Z. (2019a). A distributed data management system to support large-scale data analysis. *Journal of Systems and Software*, *148*, 105–115. DOI: 10.1016/j.jss.2018.11.007

Emara, T. Z., & Huang, J. Z. (2019b). RRPlib: A spark library for representing HDFS blocks as a set of random sample data blocks. *Science of Computer Programming*, *184*, 102301. DOI: 10.1016/j.scico.2019.102301

Emara, T. Z., & Huang, J. Z. (2020). Distributed Data Strategies to Support Large-Scale Data Analysis Across Geo-Distributed Data Centers. *IEEE Access : Practical Innovations, Open Solutions*, *8*, 178526–178538. DOI: 10.1109/ACCESS.2020.3027675

Emara, T. Z., Saleh, A. I., & Arafat, H. (2014). Power saving mechanism for VoIP services over WiMAX systems. *Wireless Networks*, *20*(5), 975–985. DOI: 10.1007/s11276-013-0650-5

Emara, T. Z., Trinh, T., & Huang, J. Z. (2023). Geographically distributed data management to support large-scale data analysis. *Scientific Reports*, *13*(1), 17783. DOI: 10.1038/s41598-023-44789-x PMID: 37853092

Emuna, R., Borowsky, A., & Biess, A. (2020). *Deep Reinforcement Learning for Human-Like Driving Policies in Collision Avoidance Tasks of Self-Driving Cars* (Version 2). arXiv. DOI: 10.48550/ARXIV.2006.04218

EquinoxsDrones. (2024, April). *Overview of the Agricultural Industry in India*. Retrieved from Equinoxsdrones.com: https://equinoxsdrones.com/importance-of-drone-technology-in-indian-agriculture-farming/

Erickson, B., & Fausti, S. W. (2021). The role of precision agriculture in food security. *Agronomy Journal, 113*(6), 4455–4462. DOI: 10.1002/agj2.20919

Er-Rousse, O., & Qafas, A. (2024). Artificial intelligence for the optimization of marine aquaculture. *E3S Web of Conferences*. https://doi.org/.DOI: 10.1051/e3sconf/202447700102

Esmaeili, N., Sharaf, E., Gomes Ataide, E. J., Illanes, A., Boese, A., Davaris, N., & Friebe, M. (2021). Deep convolution neural network for laryngeal cancer classification on contact endoscopy-narrow band imaging. *Sensors (Basel), 21*(23), 8157. DOI: 10.3390/s21238157 PMID: 34884166

Esteva, A., Kuprel, B., Novoa, R. A., Ko, J., Swetter, S. M., Blau, H. M., & Thrun, S. (2017). Dermatologist-level classification of skin cancer with deep neural networks. *Nature, 542*(7639), 115–118. DOI: 10.1038/nature21056 PMID: 28117445

Esteva, A., Robicquet, A., Ramsundar, B., Kuleshov, V., DePristo, M. A., Chou, K., Cui, C., Corrado, G. S., Thrun, S., & Dean, J. (2019). A guide to deep learning in healthcare. *Nature Medicine, 25*(1), 24–29. DOI: 10.1038/s41591-018-0316-z PMID: 30617335

Fabian. (2022, March 26). Freshwater aquarium fish diseases, symptoms & remedies. SmartAquariumGuide. https://smartaquariumguide.com/freshwater-fish-diseases-symptoms-remedies/

Fanni, S. C., Febi, M., Aghakhanyan, G., & Neri, E. (2023). Natural language processing. In *Introduction to Artificial Intelligence* (pp. 87–99). Springer International Publishing., DOI: 10.1007/978-3-031-25928-9_5

Farahani, M. S., & Ghasemi, G. (2024). How artificial intelligence plays a role in achieving sustainable development goals? *Sustainable Economies, 2*(3), 66–66. DOI: 10.62617/se.v2i3.66

Farghali, H., Canová, N. K., & Arora, M. (2021). The potential applications of artificial intelligence in drug discovery and development. *Physiological Research, 70*(1), 19-30.

Farmer, P., Kim, J. Y., Kleinman, A., & Basilico, M. (2013). *Reimagining global health: an introduction* (Vol. 26). Univ of California Press.

Faruqui, F., Yousuf, M. A., Kateb, F. A., Abdul Hamid, M., & Monowar, M. M. (2023). Healthcare As a Service (HAAS): CNN-based cloud computing model for ubiquitous access to lung cancer diagnosis. *Heliyon, 9*(11), e21520. DOI: 10.1016/j.heliyon.2023.e21520 PMID: 37942151

Faure, L.. (2022) Artificial Metabolic Networks: enabling neural computation with metabolic networks. *bioRxiv*. DOI: 10.1101/2022.01.09.475487

Fillinger, S., de la Garza, L., Peltzer, A., Kohlbacher, O., & Nahnsen, S. (2019). Challenges of big data integration in the life sciences. *Analytical and Bioanalytical Chemistry, 411*(26), 6791–6800. DOI: 10.1007/s00216-019-02074-9 PMID: 31463515

Fitzgibbon, A., & Zisserman, A. (1998, September). Automatic 3D model acquisition and generation of new images from video sequences. In *9th European Signal Processing Conference (EUSIPCO 1998)* (pp. 1-8). IEEE.

Floerl, O., Sunde, L., & Bloecher, N. (2016). Potential environmental risks associated with biofouling management in salmon aquaculture. *Aquaculture Environment Interactions, 8*, 407–417. DOI: 10.3354/aei00187

Floridi, L. (2023). The ethics of artificial intelligence: Principles, challenges, and opportunities.

Floridi, L., Cowls, J., Beltrametti, M., Chatila, R., Chazerand, P., Dignum, V., Luetge, C., Madelin, R., Pagallo, U., Rossi, F., Schafer, B., Valcke, P., & Vayena, E. (2019). AI4People—An ethical framework for a good AI society: Opportunities, risks, principles, and recommendations. *Minds and Machines, 29*(4), 611–637. DOI: 10.1007/s11023-019-09508-8 PMID: 30930541

Francis-Floyd, R., & Petty, B. D. (2020, February 27). Disorders and Diseases of Fish. MSD Veterinary Manual. https://www.msdvetmanual.com/all-other-pets/fish/disorders-and-diseases-of-fish

Freier, L., Hemmerich, J., Schöler, K., Wiechert, W., Oldiges, M., & von Lieres, E. (2016). Framework for Kriging-based iterative experimental analysis and design: Optimization of secretory protein production inCorynebacterium glutamicum. *Engineering in Life Sciences, 16*(6), 538–549. DOI: 10.1002/elsc.201500171

FSII. (n.d.). *Future of Agriculture: Drones*. Retrieved from Federation of Seed Industry of India: https://fsii.in/future-of-agriculture-drones/

Fuselage. (2020). *Solutions*. Retrieved from Fuselage: https://fuselage.co.in/solutions/

G, U., & Bv, G. (2020). A survey on plant disease prediction using machine learning and deep learning techniques. *Inteligencia Artificial, 23*(65), 136–154. DOI: 10.4114/intartif.vol23iss65pp136-154

Ganeshkumar, C., Jena, S. K., Sivakumar, A., & Nambirajan, T. (2023). Artificial intelligence in agricultural value chain: Review and future directions. *Journal of Agribusiness in Developing and Emerging Economies, 13*(3), 379–398. DOI: 10.1108/JADEE-07-2020-0140

Garg, V., . . . (Eds.). Applications of Artificial Intelligence in Business and Finance: Modern Trends. CRC Press,(2021). ISBN 9781774639658,272 Pages 8 Color & 30 B/W Illustrations,Published July 8, 2024 by Apple Academic Press.

Garlapati, A., Krishna, D. R., Garlapati, K., Rahul, U., & Narayanan, G. (2021, April). Stock price prediction using Facebook Prophet and Arima models. In 2021 6th International Conference for Convergence in Technology (I2CT) (pp. 1-7). IEEE. https://doi.org/DOI: 10.1109/I2CT51068.2021.9418057

Garrido, M.J., Villareal, M.G., Calderon, D., Muñoz, A., Macintyre, A., Mulliez, A., Goldberg, B., Felice, C.D., Salamanca, D., Martucci, D., Ubico, D., Gutnisky, G., Sipalla, H., Cornelio, I.D., Lozanoska, J., Amaya, J., Garcia, J., Araque, L., Estrada, M.S., Garita, M., Tarre, P., Gomez, R., Cornejo-Hösl, R., Ryan, R., Salojärvi, V., & Gonzalez, Z. (2015). The 2030 Agenda for Sustainable Development. *An Insider's Guide to the UN.*

GarudaAerospace. (2015). Retrieved from GarudaAerospace.com: https://www.garudaaerospace.com/

Ghani, B., & Hallerberg, S. (2021). *A Randomized Bag-of-Birds Approach to Study Robustness of Automated Audio Based Bird Species Classification.* MDPI. DOI: 10.3390/app11199226

Ghosh, P. (2021). AI Early Diagnosis Could Save Heart and Cancer Patients. *Science Correspondent. BBC News. 2018.*

Girish Kumar jha (2007). Artificial Neural Networks and its applications. *ARI, New Delhi, 48416112*

Glady, J. B. P., D'Souza, S. M., Priya, A. P., Amuthachenthiru, K., Vikram, G., & Boopathi, S. (2024). A Study on AI-ML-Driven Optimizing Energy Distribution and Sustainable Agriculture for Environmental Conservation. In *Harnessing High-Performance Computing and AI for Environmental Sustainability* (pp. 1–27). IGI Global. DOI: 10.4018/979-8-3693-1794-5.ch001

Godino-Llorente, J. I., Gómez-Vilda, P., Sáenz-Lechón, N., Blanco-Velasco, M., Cruz-Roldán, F., & Ferrer-Ballester, M. A. (2005). Support vector machines applied to the detection of voice disorders. In Nonlinear Analyses and Algorithms for Speech Processing: International Conference on Non-Linear Speech Processing, NOLISP 2005, Barcelona, Spain, April 19-22, 2005, Revised Selected Papers (pp. 219-230). Springer Berlin Heidelberg.

Gong, E. J., Bang, C. S., & Lee, J. J. (2024). Edge Artificial Intelligence Device in Real-Time Endoscopy for Classification of Gastric Neoplasms: Development and Validation Study. *Biomimetics*, *9*(12), 783. DOI: 10.3390/biomimetics9120783 PMID: 39727787

Gong, Y., & Schroeder, A. (2022). A systematic literature review of data privacy and security research on smart tourism. *Tourism Management Perspectives*, *44*, 101019. DOI: 10.1016/j.tmp.2022.101019

González, M. G., Janssen, A. P. A., & IJzerman, A. P. (2022). Oncological drug discovery: AI meets structure-based computational research. *Drug Discovery Today, 27*(9), 2404-2418.

Gul, D., & Banday, R. U. Z. (2024). *Transforming Crop Management Through Advanced AI and Machine Learning: Insights into Innovative Strategies for Sustainable Agriculture*. AI, Computer Science and Robotics Technology.

Guo, H., Cheng, Y., Liu, J., & Wang, Z. (2024). Low-cost and precise traditional Chinese medicinal tree pest and disease monitoring using UAV RGB image only. *Scientific Reports*, *14*(1), 25562. DOI: 10.1038/s41598-024-76502-x PMID: 39462013

Gupta, P., Malviya, L., & Charhate, S. V. (2019). 5G multi-element/port antenna design for wireless applications:a review. *International Journal of Microwave and Wireless Technologies*, *11*(9), 918–938. DOI: 10.1017/S1759078719000382

Gyamfi, E. K., Kropczynski, J., Johnson, J. S., & Yakubu, M. A. (2024). Internet of Things Security and Data Privacy Concerns in Smart Farming. *2024 IEEE World AI IoT Congress (AIIoT)*, 0575–0583. DOI: 10.1109/AIIoT61789.2024.10578972

Habuza, T., Navaz, A. N., Hashim, F., Alnajjar, F., Zaki, N., Serhani, M. A., & Statsenko, Y. (2021). AI applications in robotics, diagnostic image analysis and precision medicine: Current limitations, future trends, guidelines on CAD systems for medicine. *Informatics in Medicine Unlocked*, *24*, 100596. Advance online publication. DOI: 10.1016/j.imu.2021.100596

Hanamantray Karaddi, S., Dev Sharma, L., & Bhattacharya, A. (2024). Softflatten-net: A deep Convolutional neural network design for Monkeypox classification from digital skin lesions images. *IEEE Sensors Journal*, *24*(20), 32566–32576. DOI: 10.1109/JSEN.2024.3445286

Hariharan, U., Rajkumar, K., Akilan, T., & Jeyavel, J. (2021). Smart Wearable Devices for Remote Patient Monitoring in Healthcare 4.0. Internet of Medical Things: Remote Healthcare Systems and Applications, 117-135.

Hariram, N. P., Mekha, K. B., Suganthan, V., & Sudhakar, K. (2023). Sustainalism: An Integrated Socio-Economic-Environmental Model to Address Sustainable Development and Sustainability. *Sustainability (Basel)*, *15*(13), 10682. DOI: 10.3390/su151310682

Hashimoto, D. A., Rosman, G., Rus, D., & Meireles, O. R. (2018). Artificial intelligence in surgery: Promises and perils. *Annals of Surgery*, *268*(1), 70–76. DOI: 10.1097/SLA.0000000000002693 PMID: 29389679

Hassan, E., Saber, A., & Elbedwehy, S. (2024). Knowledge distillation model for acute lymphoblastic leukemia detection: Exploring the impact of Nesterov-accelerated adaptive moment estimation optimizer. *Biomedical Signal Processing and Control, 94*, 106246.

Hassan, E. (2024). Enhancing coffee bean classification: A comparative analysis of pre-trained deep learning models. *Neural Computing & Applications*, *36*(16), 9023–9052. DOI: 10.1007/s00521-024-09623-z

Hassan, E., Saber, A., & Elbedwehy, S. (2024). Knowledge distillation model for Acute Lymphoblastic Leukemia Detection: Exploring the impact of nesterov-accelerated adaptive moment estimation optimizer. *Biomedical Signal Processing and Control*, *94*, 106246. DOI: 10.1016/j.bspc.2024.106246

Hassan, E., Saber, A., El-kenawy, E. M., Bhatnagar, R., & Shams, M. Y. (2024). Early detection of Black fungus using deep learning models for efficient medical diagnosis. *2024 International Conference on Emerging Techniques in Computational Intelligence (ICETCI)*, 426-431. DOI: 10.1109/ICETCI62771.2024.10704103

Hassan, E., Shams, M. Y., Hikal, N. A., & Elmougy, S. (2022). The effect of choosing optimizer algorithms to improve computer vision tasks: A comparative study. *Multimedia Tools and Applications*, *82*(11), 16591–16633. DOI: 10.1007/s11042-022-13820-0 PMID: 36185324

Hassani, H., Silva, E. S., Unger, S., & TajMazinani, M. (2020). Artificial intelligence (AI) or intelligence augmentation (IA): What is the future? *AI, 1*(2), 143-155.

Hasselgren, C., & Oprea, T. I. (2024). Artificial intelligence for drug discovery: Are we there yet? *Annual Review of Pharmacology and Toxicology*, *64*(1), 527–550. DOI: 10.1146/annurev-pharmtox-040323-040828 PMID: 37738505

Hassija, V., Chamola, V., Mahapatra, A., Singal, A., Goel, D., Huang, K., Scardapane, S., Spinelli, I., Mahmud, M., & Hussain, A. (2024). Interpreting Black-Box Models: A Review on Explainable Artificial Intelligence. *Cognitive Computation*, *16*(1), 45–74. DOI: 10.1007/s12559-023-10179-8

Hayes, C. F., Rădulescu, R., Bargiacchi, E., Källström, J., Macfarlane, M., Reymond, M., Verstraeten, T., Zintgraf, L. M., Dazeley, R., Heintz, F., Howley, E., Irissappane, A. A., Mannion, P., Nowé, A., Ramos, G., Restelli, M., Vamplew, P., & Roijers, D. M. (2022). A practical guide to multi-objective reinforcement learning and planning. *Autonomous Agents and Multi-Agent Systems*, *36*(1), 26. DOI: 10.1007/s10458-022-09552-y

He, Y., Zhang, L., Chen, Z., & Li, C. Y. (2023). A framework of structural damage detection for civil structures using a combined multi-scale convolutional neural network and echo state network. *Engineering with Computers*, *39*(3), 1771–1789. DOI: 10.1007/s00366-021-01584-4

Hinton, G. E., Osindero, S., & Teh, Y. W. (2006). A fast learning algorithm for deep belief nets. *Neural Computation*, *18*(7), 1527–1554. DOI: 10.1162/neco.2006.18.7.1527 PMID: 16764513

Hoerl, R. W. (2020). Ridge Regression: A Historical Context. *Technometrics*, *62*(4), 420–425. DOI: 10.1080/00401706.2020.1742207

Hoffmann, U., Garcia, G., Vesin, J. M., Diserens, K., & Ebrahimi, T. (2005, March). A boosting approach to P300 detection with application to brain-computer interfaces. *In Conference Proceedings. 2nd* [IEEE.]. *International IEEE/EMBS Conference on Neural Engineering*, *2005*, 97–100.

Hollenstein, N., de la Torre, A., Langer, N., & Zhang, C. (2019, November). CogniVal: A Framework for Cognitive Word Embedding Evaluation. In Bansal, M., & Villavicencio, A. (Eds.), *Proceedings of the 23rd Conference on Computational Natural Language Learning (CoNLL)* (pp. 538–549). DOI: 10.18653/v1/K19-1050

Holzinger, A., Keiblinger, K., Holub, P., Zatloukal, K., & Müller, H. (2023). AI for life: Trends in artificial intelligence for biotechnology. *New Biotechnology*, *74*, 16–24. DOI: 10.1016/j.nbt.2023.02.001 PMID: 36754147

Hoque, A. (2024). Artificial Intelligence in Post-Harvest Drying Technologies: A Comprehensive Review on Optimization, Quality Enhancement, and Energy Efficiency. [IJSR]. *International Journal of Science and Research (Raipur, India), 13*(11), 493–502. DOI: 10.21275/SR241107163717

Hoque, A., & Padhiary, M. (2024). Automation and AI in Precision Agriculture: Innovations for Enhanced Crop Management and Sustainability. *Asian Journal of Research in Computer Science, 17*(10), 95–109. DOI: 10.9734/ajrcos/2024/v17i10512

Hoque, A., Padhiary, M., Prasad, G., & Kumar, K. (2025). Real-Time Data Processing in Agricultural Robotics. In Bora, D. J., & Bania, R. K. (Eds.), *Computer Vision Techniques for Agricultural Advancements* (pp. 431–468). IGI Global., DOI: 10.4018/979-8-3693-8019-2.ch014

Huang, H. P., Liu, Y. H., Lin, W. Z., Kang, Z. H., Cheng, C. A., & Huang, T. H. (2015). Development of a P300 brain–machine interface and design of an elastic mechanism for a rehabilitation robot. *International Journal of Automation and Smart Technology, 5*(2), 123–145.

Hu, H. C., Chang, S. Y., Wang, C. H., Li, K. J., Cho, H. Y., Chen, Y. T., & Lee, O. K. S. (2021). Deep learning application for vocal fold disease prediction through voice recognition: Preliminary development study. *Journal of Medical Internet Research, 23*(6), e25247. DOI: 10.2196/25247 PMID: 34100770

Hu, J., Li, D., Duan, Q., Han, Y., Chen, G., & Si, X. (2012). Fish species classification by color, texture and multi-class support vector machine using computer vision. *Computers and Electronics in Agriculture, 88*, 133–140. DOI: 10.1016/j.compag.2012.07.008

Hukunda, B., Rau, A., & Upadhyay, P. (2023). Reimaging Healthcare Opportunities with Artificial Intelligence. *Infosys Navigate Your Next. 2018. Available online:*https://www.infosys.com/industries/healthcare/featuresopinions/Documents/reimagining-healthcare-opportunities.pdf

Hussain, A., Draz, U., Ali, T., Tariq, S., Irfan, M., Glowacz, A., Antonino Daviu, J. A., Yasin, S., & Rahman, S. (2020). Waste Management and Prediction of Air Pollutants Using IoT and Machine Learning Approach. *Energies, 13*(15), 3930. DOI: 10.3390/en13153930

Hwang, D. K., Hsu, C. C., Chang, K. J., Chao, D., Sun, C. H., Jheng, Y. C., Yarmishyn, A. A., Wu, J.-C., Tsai, C.-Y., Wang, M.-L., Peng, C.-H., Chien, K.-H., Kao, C.-L., Lin, T.-C., Woung, L.-C., Chen, S.-J., & Chiou, S. H. (2019). Artificial intelligence-based decision-making for age-related macular degeneration. *Theranostics, 9*(1), 232–245. DOI: 10.7150/thno.28447 PMID: 30662564

Hyder, F., Kida, I., Behar, K. L., Kennan, R. P., Maciejewski, P. K., & Rothman, D. L. (2001). Quantitative functional imaging of the brain: Towards mapping neuronal activity by BOLD fMRI. NMR in Biomedicine. *NMR in Biomedicine*, *14*(7-8), 413–431. DOI: 10.1002/nbm.733 PMID: 11746934

IBEF. (2024, July 18). *Future of Drone Technology in Indian Agriculture: Enhancing Efficiency and Sustainability*. Retrieved from IBEF: https://www.ibef.org/blogs/future-of-drone-technology-in-indian-agriculture-enhancing-efficiency-and-sustainability

ideaForge. (2023, September 11). *Applications of Agriculture Drones: From Skies to Fields, Farm with Innovation!* Retrieved from ideaForge Technology: https://ideaforgetech.com/blogs/applications-of-agriculture-drones

Imade, A., Ounacer, S., El Ghoumari, M. Y., Ardchir, S., & Azzouazi, M. (2024). Agricultural Yield Prediction Using ML Algorithms in the Industry 5.0. In Chakir, A., Bansal, R., & Azzouazi, M. (Eds.), *Industry 5.0 and Emerging Technologies* (Vol. 565, pp. 135–157). Springer Nature Switzerland., DOI: 10.1007/978-3-031-70996-8_7

Indaryanto, A., Harijadi, B. D., & Sinaga, E. (2023). The growing use and impact of artificial intelligence technologies in the tourism industry. *Sustainable Engineering and Innovation*, *5*(2), 189–204. DOI: 10.37868/sei.v5i2.id238

Irrigation planning - Use every drop wisely. (2024, April 23). CropX. https://cropx.com/cropx-system/irrigation-planning/

Ishtiaq, M., Saeed, N., & Khan, M. A. (2021). Edge computing in IoT: A 6G perspective. https://doi.org/DOI: 10.36227/techrxiv.17031665

Jagani, D., & Degadwala, S. (2024). Monkeypox skin lesion classification using fine-tune CNN model. *2024 4th International Conference on Pervasive Computing and Social Networking (ICPCSN)*. DOI: 10.1109/ICPCSN62568.2024.00014

Jagdish. (2024, March 8). Artificial intelligence and its aquaculture applications. Ocean News & Technology. https://ecomagazine.com/news/fisheries-aquaculture/artificial-intelligence-and-its-aquaculture-applications/

Jamei, M., Karbasi, M., Malik, A., Jamei, M., Kisi, O., & Yaseen, Z. M. (2022). Long-term multi-step ahead forecasting of root zone soil moisture in different climates: Novel ensemble-based complementary data-intelligent paradigms. *Agricultural Water Management*, *269*, 107679. DOI: 10.1016/j.agwat.2022.107679

Jauro, F., Chiroma, H., Gital, A. Y., Almutairi, M., Abdulhamid, S. M., & Abawajy, J. H. (2020). Deep learning architectures in emerging cloud computing architectures: Recent development, challenges and next research trend. *Applied Soft Computing*, *96*, 106582. DOI: 10.1016/j.asoc.2020.106582

Javaid, M., Haleem, A., Khan, I. H., & Suman, R. (2023). Understanding the potential applications of Artificial Intelligence in Agriculture Sector. *Advanced Agrochem*, *2*(1), 15–30.

Jeffrey, L., & Bommu, R. (2024). Innovative AI Solutions for Agriculture: Enhancing CropManagement and Yield. *International Journal of Advanced Engineering Technologies and Innovations*, *1*(3), 203–221.

Jeschke, P., Alteköster, C., Hansson Mild, K., Israel, M., Ivanova, M., Schiessl, K., Shalamanova, T., Soyka, F., Stam, R., & Wilén, J. (2022). Protection of workers exposed to Radiofrequency electromagnetic fields: A perspective on open questions in the context of the new ICNIRP 2020 guidelines. *Frontiers in Public Health*, *10*, 875946. Advance online publication. DOI: 10.3389/fpubh.2022.875946 PMID: 35757616

JetayuGadgets. (n.d.). *How to Use Drones for Agriculture in India*. Retrieved from JetayuGadgets.com: https://jetayugadgets.com/blog/how-to-use-drones-for-agriculture-in-india/?srsltid=AfmBOorFqoLwMmcwo7iHjtrxQvVyEdGcs55tlVlqnZXHdtCpi9AGZjKg

Jiang, H., Peng, M., Zhong, Y., Xie, H., Hao, Z., Lin, J., Ma, X., & Hu, X. (2022). A Survey on Deep Learning-Based Change Detection from High-Resolution Remote Sensing Images. *Remote Sensing (Basel)*, *14*(7), 1552. DOI: 10.3390/rs14071552

Jiang, X., Wang, X., Chen, Y., Liu, C., & Wang, Y. (2022). A differential privacy-based scheme for sharing genomic datasets in genome-wide association studies (GWAS). *Journal of Biomedical Informatics*, *128*, 104032. DOI: 10.1016/j.jbi.2022.104032

Jiang, Y., Li, X., Luo, H., Yin, S., & Kaynak, O. (2022). Quo vadis artificial intelligence? *Discover Artificial Intelligence*, *2*(1), 4. Advance online publication. DOI: 10.1007/s44163-022-00022-8

Jin, Y., Chen, B., Lampinen, B. D., & Brown, P. H. (2020). Advancing Agricultural Production With Machine Learning Analytics: Yield Determinants for California's Almond Orchards. *Frontiers in Plant Science*, *11*, 290. DOI: 10.3389/fpls.2020.00290 PMID: 32231679

Jin, Y., Sharifi, A., Li, Z., Chen, S., Zeng, S., & Zhao, S. (2024). Carbon emission prediction models: A review. *The Science of the Total Environment*, *927*, 172319. DOI: 10.1016/j.scitotenv.2024.172319 PMID: 38599410

Jin, Z., & Chen, Y. (2015). Telemedicine in the cloud era: Prospects and challenges. *IEEE Pervasive Computing*, *14*(1), 54–61. DOI: 10.1109/MPRV.2015.19

Joshi, A., Pradhan, B., Gite, S., & Chakraborty, S. (2023). Remote-Sensing Data and Deep-Learning Techniques in Crop Mapping and Yield Prediction: A Systematic Review. *Remote Sensing (Basel)*, *15*(8), 2014. DOI: 10.3390/rs15082014

Jumper, J., Evans, R., Pritzel, A., Green, T., Figureurnov, M., Ronneberger, O., Tunyasuvunakool, K., Bates, R., Žídek, A., & Potapenko, A. (2021). Highly accurate protein structure prediction with AlphaFold. *Nature*, *596*(7873), 583–589. DOI: 10.1038/s41586-021-03819-2 PMID: 34265844

Jurgens, J. R. (2022). *Using Technology To Improve a Billion Livelihoods*. Geneva: World Economic Forum In Collaboration with Adani Group.

Kainth, G. (2023, August 28). *Drone Technology In Agriculture: Indian Experience – Analysis*. Retrieved from Eurasiareview.com: https://www.eurasiareview.com/28082023-drone-technology-in-agriculture-indian-experience-analysis/

Kalyanaraman, A., Burnett, M., Fern, A., Khot, L., & Viers, J. (2022). Special report: The AgAID AI institute for transforming workforce and decision support in agriculture. *Computers and Electronics in Agriculture*, *197*, 106944. DOI: 10.1016/j.compag.2022.106944

Kamilaris, A., & Prenafeta-Boldú, F. X. (2018). Deep learning in agriculture: A survey. *Computers and Electronics in Agriculture*, *147*, 70–90. DOI: 10.1016/j.compag.2018.02.016

Kang, S., Chen, E. C., Cifuentes, H., Cole, G., & Graham, J. (2024). Complex in vitro models positioned for impact to drug testing in pharma: A review. *Drug Testing Reports, 12*(4), 455-470.

Kanipriya, M., Sugumaran, D., Pandiyarajan, V., Yogananda, B. S., Selvakumar, P., & Mohan, S. (2024, January). Analysis of Electromyography Signals for Control of Mechanical Prosthesis using Machine Learning Techniques. In 2024 2nd International Conference on Intelligent Data Communication Technologies and Internet of Things (IDCIoT) (pp. 1059-1067). IEEE.

Karipidis, K., Mate, R., Urban, D., Tinker, R., & Wood, A. (2021). 5G mobile networks and health—A state-of-the-science review of the research into low-level RF fields above 6 GHz. *Journal of Exposure Science & Environmental Epidemiology*, *31*(4), 585–605. DOI: 10.1038/s41370-021-00297-6 PMID: 33727687

Karthika, K., Dhanalakshmi, S., Murthy, S. M., Mishra, N., Sasikala, S., & Murugan, S. (2023, October). Raspberry Pi-enabled Wearable Sensors for Personal Health Tracking and Analysis. *In 2023 International Conference on Self Sustainable Artificial Intelligence Systems* (ICSSAS) (pp. 1254-1259). IEEE.

Kashyap, B., & Kumar, R. (2021). Sensing Methodologies in Agriculture for Monitoring Biotic Stress in Plants Due to Pathogens and Pests. *Inventions (Basel, Switzerland)*, *6*(2), 29. DOI: 10.3390/inventions6020029

Kaswan, K. S., Gaur, L., Dhatterwal, J. S., & Kumar, R. (2021). AI-based natural language processing for the generation of meaningful information electronic health record (EHR) data. In Advanced AI techniques and applications in bioinformatics (pp. 41-86). CRC Press.

Katekar, V., & Cheruku, J. K. (2022). The Application of Drone Technology for Sustainable Agriculture in India. *Current Agriculture Research Journal*, *10*(3), 352–365. DOI: 10.12944/CARJ.10.3.19

Kaur, G., Rajni, , & Sivia, J. S. (2024). Integrating Data Envelopment Analysis and Machine Learning Approaches for Energy Optimization, Decreased Carbon Footprints, and Wheat Yield Prediction Across North-Western India. *Journal of Soil Science and Plant Nutrition*, *24*(1), 1424–1447. DOI: 10.1007/s42729-024-01647-7

Kazi, K. (2024). AI-Driven IoT (AIIoT) in Healthcare Monitoring. In *Using Traditional Design Methods to Enhance AI-Driven Decision Making* (pp. 77-101). IGI Global.

Keating, B. A., & Thorburn, P. J. (2018). Modelling crops and cropping systems—Evolving purpose, practice and prospects. *European Journal of Agronomy*, *100*, 163–176. DOI: 10.1016/j.eja.2018.04.007

Ke, T. T., & Sudhir, K. (2023). Privacy rights and data security: GDPR and personal data markets. *Management Science*, *69*(8), 4389–4412. DOI: 10.1287/mnsc.2022.4614

Khang, A. (Ed.). (2024). *AI-Driven Innovations in Digital Healthcare: Emerging Trends, Challenges, and Applications: Emerging Trends*. Challenges, and Applications.

Khan, M. J., & Hong, K. S. (2015). Passive BCI based on drowsiness detection: An fNIRS study. *Biomedical Optics Express*, *6*(10), 4063–4078. DOI: 10.1364/BOE.6.004063 PMID: 26504654

Kholil, M., Ismanto, I., & Fu'Ad, M. N. (2021, February). 3D reconstruction using structure from motion (SFM) algorithm and multi view stereo (MVS) based on computer vision. In *IOP Conference Series: Materials Science and Engineering* (Vol. 1073, No. 1, p. 012066). IOP Publishing. https://doi.org/DOI: 10.1088/1757-899X/1073/1/012066

Khorasani, E. S. (2008). Artificial intelligence: Structures and strategies for complex problem solving. Scalable Computing: Practice and Experience, 9(3).

Khosrowabadi, R., Quek, C., Ang, K. K., & Wahab, A. (2013). ERNN: A biologically inspired feedforward neural network to discriminate emotion from EEG signal. *IEEE Transactions on Neural Networks and Learning Systems*, *25*(3), 609–620. DOI: 10.1109/TNNLS.2013.2280271 PMID: 24807454

Khoury, M. J., & McCabe, L. L. (2015). Personalized medicine: From aspiration to reality. *Journal of the American Medical Association*, *314*(14), 1451–1452. PMID: 26461994

Kim, J., Kim, K., Kang, Y., & Jeong, S. (2024). Predicting Key Substance Levels in Aquaculture through AI-Based Water Quality Monitoring. Human Interaction and Emerging Technologies (IHIET 2024). .DOI: 10.54941/ahfe1005479

Kim, H. E., Cosa-Linan, A., Santhanam, N., Jannesari, M., Maros, M. E., & Ganslandt, T. (2022). Transfer learning for medical image classification: A literature review. *BMC Medical Imaging*, *22*(1), 69. DOI: 10.1186/s12880-022-00793-7 PMID: 35418051

Kim, S., & Nasim, I. (2020). Human electromagnetic Field exposure in 5G at 28 GHz. *IEEE Consumer Electronics Magazine*, *9*(6), 41–48. DOI: 10.1109/MCE.2019.2956223

Kim, Y. J., Bae, J. P., Chung, J. W., Park, D. K., Kim, K. G., & Kim, Y. J. (2021). New polyp image classification technique using transfer learning of network-in-network structure in endoscopic images. *Scientific Reports*, *11*(1), 3605. DOI: 10.1038/s41598-021-83199-9 PMID: 33574394

King, Z. A., O'Brien, E. J., Feist, A. M., & Palsson, B. O. (2017). Literature mining supports a next-generation modeling approach to predict cellular byproduct secretion. *Metabolic Engineering*, *39*, 220–227. DOI: 10.1016/j.ymben.2016.12.004 PMID: 27986597

Kist, A. M., Gómez, P., Dubrovskiy, D., Schlegel, P., Kunduk, M., Echternach, M., & Döllinger, M. (2021). A deep learning enhanced novel software tool for laryngeal dynamics analysis. *Journal of Speech, Language, and Hearing Research: JSLHR*, *64*(6), 1889–1903. DOI: 10.1044/2021_JSLHR-20-00498 PMID: 34000199

Kourou, K., Exarchos, T. P., Karamouzis, M. V., & Papaloukas, C. (2015). Machine learning applications in cancer prognosis and prediction. *Computational and Structural Biotechnology Journal*, *13*, 8–17. DOI: 10.1016/j.csbj.2014.11.005 PMID: 25750696

krishiviman. (2024, January 4). *Cost-Benefit Analysis: Is Investing in Agricultural Drones Worth it for Indian Farmers?* Retrieved from krishiviman: https://www.krishiviman.com/blog/cost-benefit-analysis-is-investing-in-agricultural-drones-worth-it-for-indian-farmers/

Kumar, K., & Thakur, G. S. M. (2012). Advanced applications of neural networks and artificial intelligence: A review. International journal of information technology and computer science, 4(6), 57-68.

Kumaran, S., D, S. K., A, S., & K, S. K. (2023). Crop Pest Control using IoT Real-Time Monitoring and Intervention. *2023 International Conference on Sustainable Communication Networks and Application (ICSCNA)*, 1552–1557. DOI: 10.1109/ICSCNA58489.2023.10370176

Kumar, D., Kumar, K., Roy, P., & Rabha, G. (2024). Renewable Energy in Agriculture: Enhancing Aquaculture and Post-Harvest Technologies with Solar and AI Integration. *Asian Journal of Research in Computer Science*, *17*(12), 201–219. DOI: 10.9734/ajrcos/2024/v17i12539

Kumari, A., Edla, D. R., Reddy, R. R., Jannu, S., Vidyarthi, A., Alkhayyat, A., & de Marin, M. S. G. (2024). EEG-based motor imagery channel selection and classification using hybrid optimization and two-tier deep learning. *Journal of Neuroscience Methods*, *409*, 110215. DOI: 10.1016/j.jneumeth.2024.110215 PMID: 38968976

Kumar, V., Bhalla, A., & Rathore, A. S. (2014). Design of experiments applications in bioprocessing: Concepts and approach. *Biotechnology Progress*, *30*(1), 86–99. DOI: 10.1002/btpr.1821 PMID: 24123959

Kundu, D., Rahman, M. M., Rahman, A., Das, D., Siddiqi, U. R., Alam, M. G., Dey, S. K., Muhammad, G., & Ali, Z. (2024). Federated deep learning for Monkeypox disease detection on GAN-augmented dataset. *IEEE Access : Practical Innovations, Open Solutions*, *12*, 32819–32829. DOI: 10.1109/ACCESS.2024.3370838

Kuppusamy, P., Suresh, J. K., & Shanmugananthan, S. (2023). Machine Learning-Enabled Internet of Things Solution for Smart Agriculture Operations. In Goel, N., & Yadav, R. K. (Eds.), (pp. 84–115). Advances in Systems Analysis, Software Engineering, and High Performance Computing. IGI Global., DOI: 10.4018/978-1-6684-8785-3.ch005

Langer, S. (2021). Approximating smooth functions by deep neural networks with sigmoid activation function. *Journal of Multivariate Analysis*, *182*, 104696. DOI: 10.1016/j.jmva.2020.104696

Latif, S., Qadir, J., Farooq, S., & Imran, M. (2017). How 5G wireless (and concomitant technologies) will revolutionize healthcare? *Future Internet*, *9*(4), 93. DOI: 10.3390/fi9040093

Laves, M.-H., Bicker, J., Kahrs, L. A., & Ortmaier, T. (2019). A dataset of laryngeal endoscopic images with comparative study on convolution neural network based semantic segmentation. *Medical Image Analysis*, *54*, 158–170. PMID: 30649670

Le Quy, T., Roy, A., Iosifidis, V., Zhang, W., & Ntoutsi, E. (2022). A survey on datasets for fairness-aware machine learning. *Wiley Interdisciplinary Reviews. Data Mining and Knowledge Discovery*, *12*(3), e1452. DOI: 10.1002/widm.1452

LeCun, Y., Bengio, Y., & Hinton, G. (2015). Deep learning. *Nature*, *521*(7553), 436–444. DOI: 10.1038/nature14539 PMID: 26017442

Lenhardt, A., Kaper, M., & Ritter, H. J. (2008). An adaptive P300-based online brain–computer interface. *IEEE Transactions on Neural Systems and Rehabilitation Engineering*, *16*(2), 121–130. DOI: 10.1109/TNSRE.2007.912816 PMID: 18403280

Leung, W. K., Liu, X., & Meng, H. (2019). CNN-RNN-CTC based end-to-end mispronunciation detection and diagnosis. In ICASSP 2019-2019 IEEE International Conference on Acoustics, Speech and Signal Processing (ICASSP) (pp. 8132-8136). IEEE. DOI: 10.1109/ICASSP.2019.8682654

Levchenko, O., Kolev, B., Yagoubi, D.-E., Akbarinia, R., Masseglia, F., Palpanas, T., Shasha, D., & Valduriez, P. (2021). BestNeighbor: Efficient evaluation of kNN queries on large time series databases. *Knowledge and Information Systems*, *63*(2), 349–378. DOI: 10.1007/s10115-020-01518-4

Li, B., Han, X., & Wu, D. (2018, January). Real-time crowd density estimation based on convolutional neural networks. In *2018 International Conference on Intelligent Transportation, Big Data & Smart City (ICITBS)* (pp. 690-694). IEEE. https://doi.org/DOI: 10.1109/ICITBS.2018.00179

Li, D., Li, X., Wang, Q., & Hao, Y. (2022). Advanced techniques for the intelligent diagnosis of fish diseases: A review. *Animals (Basel)*, *12*(21), 2938. DOI: 10.3390/ani12212938 PMID: 36359061

Li, G., Rabe, K. S., Nielsen, J., & Engqvist, M. K. M. (2019). Machine Learning Applied to Predicting Microorganism Growth Temperatures and Enzyme Catalytic Optima. *ACS Synthetic Biology*, *8*(6), 1411–1420. DOI: 10.1021/acssynbio.9b00099 PMID: 31117361

Li, H., Zhang, Q., Liu, X., Zhou, J., & Wang, Y. (2024). Evaluation of federated learning models for Parkinson's disease prediction using genomic and transcriptomic data. *Bioinformatics (Oxford, England)*, *40*(1), bta009. Advance online publication. DOI: 10.1093/bioinformatics/bta009 PMID: 38141210

Litjens, G., Kooi, T., Bejnordi, B. E., Setio, A. A., Ciompi, F., Ghafoorian, M., Van Der Laak, J. A., Van Ginneken, B., & Sánchez, C. I. (2017). A survey on deep learning in medical image analysis. *Medical Image Analysis*, *42*, 60–88. DOI: 10.1016/j.media.2017.07.005 PMID: 28778026

Liu, J., Du, H., Huang, L., Xie, W., & Liu, K. (2024). AI-powered microfluidics: Shaping the future of phenotypic drug discovery. *ACS Applied Materials & Interfaces, 16*(12), 2204-2212.

Liu, G. S., Hodges, J. M., Yu, J., Sung, C. K., Erickson-DiRenzo, E., & Doyle, P. C. (2023). End-to-end deep learning classification of vocal pathology using stacked vowels. *Laryngoscope Investigative Otolaryngology*, *8*(5), 1312–1318. DOI: 10.1002/lio2.1144 PMID: 37899847

Liu, X., Mehraliyev, F., Liu, C., & Schuckert, M. (2020). The roles of social media in tourists' choices of travel components. *Tourist Studies*, *20*(1), 27–48. DOI: 10.1177/1468797619873107

Liu, Y. H., Wu, C. T., Cheng, W. T., Hsiao, Y. T., Chen, P. M., & Teng, J. T. (2014). Emotion recognition from single-trial EEG based on kernel Fisher's emotion pattern and imbalanced quasiconformal kernel support vector machine. *Sensors (Basel)*, *14*(8), 13361–13388. DOI: 10.3390/s140813361 PMID: 25061837

Liu, Y., & Gunawan, R. (2017). Bioprocess optimization under uncertainty using ensemble modeling. *Journal of Biotechnology*, *244*, 34–44. DOI: 10.1016/j.jbiotec.2017.01.013 PMID: 28137617

Li, Y., Gu, W., Yue, H., Lei, G., Guo, W., Wen, Y., & Lei, W. (2023). Real-time detection of laryngopharyngeal cancer using an artificial intelligence-assisted system with multimodal data. *Journal of Translational Medicine*, *21*(1), 698. DOI: 10.1186/s12967-023-04572-y PMID: 37805551

Li, Y., Long, J., Yu, T., Yu, Z., Wang, C., Zhang, H., & Guan, C. (2010). An EEG-based BCI system for 2-D cursor control by combining Mu/Beta rhythm and P300 potential. *IEEE Transactions on Biomedical Engineering*, *57*(10), 2495–2505. DOI: 10.1109/TBME.2010.2055564 PMID: 20615806

Li, Y., Xu, Y., Ma, Z., Ye, Y., Gao, L., & Sun, Y. (2022). An XGBoost-based model for assessment of aortic stiffness from wrist photoplethysmogram. *Computer Methods and Programs in Biomedicine*, *226*, 107128. DOI: 10.1016/j.cmpb.2022.107128 PMID: 36150230

Liyakat, K. K. S. (2024). Smart Agriculture based on AI-Driven-IoT (AIIoT): A KSK Approach. *labour, 18*, 19.

Lodato, F., Garzia, A., Valbonesi, S., Ruello, G., Iodice, A., Matera, F., Salvo, P., & Massa, R. (2024). Ray tracing tools assessment for the evaluation of EMF levels generated by 5G NR systems: An overview. *2024 IEEE International Symposium on Measurements & Networking (M&N), 120*, 1-6. https://doi.org/DOI: 10.1109/MN60932.2024.10615802

Londhe, K., Dharmadhikari, N., Zaveri, P., & Sakoglu, U. (2024). Enhanced Travel Experience using Artificial Intelligence: A Data-driven Approach. *Procedia Computer Science*, *235*, 1920–1928. DOI: 10.1016/j.procs.2024.04.182

Lou, D., Chen, M., & Ye, J. (2007). Study on a fish disease case reasoning system based on image retrieval. *New Zealand Journal of Agricultural Research*, *50*(5), 887–893. DOI: 10.1080/00288230709510364

Loureiro, S. M. C., & Nascimento, J. (2021). Shaping a view on the influence of technologies on sustainable tourism. *Sustainability (Basel)*, *13*(22), 12691. DOI: 10.3390/su132212691

Lukita, C., Pangilinan, G. A., Chakim, M. H. R., & Saputra, D. B. (2023). Examining the impact of artificial intelligence and internet of things on smart tourism destinations: A comprehensive study. [ATT]. *Aptisi Transactions on Technopreneurship*, *5*(2sp), 135–145. DOI: 10.34306/att.v5i2sp.332

Lutz, C. G. (2023, March 1). The rise of AI in aquaculture. The Fish Site. https://thefishsite.com/articles/the-rise-of-ai-in-aquaculture-artificial-intelligence

Luxton, D. D., & Riek, L. D. (2019). Artificial intelligence and robotics in rehabilitation. In Brenner, L. A., Reid-Arndt, S. A., Elliott, T. R., Frank, R. G., & Caplan, B. (Eds.), *Handbook of rehabilitation psychology* (3rd ed., pp. 507–520). American Psychological Association., DOI: 10.1037/0000129-031

Mahadev Madgule, N. (2023). Vinayaka, Yeshwant M. Sonkhaskar, Dhiren Ramanbhai Patel, R. Karthikeyan, P. Selvakumar,(2023). Mechanical properties and microstructure of activated TIG welded similar joints of Inconel alloys by desirability approaches. *Materials Today: Proceedings*, *77*, 528–533. DOI: 10.1016/j.matpr.2022.12.250

Mahfouz, N. B., Moustafa, E. M., & El-Baz, N. F. (2016). Studies on Some Bacterial Diseases in Some Cultured Freshwater Fishes in Kafr El-Sheikh Governorate. *Kafrelsheikh Veterinary Medical Journal*, *14*(1), 149–161. DOI: 10.21608/kvmj.2016.108686

Mahmoud, A., Alhussein, M., Aurangzeb, K., & Takaoka, E. (2024). Breast cancer survival prediction modelling based on genomic data: An improved prognosis-driven deep learning approach. *IEEE Access, 12*(6), 455-470.

Mahmoud, A., Awad, N. A., Alsubaie, N., Ansarullah, S. I., Alqahtani, M. S., Abbas, M., Usman, M., Soufiene, B. O., & Saber, A. (2023). Advanced deep learning approaches for accurate brain tumor classification in medical imaging. *Symmetry, 15*(3), 571.

Mahmoud, A., Hamza, T., & Rashad, M. Z. (2020). Prediction of chemical toxicity for drug design using AIRS algorithms and hybrid classifiers. *International Journal of Applied Mathematics & Information Sciences, 14*(2), 133-144.

Mahmoud, A., Awad, N. A., Alsubaie, N., Ansarullah, S. I., Alqahtani, M. S., Abbas, M., Usman, M., Soufiene, B. O., & Saber, A. (2023). Advanced deep learning approaches for accurate brain tumor classification in medical imaging. *Symmetry*, *15*(3), 571. DOI: 10.3390/sym15030571

Maiseli, B. J., Abdalla, A. T., Massawe, L. V., Mbise, M., Mkocha, K., Nassor, N. A., Ismail, M., Michael, J., & Kimambo, S. (2023). Brain–computer interface: Trend, challenges, and threats. *Brain Informatics*, *10*(1), 20. DOI: 10.1186/s40708-023-00199-3 PMID: 37540385

Makin, J. G., Moses, D. A., & Chang, E. F. (2019). Machine translation of cortical activity to text with an encoder–decoder framework. *Nature Neuroscience*, *23*(4), 575–582. DOI: 10.1038/s41593-020-0608-8 PMID: 32231340

Mak, K. K., Wong, Y. H., & Pichika, M. R. (2023). Artificial intelligence in drug discovery and development. In *Drug Discovery and Evaluation [Springer.]. *Safety (Basel, Switzerland)*, •••, 345–378.

Malone, G. (2024, January 30). *Early Pest Detection: The Drone Advantage in Farming*. Retrieved from Medium.com: https://medium.com/@gmalone_38909/early-pest-detection-the-drone-advantage-in-farming-d3214fc088ad

Mandal, A. K. (2023). Usage of particle swarm optimization in digital images selection for Monkeypox virus prediction and diagnosis. DOI: 10.21203/rs.3.rs-2421266/v1

Mandal, A., & Ghosh, A. R. (2023). Role of artificial intelligence in fish growth and health status monitoring: A review on sustainable aquaculture. *Aquaculture International*, *32*(3), 2791–2820. DOI: 10.1007/s10499-023-01297-z

Ma, R., Aghasadeghi, N., Jarzebowski, J., Bretl, T., & Coleman, T. P. (2011). A stochastic control approach to optimally designing hierarchical flash sets in P300 communication prostheses. *IEEE Transactions on Neural Systems and Rehabilitation Engineering*, *20*(1), 102–112. DOI: 10.1109/TNSRE.2011.2179560 PMID: 22203722

Marchetti, N. (2017). Towards the 5th generation of wireless communication systems. *arXiv preprint arXiv:1702.00370.*

Mardis, E. R. (2017). DNA sequencing technologies: 2006–2016. *Nature Protocols*, *12*(2), 365–368. DOI: 10.1038/nprot.2016.182 PMID: 28055035

Marescotti, D., Narayanamoorthy, C., Bonjour, F., Kuwae, K., et al. (2022). AI-driven laboratory workflows enable operation in the age of social distancing. *SLAS Technology, 27*(3), 129-138.

Martinsson, J. (2017). Bird species identification using convolutional neural networks.

Marutdrones. (2019). Retrieved from Marutdrones.com: https://marutdrones.com/products/

Marx, V. (2013). The big challenges of big data. *Nature*, *498*(7453), 255–260. DOI: 10.1038/498255a PMID: 23765498

Matrouk, K., Trabelsi, Y., Gomathy, V., Arun Kumar, U., Rathish, C., & Parthasarathy, P. (2023). Energy efficient data transmission in intelligent transportation system (ITS): Millimeter (Mm wave) based routing algorithm for connected vehicles. *Optik (Stuttgart)*, *273*, 170374. DOI: 10.1016/j.ijleo.2022.170374

Mattei, E., Lozzi, D., Matteo, A. D., Polsinelli, M., Manes, C., Mignosi, F., & Placidi, G. (2024). Deep Learning Architecture analysis for EEG-Based BCI Classification under Motor Execution. *2024 IEEE 37th International Symposium on Computer-Based Medical Systems (CBMS)*, 549-555.

Ma, Y., Zhang, Z., Kang, Y., & Özdoğan, M. (2021). Corn yield prediction and uncertainty analysis based on remotely sensed variables using a Bayesian neural network approach. *Remote Sensing of Environment*, *259*, 112408. DOI: 10.1016/j.rse.2021.112408

Mbunge, E., & Muchemwa, B. (2022). Deep learning and machine learning techniques for analyzing travelers' online reviews: a review. Optimizing digital solutions for hyper-personalization in tourism and hospitality, 20-39. https://doi.org/DOI: 10.4018/978-1-7998-8306-7.ch002

McCarthy, J. (1960). Recursive Functions of Symbolic Expressions and Their Computation by Machine, Part I. *Communications of the ACM*, *3*(4), 184–195. DOI: 10.1145/367177.367199

McCarthy, J., Minsky, M. L., Rochester, A., & Shannon, C. E. (1956). *A Proposal for the Dartmouth Summer Research Project on Artificial Intelligence*. Dartmouth College.

McCoy, L. G., Brenna, C. T. A., Chen, S. S., Vold, K., & Das, S. (2022). Believing in black boxes: Machine learning for healthcare does not need explainability to be evidence-based. *Journal of Clinical Epidemiology*, *142*, 252–257. DOI: 10.1016/j.jclinepi.2021.11.001 PMID: 34748907

McCredden, J. E., Weller, S., & Leach, V. (2023). The assumption of safety is being used to justify the rollout of 5G technologies. *Frontiers in Public Health*, *11*, 1058454. Advance online publication. DOI: 10.3389/fpubh.2023.1058454 PMID: 36815158

Mecha, N. J. M. F., & Gonzales-Plasus, M. M. (2023). The common diseases of freshwater ornamental fishes and the treatments applied by local fish owners in Puerto Princesa City, Palawan, Philippines. *Journal of Fisheries*, *11*(1), 111205–111205.

Meena, B. S. (n.d.). *Drones in Crop Monitoring and Management: Uses, Types, Advantages and Limitations*. Retrieved from WikiFarmer.com: https://wikifarmer.com/library/en/article/drones-in-crop-monitoring-and-management-uses-types-advantages-and-limitations

Mehyadin, A. E., Abdulazeez, A. M., Hasan, D. A., & Saeed, J. N. (2021). Birds sound classification based on machine learning algorithms. *Asian Journal of Research in Computer Science*, *9*(4), 1–11.

Mei, Z., Vatsavai, R., & Chirkova, R. (2023). Q-learning Based Simulation Tool for Studying Effectiveness of Dynamic Application of Fertilizer on Crop Productivity. *Proceedings of the 11th ACM SIGSPATIAL International Workshop on Analytics for Big Geospatial Data*, 13–22. DOI: 10.1145/3615833.3628591

Mekonnen, Y., Namuduri, S., Burton, L., Sarwat, A., & Bhansali, S. (2020). Review—Machine Learning Techniques in Wireless Sensor Network Based Precision Agriculture. *Journal of the Electrochemical Society*, *167*(3), 037522. DOI: 10.1149/2.0222003JES

Melhart, D., Togelius, J., Mikkelsen, B., Holmgård, C., & Yannakakis, G. N. (2023). The ethics of AI in games. *IEEE Transactions on Affective Computing*, *15*(1), 1–14. DOI: 10.1109/TAFFC.2023.3276425

Melnychenko, S., Boiko, M., Okhrimenko, A., Bosovska, M., & Mazaraki, N. (2020). Foresight technologies of economic systems: Evidence from the tourism sector of Ukraine. *Problems and Perspectives in Management*, *18*(4), 303–318. DOI: 10.21511/ppm.18(4).2020.25

Meng, L., Jin, J., & Wang, X. (2011, October). A comparison of three electrode channels selection methods applied to SSVEP BCI. *In 2011 4th international conference on biomedical engineering and informatics* (BMEI) (Vol. 1, pp. 584-587). IEEE.

Miller, A. B., Sears, M. E., Morgan, L. L., Davis, D. L., Hardell, L., Oremus, M., & Soskolne, C. L. (2019). Risks to health and well-being from radio-frequency radiation emitted by cell phones and other wireless devices. *Frontiers in Public Health*, *7*, 223. Advance online publication. DOI: 10.3389/fpubh.2019.00223 PMID: 31457001

Milletari, F., Frei, J., Aboulatta, M., Vivar, G., & Ahmadi, S. A. (2018). Cloud deployment of high-resolution medical image analysis with TOMAAT. *IEEE Journal of Biomedical and Health Informatics*, *23*(3), 969–977. DOI: 10.1109/JBHI.2018.2885214 PMID: 30530377

Millington, I. (2019). *AI for Games*. CRC Press. DOI: 10.1201/9781351053303

Mishra, S. (2023, August 5). *Revolutionizing Agriculture: Punjab Farmers Embrace Drones for Pesticide Spraying*. Retrieved from TractorNews: https://tractornews.in/news/revolutionizing-agriculture-punjab-farmers-embrace-drones-for-pesticide-spraying/

Mishra, S. (2023, September 6). *Assam Agriculture Minister Atul Bora Inaugurates Drone Technology Demonstration in Agriculture: VIDEO*. Retrieved from TractorNews: https://tractornews.in/news/assam-agriculture-minister-atul-bora-inaugurates-drone-technology-demonstration-in-agriculture-video/

Mistry, H. K., Mavani, C., Goswami, A., & Patel, R. (2024). The Impact Of Cloud Computing And Ai On Industry Dynamics And Competition. *Educational Administration: Theory and Practice*, *30*(7), 797–804.

Mitchell, T. M. (1997). *Machine Learning*. McGraw-Hill.

Mitra, S. (2023, July 6). *Himachal Pradesh Signs INR 200 Cr MoU With FICCI To Promote Drone Production*. Retrieved from Inc42.com: https://inc42.com/buzz/himachal-pradesh-signs-inr-200-cr-mou-with-ficci-to-promote-drone-production/

Mitra, S., Datta, S., Perkins, T., & Michailidis, G. (2008). *Introduction to machine learning and bioinformatics*. CRC Press. DOI: 10.1201/b17186

Mittelstadt, B. D., Allo, P., Taddeo, M., Wachter, S., & Floridi, L. (2016). The ethics of algorithms: Mapping the debate. *Big Data & Society*, *3*(2), 2053951716679679. DOI: 10.1177/2053951716679679

Mizik, T. (2023). How can precision farming work on a small scale? A systematic literature review. *Precision Agriculture*, *24*(1), 384–406. DOI: 10.1007/s11119-022-09934-y

Moghimi, S., Kushki, A., Marie Guerguerian, A., & Chau, T. (2013). A review of EEG-based brain-computer interfaces as access pathways for individuals with severe disabilities. *Assistive Technology*, *25*(2), 99–110. DOI: 10.1080/10400435.2012.723298 PMID: 23923692

Mohale, H., Narsale, S., Kadam, R., Prakash, P., Sheikh, S., Mansukhbhai, C., Kirtikumar, P., & Baraiya, R. (2024). Artificial Intelligence in Fisheries and Aquaculture: Enhancing Sustainability and Productivity. *Archives of Current Research International*. https://doi.org/.DOI: 10.9734/acri/2024/v24i3650

Mohamed, N., Almutairi, R. L., Abdelrahim, S., Alharbi, R., Alhomayani, F. M., Elamin Elnaim, B. M., Elhag, A. A., & Dhakal, R. (2023). Automated Laryngeal cancer detection and classification using dwarf mongoose optimization algorithm with deep learning. *Cancers (Basel)*, *16*(1), 181. DOI: 10.3390/cancers16010181 PMID: 38201608

Mohammed Yousef Shaheen. (2021). Applications of Artificial Intelligence (AI) in healthcare: A review. *Science OpenPreprints. DOI:* DOI: 10.14293/S2199-1006.1.SOR-.PPVRY8K.v1

Mohammed, M. A., Abdulkareem, K. H., Mostafa, S. A., Khanapi Abd Ghani, M., Maashi, M. S., Garcia-Zapirain, B., & Al-Dhief, F. T. (2020). Voice pathology detection and classification using convolutional neural network model. *Applied Sciences (Basel, Switzerland)*, *10*(11), 3723. DOI: 10.3390/app10113723

Mohanta, B., Das, P., & Patnaik, S. (2019). Healthcare 5.0: A paradigm shift in digital healthcare system using artificial intelligence, IOT and 5G communication. *2019 International Conference on Applied Machine Learning (ICAML)*. https://doi.org/DOI: 10.1109/ICAML48257.2019.00044

Mohanty, S. P., Hughes, D. P., & Salathé, M. (2016). Using deep learning for image-based plant disease detection. *Frontiers in Plant Science*, *7*, 1419. DOI: 10.3389/fpls.2016.01419 PMID: 27713752

Mohbey, K. K., Meena, G., Kumar, S., & Lokesh, K. (2023). A CNN-LSTM-Based hybrid deep learning approach for sentiment analysis on Monkeypox tweets. *New Generation Computing*, *42*(1), 89–107. DOI: 10.1007/s00354-023-00227-0

Mohyuddin, G., Khan, M. A., Haseeb, A., Mahpara, S., Waseem, M., & Saleh, A. M. (2024). Evaluation of Machine Learning Approaches for Precision Farming in Smart Agriculture System: A Comprehensive Review. *IEEE Access : Practical Innovations, Open Solutions*, *12*, 60155–60184. DOI: 10.1109/ACCESS.2024.3390581

Morales-García, J., Bueno-Crespo, A., Terroso-Sáenz, F., Arcas-Túnez, F., Martínez-España, R., & Cecilia, J. M. (2023). Evaluation of synthetic data generation for intelligent climate control in greenhouses. *Applied Intelligence*, *53*(21), 24765–24781. DOI: 10.1007/s10489-023-04783-2

Morley, J., Machado, C. C., Burr, C., Cowls, J., Joshi, I., Taddeo, M., & Floridi, L. (2020). The ethics of AI in health care: A mapping review. *Social Science & Medicine*, *260*, 113172. DOI: 10.1016/j.socscimed.2020.113172 PMID: 32702587

Mount, D. W. (2004). *Bioinformatics: Sequence and Genome Analysis* (2nd ed.). Cold Spring Harbor Laboratory Press.

Mullowney, M. W., Duncan, K. R., Elsayed, S. S., Garg, N., van der Hooft, J. J., Martin, N. I., Meijer, D., Terlouw, B. R., Biermann, F., Blin, K., Durairaj, J., Gorostiola González, M., Helfrich, E. J. N., Huber, F., Leopold-Messer, S., Rajan, K., de Rond, T., van Santen, J. A., Sorokina, M., & Medema, M. H. (2023). Artificial intelligence for natural product drug discovery. *Nature Reviews. Drug Discovery*, *22*(11), 895–916. DOI: 10.1038/s41573-023-00774-7 PMID: 37697042

Munif, M., Nasir, H., & Ahmad, M. (2024). Optimizing ant colony system algorithm with rule-based data classification for smart aquaculture. *Indonesian Journal of Electrical Engineering and Computer Science*, *33*(1), 261. Advance online publication. DOI: 10.11591/ijeecs.v33.i1.pp261-268

Murugan, R., Thomas, F. S., Geetha Shree, G., Glory, S., & Shilpa, A. (2020). Linear Regression Approach to Predict Crop Yield. *Asian Journal of Computer Science and Technology*, *9*(1), 40–44. DOI: 10.51983/ajcst-2020.9.1.2152

Nagi, A. T., Awan, M. J., Mohammed, M. A., Mahmoud, A., Majumdar, A., & Thinnukool, O. (2022). Performance analysis for COVID-19 diagnosis using custom and state-of-the-art deep learning models. *Applied Sciences (Basel, Switzerland)*, *12*(13), 6364. DOI: 10.3390/app12136364

Najjar, R. (2023). Redefining radiology: A review of artificial intelligence integration in medical imaging. *Diagnostics (Basel)*, *13*(17), 2760. DOI: 10.3390/diagnostics13172760 PMID: 37685300

Naseer, I. (2023). AWS Cloud Computing Solutions: Optimizing Implementation for Businesses. Statistics. *Computing and Interdisciplinary Research*, *5*(2), 121–132. DOI: 10.52700/scir.v5i2.138

Naseer, N., & Hong, K. S. (2015). fNIRS-based brain-computer interfaces: A review. *Frontiers in Human Neuroscience*, *9*, 3. PMID: 25674060

Nasser, N., Emad-ul-Haq, Q., Imran, M., Ali, A., Razzak, I., & Al-Helali, A. (2023). A smart healthcare framework for detection and monitoring of COVID-19 using IoT and cloud computing. *Neural Computing & Applications*, *35*(19), 1–15. DOI: 10.1007/s00521-021-06396-7 PMID: 34522068

Navidan, H., Moshiri, P. F., Nabati, M., Shahbazian, R., Ghorashi, S. A., Shah-Mansouri, V., & Windridge, D. (2021). Generative Adversarial Networks (GANs) in networking: A comprehensive survey & evaluation. *Computer Networks*, *194*, 108149. DOI: 10.1016/j.comnet.2021.108149

Neethirajan, S. (2024). Artificial intelligence and sensor innovations: Enhancing livestock welfare with a human-centric approach. *Human-Centric Intelligent Systems*, *4*(1), 77–92. DOI: 10.1007/s44230-023-00050-2

Ness, S., Shepherd, N., & Xuan, T. (2023). Synergy Between AI and Robotics: A Comprehensive Integration. *Asian Journal of Research in Computer Science.*, *16*(4), 80–94. DOI: 10.9734/ajrcos/2023/v16i4372

Newell, A., & Simon, H. A. (1972). *Human Problem Solving*. Prentice-Hall.

Ngugi, L. C., Abelwahab, M., & Abo-Zahhad, M. (2021). Recent advances in image processing techniques for automated leaf pest and disease recognition – A review. *Information Processing in Agriculture*, *8*(1), 27–51. DOI: 10.1016/j.inpa.2020.04.004

Nicolae, R. A., Constantinescu, A. M., Costea, A. C., Girbea, G., Constantinescu, C. M., Neagu, E., & Barbarii, L. E. (2019). A rare genetic gender anomaly identified in a paternity case presenting AMEL-Y dropout. *Forensic Science International. Genetics Supplement Series*, *7*(1), 555–556. DOI: 10.1016/j.fsigss.2019.10.088

Nobel, S. N., Swapno, S. M. R., Islam, M. R., Safran, M., Alfarhood, S., & Mridha, M. F. (2024). A machine learning approach for vocal fold segmentation and disorder classification based on ensemble method. *Scientific Reports*, *14*(1), 14435. DOI: 10.1038/s41598-024-64987-5 PMID: 38910146

Noga, E. J. (2010). *Fish Disease: Diagnosis and Treatment* (2nd ed.). Wiley-Blackwell., DOI: 10.1002/9781118786758

Nwankwo, W., Nwankwo, C. P., & Wilfred, A. (2022). Leveraging on Artificial Intelligence to Accelerate Sustainable Bioeconomy. *IUP Journal of Knowledge Management, 20*(2).

Obermeyer, Z., Powers, B. W., Vogeli, C., & Mullainathan, S. (2019). Dissecting racial bias in an algorithm used to manage the health of populations. *Science*, *366*(6464), 447–453. DOI: 10.1126/science.aax2342 PMID: 31649194

Onmanorama. (2024, July 10). *In a First, Drones Replace Tractors to Sow Paddy Seeds in Kerala*. Retrieved from Onmanorama.com: https://www.onmanorama.com/news/kerala/2024/07/10/drones-sows-paddy-seeds-in-kuttanad-for-the-first-time-in-kerala.html

Owen, J. (2024). The Pinnacle of AI-Driven Robotics: Achieving New Heights in Technology.

Oyetunde, T., Liu, D., Martin, H. G., & Tang, Y. J. (2019). Machine learning framework for assessment of microbial factory performance. *PLoS One*, *14*(1), e0210558. DOI: 10.1371/journal.pone.0210558 PMID: 30645629

Pacal, I. (2024). *Improved vision transformer with lion optimizer for lung diseases detection*. Uluslararası Muhendislik Arastirma ve Gelistirme Dergisi., DOI: 10.29137/umagd.1469472

Pachiappan, K., Anitha, K., Pitchai, R., Sangeetha, S., Satyanarayana, T. V. V., & Boopathi, S. (2024). Intelligent Machines, IoT, and AI in Revolutionizing Agriculture for Water Processing. In *Handbook of Research on AI and ML for Intelligent Machines and Systems* (pp. 374–399). IGI Global.

Padhiary, M. (2024). The Convergence of Deep Learning, IoT, Sensors, and Farm Machinery in Agriculture. In *Designing Sustainable Internet of Things Solutions for Smart Industries* (pp. 109–142). IGI Global., DOI: 10.4018/979-8-3693-5498-8.ch005

Padhiary, M., Barbhuiya, J. A., Roy, D., & Roy, P. (2024). 3D Printing Applications in Smart Farming and Food Processing. *Smart Agricultural Technology, 100553*, 100553. Advance online publication. DOI: 10.1016/j.atech.2024.100553

Padhiary, M., & Kumar, R. (2024a). Assessing the Environmental Impacts of Agriculture, Industrial Operations, and Mining on Agro-Ecosystems. In Azrour, M., Mabrouki, J., Alabdulatif, A., Guezzaz, A., & Amounas, F. (Eds.), *Smart Internet of Things for Environment and Healthcare* (Vol. 1165, pp. 107–126). Springer Nature Switzerland., DOI: 10.1007/978-3-031-70102-3_8

Padhiary, M., & Kumar, R. (2024b). Enhancing Agriculture Through AI Vision and Machine Learning: The Evolution of Smart Farming. In Thangam, D. (Ed.), (pp. 295–324). Advances in Computational Intelligence and Robotics. IGI Global., DOI: 10.4018/979-8-3693-5380-6.ch012

Padhiary, M., Kumar, R., & Sethi, L. N. (2024). *Navigating the Future of Agriculture: A Comprehensive Review of Automatic All-Terrain Vehicles in Precision Farming. Journal of The Institution of Engineers.* Series A., DOI: 10.1007/s40030-024-00816-2

Padhiary, M., Roy, D., & Dey, P. (2024). Mapping the Landscape of Biogenic Nanoparticles in Bioinformatics and Nanobiotechnology: AI-Driven Insights. In Das, S., Khade, S. M., Roy, D. B., & Trivedi, K. (Eds.), *Synthesizing and Characterizing Plant-Mediated Biocompatible Metal Nanoparticles* (pp. 337–376). IGI Global., DOI: 10.4018/979-8-3693-6240-2.ch014

Padhiary, M., & Roy, P. (2024). Advancements in Precision Agriculture: Exploring the Role of 3D Printing in Designing All-Terrain Vehicles for Farming Applications. [IJSR]. *International Journal of Science and Research (Raipur, India), 13*(5), 861–868. DOI: 10.21275/SR24511105508

Padhiary, M., Roy, P., Dey, P., & Sahu, B. (2024). Harnessing AI for Automated Decision-Making in Farm Machinery and Operations: Optimizing Agriculture. In Hai-Jew, S. (Ed.), (pp. 249–282). Advances in Computational Intelligence and Robotics. IGI Global., DOI: 10.4018/979-8-3693-6230-3.ch008

Padhiary, M., Tikute, S. V., Saha, D., Barbhuiya, J. A., & Sethi, L. N. (2024). Development of an IOT-Based Semi-Autonomous Vehicle Sprayer. *Agricultural Research.* Advance online publication. DOI: 10.1007/s40003-024-00760-4

Padyana, U. K., Rai, H. P., Ogeti, P., & Fadnavis, N. S. (2023). AI and machine learning in cloud-based internet of things (IoT) solutions: A comprehensive review and analysis. *Integrated Journal for AI Solutions, 15*(7), 45-62.

Pandey, S. S., Kovaleva, M., & Barelle, C. J. (2022). Overview, generation, and significance of variable new antigen receptors (VNARs) as a platform for drug and diagnostic development. In *Single-Domain Antibodies* (pp. 223–240). Springer. DOI: 10.1007/978-1-0716-2075-5_2

Panicker, R. C., Puthusserypady, S., & Sun, Y. (2011). An asynchronous P300 BCI with SSVEP-based control state detection. *IEEE Transactions on Biomedical Engineering*, *58*(6), 1781–1788. DOI: 10.1109/TBME.2011.2116018 PMID: 21335304

Pappu, S. M. J., & Gummadi, S. N. (2017). Artificial neural network and regression coupled genetic algorithm to optimize parameters for enhanced xylitol production by Debaryomyces nepalensis in bioreactor. *Biochemical Engineering Journal*, *120*, 136–145. DOI: 10.1016/j.bej.2017.01.010

Park, J., Oh, M., & Han, S. (2007). Fish Disease Diagnosis System Based on Image Processing of Pathogens' Microscopic Images. 2007 Frontiers in the Convergence of Bioscience and Information Technologies, 878-883.

Patel, V., & Shah, M. (2022). Artificial intelligence and machine learning in drug discovery and development. *Intelligent Medicine, 3*(4), 295-309.

Patel, D., & Tarakji, K. G. (2021). Smartwatch diagnosis of atrial fibrillation in patient with embolic stroke of unknown source: A case report. *Cardiovascular Digital Health Journal*, *2021*(2), 84–87. DOI: 10.1016/j.cvdhj.2021.01.001 PMID: 35265894

Paul, J. (2024). Introduction to infectious diseases. *Disease Causing Microbes*, 1-63. https://doi.org/DOI: 10.1007/978-3-031-28567-7_1

Peddina, K., & Mandava, A. K. (2024). An optimized deep network-based fish tracking and classification from underwater images. *Multimedia Tools and Applications*. Advance online publication. DOI: 10.1007/s11042-024-19417-z

Pedersen, M., Larsen, C. F., Madsen, B., & Eeg, M. (2023). Localization and quantification of glottal gaps on deep learning segmentation of vocal folds. *Scientific Reports*, *13*(1), 878. DOI: 10.1038/s41598-023-27980-y PMID: 36650265

Pedregosa, F.. (2011). Scikit-learn: Machine Learning in Python. *Journal of Machine Learning Research*, *12*, 2825–2830.

Pennanen, H., Hänninen, T., Tervo, O., Tölli, A., & Latva-aho, M. (2024). 6G: The Intelligent Network of Everything—A Comprehensive Vision, Survey, and Tutorial. *arXiv preprint arXiv:2407.09398.*

Pérez-Granados, C., & Traba, J. (2021). Estimating bird density using passive acoustic monitoring: A review of methods and suggestions for further research. *The Ibis*, *163*(3), 765–783.

Petrovic, B., Kononets, Y., & Csambalik, L. (2025). Adoption of drone, sensor, and robotic technologies in organic farming systems of Visegrad countries. *Heliyon*, *11*(1), e41408. DOI: 10.1016/j.heliyon.2024.e41408 PMID: 39958722

Pham, M., Lin, J., & Zhang, Y. (2018). Diagnosing voice disorder with machine learning. In 2018 IEEE International Conference on Big Data (Big Data) (pp. 5263-5266). IEEE. DOI: 10.1109/BigData.2018.8622250

Pinto, J., de Azevedo, C. R., Oliveira, R., & von Stosch, M. (2019). A bootstrap-aggregated hybrid semi-parametric modeling framework for bioprocess development. *Bioprocess and Biosystems Engineering*, *42*(11), 1853–1865. DOI: 10.1007/s00449-019-02181-y PMID: 31375965

Popescu, S. M., Mansoor, S., Wani, O. A., Kumar, S. S., Sharma, V., Sharma, A., Arya, V. M., Kirkham, M. B., Hou, D., Bolan, N., & Chung, Y. S. (2024). Artificial intelligence and IoT driven technologies for environmental pollution monitoring and management. *Frontiers in Environmental Science*, *12*, 1336088. DOI: 10.3389/fenvs.2024.1336088

Pore, A., Finocchiaro, M., Dall'Alba, D., Hernansanz, A., Ciuti, G., Arezzo, A., & Fiorini, P. (2022). Colonoscopy navigation using end-to-end deep visuomotor control: A user study. In *2022 IEEE/RSJ International Conference on Intelligent Robots and Systems (IROS)* (pp. 9582-9588). IEEE. DOI: 10.1109/IROS47612.2022.9981480

Pourqasem, J. (2018). Cloud-based IoT: Integration cloud computing with Internet of things. *International Journal of Research in Industrial Engineering*, *7*(4), 482–494.

Pradhan, B., Das, S., Roy, D. S., Routray, S., Benedetto, F., & Jhaveri, R. H. (2023). An AI-assisted smart healthcare system using 5G communication. *IEEE Access : Practical Innovations, Open Solutions*, *11*, 108339–108355. DOI: 10.1109/ACCESS.2023.3317174

Prospera. (2020). *AI for precision agriculture*. Retrieved from https://www.prospera.ag

Pun, F. W., Ozerov, I. V., & Zhavoronkov, A. (2023). AI-powered therapeutic target discovery. *Trends in Pharmacological Sciences, 44*(2), 166-177.

Qureshi, H. N., Manalastas, M., Ijaz, A., Imran, A., Liu, Y., & Al Kalaa, M. O. (2022). Communication requirements in 5G-Enabled healthcare applications: Review and considerations. *Health Care*, *10*(2), 293. DOI: 10.3390/healthcare10020293 PMID: 35206907

Rabha, G., Kumar, K., Kumar, D., & Kumar, D. (2024). A Comprehensive Review of Integrating AI and IoT in Farm Machinery: Advancements, Applications, and Sustainability. *International Journal of Research and Analytical Reviews, 11*(4).

Radivojevic, T., Costello, Z., Workman, K., & Garcia Martin, H. (2020). A machine learning Automated Recommendation Tool for synthetic biology. *Nature Communications, 11*(1), 4879. DOI: 10.1038/s41467-020-18008-4 PMID: 32978379

Radwan, T., Azer, M. A., & Abdelbaki, N. (2017). Cloud computing security: Challenges and future trends. *International Journal of Computer Applications in Technology, 55*(2), 158–172. DOI: 10.1504/IJCAT.2017.082865

Ragab, S., Hoseinifar, S., Doan, H., Rossi, W., Davies, S., Ashour, M., & El-Haroun, E. (2024). Overview of aquaculture Artificial Intelligence (AAI) applications: Enhance sustainability and productivity, reduce labor costs, and increase the quality of aquatic products. *Annals of Animal Science*. Advance online publication. DOI: 10.2478/aoas-2024-0075

Raghavendra, V. (2024, November 12). *Andhra Pradesh Govt. to Deploy 'Kisan Drones' in Agriculture Sector this Year*. Retrieved from TheHindu: https://www.thehindu.com/news/national/andhra-pradesh/andhra-pradesh-govt-to-deploy-kisan-drones-in-agriculture-sector-this-year/article68856176.ece

Raha, A. D., Gain, M., Debnath, R., Adhikary, A., Qiao, Y., Hassan, M. M., Bairagi, A. K., & Islam, S. M. (2024). Attention to Monkeypox: An interpretable Monkeypox detection technique using attention mechanism. *IEEE Access : Practical Innovations, Open Solutions, 12*, 51942–51965. DOI: 10.1109/ACCESS.2024.3385099

Rajendiran, G., & Rethnaraj, J. (2024). IoT-Integrated Machine Learning-Based Automated Precision Agriculture-Indoor Farming Techniques. In Nguyen, T. V. T., & Vo, N. T. M. (Eds.), (pp. 289–317). Advances in Media, Entertainment, and the Arts. IGI Global., DOI: 10.4018/979-8-3693-0639-0.ch013

Rajendran, G. a. (2024, September 16). *TN Farmers Spray, Sow and Soar into the Drone Age*. Retrieved from Times of India: https://timesofindia.indiatimes.com/city/chennai/tamil-nadu-farmers-embrace-drones-for-efficient-agriculture/articleshow/113373943.cms

Raj, N., Masih, J., Selvaraj, K., Pavithra, G., & Ramkumar, E. V. (2024, April). IoT & AI Enabled Information and Communication Technologies in Transforming Rural Communities via Digital Agriculture. In *2024 Ninth International Conference on Science Technology Engineering and Mathematics (ICONSTEM)* (pp. 1-6). IEEE. DOI: 10.1109/ICONSTEM60960.2024.10568764

Raj, R., & Kos, A. (2023). Artificial Intelligence: Evolution, Developments, Applications, and Future Scope. *Przeglad Elektrotechniczny*, *99*(2).

Raliya, R. (2024). Artificial Intelligence for Precision and Sustainable Agricultural. *ACS Agricultural Science & Technology*, *4*(6), 628–630. DOI: 10.1021/acsagscitech.4c00290

Ramakrishnan, T., Mohan Gift, M. D., Chitradevi, S., Jegan, R., & Subha Hency Jose, P. Nagaraja, Rajneesh Sharma H.N., Selvakumar P., Sintayehu Mekuria Hailegiorgis, (2022). "Study of Numerous Resins Used in Polymer Matrix Composite Materials", Advances in Materials Science and Engineering, vol. 2022, Article ID 1088926, 8 pages, 2022. https://doi.org/DOI: 10.1155/2022/1088926

Rane, N., Choudhary, S., & Rane, J. (2023). Sustainable tourism development using leading-edge Artificial Intelligence (AI), Blockchain, Internet of Things (IoT), Augmented Reality (AR) and Virtual Reality (VR) technologies. Blockchain, Internet of Things (IoT), Augmented Reality (AR) and Virtual Reality (VR) technologies (October 31, 2023). http://dx.doi.org/DOI: 10.2139/ssrn.4642605

Rao, S. K., & Prasad, R. (2018). Impact of 5G technologies on industry 4.0. *Wireless Personal Communications*, *100*(1), 145–159. DOI: 10.1007/s11277-018-5615-7

Rathore, A. K., Chaurasia, R. K., Mishra, R., & Kumar, H. (2012). Road map and challenges in 4G wireless system. *Journal of Electrical & Electronics*, *01*(03). Advance online publication. DOI: 10.4172/2167-101X.1000104

Rebsamen, B., Guan, C., Zhang, H., Wang, C., Teo, C., Ang, M. H., & Burdet, E. (2010). A brain controlled wheelchair to navigate in familiar environments. *IEEE Transactions on Neural Systems and Rehabilitation Engineering*, *18*(6), 590–598. DOI: 10.1109/TNSRE.2010.2049862 PMID: 20460212

Ren, H., Guo, J., Cheng, S., & Li, Y. (2024). Pooling-based visual transformer with low complexity attention hashing for image retrieval. *Expert Systems with Applications*, *241*, 122745. DOI: 10.1016/j.eswa.2023.122745

Resendis-Antonio, O. (2013) Constraint-based Modeling. In Encyclopedia of Systems Biology (Dubitzky, W. et al., eds), pp. 494-498, Springer New York. DOI: 10.1007/978-1-4419-9863-7_1143

Richard, B. (2017). Hotel chains: Survival strategies for a dynamic future. *Journal of Tourism Futures*, *3*(1), 56–65. DOI: 10.1108/JTF-06-2016-0018

Rokade, S., & Mishra, N. (2024, January). Intelligent Health: Empowering Illness and Prognosis with Machine Learning-Enabled Electronic Health Records on Blockchain. *In 2024 International Conference on Advancements in Smart, Secure and Intelligent Computing* (ASSIC) (pp. 1-5). IEEE.

Rokade, S., & Mishra, N. (2024). Intelligent health: Empowering illness and prognosis with machine learning-enabled electronic health records on blockchain. *2024 International Conference on Advancements in Smart, Secure and Intelligent Computing (ASSIC), 23*, 1-5. https://doi.org/DOI: 10.1109/ASSIC60049.2024.10508019

Rome, B. N., Egilman, A. C., & Kesselheim, A. S. (2022). Trends in prescription drug launch prices, 2008-2021. *JAMA, 327*(21), 2145-2150.

Roriz, R., Cabral, J., & Gomes, T. (2021). Automotive LiDAR Technology: A Survey. *IEEE Transactions on Intelligent Transportation Systems, 23*(7), 6282–6297. DOI: 10.1109/TITS.2021.3086804

Rosenblatt, F. (1958). The Perceptron: A Probabilistic Model for Information Storage and Organization in the Brain. *Psychological Review, 65*(6), 386–408. DOI: 10.1037/h0042519 PMID: 13602029

Routray, S. K., Jha, M. K., Sharma, L., Sarkar, S., Javali, A., & Tengshe, R. (2018). Energy consumption aspects of 5G waveforms. *2018 International Conference on Wireless Communications, Signal Processing and Networking (WiSPNET)*, 1-5. https://doi.org/DOI: 10.1109/WiSPNET.2018.8538674

Ruangurai, P., Dailey, M. N., Ekpanyapong, M., & Soni, P. (2022). Optimal vision-based guidance row locating for autonomous agricultural machines. *Precision Agriculture, 23*(4), 1205–1225. DOI: 10.1007/s11119-022-09881-8

Rudy, S. H., & Sapsis, T. P. (2023). Output-weighted and relative entropy loss functions for deep learning precursors of extreme events. *Physica D. Nonlinear Phenomena, 443*, 133570. DOI: 10.1016/j.physd.2022.133570

Rumelhart, D. E., Hinton, G. E., & Williams, R. J. (1986). Learning representations by backpropagating errors. *Nature, 323*(6088), 533–536. DOI: 10.1038/323533a0

Russell, C. L. (2018). 5 G wireless telecommunications expansion: Public health and environmental implications. *Environmental Research, 165*, 484–495. DOI: 10.1016/j.envres.2018.01.016 PMID: 29655646

Saber, A., Hussien, A. G., Awad, W. A., & Mahmoud, A. (2023). Adapting the pre-trained convolutional neural networks to improve the anomaly detection and classification in mammographic images. *Scientific Reports, 13*(1), 14877.

Saber, A., Sakr, M., Abo-Seida, O. M., & Keshk, A. (2021, May). Automated Breast Cancer Detection and Classification Techniques–A survey. In 2021 International Mobile, Intelligent, and Ubiquitous Computing Conference (MIUCC) (pp. 200-207). IEEE.

Saber, A., Sakr, M., Abo-Seida, O. M., & Keshk, A. (2021b, May). Automated Breast Cancer Detection and Classification Techniques–A survey. In 2021 International Mobile, Intelligent, and Ubiquitous Computing Conference (MIUCC) (pp. 200-207). IEEE.

Saber, A., Hussien, A. G., Awad, W. A., Mahmoud, A., & Allakany, A. (2023). Adapting the pre-trained convolutional neural networks to improve the anomaly detection and classification in mammographic images. *Scientific Reports*, *13*(1), 14877. DOI: 10.1038/s41598-023-41633-0 PMID: 37689757

Saber, A., Sakr, M., Abou-Seida, O., & Keshk, A. (2021a). A novel transfer-learning model for automatic detection and classification of breast cancer based deep CNN. *Kafrelsheikh Journal of Information Sciences*, *2*(1), 1–9. DOI: 10.21608/kjis.2021.192207

Sabir, R. M., Mehmood, K., Sarwar, A., Safdar, M., Muhammad, N. E., Gul, N., Athar, F., Majeed, M. D., Sattar, J., Khan, Z., Fatima, M., Hussain, M. A., & Akram, H. M. B. (2024). Remote Sensing and Precision Agriculture: A Sustainable Future. In Kanga, S., Singh, S. K., Shevkani, K., Pathak, V., & Sajan, B. (Eds.), *Transforming Agricultural Management for a Sustainable Future* (pp. 75–103). Springer Nature Switzerland., DOI: 10.1007/978-3-031-63430-7_4

Sabzevari, M., Szedmak, S., Penttilä, M., Jouhten, P., & Rousu, J. (2022). Strain design optimization using reinforcement learning. *PLoS Computational Biology*, *18*(6), e1010177. DOI: 10.1371/journal.pcbi.1010177 PMID: 35658018

Sachane, M. N., & Patil, S. A. (2024). Hybrid QDCNN-DNFN for Laryngeal Cancer Detection Using Image and Voice Analysis in Federated Learning. *Sensing and Imaging*, *26*(1), 1. DOI: 10.1007/s11220-024-00531-z

Sadybekov, A. V., & Katritch, V. (2023). Computational approaches streamlining drug discovery. *Nature, 621*(5), 243-250.

Sahu, B. (2024). Artificial Intelligence and Automation in Smart Agriculture: A Comprehensive Review of Precision Farming, All-Terrain Vehicles, IoT Innovations, and Environmental Impact Mitigation. [IJSR]. *International Journal of Science and Research (Raipur, India)*, *13*(11), 656–665. DOI: 10.21275/SR241110184009

Saikanth, D. R. K., Supriya, , Singh, B. V., Rai, A. K., Bana, S. R., Sachan, D. S., & Singh, B. (2023). Advancing Sustainable Agriculture: A Comprehensive Review for Optimizing Food Production and Environmental Conservation. *International Journal of Plant and Soil Science, 35*(16), 417–425. DOI: 10.9734/ijpss/2023/v35i163169

Saiwa. (2025, February 5). *The Rise of Drone Livestock Monitoring | A Look at Cutting-Edge Tech for Modern Farmers.* Retrieved from Saiwa: https://saiwa.ai/blog/Drone-Livestock-Monitoring/

Sajid, M. Z., Qureshi, I., Abbas, Q., Albathan, M., Shaheed, K., Youssef, A., Ferdous, S., & Hussain, A. (2023). Mobile-HR: An Ophthalmologic-Based Classification System for Diagnosis of Hypertensive Retinopathy Using Optimized MobileNet Architecture. *Diagnostics (Basel), 13*(8), 1439. DOI: 10.3390/diagnostics13081439 PMID: 37189539

Sakr, M., Saber, A. M., Abo-Seida, O., & Keshk, A. (2020). Machine learning for breast cancer classification using k-star algorithm. *Applied Mathematics & Information Sciences, 14*(5), 855–863. DOI: 10.18576/amis/140513

Sakthivel, S., & Prabhu, V. (2022). Optimal Deep Learning-Based Vocal Fold Disorder Detection and Classification Model on High-Speed Video Endoscopy. *Journal of Healthcare Engineering, 2022*(1), 4248938. DOI: 10.1155/2022/4248938 PMID: 36353680

Saleem, M. H., Potgieter, J., & Arif, K. M. (2021). Automation in Agriculture by Machine and Deep Learning Techniques: A Review of Recent Developments. *Precision Agriculture, 22*(6), 2053–2091. DOI: 10.1007/s11119-021-09806-x

Salehi, F. (2024). The role of artificial intelligence in revolutionizing the agriculture industry in Canada. *Asian Journal of Research and Review in Agriculture*, 70-78.

Samal, P., & Hashmi, M. F. (2024). Role of machine learning and deep learning techniques in EEG-based BCI emotion recognition system: A review. *Artificial Intelligence Review, 57*(3), 50. DOI: 10.1007/s10462-023-10690-2

Samuel, P. K. (2024). AI for addressing challenges in agriculture. *Agricultural Engineering Today, 48*(1), 5–8.

Sana, J. (2024, November 25). *How to Use Agricultural Drones for Soil Health and Fertility Management.* Retrieved from Agribizdrones.com: https://agribizdrones.com/how-to-use-agricultural-drones-for-soil-health/

Sanyaolu, M., & Sadowski, A. (2024). The Role of Precision Agriculture Technologies in Enhancing Sustainable Agriculture. *Sustainability (Basel), 16*(15), 6668. DOI: 10.3390/su16156668

Saquete, E., Tomás, D., Moreda, P., Martínez-Barco, P., & Palomar, M. (2020). Fighting post-truth using natural language processing: A review and open challenges. *Expert Systems with Applications, 141*, 112943. DOI: 10.1016/j.eswa.2019.112943

Saraf, A. P.. (2020). *"Explainable artificial intelligence for aviation safety applications." AIAAAviation Forum*. Using Big Data to Increase Safety., DOI: 10.2514/6.2020-2881

Sayad, Y. O., Mousannif, H., & Al Moatassime, H. (2019). Predictive modeling of wildfires: A new dataset and machine learning approach. *Fire Safety Journal, 104*, 130–146. DOI: 10.1016/j.firesaf.2019.01.006

Schwartz, R., Schwartz, R., Vassilev, A., Greene, K., Perine, L., Burt, A., & Hall, P. (2022). Towards a standard for identifying and managing bias in artificial intelligence (Vol. 3, p. 00). US Department of Commerce, National Institute of Standards and Technology. https://doi.org/DOI: 10.6028/NIST.SP.1270

Sebayang, A. H., Masjuki, H. H., Ong, H. C., Dharma, S., Silitonga, A. S., Kusumo, F., & Milano, J. (2017). Optimization of bioethanol production from sorghum grains using artificial neural networks integrated with ant colony. *Industrial Crops and Products*, *97*, 146–155. DOI: 10.1016/j.indcrop.2016.11.064

Serey, J., Quezada, L., Alfaro, M., Fuertes, G., Vargas, M., Ternero, R., & Gutierrez, S. (2021). Artificial intelligence methodologies for data management. *Symmetry*, *13*(11), 2040. DOI: 10.3390/sym13112040

Shad, E. H. T., Molinas, M., & Ytterdal, T. (2020). Impedance and noise of passive and active dry EEG electrodes: A review. *IEEE Sensors Journal*, *20*(24), 14565–14577. DOI: 10.1109/JSEN.2020.3012394

Shah, T. (2023, March 17). *Drone Technology in Agriculture: Benefits and Challenges for Rural India*. Retrieved from The Times of India: https://timesofindia.indiatimes.com/blogs/voices/drone-technology-in-agriculture-benefits-and-challenges-for-rural-india/

Shah, V., & Konda, S. R. (2022). Cloud Computing in Healthcare: Opportunities, Risks, and Compliance. *Revista Española de Documentación Cientifica*, *16*(3), 50–71.

Shaikh, T. A., Rasool, T., & Lone, F. R. (2022). Towards leveraging the role of machine learning and artificial intelligence in precision agriculture and smart farming. *Computers and Electronics in Agriculture*, *198*, 107119. DOI: 10.1016/j.compag.2022.107119

Shakeel, P. M., Tolba, A., Al-Makhadmeh, Z., & Jaber, M. M. (2020). RETRACTED ARTICLE: Automatic detection of lung cancer from biomedical data set using discrete AdaBoost optimized ensemble learning generalized neural networks. *Neural Computing & Applications*, *32*(3), 777–790. DOI: 10.1007/s00521-018-03972-2

Shamsuddin, D., Danilevicz, M. F., Al-Mamun, H. A., Bennamoun, M., & Edwards, D. (2024). Multimodal Deep Learning Integration of Image, Weather, and Phenotypic Data Under Temporal Effects for Early Prediction of Maize Yield. *Remote Sensing (Basel)*, *16*(21), 4043. DOI: 10.3390/rs16214043

Sharma, M., Mahajan, Y., & Saber, A. (2024). Using Augmented Reality (AR) and the Internet of Things (IoT) to Improve Water Management Maintenance and Training. In Innovations in Machine Learning and IoT for Water Management (pp. 252-270). IGI Global. https://doi.org/DOI: 10.4018/979-8-3693-1194-3.ch013

Sharma, S., Gupta, S., Gupta, D., Juneja, S., Mahmoud, A., El-sappagh, S., & Kwak, K. S. (2022). Transfer learning based modified inception model for diagnosis of Alzheimer's disease. *Frontiers in Computational Neuroscience, 16*(119), 1-13.

Sharma, A., Jain, A., Gupta, P., & Chowdary, V. (2021a). Machine Learning Applications for Precision Agriculture: A Comprehensive Review. *IEEE Access : Practical Innovations, Open Solutions*, *9*, 4843–4873. DOI: 10.1109/ACCESS.2020.3048415

Sharma, I., & Aggarwal, A. (2024). Revolutionizing Creative Tourism: Integrating AI in Marketing for Immersive Travel Experiences. In Manohar, S., Mittal, A., Raju, S., & Nair, A. (Eds.), *Innovative Technologies for Increasing Service Productivity* (pp. 179–197). IGI Global., DOI: 10.4018/979-8-3693-2019-8.ch011

Sharma, S., Sato, K., & Gautam, B. P. (2023). A Methodological Literature Review of Acoustic Wildlife Monitoring Using Artificial Intelligence Tools and Techniques. *Sustainability (Basel)*, *15*(9), 7128. DOI: 10.3390/su15097128

Sharma, V., Tripathi, A. K., & Mittal, H. (2022). Technological revolutions in smart farming: Current trends, challenges & future directions. *Computers and Electronics in Agriculture*, *201*, 107217. DOI: 10.1016/j.compag.2022.107217

Sherstinsky, A. (2020). Fundamentals of Recurrent Neural Network (RNN) and Long Short-Term Memory (LSTM) network. *Physica D. Nonlinear Phenomena*, *404*, 132306. DOI: 10.1016/j.physd.2019.132306

Shi, L. (2022). Analysis of the security of 5G technology from the network level. *SHS Web of Conferences*, *144*, 02007. https://doi.org/DOI: 10.1051/shsconf/202214402007

Shmelova, T.. (2023). Artificial intelligence methods and applications in aviation. In *Handbook of Research on AI Methods and Applications in Computer Engineering* (pp. 108–140). IGI Global., DOI: 10.4018/978-1-6684-6937-8.ch006

Shortliffe, E. H. (1976). *Computer-Based Medical Consultations: MYCIN*. Elsevier.

Shukla, S. (2024, April 3). *Rising Impact: How Drones Are Leading an Agricultural Revolution in India*. Retrieved from KrishiJagran.com: https://krishijagran.com/explainers/how-drones-are-leading-an-agricultural-revolution-in-india/

Shukla, A. K., & Kumar, V. S. (2023). Cloud computing with artificial intelligence techniques for effective disease detection. *International Journal of Data Informatics and Intelligent Computing*, *2*(1), 32–41. DOI: 10.59461/ijdiic.v2i1.45

Shukla, V., Rana, S., & Prashar, S. (2024). Examining the potential of virtual and augmented reality in enhancing tourism experiences. *The Bottom Line (New York, N.Y.)*. Advance online publication. DOI: 10.1108/BL-05-2023-0139

Shulman, R. G., Rothman, D. L., & Hyder, F. (2007). A BOLD search for baseline. *NeuroImage*, *36*(2), 277–281. DOI: 10.1016/j.neuroimage.2006.11.035 PMID: 17223362

Silva, M. M. D., & Guerreiro, J. (2020). On the 5G and Beyond. *Applied Sciences (Basel, Switzerland)*, *10*(20), 7091. DOI: 10.3390/app10207091

Silveira, E. C., Corrêa, C. F. S., Silva, L. M., Santos, B. A., Pretti, S. M., & de Melo, F. F. (2022). Recognition of esophagitis in endoscopic images using transfer learning. *World Journal of Gastrointestinal Endoscopy*, *14*(5), 311–319. DOI: 10.4253/wjge.v14.i5.311 PMID: 35719896

Simkó, M., & Mattsson, M. (2019). 5G wireless communication and health effects—A pragmatic review based on available studies regarding 6 to 100 GHz. *International Journal of Environmental Research and Public Health*, *16*(18), 3406. DOI: 10.3390/ijerph16183406 PMID: 31540320

Singh, M., Joshi, M., Tyagi, K. D., & Tyagi, V. B. (2024). Future Professions in Agriculture, Medicine, Education, Fitness, Research and Development, Transport, and Communication. *Topics in Artificial Intelligence Applied to Industry*, *4*(0), 181–202. DOI: 10.1002/9781394216147.ch10

Singh, R. K., Berkvens, R., & Weyn, M. (2021). AgriFusion: An Architecture for IoT and Emerging Technologies Based on a Precision Agriculture Survey. *IEEE Access : Practical Innovations, Open Solutions*, *9*, 136253–136283. DOI: 10.1109/ACCESS.2021.3116814

Singh, R., & Gill, S. S. (2023). Edge AI: A survey. *Internet of Things and Cyber-Physical Systems, 3*, 71–92. DOI: 10.1016/j.iotcps.2023.02.004

Singh, R., Kurukuru, V., & Khan, M. (2023). Advanced Power Converters and Learning in Diverse Robotic Innovation: A Review. *Energies, 16*(20), 7156. DOI: 10.3390/en16207156

Singh, V., Singh, S., & Gupta, P. (2020). Real-time anomaly recognition through CCTV using neural networks. *Procedia Computer Science, 173*, 254–263. DOI: 10.1016/j.procs.2020.06.030

Sinha, S., & Vohora, D. (2018). Drug discovery and development: An overview. *Pharmaceutical medicine and translational clinical research*, 19-32.

Sishodia, R. P., Ray, R. L., & Singh, S. K. (2020). Applications of Remote Sensing in Precision Agriculture: A Review. *Remote Sensing (Basel), 12*(19), 3136. DOI: 10.3390/rs12193136

Soice, E., & Johnston, J. (2021). Immortalizing cells for human consumption. *International Journal of Molecular Sciences, 22*(5), 2315.

Soleymani, F., Paquet, E., Viktor, H., & Michalowski, W. (2022). Protein–protein interaction prediction with deep learning: A comprehensive review. *Computational Biology, 37*(4), 123-137.

Song, X.-P., Li, H., Potapov, P., & Hansen, M. C. (2022). Annual 30 m soybean yield mapping in Brazil using long-term satellite observations, climate data and machine learning. *Agricultural and Forest Meteorology, 326*, 109186. DOI: 10.1016/j.agrformet.2022.109186

Srivastava, R., Kumar, N., & Sandhan, T. (2024). Binary Classification of Laryngeal Images Utilising ResNet-50 CNN Architecture. *Indian Journal of Otolaryngology and Head and Neck Surgery : Official Publication of the Association of Otolaryngologists of India*, •••, 1–8. DOI: 10.1007/s12070-024-05202-9

Stephens, Z. D., Lee, S. Y., Faghri, F., Campbell, R. H., Zhai, C., Efron, M. J., Iyer, R., Schatz, M. C., Sinha, S., & Robinson, G. E. (2015). Big data: Astronomical or genomical? *PLoS Biology, 13*(7), e1002195. DOI: 10.1371/journal.pbio.1002195 PMID: 26151137

Stoner, J. (2023, July 14). *The Role of Drones in Sustainable Agriculture*. Retrieved from Flyeye.io: https://www.flyeye.io/the-role-of-drones-in-sustainable-agriculture/

Stoner, J. (2025, January 18). *Agricultural Drones: Farming with AI Integration (2025)*. Retrieved from Flyeyc.io: https://www.flyeye.io/agricultural-drones-ai/

Talaat, F. M., El-Sappagh, S., Alnowaiser, K., & Hassan, E. (2024). Improved prostate cancer diagnosis using a modified resnet50-based deep learning architecture. *BMC Medical Informatics and Decision Making*, *24*(1), 23. Advance online publication. DOI: 10.1186/s12911-024-02419-0 PMID: 38267994

Talebi, E., & Nezhad, M. K. (2024). Revolutionizing animal sciences: Multifaceted solutions and transformative impact of AI technologies. *CABI Reviews, 19*(1).

Tang, M., Mehrotra, A., & Stern, A. D. (2022). Rapid Growth Of Remote Patient Monitoring Is Driven By A Small Number Of Primary Care Providers: Study examines the growth of remote patient monitoring. *Health Affairs*, *41*(9), 1248–1254. DOI: 10.1377/hlthaff.2021.02026 PMID: 36067430

Tang, Y., Liu, C., & Yuan, X. (2024). Recognition of bird species with birdsong records using machine learning methods. *PLoS One*, *19*(2), e0297988.

Taranis: Leading Crop Management Software. (2023, October 30). Taranis. https://www.taranis.com/

Tariq, M. U. (2024a). Multi-Agent Models in Healthcare System Design. In Dall'Acqua, L. (Ed.), *Bioethics of Cognitive Ergonomics and Digital Transition* (pp. 143–170). IGI Global., DOI: 10.4018/979-8-3693-2667-1.ch008

Tariq, M. U. (2024b). Social Innovations for Improving Healthcare. In Chandan, H. (Ed.), *Social Innovations in Education, Environment, and Healthcare* (pp. 302–317). IGI Global., DOI: 10.4018/979-8-3693-2569-8.ch015

Tariq, M. U. (2024c). Leveraging AI for Entrepreneurial Innovation in Healthcare. In Özsungur, F. (Ed.), *Generating Entrepreneurial Ideas With AI* (pp. 192–216). IGI Global., DOI: 10.4018/979-8-3693-3498-0.ch009

Tariq, M. U. (2025a). Navigating the Convergence of IoMT and Generative AI in Global Healthcare: Opportunities, Challenges, and Ethical Considerations. In Kumar, V., Katina, P., & Zhao, J. (Eds.), *Convergence of Internet of Medical Things (IoMT) and Generative AI* (pp. 513–534). IGI Global Scientific Publishing., DOI: 10.4018/979-8-3693-6180-1.ch022

Tariq, M. U. (2025b). AI-Powered Breakthroughs: Revolutionizing Cognitive Psychology and Neuropsychology With Machine Learning. In Bansal, R., Maqableh, T., Shuklaa, G., Rabby, F., & Lathabhavan, R. (Eds.), *Transforming Neuropsychology and Cognitive Psychology With AI and Machine Learning* (pp. 65–92). IGI Global Scientific Publishing., DOI: 10.4018/979-8-3693-9341-3.ch003

Tariq, M. U. (2025c). Ethics and Conduct: Navigating the Moral Landscape of Virtual. In Darwish, D. (Ed.), *Building Power, Safety, and Trust in Virtual Communities* (pp. 297–332). IGI Global., DOI: 10.4018/979-8-3693-3868-1.ch014

Taseer, A., & Han, X. (2024). Advancements in variable rate spraying for precise spray requirements in precision agriculture using Unmanned aerial spraying Systems: A review. *Computers and Electronics in Agriculture*, *219*, 108841. DOI: 10.1016/j.compag.2024.108841

Terentev, A., Dolzhenko, V., Fedotov, A., & Eremenko, D. (2022). Current State of Hyperspectral Remote Sensing for Early Plant Disease Detection: A Review. *Sensors (Basel)*, *22*(3), 757. DOI: 10.3390/s22030757 PMID: 35161504

Thangamani, R., Suguna, R. K., & Kamalam, G. K. (2024). Drones and Autonomous Robotics Incorporating Computational Intelligence. In Prakash, K. B., Peddapelli, S. K., Tam, I. C. K., Woo, W. L., & Jain, V. (Eds.), *Computational Intelligent Techniques in Mechatronics* (1st ed., pp. 243–296). Wiley., DOI: 10.1002/9781394175437.ch8

TheEconomicTimes. (2023, July 5). *Himachal Govt Signs MoUs for Production of Drones to Promote Technology-Driven Governance*. Retrieved from TheEconomicTimes: https://economictimes.indiatimes.com/news/economy/policy/himachal-govt-sign-mous-for-production-of-drones-to-promote-technology-driven-governance/articleshow/101518242.cms

TheHillsTimes. (2024, September 13). *Youths Use Drone Technology for Winter Paddy Harvest in Assam*. Retrieved from TheHillsTimes: https://thehillstimes.in/assam/youths-use-drone-technology-for-winter-paddy-harvest-in-assam

Thorat, R., & Gupta, A. (2024). Transfer learning-enabled skin disease classification: The case of monkeypox detection. *Multimedia Tools and Applications*, *83*(35), 82925–82943. DOI: 10.1007/s11042-024-18750-7

Thulasidas, M., Guan, C., & Wu, J. (2006). Robust classification of EEG signal for brain-computer interface. *IEEE Transactions on Neural Systems and Rehabilitation Engineering*, *14*(1), 24–29. DOI: 10.1109/TNSRE.2005.862695 PMID: 16562628

Tokuyama, K., Shimodaira, Y., Terawaki, T., Kusunose, Y., Nakai, H., Tsuji, Y., Toya, Y., Matsuda, F., & Shimizu, H. (2020). Data science-based modeling of the lysine fermentation process. *Journal of Bioscience and Bioengineering*, *130*(4), 409–415. DOI: 10.1016/j.jbiosc.2020.06.011 PMID: 32709563

Tolentino-Zondervan, F., Ngoc, P. T. A., & Roskam, J. L. (2023). Use cases and future prospects of blockchain applications in global fishery and aquaculture value chains. *Aquaculture (Amsterdam, Netherlands)*, *565*, 739158. DOI: 10.1016/j.aquaculture.2022.739158

Torii, T. (2000). Research in autonomous agriculture vehicles in Japan. *Computers and Electronics in Agriculture*, *25*(1-2), 133–153. DOI: 10.1016/S0168-1699(99)00060-5

Torkzadehmahani, R., Schöttle, P., & Wachter, S. (2020). "Privacy-preserving AI techniques in biomedicine: An overview of federated learning and differential privacy." *npj. Digital Medicine*, *3*(1), 135. DOI: 10.1038/s41746-020-00333-z

Tran, T. T. V., Wibowo, A. S., & Tayara, H. (2023). Artificial intelligence in drug toxicity prediction: Recent advances, challenges, and future perspectives. *Journal of Chemical Biology, 56*(2), 411-429.

Tran, B. A., Dao, T. T. P., Dung, H. D. Q., Van, N. B., Ha, C. C., Pham, N. H., & Tran, M. T. (2023). Support of deep learning to classify vocal fold images in flexible laryngoscopy. *American Journal of Otolaryngology*, *44*(3), 103800. DOI: 10.1016/j.amjoto.2023.103800 PMID: 36905912

Treulle, S. (2025, January 14). Aquarium fish diseases & treatments (Complete guide with diagnostic chart). PangoVet. https://pangovet.com/pet-health-wellness/aquarium/aquarium-fish-diseases-and-treatments/

Trinh, T., Duc, L. P., Tran, C. T., Duy, T. T., & Emara, T. Z. (2021). A new stratified block model to process large-scale data for a small cluster. *International Conference on Artificial Intelligence and Big Data in Digital Era*, 253–263. Springer International Publishing Cham. https://doi.org/DOI: 10.1007/978-3-030-97610-1_21

TrinhHuy., T., Trinh, T., NguyenViet, H., Emara, T. Z., & VuongThi, N. (2025, February). *Event scale prediction in event-based social networks.* 11th EAI International Conference on Industrial Networks and Intelligent Systems, Da Nang, Vietnam.

Trinh, T., Nguyen, N.-T., Wu, D., Huang, J. Z., & Emara, T. Z. (2019). A new location-based topic model for event attendees recommendation. *2019 IEEE-RIVF International Conference on Computing and Communication Technologies (RIVF)*, 1–6. IEEE. https://doi.org/DOI: 10.1109/RIVF.2019.8713716

Tsung, C. K., & Tso, Y. A. (2022). Recognizing edge-based diseases of vocal cords by using convolutional neural networks. *IEEE Access : Practical Innovations, Open Solutions*, *10*, 120383–120397. DOI: 10.1109/ACCESS.2022.3221457

Turing, A. M. (1936). On Computable Numbers, with an Application to the Entscheidungsproblem. *Proceedings of the London Mathematical Society*, *42*(1), 230–265. DOI: 10.1112/plms/s2-42.1.230

Turing, A. M. (1950). Computing Machinery and Intelligence. *Mind*, *59*(236), 433–460. DOI: 10.1093/mind/LIX.236.433

Turnip, A., & Hong, K. S. (2012). Classifying mental activities from EEG-P300 signals using adaptive neural network. Int. *J. Innov. Comp. Inf. Control*, *8*(7), 5839–5850.

Tzounis, A., Katsoulas, N., Bartzanas, T., & Kittas, C. (2017). Internet of Things in agriculture, recent advances and future challenges. *Biosystems Engineering*, *164*, 31–48. DOI: 10.1016/j.biosystemseng.2017.09.007

Ullah, H., Gopalakrishnan Nair, N., Moore, A., Nugent, C., Muschamp, P., & Cuevas, M. (2019). 5G communication: An overview of vehicle-to-Everything, drones, and healthcare use-cases. *IEEE Access : Practical Innovations, Open Solutions*, *7*, 37251–37268. DOI: 10.1109/ACCESS.2019.2905347

Umar, Y. R. M., & Mishra, N. (2024, January). GeneOntology's Disease Gene Identification Assessment by DisGeNET. *In 2024 International Conference on Advancements in Smart, Secure and Intelligent Computing (ASSIC)* (pp. 1-7). IEEE.

Urli, T., Brotánková, J., Kilby, P., & Hentenryck, P. V. (2016). Intelligent Habitat Restoration Under Uncertainty. *AAAI Conference on Artificial Intelligence.*

Üstündağ, B. (2017). An adaptive Mealy machine model for monitoring crop status. *Journal of Integrative Agriculture*, *16*(2), 252–265. DOI: 10.1016/S2095-3119(16)61430-2

Vamathevan, J., Clark, D., Czodrowski, P., Dunham, I., Ferran, E., Lee, G., Li, B., Madabhushi, A., Shah, P., & Sumbayev, V. (2019). Applications of machine learning in drug discovery and development. *Nature Reviews. Drug Discovery*, *18*(6), 463–477. DOI: 10.1038/s41573-019-0024-5 PMID: 30976107

Van Rosmalen, R. P., Smith, R. W., Martins dos Santos, V. A. P., Fleck, C., & Suarez-Diez, M. (2021). Model reduction of genome-scale metabolic models as a basis for targeted kinetic models. *Metabolic Engineering*, *64*, 74–84. DOI: 10.1016/j.ymben.2021.01.008 PMID: 33486094

Vanegas, J., Valencia, M., Restrepo, J., & Muñeton, G. (2020). Modeling determinants of tourism demand in Colombia. *Tourism and Hospitality Management*, *26*(1), 49–67. DOI: 10.20867/thm.26.1.4

Vansteensel, M. J., Klein, E., van Thiel, G., Gaytant, M., Simmons, Z., Wolpaw, J. R., & Vaughan, T. M. (2023). Towards clinical application of implantable brain–computer interfaces for people with late-stage ALS: Medical and ethical considerations. *Journal of Neurology*, *270*(3), 1323–1336. DOI: 10.1007/s00415-022-11464-6 PMID: 36450968

Veeramanju, K. T. (2023). Revolutionizing agriculture: A case study of IBM's AI innovations. [IJAEML]. *International Journal of Applied Engineering and Management Letters*, *7*(4), 95–114.

Vega-Ramon, F., Zhu, X., Savage, T. R., Petsagkourakis, P., Jing, K., & Zhang, D. (2021). Kinetic and hybrid modeling for yeast astaxanthin production under uncertainty. *Biotechnology and Bioengineering*, *118*(12), 4854–4866. DOI: 10.1002/bit.27950 PMID: 34612511

Venkatesh, K., & Naik, K. J. (2024). An ensemble transfer learning for nutrient deficiency identification and yield-loss prediction in crop. *Multimedia Tools and Applications*, *83*(32), 78535–78561. DOI: 10.1007/s11042-024-18592-3

Viana-Ferreira, C., Guerra, A., Silva, J. F., Matos, S., & Costa, C. (2017). An intelligent cloud storage gateway for medical imaging. *Journal of Medical Systems*, *41*(9), 141. DOI: 10.1007/s10916-017-0790-8 PMID: 28780714

Vijayakumar, G., Rajkumar, M., Rajiv Chandar, N., Selvakumar, P., & Duraisamy, R. (2024). Muniyandi Rajkumar, Rajiv Chandar N, Selvakumar P and Ramesh Duraisamy, (2024). Environmentally friendly TDS removal from waste water by electrochemical ion exchange batch-type recirculation (EIR) technique. *Environmental Science. Water Research & Technology*, *10*(4), 826–835. DOI: 10.1039/D3EW00793F

Vornoli, A., Falcioni, L., Mandrioli, D., Bua, L., & Belpoggi, F. (2019). The contribution of in vivo mammalian studies to the knowledge of adverse effects of Radiofrequency radiation on human health. *International Journal of Environmental Research and Public Health*, *16*(18), 3379. DOI: 10.3390/ijerph16183379 PMID: 31547363

VuThi., K., VuongThi, N., NguyenXuan, H., Emara, T. Z., & Trinh, T. (2024). Feature Selection Based on Top Frequency. *The International Conference on Intelligent Systems & Networks*, 58–65. Springer Nature Singapore. https://doi.org/ DOI: 10.1007/978-981-97-5504-2_7

Wahab, A., Khan, U. S., Nawaz, T., Akbar, H., Shah, S. T., Khalid, A., Ansari, A. R., & Nawaz, R. (2024). Improved Accuracy for Subject-Dependent and Subject-Independent Deep Learning-Based SSVEP BCI Classification: A User-Friendly Approach. *IEEE Access : Practical Innovations, Open Solutions, 12*, 115935–115950. DOI: 10.1109/ACCESS.2024.3442235

Wang, D., Cao, W., Zhang, F., Li, Z., Xu, S., & Wu, X. (2022). A Review of Deep Learning in Multiscale Agricultural Sensing. *Remote Sensing (Basel), 14*(3), 559. DOI: 10.3390/rs14030559

Wang, Y., Zhang, Z., Feng, L., Du, Q., & Runge, T. (2020). Combining Multi-Source Data and Machine Learning Approaches to Predict Winter Wheat Yield in the Conterminous United States. *Remote Sensing (Basel), 12*(8), 1232. DOI: 10.3390/rs12081232

Wani, S. U. D., Khan, N. A., Thakur, G., Gautam, S. P., Ali, M., Alam, P., Alshehri, S., Ghoneim, M. M., & Shakeel, F. (2022). Utilization of Artificial Intelligence in Disease Prevention: Diagnosis, Treatment, and Implications for the Healthcare Workforce. *Health Care, 10*(4), 608. DOI: 10.3390/healthcare10040608 PMID: 35455786

Wei, C., Salloum, S., Emara, T. Z., Zhang, X., Huang, J. Z., & He, Y. (2018). A Two-Stage Data Processing Algorithm to Generate Random Sample Partitions for Big Data Analysis. In M. Luo & L.-J. Zhang (Eds.), *Cloud Computing -- CLOUD 2018* (Vol. 10967, Issue 2018060, pp. 347–364). Springer International Publishing. DOI: 10.1007/978-3-319-94295-7_24

Wei, P., Wang, D., Zhao, Y., Tyagi, S. K. S., & Kumar, N. (2020). Blockchain data-based cloud data integrity protection mechanism. *Future Generation Computer Systems, 102*, 902–911. DOI: 10.1016/j.future.2019.09.028

Weissler, E. H., Naumann, T., Andersson, T., Ranganath, R., Elemento, O., Luo, Y., Freitag, D. F., Benoit, J., Hughes, M. C., Khan, F., Slater, P., Shameer, K., Roe, M., Hutchison, E., Kollins, S. H., Broedl, U., Meng, Z., Wong, J. L., Curtis, L., & Ghassemi, M. (2021). The role of machine learning in clinical research: Transforming the future of evidence generation. *Trials, 22*(1), 537. DOI: 10.1186/s13063-021-05489-x PMID: 34399832

West, D. M. (2016). How 5G technology enables the health internet of things. *Brookings Center for Technology Innovation, 3*(1), 20.

Weyand, S., Takehara-Nishiuchi, K., & Chau, T. (2015). Weaning off mental tasks to achieve voluntary self-regulatory control of a near-infrared spectroscopy brain-computer interface. *IEEE Transactions on Neural Systems and Rehabilitation Engineering, 23*(4), 548–561. DOI: 10.1109/TNSRE.2015.2399392 PMID: 25667356

Whig, P., Kasula, B. Y., Yathiraju, N., Jain, A., & Sharma, S. (2024). *Transforming Aviation: The Role of Artificial Intelligence in Air Traffic Management. New Innovations in AI, Aviation, and Air Traffic Technology*. IGI Global., DOI: 10.4018/979-8-3693-1954-3.ch004

White, T., Blok, E., & Calhoun, V. D. (2022). Data sharing and privacy issues in neuroimaging research: Opportunities, obstacles, challenges, and monsters under the bed. *Human Brain Mapping, 43*(6), 1830-1850.

Wick, M. R., & Thompson, W. B. (1992, March). Reconstructive expert system explanation. *Artificial Intelligence, 54*(1–2), 33–70. DOI: 10.1016/0004-3702(92)90087-E

Wilkinson, M. D., Dumontier, M., Aalbersberg, I. J. J., Appleton, G., Axton, M., Baak, A., Blomberg, N., Boiten, J.-W., da Silva Santos, L. B., Bourne, P. E., Bouwman, J., Brookes, A. J., Clark, T., Crosas, M., Dillo, I., Dumon, O., Edmunds, S., Evelo, C. T., Finkers, R., & Mons, B. (2016). The FAIR Guiding Principles for scientific data management and stewardship. *Scientific Data, 3*(1), 160018. DOI: 10.1038/sdata.2016.18 PMID: 26978244

Wolfert, S., Ge, L., Verdouw, C., & Bogaardt, M. J. (2017). Big data in smart farming – A review. *Agricultural Systems, 153*, 69–80. DOI: 10.1016/j.agsy.2017.01.023

Wolpaw, J. R., Birbaumer, N., Heetderks, W. J., McFarland, D. J., Peckham, P. H., Schalk, G., Donchin, E., Quatrano, L. A., Robinson, C. J., & Vaughan, T. M. (2000). Brain-computer interface technology: A review of the first international meeting. *IEEE Transactions on Rehabilitation Engineering, 8*(2), 164–173. DOI: 10.1109/TRE.2000.847807 PMID: 10896178

Wolpaw, J. R., Birbaumer, N., McFarland, D. J., Pfurtscheller, G., & Vaughan, T. M. (2002). Brain–computer interfaces for communication and control. *Clinical Neurophysiology, 113*(6), 767–791. DOI: 10.1016/S1388-2457(02)00057-3 PMID: 12048038

Wongchai, A., Shukla, S., Ahmed, A., Sakthi, U., Jagdish, M., & Kumar, R. (2022). Artificial intelligence - enabled soft sensor and internet of things for sustainable agriculture using ensemble deep learning architecture. *Computers & Electrical Engineering, 102*, 108128. Advance online publication. DOI: 10.1016/j.compeleceng.2022.108128

Wu, B., Wang, C., & Du, K. (2023). A Smart Aquaculture System Exploiting IoT, AI and Cloud Computing. *2023 IEEE International Conference on Smart Internet of Things (SmartIoT)*, 251-256. https://doi.org/DOI: 10.1109/SmartIoT58732.2023.00045

Wu, J., & Yang, H. (2015). Linear Regression-Based Efficient SVM Learning for Large-Scale Classification. *IEEE Transactions on Neural Networks and Learning Systems, 26*(10), 2357–2369. DOI: 10.1109/TNNLS.2014.2382123 PMID: 25576581

Wu, Q., Xu, J., Zeng, Y., Ng, D. W. K., Al-Dhahir, N., Schober, R., & Swindlehurst, A. L. (2021). A comprehensive overview on 5G-and-beyond networks with UAVs: From communications to sensing and intelligence. *IEEE Journal on Selected Areas in Communications, 39*(10), 2912–2945. DOI: 10.1109/JSAC.2021.3088681

Wu, S. G., Wang, Y., Jiang, W., Oyetunde, T., Yao, R., Zhang, X., Shimizu, K., Tang, Y. J., & Bao, F. S. (2016). Rapid Prediction of Bacterial Heterotrophic Fluxomics Using Machine Learning and Constraint Programming. *PLoS Computational Biology, 12*(4), e1004838. DOI: 10.1371/journal.pcbi.1004838 PMID: 27092947

Xiao Wen, E. L., Kang, H. M., & Vistro, D. M. (2022). The Application of Machine Learning in Agriculture Sustainability: A Review. *International Journal of Data Science and Advanced Analytics, 4*(4), 66–70. DOI: 10.69511/ijdsaa.v4i4.94

Xiaoshuan, Z., Zetian, F., & Ruimei, W. (2004). Development of the ES-FDD: an expert system for fish disease diagnosis. Oceans '04 MTS/IEEE Techno-Ocean '04 (IEEE Cat. No.04CH37600), 1, 482-487.

Xuan, A., & Shen, L. (2023). Reconstruction of three-dimensional turbulent flow structures using surface measurements for free-surface flows based on a convolutional neural network. *Journal of Fluid Mechanics, 959*, A34. DOI: 10.1017/jfm.2023.154

Yadav, P., Rishiwal, V., Yadav, M., Alotaibi, A., Maurya, V., Agarwal, U., & Sharma, S. (2024). Investigation and Empirical Analysis of Transfer Learning for Industrial IoT Networks. *IEEE Access : Practical Innovations, Open Solutions, 12*, 173351–173379. DOI: 10.1109/ACCESS.2024.3499741

Yadav, S., & Qidwai, T. (2024). Machine learning-based monkeypox virus image prognosis with feature selection and advanced statistical loss function. *Medicine in Microecology, 19*, 100098. DOI: 10.1016/j.medmic.2024.100098

Yang, H. M., Duan, T., Ding, D., Bagul, A., Langlotz, C., & Shpanskaya, K. (2017). CheXNet: radiologist-level pneumonia detection on chest x-rays with deep learning. arXiv preprint arXiv:1711.05225.

Yang, J., Chen, F., Das, R. K., Zhu, Z., & Zhang, S. (2024). Adaptive-avg-Pooling based attention vision transformer for face anti-spoofing. *ICASSP 2024 - 2024 IEEE International Conference on Acoustics, Speech and Signal Processing (ICASSP)*, 3875-3879. DOI: 10.1109/ICASSP48485.2024.10446940

Yang, C., Wang, Y., Lan, S., Wang, L., Shen, W., & Huang, G. Q. (2022). Cloud-edge-device collaboration mechanisms of deep learning models for smart robots in mass personalization. *Robotics and Computer-integrated Manufacturing*, *77*, 102351. DOI: 10.1016/j.rcim.2022.102351

Yang, S., Frier, R., & Shi, Q. (2021). *Acoustic Classification of Bird Species Using Wavelets and Learning Algorithms*. ACM Digital Library. DOI: 10.1145/3457682.3457692

Yang, W., & Jeon, J. Y. (2020). Design strategies and elements of building envelope for urban acoustic environment. *Building and Environment*, *182*, 107121. DOI: 10.1016/j.buildenv.2020.107121

Yao, L., Ma, R., & Wang, H. (2021). Baidu index-based forecast of daily tourist arrivals through rescaled range analysis, support vector regression, and autoregressive integrated moving average. *Alexandria Engineering Journal*, *60*(1), 365–372. DOI: 10.1016/j.aej.2020.08.037

Yépez-Ponce, D. F., Salcedo, J. V., Rosero-Montalvo, P. D., & Sanchis, J. (2023). Mobile robotics in smart farming: Current trends and applications. *Frontiers in Artificial Intelligence*, *6*, 1213330. DOI: 10.3389/frai.2023.1213330 PMID: 37719082

Yin, Q. Y., Yang, J., Huang, K. Q., Zhao, M. J., Ni, W. C., Liang, B., ... & Wang, L. (2023). AI in human-computer gaming: Techniques, challenges and opportunities. Machine intelligence research, 20(3), 299-317.

Yin, E., Zhou, Z., Jiang, J., Chen, F., Liu, Y., & Hu, D. (2013). A speedy hybrid BCI spelling approach combining P300 and SSVEP. *IEEE Transactions on Biomedical Engineering*, *61*(2), 473–483. PMID: 24058009

Yolcu Oztel, G. (2024). Vision transformer and CNN-based skin lesion analysis: Classification of monkeypox. *Multimedia Tools and Applications*, *83*(28), 71909–71923. DOI: 10.1007/s11042-024-19757-w

Yousef, A. M., Deliyski, D. D., Zacharias, S. R., de Alarcon, A., Orlikoff, R. F., & Naghibolhosseini, M. (2022). A deep learning approach for quantifying vocal fold dynamics during connected speech using laryngeal high-speed videoendoscopy. *Journal of Speech, Language, and Hearing Research: JSLHR*, *65*(6), 2098–2113. DOI: 10.1044/2022_JSLHR-21-00540 PMID: 35605603

Yue, Y., Zeng, X., Lin, H., Xu, J., Zhang, F., Zhou, K., & Li, Z. (2024). A deep learning based smartphone application for early detection of nasopharyngeal carcinoma using endoscopic images. npj. *Digital Medicine*, *7*(1), 384. PMID: 39738998

Zafar, A., Aamir, M., Mohd Nawi, N., Arshad, A., Riaz, S., Alruban, A., Dutta, A. K., & Almotairi, S. (2022). A Comparison of Pooling Methods for Convolutional Neural Networks. *Applied Sciences (Basel, Switzerland)*, *12*(17), 8643. DOI: 10.3390/app12178643

Zaitseva, N., Shur, P., Chetverkina, K., & Khasanova, A. (2020). Developing methodical approaches to substantiating average annual maximum permissible concentrations of hazardous substances in ambient air in settlements as per acceptable health risk. *Health Risk Analysis*, (3), 39–48. DOI: 10.21668/health.risk/2020.3.05.eng

Zampieri, G., Vijayakumar, S., Yaneske, E., & Angione, C. (2019). Machine and deep learning meet genome-scale metabolic modeling. *PLoS Computational Biology*, *15*(7), e1007084. DOI: 10.1371/journal.pcbi.1007084 PMID: 31295267

Zhang, J., & Li, D. (2008). A Call Center Oriented Consultant System for Fish Disease Diagnosis in China. In: Li, D. (eds) Computer And Computing Technologies In Agriculture, Volume II. CCTA 2007. The International Federation for Information Processing, vol 259. Springer, Boston, MA. DOI: 10.1007/978-0-387-77253-0_96

Zhang, C., Ueng, Y., Studer, C., & Burg, A. (2020). Artificial intelligence for 5G and beyond 5G: Implementations, algorithms, and optimizations. *IEEE Journal on Emerging and Selected Topics in Circuits and Systems*, *10*(2), 145–148. DOI: 10.1109/JETCAS.2020.2999944

Zhang, D., Rodrigues, J. J., Zhai, Y., & Sato, T. (2021). Design and implementation of 5G e-Health systems: Technologies, use cases, and future challenges. *IEEE Communications Magazine*, *59*(9), 80–85. DOI: 10.1109/MCOM.001.2100035

Zhang, J., Petersen, S. D., Radivojevic, T., Ramirez, A., Pérez-Manríquez, A., Abeliuk, E., Sánchez, B. J., Costello, Z., Chen, Y., Fero, M. J., Martin, H. G., Nielsen, J., Keasling, J. D., & Jensen, M. K. (2020). Combining mechanistic and machine learning models for predictive engineering and optimization of tryptophan metabolism. *Nature Communications*, *11*(1), 4880. DOI: 10.1038/s41467-020-17910-1 PMID: 32978375

Zhang, K., Wang, H., Cheng, Y., Liu, H., Gong, Q., Zeng, Q., & Chen, D. (2024). Early gastric cancer detection and lesion segmentation based on deep learning and gastroscopic images. *Scientific Reports*, *14*(1), 7847. DOI: 10.1038/s41598-024-58361-8 PMID: 38570595

Zhang, S., Gao, Y., Cai, J., Yang, H., Zhao, Q., & Pan, F. (2023). *A Novel Bird Sound Recognition Method Based on Multifeature Fusion and a Transformer Encoder.* Intelligent Voice Ltd. DOI: 10.3390/s23198099

Zhang, W., Chien, J., Yong, J., & Kuang, R. (2017). Network-based machine learning and graph theory algorithms for precision oncology. *NPJ Precision Oncology*, *1*(1), 25. DOI: 10.1038/s41698-017-0029-7 PMID: 29872707

Zhavoronkov, A., Ivanenkov, Y. A., Aliper, A., Veselov, M. S., Aladinskiy, V., Aladinskaya, A. V., Terentiev, V., Polykovskiy, D., Kuznetsov, M. D., Asadulaev, A., Volkov, Y., Zholus, A., Shayakhmetov, R. R., Zhebrak, A., Minaeva, L. I., Zagribelnyy, B. A., Lee, L. H., Soll, R., Madge, D., & Aspuru-Guzik, A. (2019). Deep learning enables rapid identification of potent DDR1 kinase inhibitors. *Nature Biotechnology*, *37*(9), 1038–1040. DOI: 10.1038/s41587-019-0224-x PMID: 31477924

Zheng, Z. Y., Guo, X.-N., Zhu, K.-X., Peng, W., & Zhou, H.-M. (2017). Artificial neural network - Genetic algorithm to optimize wheat germ fermentation condition: Application to the production of two anti-tumor benzoquinones. *Food Chemistry*, *227*, 264–270. DOI: 10.1016/j.foodchem.2017.01.077 PMID: 28274431

Zhong, S., Zhang, K., Bagheri, M., Burken, J. G., Gu, A., Li, B., Ma, X., Marrone, B. L., Ren, Z. J., Schrier, J., Shi, W., Tan, H., Wang, T., Wang, X., Wong, B. M., Xiao, X., Yu, X., Zhu, J.-J., & Zhang, H. (2021). Machine learning: New ideas and tools in environmental science and engineering. *Environmental Science & Technology*, *55*(19), 12741–12754. DOI: 10.1021/acs.est.1c01339 PMID: 34403250

Zhong, Y., Zhang, J., Liu, S., Ding, C., & Triantafyllopoulos, A. (2023). A review of automatic recognition technology for bird vocalizations in the deep learning era. *Ecological Informatics*.

Zhou, Y., Qian, C., Guo, Y., Wang, Z., Wang, J., Qu, B., . . . Qu, X. (2021, November). XCloud-pFISTA: A medical intelligence cloud for accelerated MRI. In 2021 43rd Annual International Conference of the IEEE Engineering in Medicine & Biology Society (EMBC) (pp. 3289-3292). IEEE. DOI: 10.1109/EMBC46164.2021.9630813

Zouin, B., Zahir, J., Baletaud, F., Vigliola, L., & Villon, S. (2024). Improving CNN Fish Detection and Classification with Tracking. *Applied Sciences (Basel, Switzerland)*, *14*(22), 10122. DOI: 10.3390/app142210122

About the Contributors

Tamer Emara holds a Ph.D. in Information and Communication Engineering from Shenzhen University, China, and is an Assistant Professor at the Faculty of Computers and Artificial Intelligence, Damietta University, Egypt. He earned his M.Sc. and B.Sc. degrees from Mansoura and Tanta Universities, respectively. Dr. Emara has received numerous awards, including the Excellent Paper Award from the Big Data Mining and Analytics journal in 2021, the Outstanding Scientific Award from the Higher Institute of Engineering and Technology Kafrelsheikh, the Shenzhen Universiade International Scholarship, and recognition as the Outstanding Doctoral Student of Shenzhen University in 2016. He actively contributes to the academic community as a member of the TPC and Organization Committees for various conferences and serves as a reviewer for multiple journals. Dr. Emara is a senior member of the IEEE and a member of the ACM. His research interests span big data management and applications, machine learning, distributed and parallel computing, bioinformatics, and IoT.

Esraa Hassan was born in Kafrelsheikh, Kafrelsheikh, Egypt, in 1992. She received a B.Sc. Degree in Computer Science and a M.Sc. Degree in Computer Science from Mansoura University, Egypt, in 2013 and 2018, respectively. She received her Ph.D. in computer science from Mansoura University, Egypt, in 2022. She is currently a lecturer at the Faculty of Artificial Intelligence, Kafrelsheikh University, Egypt. She has published many research articles in prestigious international conferences and reputable journals. She is also a reviewer for many journals. Her research interests include machine learning, deep learning, optimization, and bioinformatics.

Thanh Trinh received the PhD degree in computer science from Shenzhen University, China, and the MSc degree in information systems design from University of Central Lancashire, UK. He is currently a lecturer with the faculty of computer science, Phenikaa University. He published many papers on his research topics. His research includes efficient query, big data, social network, classification, AI/ML for disasters, and climate change. He participated in many projects, from the institute level to government level. In 2020, he and his team awarded with the price in Science and Technology in Ministry of Natural Resources and Environment. Currently, he leads one course of Eramus+ project, between Vietnamese universities and European universities, about Climate change Adaptation using Digital geospatial twins and Earth Observation.

Genghui Li received the M.Sc. degree in computer science and technology from Shenzhen University, Shenzhen, China, in 2016, and the Ph.D. degree in computer science from City University of Hong Kong, Hong Kong, China, in 2021. From 2022 to 2023, he was a Post-Doctoral Research Fellow with the School of System Design and Intelligent Manufacturing, Southern University of Science and Technology, Shenzhen, China. He is currently an Assistant Professor with the College of Computer Science and Software Engineering, Shenzhen University, Shenzhen, China. His research interests include evolutionary computation, computational intelligence, machine learning, and their applications. He has published over 50 SCI papers on computational intelligence/artificial intelligence in reputable journals.

Abeer Saber was born in Damietta, Dumyat, Egypt, in 1992. She received the B.Sc. Degree in Computer Science and the M.Sc. Degree in Computer Science from Mansoura University, Egypt, in 2013 and 2018, respectively. She is received the Ph.D. in computer science from Menoufia University, Egypt, in 2022. She is currently a Lecturer in information technology with the Faculty of Computers and Artificial intelligence, Damietta University, Egypt. She has published many research articles in prestigious international conferences and reputable journals. She is also a reviewer for many journals. Her current research interests include big data analysis, semantic web, linked open data, Optimization, machine learning, Deep learning, bioinformatics, and IoT.

***

J. Arthy is working as an Assistant Professor in the Department of CSE at SRMIST, Ramapuram Campus. She is pursuing Ph. D degree in the area of Web Mining at SRMIST from 2022.She has 11 years of teaching experience with good programming skills. She completed her M.E. (CSE) from Annamalai University in 2012. She is a life member of ISTE. She has guided many projects for UG students and also for many project competitions and hackathons. She is also certified in various courses from NPTEL, ISRO, MATLAB, Udemy, Great Learning, Coursera, etc. She has authored or co-authored research papers that have been published in international journals, conferences, and symposiums. She is also a co-inventor for five patents. Her area of research includes Knowledge Based System and cybersecurity.

Nithya B is an esteemed Assistant Professor at B.S Abdur Rahman Crescent institute of science and technology . With a robust background in these dynamic areas, she is dedicated to guiding and mentoring students, helping them navigate the complexities of modern data science. Her scholarly contributions include numerous publications in both international and national conferences. Nithya's dedication to teaching and research has profoundly impacted the academic community. Her ongoing commitment to excellence continues to make her a valuable asset to the institution and the broader academic landscape.

Mohamed Badouch is a PhD student in computer science at Faculty of Sciences, Ibn Zohr University, Agadir, Morocco. He received the master degree in information systems engineering from Cadi Ayyad University in 2017. His main research is in the areas of artificial intelligence, and machine learning.

J.Balaji is Currently working as an Assistant Professor in the Department of Information Technology at Nandha College of Technology, Erode - 638052. His area of interest is Cloud Computing, Artificial Intelligence and Data Mining.

Rahul Bundele I am a software engineering student specializing in real-time web and desktop applications. With experience in technologies like React, Go,java script, python,postgresql and Redis, I have developed several projects, including a ride-sharing app that integrates mapping and WebSocket functionalities for live user tracking. My interests include scalable server design and optimizing performance for high-concurrency applications.

Jetlin C P received her BE degree in computer science and Engineering from TJ institute of Technology, chennai, India in 2012 and her ME degree in computer science and Engineering from Agni college of Technology, Chennai, India, in 2014. She is currently working as an Assistant Professor in St. Joseph's Institute of Technology, chennai. Her area of interest include Artificial Intelligence, Machine Learning, Deep Learning, Big Data Analytical etc.

Naglaa Elbaz holds a Ph.D. in Fish Diseases and Aquatic Animal Medicine from the Faculty of Veterinary Medicine, Damanhour University, Egypt (2024). She earned her M.Sc. from Kafrelsheikh University and a bachelor's degree in Veterinary Medicine from South Valley University in 2011. Currently, she serves as the Head of the Department of Fish Diseases and previously led the Research Laboratory at the Meat and Fish Center in the Kafr El-Sheikh Directorate. Her research interests focus on marine and fish aquaculture, particularly in disease management and sustainable aquaculture practices.

Nagwa Elmobark is a dedicated Ph.D. researcher in the Department of Information Systems, Faculty of Computer and Information Sciences at Mansoura University, Egypt. With a profound passion for technology and a strong academic background, Nagwa has emerged as an expert in the fields of Big Data, Data Analysis, Mobile Application Development, Fog Computing, Hadoop, GIS (Geographic Information Systems), and Python programming. Also a reviewer of different academic journals.

Sohaila Elshamey is a researcher and academic at Damietta University, Egypt. She specializes in IT, CS and Environmental Studies. With a strong commitment to education and research, Dr. Elshamey has actively participated in teaching, mentoring, and scholarly collaborations with national and international institutions. Her research focuses on specific topics, such as machine learning, sustainable development. Dr. Elshamey is also participated in various academic conferences, workshops, and projects. Her work aims to bridge the gap between theoretical research and real-world applications, contributing to advancements in her field.

G.David Raj is currently working as Assistant Professor in Department of Computer Application at B.S. Abdur Rahman Crescent Institute of Science and Technology, Chennai. His research interests include nature inspired algorithms, soft computing, and computational intelligence algorithms.

Azmirul Hoque is a research scholar at the Department of Agricultural Engineering, Triguna Sen School of Technology, Assam University, Silchar, India. His research areas are artificial intelligence, machine learning, and refractance window drying and food processing.

K. Kavitha received her Bachelor's degree in Electronics and Communication Engineering from Madurai Kamaraj University, Madurai, India and her Master's degree in Microwave & Optical Engineering from Madurai Kamaraj University, Madurai, India in She completed her doctoral degree at Anna University, Chennai, specializing in Hyperspectral Image Classification. Currently, she holds the position of Professor and Head in the Department of Electronics and Communication Engineering at Velammal College of Engineering and Technology in Madurai, Tamil Nadu. Her career spans over 23 years, during which she has made significant contributions in the realms of teaching and research. Her research interests encompass diverse areas, including Antenna Design, Hyperspectral Image Classification and

Optical System Design. She has published and presented over 100 research articles in renowned international journals and international and national conferences. Presently, she is supervising the research works of ten Ph.D. scholars in various areas of her expertise. In recognition of her outstanding contributions to education, Dr.K.Kavitha was honored with Dr.A.P.K Abdul Kalam Teachers Excellence Award.

K. Raja, a distinguished academician, currently serves as Professor and Head of the Department at SRM Institute of Science and Technology, Ramapuram Campus, Chennai. He earned a Master's degree (M.E) from Madras University in 2001 and a Doctorate (Ph.D.) in Knowledge-Based Systems from Sathyabama University, Chennai in 2006. His passion for knowledge drove him to attain a second Ph.D. degree in Management from Periyar University in 2020. With 28 years of experience in academia, administration, research, and teaching, he has been associated with reputable institutions like Madras Institute of Technology, Anna University. Membership in esteemed organizations such as IEI, CSI, ISTE, and IACSIT highlights his commitment to education and professional development. Whose research contributions are commendable, with 75 published papers in prestigious international and national journals. He presented 54 papers at international conferences in various countries, including Singapore, Malaysia, Sri Lanka, Doha Qatar, and India. As an accomplished author, he has written 5 books on diverse subjects. As a highly regarded expert in his field, he actively reviews for international and national journals. His dedication to nurturing young minds is evident through the successful supervision of 6 Ph.D. candidates and ongoing mentorship of more than 4 research scholars. His expertise spans various subjects, encompassing Knowledge-Based Systems, Knowledge Management, Technology Management, Computer Networks, Network Security, Data Structures, Algorithms, HR, and Quality Systems. Throughout his career, he has received well-deserved accolades, including the Best Teacher Award, Academic Researcher Award, and Head of the Department Award.

Kundan Kumar is a researcher at the Department of Agricultural Engineering, Triguna Sen School of Technology, Assam University, Silchar, India. His research areas are artificial intelligence, machine learning, automation, and agriculture.

Hasna Mahmoud is a PhD researcher in Computer Science at the Faculty of Sciences University Ibn Zohr. I graduated with a Master's degree in Big Data and Artificial Intelligence from the Faculty of Taroudant in 2023. My research centers on advancements in Artificial Intelligence.

V. Mathumitha is currently working as an Assistant Professor in Bharath Institute of Higher Education and Research, Chennai. Her Research area includes Cloud Computing, Artificial Intelligence, Network Security.

Nilamadhab Mishra is an Associate Professor in the School of Computer Science and Engineering, AI-ML Division, at VIT Bhopal University, India. He received his Doctor of Philosophy (Ph.D.) in Computer Science & Information Engineering with a specialization in Data Science & Machine Learning from Chang Gung University, Taiwan. He has more than 22 years of national and international involvement in academic teaching and research at recognized Indian, Taiwanese, and African universities. He has over 50 publications in SCI/SCIE and SCOPUS-indexed journals, ISBN books and chapters, Indian and Australian patents, IEEE conference proceedings, and others. He has served as a reviewer, associate editor, and editorial board member for SCI/SCIE-indexed journals and conferences. Dr. Mishra worked on multiple funded research projects from the MOST, NSC, and CGU Memorial Hospital, Taiwan, during his Ph.D. and has been involved with several professional bodies. He is currently an Associate Editor for two SCOPUS-indexed journals, and his research interests span the areas of AI, Data Science, Machine Learning, and Cognitive Analytics & Applications.

Rudra Kalyan Nayak serves as a Senior Assistant Professor in the School of Computing Science & Engineering at VIT Bhopal University. He has more than 14 years of expertise in teaching, mentoring and research. His Doctorate in Computer Science & Engineering was awarded by Siksha 'O' Anusandhan University in Bhubaneswar, Odisha, where he conducted research in the field of Financial Engineering. At the same university, he earned his M. Tech. in information technology. Dr. Nayak possesses nine patents total, two of which are granted Australian design patents and two of which are granted Indian design patents. In addition, he has written three book chapters and two books. He boasts over 45 international papers to his name, appearing in reputable journals and IEEE/Springer conference proceedings. Additionally, he wrote an article for the IEEE India Council Newsletter. Speaking with expertise, Dr. Nayak has participated in faculty development programs. He has reviewed papers for a number of conferences and peer-reviewed journals. His areas of interest in research are Artificial Intelligence, Machine Learning, Financial Engineering, Bioinformatics and Computer Vision. A number of honors have been bestowed upon him, including the I2OR International Teaching Excellence Award 2023, I2OR National Elite Teacher Award 2022, 100 Eminent Academicians of 2021, Dynamic Teacher of the Year 2019, I2OR Preeminent Researcher Award 2019 and Green ThinkerZ Bright Educator Award 2019. A professional member of ACM, CSTA, IAENG, ICSES, theIRED, SDIWC, and a life member of ISTE, Dr. Nayak holds these memberships. He was also a London Journals Press Honorary Rosalind Member.

S. Ninisha Nels is Currently working as an Assistant Professor in the Department of Computer Science and Engineering at Noorul Islam Centre for Higher Education, Kumaracoil, Thuckalay, India. Her Research area includes Cyber security, Artificial Intelligence, and Cloud Computing.

Nehru P is an esteemed Assistant Professor at B.S Abdur Rahman Crescent Institute of Science and Technology. With a strong academic background and a passion for education, Nehru has been inspiring and guiding students in the field of Generative AI and Blockchain. His research interests include Plant disease prevention and he has published several papers in renowned journals and conferences. Nehru's dedication to teaching and research has made a significant impact on the academic community, and he continues to be a valuable asset to the institution.

P. Selvakumar is currently an Associate Professor in the Department of Science and Humanities at Nehru Institute of Technology, Coimbatore, Tamil Nadu, India. Since 2023, he has served as the Department IQAC Coordinator at the institute. Prior to joining Nehru Institute of Technology, he held positions as an Assistant Professor in the Department of Science and Humanities various engineering colleges in Coimbatore, Tamil Nadu, India, spanning from 2011 to 2023. Dr.P. Selvakumar's academic and research interests lie primarily in the field of phytochemistry. He has made significant contributions to his field, with an impressive record that includes 30 journal publications, 12 patents, and 20 conference papers. Additionally, he has organized 5 seminars and conferences, further demonstrating his commitment to academic and scientific exchange. Throughout his career, Dr. P.Selvakumar has been recognized for his dedication and excellence in teaching, having been honored with several awards, including the Best Teacher Awards. He holds a prominent position in the academic community, contributing actively to research.

Mrutyunjay Padhiary is an Assistant Professor in the Department of Agricultural Engineering at Assam University, Silchar, India, with a deep academic background. His previous positions included significant teaching and research duties at the Faculty of Technology, UBKV in West Bengal, and Tribhuvan University in Nepal, as well as research experience at IIT Kharagpur. He has been awarded multiple awards, including an International NUS Research Fellowship and an excellent score on the 2001 GATE Exam, demonstrating his proficiency in precision agriculture. He has made substantial contributions to biofuel production, artificial intelligence vision, aquaculture, precision agriculture, agricultural automation, and artificial intelligence. He is currently working on several research projects, which range from autonomous surface vessels to precision agriculture all-terrain vehicles.

P. Venkadesh M.E, Ph.D., is currently working as a Professor in the Department of Artificial Intelligence & Data Science in V.S.B College of Engineering Technical Campus, Coimbatore, TamilNadu, India . He received his M.E Degree in Computer Science & Engineering from Sathyabama University, Chennai, TamilNadu, India in 2007. He completed his Ph.D at Noorul Islam Centre for Higher Education in 2017 under the area of Network Security . He is guiding five research scholars and three scholars received their Ph.D degree under his guidance. His research area includes Network Security, Image Processing, Wireless Communications and Cloud Computing. He has a teaching experience of more than 22 years and had published more than 27 research papers in SCI/Scopus indexed journals . He has published 12 Patents and two patents were granted. He has published more than 3 text books and 3 Scopus indexed book chapters He is a life member of ISTE.

Vignesh R is a data science and machine learning enthusiast with a focus on anomaly detection and cybersecurity. I am currently interning at CDAC Chennai, where I work on a high-impact anomaly detection project for large network datasets. My approach leverages advanced machine learning techniques, including ensemble methods that combine models like autoencoders, Isolation Forest, and One-Class SVM, to enhance the detection of anomalies. One of my recent projects was the Student Exam Performance Indicator, an end-to-end web application built using Flask, Pandas, Scikit-learn, and other data science libraries. Through this project, I gained hands-on experience with data ingestion, exploratory data analysis, model building, and hyperparameter tuning. My work with TensorFlow has deepened my understanding of deep learning, and I am currently exploring real-time data processing to enable live anomaly detection. I am passionate about cybersecurity, with a particular interest in identifying data exfiltration threats within network logs. As I continue to develop my skills, I am dedicated to creating innovative, data-driven solutions to complex challenges in machine learning and cybersecurity.

Omnia Raied Accomplished Data Scientist, Teaching Assistant, and Business Developer with expertise in data analysis, machine learning, market research, and strategic planning. Currently serving as a Teaching Assistant in the Communication & Electronics Engineering Department, contributing to academic instruction, curriculum development, and student mentorship. Proficient in analyzing complex datasets using tools such as Python, SQL, and Excel to deliver actionable insights, enhance operational efficiency, and improve decision-making processes.

Aiswarya Ramesh is currently doing her B.Tech Artificial Intelligence and Data Science in V.S.B College of Engineering Technical Campus, Coimbatore. Her Research Area Includes Bio informatics, Data mining and Cyber Security

Ashwath S is a B.Tech student in Computer Science at SRM University, maintaining a CGPA of 9.5. My interests lie primarily in artificial intelligence and machine learning, and I've worked on various projects that fuel my passion for these fields. One of my recent projects involved using a Convolutional Neural Network to classify bird species from audio recordings. This project combined my interest in machine learning with a focus on environmental science, utilizing spectrographic analysis to aid in automated wildlife monitoring. Born and raised in Chennai, I completed my schooling at Bala Vidya Mandir and Sairam Vidyalaya, achieving 92% in my 10th grade and 85% in my 12th. Outside of academics, I enjoy playing chess and badminton with friends. I am dedicated to using technology to make a positive impact and look forward to contributing further to the fields of AI and machine learning.

S. Raj Kumar received his bachelor's degree in Electronics and Communication Engineering from Anna University, Chennai, and his master's degree in VLSI Design from the same institution. He completed his doctoral degree at Anna University, Chennai, specializing in VLSI device modeling. Currently, he serves as an Assistant Professor in the Department of Electronics Engineering (VLSI Design and Technology) at Velammal College of Engineering and Technology, Madurai, Tamil Nadu, India. His career spans over 11 years, during which he has made significant contributions to teaching and research. His research interests include diverse areas such as modeling and simulation of analog VLSI. He has published over 10 research papers in international journals and presented 20 papers at international conferences. He has also received the award for "Best Outcome in Academic Performance"..

Allwyn Sundarraj is an esteemed Associate Professor at Sri Shakthi Institute of Engineering and technology . With a robust background in these dynamic areas, He is dedicated to guiding and mentoring students, helping them navigate the complexities of modern research field.His scholarly contributions include numerous publications in both international and national conferences. Allwyn Sundarraj dedication to teaching and research has profoundly impacted the academic community. His ongoing commitment to excellence continues to make her a valuable asset to the institution and the broader academic landscape.

Sneh S is a UG research scholar at Chitkara University, Himachal Pradesh, India.

S.V. Divya M.E, Ph.D., is currently working as a Professor in the Department of Computer Science & Engineering in V.S.B College of Engineering Technical Campus, Coimbatore, TamilNadu, India . She has the teaching experience of more than 16 years and had published more than 20 research papers in SCI/Scopus indexed journals, 2 Patents and 2 text books. She had published more than 15 papers in National/International Conferences. She received her B.Tech Degree in Information Technology from Jayamatha Engineering College, Aralvoimozhi and M.E Degree in Computer Science and Engineering from Mepco Schlenk Engineering College, Sivakasi in 2005 and 2007 respectively. She received her Ph.D in Computer Science and Engineering at Noorul Islam Centre for Higher Education in 2017 under the area of Cloud Computing and guiding four research scholars. Her research area includes Wireless Communications, Cloud Computing, Big Data and Cyber Security. Acted a s reviewer in many journals and IEEE conferences. She is a Life member of ISTE.

Muhammad Usman Tariq has more than 16+ year's experience in industry and academia. He has authored more than 200+ research articles, 110+ case studies, 105+ book chapters and several books other than 4 patents. He is founder and CEO of The Case HQ, a unique repository for courses, narrative and video case studies. He has been working as a consultant and trainer for industries representing six sigma, quality, health and safety, environmental systems, project management, and information security standards. His work has encompassed sectors in aviation, manufacturing, food, hospitality, education, finance, research, software and transportation. He has diverse and significant experience working with accreditation agencies of ABET, ACBSP, AACSB, WASC, CAA, EFQM and NCEAC. Additionally, Dr. Tariq has operational experience in incubators, research labs, government research projects, private sector startups, program creation and management at various industrial and academic levels. He is Certified Higher Education Teacher from Harvard University, USA, Certified Online Educator from HMBSU, Certified Six Sigma Master Black Belt and has been awarded PFHEA, SFSEDA, SMIEEE, and CMBE.

Amit Thakur is a passionate scholar currently pursuing a Ph.D. in Machine Learning at VIT BHOPAL University. With a background in computer science and a fascination for artificial intelligence, Amit's research focuses on developing innovative algorithms for natural language processing and computer vision applications. Throughout his academic journey, Amit has demonstrated a strong aptitude for theoretical and applied research in machine learning. His work combines advanced mathematical techniques with practical implementations to address real-world challenges in data analysis and pattern recognition. In addition to his research endeavors, Amit is actively involved in teaching and mentoring undergraduate students, sharing his knowledge and enthusiasm for machine learning. He volunteers his time to organize workshops and seminars, fostering a collaborative learning environment within the academic community. Driven by a desire to harness the power of machine learning for societal benefit, Amit is dedicated to pushing the boundaries of AI research and developing intelligent systems that can positively impact people's lives. Through his interdisciplinary approach and commitment to excellence, Amit strives to be a leading figure in the field of machine learning and contribute to its continued advancement.

Anusha Thakur is an "Independent Researcher" with nearly "45 solo research papers" in the World's reputed books and journals. Anusha's exemplary contributions without Ph.D. degree, and in nearly three years, have earned her the title of the "Best Woman Researcher of the Year" - (Category - Below 35 years) during the 4th International Inspirational Women Awards - (IIWA), 2023, held in New Delhi, in June 2023. Anusha has been conferred the prestigious "Shiksha Bharati Award", in October 2023 by the "Indian Achievers Forum" in recognition of her extraordinary professional achievements and contribution towards nation-building. She has been awarded the "Young Turks Award" at the 3AI GCC-X Summit in July 2024. Anusha has also won the "CEO Award" at the Genai Max Conclave 2024 by 3AI. One of her published Chapter titled "Impact of Mood of the Millennial Customers on Purchase of Apparels Online", published by IGI Global, USA was chosen to be included as a "reprinted chapter" in 2021 for the book "Research Anthology on E-Commerce Adoption, Models, and Applications for Modern Business (3 Volumes)" by the "Information Resources Management Association (USA). One of her published Chapter includes "Market for Plant-Based Meat Alternatives" for the Book titled "Environmental, Health, and Business Opportunities in the New Meat Alternatives Market". The Book won the #Best Book in the World in the Prestigious 25th Gourmand Awards, in 2020. Gourmand Awards is considered the Oscars of Food and Wine Publications in the world. Also, this book won "National Winner in Australia" in 2019, in the "Vegetarian Writing" category. Anusha serves as the Associate Editor of IJSMI, IGI Global, USA. She is an "Editorial Review Board Member" for Journals "IJEIS, IJAMTR, IJBDAH, and IJAIBM (IGI Global, USA)". She is also a Peer Reviewer for IGI Global, USA, and Common Ground Research Networks, USA. Anusha has pursued B.Tech in "Electronics and Instrumentation Technology" from Tumkur, Karnataka, and MBA in "The Department of Economics and International Business" from the University of Petroleum & Energy Studies, (UPES) Dehradun. She has work experience of approximately 4+ years in the Market Research Industry as a Senior Research Analyst with extensive knowledge of market research in diversified industries.

Index

Symbols

A

B

C

D

E

F

G

H

I

L

M

N

P

R

S

V

www.ingramcontent.com/pod-product-compliance
Lightning Source LLC
LaVergne TN
LVHW062243200926
848484LV00006B/1602